oked Potential Primer

Rainer Spehlmann died unexpectedly as this book went to press. The publishers join his family and colleagues in mourning the loss of a respected physician and prolific medical scientist. This book stands in testimony to Dr. Spehlmann's constant search for understanding in his specialty, his eagerness to convey that knowledge to his peers, and his commitment to discover better diagnostic and treatment techniques for the benefit of countless patients.

Evoked Potential Primer

Visual, auditory, and somatosensory evoked potentials in clinical diagnosis

Rainer Spehlmann, M.D.

Professor of Neurology
Northwestern University Medical School
EEG Laboratories of the VA Lakeside Medical Center,
Chicago, Illinois, and the Evanston Hospital, Evanston,
Illinois, and Retina Laboratory of the Department of
Ophthalmology of Northwestern University Medical School

Butterworth Publishers
Boston • London
Sydney • Wellington • Durban • Toronto

Copyright © 1985 by Butterworth Publishers.
All rights reserved.

No part of this publication may be reproduced, stored in a retrieval system, or transmitted, in any form or by any means, electronic, mechanical, photocopying, recording, or otherwise, without the prior written permission of the publisher.

Every effort has been made to ensure that the drug dosage schedules within this text are accurate and conform to standards accepted at time of publication. However, as treatment recommendations vary in the light of continuing research and clinical experience, the reader is advised to verify drug dosage schedules herein with information found on product information sheets. This is especially true in cases of new or infrequently used drugs.

Library of Congress Cataloging in Publication Data

Spehlmann, Rainer, 1931–
   Evoked potential primer.

   Includes bibliographies and index.
   1. Evoked potentials (Electrophysiology)
2. Nervous system—Diseases—Diagnosis. I. Title.
[DNLM: 1. Evoked Potentials. WL 102 S742e]
RC386.6.E86S63 1985   616.07'547   84–17500
ISBN 0–409–95158–7 (casebound)
ISBN 0–409–90005–2 (paperbound)

Butterworth Publishers
80 Montvale Avenue
Stoneham, MA 02180

10 9 8 7 6 5 4

Printed in the United States of America

Contents

Preface ix
Abbreviations xiii

Part A
Evoked Potentials

1. General Description of Evoked Potentials 6
   1.1 Definition 7
   1.2 Clinical use of EPs 7
   1.3 Types of EPs 8
   1.4 The generators of EPs 12

2. General Methods of Stimulating and Recording 15
   2.1 Stimulus methods 16
   2.2 Recording methods 17
   2.3 Amplifiers 21
   2.4 Filters 27

3. Averaging 35
   3.1 Principle of averaging 36
   3.2 Computer types 38
   3.3 Acquisition parameters 38
   3.4 Averaging devices 47
   3.5 Magnetic tape recording 51
   3.6 Other methods: Frequency and power spectral analysis 52

4. General Characteristics of Evoked Potentials 53
   4.1 General description of EPs 54
   4.2 Variability of EPs 61
   4.3 Selection of EP type and of EP parameters to be measured 62

    4.4   Labeling EP records   63

5. Clinical Interpretation   65
    5.1   Principles of clinical interpretation   66
    5.2   Definition of normal: Control groups   66
    5.3   Types and clinical implications of EP abnormalities   68
    5.4   Strategy of localization and lateralization   71
    5.5   Report   71

References for Part A   73

Part B
Visual Evoked Potentials

6. VEP Types, Principles, and General Methods of Stimulating and Recording   85
    6.1   VEP types   86
    6.2   The subject   88
    6.3   General description of stimulus parameters   89
    6.4   General description of recording parameters   93

7. The Normal Transient VEP to Checkerboard Pattern Reversal   95
    7.1   Description of the normal transient pattern reversal VEP   96
    7.2   Subject variables   98
    7.3   Effect of stimulus characteristics   99
    7.4   Effect of recording parameters   103
    7.5   Prechiasmal, chiasmal, retrochiasmal, and ophthalmological strategies   109

8. The Abnormal Transient VEP to Checkerboard Pattern Reversal   116
    8.1   Criteria distinguishing abnormal pattern reversal VEPs   117
    8.2   General clinical interpretation of abnormal VEPs to checkerboard pattern reversal   117
    8.3   Neurological disorders that cause abnormal pattern reversal VEPs   122
    8.4   Ophthalmological disorders that cause abnormal pattern reversal VEPs   132

9. The Transient VEP to Diffuse Light Stimuli   135
    9.1   Clinical usefulness of the flash VEP   136
    9.2   The normal transient flash VEP   136
    9.3   Subject variables   137
    9.4   Stimulus characteristics   138
    9.5   Recording parameters: Electrode placements and combinations   139
    9.6   Criteria distinguishing abnormal flash VEPs   140
    9.7   Neurological disorders that cause abnormal flash VEPs   140
    9.8   Ophthalmological disorders that cause abnormal flash VEPs   142

10. VEPs to Other Stimuli   144
    10.1   Transient VEPs to checkerboard appearance, disappearance, and flash   145
    10.2   Steady-state VEPs to checkerboard pattern reversal, appearance, disappearance, and flash   146
    10.3   Steady-state VEPs to diffuse light stimuli   149
    10.4   VEPs to sine wave gratings   151
    10.5   VEPs to bar gratings   155

10.6 VEPs to small macular light spots 155
10.7 VEPs to moving and stereoscopic random dot patterns 157
10.8 VEPs to other patterned stimuli 157
10.9 Photopic versus scotopic VEPs 157
10.10 Color VEPs 157
10.11 VEPs to blinking and shift of gaze 158
10.12 Electric stimulation of the eye 158
10.13 The ERG to diffuse and patterned light 158

References for Part B 160

Part C
Auditory Evoked Potentials

11. AEP Types, Principles, and General Methods of Stimulating and Recording 194
    11.1 AEP types 195
    11.2 The subject 195
    11.3 Stimulation methods 197
    11.4 Recording electrode placements, montages, and polarity convention 202

12. The Normal BAEP 204
    12.1 Description of the normal BAEP 205
    12.2 Subject variables 208
    12.3 Stimulus characteristics 209
    12.4 Recording parameters 211
    12.5 General strategy 212

13. The Abnormal BAEP 217
    13.1 Criteria distinguishing abnormal BAEPs 218
    13.2 General clinical interpretation of abnormal BAEPs 219
    13.3 Neurological disorders that cause abnormal transient BAEPs 223
    13.4 Audiological disorders that cause abnormal BAEPs 232

14. Other AEPs 236
    14.1 The slow brain stem AEP 237
    14.2 The frequency following potential 237
    14.3 The middle latency AEP 240
    14.4 The 40-Hz AEP 242
    14.5 The long-latency AEP 242
    14.6 The electrocochleogram 246
    14.7 Sonomotor AEPs 249

References for Part C 252

Part D
Somatosensory Evoked Potentials

15. SEP Types, Principles, and General Methods of Stimulating and Recording 282
    15.1 SEP types 283
    15.2 The subject 284
    15.3 Stimulation methods 285
    15.4 General recording parameters 287

16. Normal SEPs to Arm Stimulation 289
    16.1 Normal SEPs at different recording sites 290
    16.2 Subject variables 295
    16.3 Stimulus characteristics 297
    16.4 Recording parameters 298
    16.5 General strategy of stimulating and recording 299

17. Abnormal SEPs to Arm Stimulation   300
    17.1 Criteria distinguishing abnormal SEPs   301
    17.2 General clinical interpretation of abnormal SEPs to arm stimulation   302
    17.3 Peripheral lesions that cause abnormal SEPs to arm stimulation   304
    17.4 Lesions of cervical roots and of the cervical cord that cause abnormal SEPs to arm stimulation   306
    17.5 Lesions of the brain stem and cerebral hemispheres that cause abnormal SEPs to arm stimulation   308

18. Normal SEPs to Leg Stimulation   319
    18.1 Normal SEPs at different recording sites   320
    18.2 Subject variables   326
    18.3 Effect of stimulus parameters   326
    18.4 Recording parameters   327
    18.5 General strategy of stimulating and recording   327

19. Abnormal SEPs to Leg Stimulation   329
    19.1 Criteria distinguishing abnormal SEPs   330
    19.2 General clinical interpretation of abnormal SEPs to leg stimulation   331
    19.3 Peripheral nerve and root lesions that cause abnormal SEPs to leg stimulation   333
    19.4 Lesions of spinal cord and brain stem that cause abnormal SEPs to leg stimulation   334
    19.5 Cerebral lesions that cause abnormal SEPs to leg stimulation   338

20. Other SEPs   339
    20.1 Sensory nerve action potentials   340
    20.2 SEPs to trigeminal nerve stimulation   340
    20.3 SEPs to pudendal and bladder stimulation   340
    20.4 SEPs to adequate stimuli   340
    20.5 Somatomotor EPs   343

References for Part D   344

Part E
Event-Related and Other Potentials

21. Event-Related Potentials, Olfactory Evoked Potentials, and Magnetic Evoked Fields   367
    21.1 The $\overline{P300}$ 368
    21.2 The contingent negative variation   370
    21.3 Readiness potential (Bereitschaftspotential) and other movement-related potentials   372
    21.4 Potentials preceding speech and accompanying writing   374
    21.5 Olfactory EPs   374
    21.6 Magnetic evoked fields   374

References for Part E   375

Appendix   382

Index   383

Preface

This book is an introduction to the use of evoked potentials in clinical diagnosis. It addresses itself primarily to residents in neurology, neurosurgery, and related fields; to other physicians, neuroscientists, and technicians; and to anyone else who wants to learn how to record and interpret evoked potentials. For this group, the book is meant to be used in addition to practical experience; it cannot replace learning experience in the laboratory, but it can serve as a guide and a reference for this experience. The book can also fill the need of those who, although not directly involved in the recording and interpretation of evoked potentials, want to become familiar with the methods and diagnostic abilities of the test. Even persons with some experience in evoked potentials may find the book useful because it presents a systematic review and includes topics that are rarely dealt with in simple terms.

Although evoked potentials are being used more and more widely, it is not easy to find sources of information that cover the entire field evenly and completely. A variety of methods has been developed in different laboratories, each method producing different kinds of evoked potentials. Uniform standards have not been adopted yet, although a few recommendations[1,2,3,4,5] have been published. Under these conditions, it seems best to compromise between the extremes of either giving brief reviews of a great variety of evoked potentials produced with different methods or providing long descriptions of a few evoked potentials elicited with the author's preferred methods. To give the reader a practical starting point, the most commonly used methods for visual, auditory, and somatosensory evoked potentials are here described in sufficient detail to help the beginner get started with the recording and the interpretation of evoked potentials while less commonly used methods are presented only briefly. The guidelines of the American EEG Society[1] are followed wherever they apply.

The book has five parts. Part A gives a general overview of evoked potentials, discussing definitions of evoked potential types, methods of stimulation and recording, use of computer averaging, characteristic features and nomenclature of evoked potentials, criteria distinguishing abnormal evoked potentials, and principles of clinical interpretation. Parts B, C, and D deal with visual, auditory, and somatosensory evoked potentials, respectively. Each part starts with a general chapter describing evoked potential types and stimulus and recording methods for the particular modality. Subsequent chapters focus on the diagnostically most important type of evoked potential and describe the normal evoked potential and its variations, strategy of detecting and localizing lesions, criteria distinguishing abnormal evoked potentials, principles of clinical interpretation of specific evoked potential abnormalities, and particular clinical disorders characterized by abnormal evoked potentials. Less important evoked potential types are described by their essential features only. Part E gives a short description of event-related and a few rarely used evoked potentials.

To facilitate reading and review, a detailed table of contents is placed at the beginning of each part. Each chapter starts with a summary. The numbered items of the summary refer to numbered sections of the text which expand on the material. The general characteristics are dealt with in Part A; special features are discussed separately with each evoked potential type in the subsequent parts of the book. Related features of different evoked potentials can be found through cross-references in the text and through the index.

References to the literature favor more recent publications through which the older ones can be found. References are selected not very critically for topics on which only one or a few investigations have been published. References to animal studies are kept to a minimum because they rarely help to explain the results of clinical evoked potential recordings.

Figures are used extensively to illustrate the technical background and to show actual recordings of normal and abnormal evoked potentials. Many figures of evoked potentials are selected from the literature to illustrate the wide range of methods used in different laboratories; I thank the many authors and publishers who kindly permitted me to reproduce figures from their articles. Several tables summarize related materials but are not meant to invite memorization.

In writing this book, I have been greatly helped by the experience of teaching residents, other physicians, and technicians: Their questions have indicated to me which problems need discussion and how to discuss them. Special thanks are due to Dr. Karyl Norcross, Dr. Cynthia Stack,

Dr. Sam Ho, Dr. Glen Ackerman, Dr. Bruce Cohen, and Dr. Dan Johnson for reviewing the manuscript at different stages and to Dr. Ho for the permission to use some of his recordings for illustrations. Mr. C.C. Smathers helped me design many of the illustrations.

5. International Federation of Societies for Electroencephalography and Clinical Neurophysiology. 1983. Recommendations for the practice of clinical neurophysiology. Amsterdam: Elsevier.

References

1. American Electroencephalographic Society. 1984. Guidelines for clinical evoked potential studies. J. Clin. Neurophysiol. 1:3–53.
2. Arden, G.B., Bodis-Wollner, I., Halliday, A.M., Jeffreys, A., Kulikowski, J.J., Spekreijse, H., and Regan, D. 1977. Methodology of patterned visual stimulation (report of the Brussels symposium ad-hoc committee). In: Visual evoked potentials in man: New developments, ed. J.E. Desmedt, pp. 3–15. Oxford: Clarendon Press.
3. Desmedt, J.E. 1977. Some observations on the methodology of cerebral evoked potentials in man. Prog. Clin. Neurophysiol. 1:12–29.
4. Donchin, E., Callaway, E., Cooper, R., Desmedt, J.E., Goff, W.R., Hillyard, S.A., and Sutton, S. 1977. Publication criteria for studies of evoked potentials (EP) in man. Report of a committee. Prog. Clin. Neurophysiol. 1:1–11.

Abbreviations

AEP	Auditory evoked potential
BAEP	Brain stem auditory evoked potential
CNS	Central nervous system
CFPD	Critical frequency of photic driving
CM	Cochlear microphonic
CNV	Contingent negative variation
ECochG	Electrocochleogram
ECG	Electrocardiogram
EEG	Electroencephalogram, electroencephalographic
EMG	Electromyogram, electromyographic
EP	Evoked potential
ERA	Electric response audiometry
ERG	Electroretinogram
ERP	Event-related potential
FFP	Frequency following potential
HL	Hearing level
IPL	Interpeak latency
LLAEP	Long-latency auditory evoked potential
MLAEP	Middle latency auditory evoked potential
NAP	Nerve action potential
nHL	Normal hearing level
peSPL	Peak equivalent sound pressure level
SEP	Somatosensory evoked potential
SL	Sensory level
SNAP	Sensory nerve action potential
SP	Summating potential
SPL	Sound pressure level
VEP	Visual evoked potential

Evoked Potential Primer

A
Evoked potentials

1. GENERAL DESCRIPTION OF EVOKED POTENTIALS
   1.1 Definition
   1.2 Clinical use of EPs
   1.3 Types of EPs
      1.3.1 EPs defined by stimulus modality and type
      1.3.2 EPs defined by generator
         1.3.2.1 Cortical EPs
         1.3.2.2 Subcortical EPs
      1.3.3 EPs defined by recording site
      1.3.4 EPs defined by recording method
         1.3.4.1 Near-field recordings
         1.3.4.2 Far-field recordings
         1.3.4.3 Mixed far-field and near-field recordings
      1.3.5 EPs defined by stimulus rate
         1.3.5.1 Transient EPs
         1.3.5.2 Steady-state EPs
      1.3.6 EPs defined by stimulus duration
      1.3.7 EPs to unilateral and bilateral stimuli
      1.3.8 EPs recorded in the midline and laterally
   1.4 The generators of EPs

2. GENERAL METHODS OF STIMULATING AND RECORDING
   2.1 Stimulus methods
      2.1.1 Stimulus type
      2.1.2 Stimulus intensity
      2.1.3 Stimulus duration
      2.1.4 Stimulus rate
      2.1.5 Number of stimuli
      2.1.6 Size and location of stimulated area

2.2 Recording methods
   2.2.1 Electrode types and application methods
      2.2.1.1 EEG electrodes
      2.2.1.2 Clip electrodes
      2.2.1.3 Adhesive ECG electrodes
      2.2.1.4 Needle electrodes
      2.2.1.5 Electrocochleographic electrodes
      2.2.1.6 Electroretinographic electrodes
      2.2.1.7 Electrocorticographic electrodes
      2.2.1.8 Intracerebral electrodes
   2.2.2 Electric properties of recording electrodes
      2.2.2.1 Electrode materials
      2.2.2.2 Electrode resistance
      2.2.2.3 Electrode impedance
      2.2.2.4 Electrode polarization and bias potentials
   2.2.3 Electrode placement
2.3 Amplifiers
   2.3.1 Input board
   2.3.2 Calibration
   2.3.3 Input impedance
   2.3.4 Differential amplification and polarity convention
      2.3.4.1 Discrimination of cerebral potentials and polarity convention
      2.3.4.2 Rejection of artifacts
      2.3.4.3 Amplification
         2.3.4.3.1 Gain and sensitivity
         2.3.4.3.2 Recording and display gain
2.4 Filters
   2.4.1 Analog filters
      2.4.1.1 Low frequency filters or high-pass filters
      2.4.1.2 High frequency filters or low-pass filters
      2.4.1.3 60-Hz notch filter
      2.4.1.4 Phase shift
   2.4.2 Digital filters

3. AVERAGING
3.1 Principle of averaging
3.2 Computer types
3.3 Acquisition parameters
   3.3.1 Number of channels
   3.3.2 Triggering
   3.3.3 Number of responses to be averaged: Noise reduction
   3.3.4 Horizontal parameters of analysis: Analysis period, number of points, sampling rate, and dwell time
   3.3.5 Vertical parameters of analysis: Number of bits, resolution, and recording gain
3.4 Averaging devices
   3.4.1 Artifact rejection option
   3.4.2 Oscilloscope displays
   3.4.3 Smoothing
   3.4.4 Cursors
   3.4.5 Addition, inversion, and subtraction of EPs

3.4.6 Copies of EPs
    3.4.6.1 Hard copies
        3.4.6.1.1 XY plots
        3.4.6.1.2 Photographs
        3.4.6.1.3 Photosensitive paper copies
    3.4.6.2 Magnetic copies
3.5 Magnetic tape recording
3.6 Other methods: Frequency and power spectral analysis

4. GENERAL CHARACTERISTICS OF EVOKED POTENTIALS
  4.1 General description of EPs
    4.1.1 Polarity
    4.1.2 Number in sequence
    4.1.3 Latency
      4.1.3.1 Peak identification
      4.1.3.2 Latency of transient EPs
        4.1.3.2.1 Peak latency
        4.1.3.2.2 Interpeak latency and central conduction time
      4.1.3.3 Latency of steady-state EPs
    4.1.4 Amplitude
      4.1.4.1 Absolute amplitude measurements
        4.1.4.1.1 Peak amplitude
        4.1.4.1.2 Peak-to-peak amplitude
      4.1.4.2 Amplitude ratios
    4.1.5 Waveshape
    4.1.6 Distribution
    4.1.7 Naming of peaks
      4.1.7.1 Nomenclatures
        4.1.7.1.1 Polarity-latency
        4.1.7.1.2 Number of a peak in sequence
        4.1.7.1.3 Polarity-number in sequence
      4.1.7.2 Observed versus characteristic peaks
  4.2 Variability of EPs
    4.2.1 Intraindividual and interindividual variability
      4.2.1.1 Intraindividual variability
        4.2.1.1.1 Changes of attention and alertness
        4.2.1.1.2 Background activity
        4.2.1.1.3 Heart rate and blood pressure
        4.2.1.1.4 Time between averages
        4.2.1.1.5 Age
        4.2.1.1.6 Difference between laboratories
        4.2.1.1.7 Abnormal cerebral function
      4.2.1.2 Interindividual variability
  4.3 Selection of EP type and of EP parameters to be measured
    4.3.1 Selection of EP type
    4.3.2 Selection of EP parameters to be measured
  4.4 Labeling EP records

5. CLINICAL INTERPRETATION
  5.1 Principles of clinical interpretation
  5.2 Definition of normal: Control groups
    5.2.1 Characteristics of normal subjects
    5.2.2 Statistical definition of normal ranges
  5.3 Types and clinical implications of EP abnormalities
    5.3.1 Abnormally long latency

5.3.1.1 Abnormally long peak latency
5.3.1.2 Abnormally long interpeak latency
5.3.1.3 Abnormal differences of peak or interpeak latencies
5.3.2 Abnormal amplitude
   5.3.2.1 Reduced amplitude
   5.3.2.2 Abnormal amplitude ratios
   5.3.2.3 Complete absence of an entire EP
5.3.3 Abnormal shape

5.4 Strategy of localization and lateralization
   5.4.1 Correlation between EP peaks and anatomical structures
   5.4.2 Localization by combinations of unilateral stimulation and recording in partially or completely crossed pathways
5.5 Report

References for part A

1

General description of evoked potentials

SUMMARY

1.1. Evoked potentials (EPs) are here defined as computer averaged electric responses of the nervous system to sensory stimulation.

1.2. EPs are used to test conduction through sensory pathways.

1.3. Different types of EPs can be distinguished by stimulus modality: Visual EPs (VEPs), auditory EPs (AEPs), and somatosensory EPs (SEPs); these are further divided into VEPs to checkerboard pattern, flash, and other stimuli, AEPs of short, middle, and long latency, and SEPs to stimulation of arm and leg nerves. EPs can also be classified by their source as cortical and subcortical, by the recording site as scalp, neck, clavicular, and lumbosacral SEPs, by specific recording methods as near-field or far-field EPs, and by other variations of stimulus and recording conditions.

1.4. The generators differ depending on the EP type: Cortical EPs are mainly generated by the spatial and temporal addition of postsynaptic potentials of cortical neurons; peripheral nerve and plexus EPs are produced by nerve impulses traveling in nerve fibers; subcortical potentials may be produced by varying combinations of both types of elements.

1.1 DEFINITION

Evoked potentials (EPs) are here defined as electric responses of the nervous system to sensory stimulation. They consist of a sequence of deflections, or waves, each characterized by latency, amplitude, and other features described in detail elsewhere (4.1). In clinical testing, EPs are elicited by visual or auditory stimulation or by electric stimulation of sensory nerves. These EPs are recorded from the surface of the body with electrodes on the scalp or on the skin over the spinal cord or peripheral nerves. Rarely are EPs in humans recorded with electrodes placed in the depth or on the surface of the brain. In research studies, EPs may be elicited by electrically stimulating points within the central nervous system (CNS) with surgically inserted electrodes.

A single response to a stimulus usually has low amplitude and may be partly or completely obscured by ongoing spontaneous EEG activity. To extract EPs from the EEG, single responses are elicited repeatedly and averaged with a computer; this method reduces the amplitude of the EEG component that is not related to the stimulus and thereby enhances the features of the response component that is time-locked to the stimulus. The term *evoked potential* is here arbitrarily used to indicate only the average of individual responses; the term *response* is used for any recording following a single stimulus regardless of whether stimulus-related deflections can be distinguished without averaging. The term *peak*, or *wave*, is used to denote the upgoing or downgoing deflections that make up an EP. The word *component*, which is commonly used in the same sense, is here reserved to describe separable contributions to the same peak or wave as in *noise component* or *low-frequency component*.

The name *event-related potential* (ERP) is now sometimes used to denote both EPs and other kinds of potentials that are the result of cognitive processes following a stimulus or of pregnancy processes preceding a movement or speech. Such potentials, not yet used widely in clinical diagnosis, are here described only briefly (Chapter 21).

1.2 CLINICAL USE OF EPs

EPs are used mainly to test conduction in the visual, auditory, and somatosensory systems, especially in the central parts of these systems. EPs are so sensitive that they can detect lesions not discovered by clinical or other laboratory techniques. Furthermore, EPs often help to localize lesions to certain segments of a central sensory pathway. EPs therefore have become valuable diagnostic tools in clinical neurology. They are especially helpful in the diagnosis of multiple sclerosis where the demonstration of clinically silent defects may prove the presence of multiple lesions. EPs may be useful in the diagnosis of other structural lesions, some degenerative diseases, and even a few metabolic encephalopathies. EPs are also abnormal in many other disorders that are more conveniently diagnosed by other means. In cases of diffuse disorders, EPs may be used to show involvement of a particular sensory pathway. In some medical centers, EPs are used during surgical procedures to monitor the condition of structures at the operative site, for instance, the condition of spinal sensory pathways during operations on the spinal column or that of the brain stem and of the acoustic nerve during the removal of acoustic neurinomas. EPs may be used in ophthalmological and audiological studies, especially in the evaluation of infants and other noncommunicative subjects.

1.3 TYPES OF EPs

Many types of EPs can be distinguished according to stimulation and recording methods. Some EP types are listed in Table 1.1; for clinical purposes, EPs are further broken down as shown in Tables 6.1, 11.1, and 15.1

1.3.1 EPs defined by stimulus modality and type

EPs are divided by stimulus modality into visual evoked potentials (VEPs), auditory evoked potentials (AEPs), and somatosensory evoked potentials (SEPs). VEPs are subdivided by stimulus content into VEPs to checkerboard patterns, diffuse light, and other types of stimuli. AEPs are subdivided mainly by latency into short-latency AEPs, including brain stem AEPs (BAEPs), middle latency AEPs (MLAEPs), and long-latency AEPs (LLAEPs). SEPs are subdivided chiefly by the location of the stimulus into SEPs to arm and leg nerve stimulation.

1.3.2 EPs defined by generator

1.3.2.1 Cortical EPs

Cortical EPs are generated by primary sensory and higher cortical areas. They have latencies of over 10–20 msec and amplitudes of up to 10 µV or more. Cortical EPs generally are recorded with near-field recording methods (1.3.4.1). Cortical VEPs and SEPs are derived with electrodes placed near the primary receiving areas in the occipital and parietal areas, respectively. Cortical AEPs are recorded with electrodes not directly overlying the auditory cortex. In all three modalities, cortical EP peaks may be preceded by peaks generated by subcortical structures.

Cortical EPs are of different importance for each sensory modality. The usual clinical VEP test records only cortical EPs. In SEP studies, cortical SEPs are recorded simultaneously with subcortical SEPs derived with electrodes over the spine and peripheral nerves. Studies of AEPs do not record cortical EPs routinely because they are of much less practical use than the earlier brain stem AEPs.

1.3.2.2 Subcortical EPs

Subcortical EPs are generated by the chains of neurons in a sensory pathway to the cortical receiving area. These EPs

TABLE 1.1. EP types

Characteristic	Type

A. Stimulus modality and type	1. Visual evoked potentials (VEPs)
	   a. VEP to checkerboard pattern stimulation
	   b. VEP to diffuse light stimulation
	   c. VEPs to other stimuli
	2. Auditory evoked potentials (AEPs)
	   a. Short-latency AEP
	   b. Middle latency AEP
	   c. Long-latency AEP
	   d. Other AEPs
	3. Somatosensory evoked potentials (SEPs)
	   a. SEP to arm nerve stimulation
	   b. SEP to leg nerve stimulation
	   c. Other SEPs

B. EP origin	1. Cortical EP
	2. Subcortical EPs
	   a. Brain stem EP
	   b. Spinal EP
	   c. Brachial plexus EP
	   d. Cauda equina EP
	   e. Sensory nerve action potential (SNAP)

C. Recording site	1. Scalp EP 2. Neck EP 3. Clavicular EP 4. Lumbosacral EP
D. Recording method	1. Near-field EP 2. Far-field EP

TABLE 1.1. *(continued)*

Characteristic	Type
E. Stimulus rate	1. Transient EP
	2. Steady-state EP
F. Stimulus duration	1. EP to brief stimuli
	2. EP to onset or end of long stimuli
	3. EP to gradually changing stimuli
G. Unilateral and bilateral stimuli	1. EP to stimulation of one side
	2. EP to stimulation of both sides
H. Midline and lateral recordings	1. EP recorded in the midline
	2. EPs recorded unilaterally
	a. EP ipsilateral to the stimulus
	b. EP contralateral to the stimulus
	3. EPs recorded bilaterally

have latencies of less than 10–20 msec. Because the brain stem and spinal cord are relatively far away from recording electrodes on the head and neck, potentials generated in the auditory and somatosensory afferent pathways are much attenuated by the intervening tissues, have amplitudes of usually less than 1 μV at surface recording electrodes, and must be recorded with far-field methods (1.3.4.2).

1.3.3 EPs defined by recording site

Scalp EPs are recorded with an electrode placed near the cortical receiving areas of the stimulated sensory system. While recordings between such an electrode and a closely spaced reference electrode yields only the cortical EP, recordings of AEPs and SEPs between a scalp electrode and a distant reference electrode may show a sequence of subcortical and cortical EPs. Such reference electrodes may be placed on the ear, shoulder, elbow, hand, or knee. In most laboratories, subcortical SEPs are instead recorded locally with electrodes placed on cervical, clavicular, and lumbosacral locations. Electrodes may be placed in the ear canal or middle ear to record action potentials of the acoustic nerve. Electrodes at or near the eye may be used to record the electroretinogram.

1.3.4 EPs defined by recording method

1.3.4.1 Near-field recordings

Near-field recording methods are used to record cortical EPs. One electrode is placed close to the area under study, and the other electrode is placed over an electrically more quiet area several centimeters away. A recording between these electrodes yields responses of 1–10 µV and requires collection of only about 100 responses for a clear definition of the cortical EP. Repetition rates of 1–2/sec are usually used for transient cortical EPs because these EPs have relatively long latencies and durations and may interact with each other at higher rates (1.3.5).

1.3.4.2 Far-field recordings

Far-field recording methods are used mainly for recording of potentials produced in the brain stem and spinal cord, i.e., far away from surface electrodes. The electric field generated deep in the brain has a wide distribution at the surface so that the exact location of recording electrodes is not critical, although they must be fairly far apart to pick up the small voltage differences on the surface. The amplitude of the potentials is much attenuated at the surface and usually measures less than 1 µV, requiring averages of 1,000 or more responses for a clearly defined EP. Because subcortical EPs have short latency and duration and are fairly resistant to fast repetition, stimulus rates of 5–10/sec or more may be used. Far-field recordings may contain peaks generated by structures separated by relatively long distances: AEP recordings between electrodes on vertex and ear can show peaks that are generated by the acoustic nerve and its relays in the brain stem. SEPs recorded between an electrode on the scalp and another electrode on the shoulder or knee may reflect subcortical EPs from the entire afferent pathway.

1.3.4.3 Mixed far-field and near-field recordings

Recordings of AEPs and SEPs between a scalp electrode near the cortical sensory area and a distant electrode may show small early peaks which represent far-field recordings from distant, subcortical structures and precede the larger cortical potentials generated near the scalp electrode (1.3.3).

1.3.5 EPs defined by stimulus rate

1.3.5.1 Transient EPs

Separate responses are elicited by stimuli spaced so far apart in time that each response is completed before the beginning of the next one. This requires stimulus rates no faster than 1–2/sec for the relatively long-lasting VEPs and long-latency cortical AEPs and SEPs. Short-latency cortical AEPs and SEPs and subcortical EPs have much shorter duration and can be elicited as transient EPs at rates of 10/sec or more. Transient EPs are the most commonly used clinical EPs.

1.3.5.2 Steady-state EPs

If repetitive stimuli are spaced so closely that each response interacts with the next one, the EP becomes a rhythmical wave which has peaks at the same frequency as the stimuli and may contain harmonic and subharmonic components at multiples of that frequency; at higher stimulus rates, EPs to individual stimuli can no longer be distinguished. Stimulus rates producing steady-state EPs vary with the EP type. The long-lasting VEPs develop into steady state at rates beginning at about 5/sec; steady-state VEPs may be obtained at rates as high as 70/sec (10.2, 10.3). In contrast, the brief short-latency AEP can be driven to become a steady-state EP, or frequency following potential (14.2), at rates of about 250–1,000/sec. The middle latency AEP can be driven into steady state at rates of about 40/sec (14.4).

1.3.6 EPs defined by stimulus duration

A brief stimulus such as a light flash, a click, or a tone pip elicits an EP that is the compound of an EP to the onset and an EP to the end of the stimulus. These EPs can be separated with longer stimuli. A frequently used prolonged stimulus is the alternation of complementary checkerboard patterns. Gradually changing stimuli such as the modulation of light or tone intensity may also serve as a stimulus for EPs.

1.3.7 EPs to unilateral and bilateral stimuli

Auditory and somatosensory stimuli are usually applied to one side of the body at a time. Visual testing uses separate stimulation of each eye and of each visual half-field.

Bilateral stimulation produces EPs that differ from the sum of the EPs to unilateral stimulation to an extent depending on the cortical and subcortical level of recording and on the anatomy of the sensory system under study. This difference reveals the amount of interaction between responses.

1.3.8 EPs recorded in the midline and laterally

Both midline and lateral occipital recordings are usually used in VEP recordings, although midline recordings would suffice in most cases of prechiasmal lesions. Recordings of AEPs are made between vertex and one or both ears or mastoids. Electrode placements for SEPs are in the midline except in arm nerve stimulation where lateral recordings are made from Erb's point on the stimulated side and from the contralateral parietal scalp.

1.4 THE GENERATORS OF EPs

The features of an EP only very grossly suggest its origin. In general, peaks of relatively high amplitude and restricted distribution on the scalp are likely to be generated in the cortex under one of the recording electrodes, especially if the latency is long enough to be accounted for by conduction through the afferent pathway (1.3.4.1). Peaks of low amplitude and wide distribution are more likely to be generated in subcortical structures, especially if they have short latency (1.3.4.2).

The location of EP generators may be studied by delineating the distribution of the peaks in simultaneous recordings from many scalp electrode positions. In a common method of analysis, the magnitude of an EP deflection at a certain latency is plotted for each recording point on a head diagram; points of equal potential are connected to give a map of concentric isopotential lines which outline the maximum of the scalp potential and its gradient (Figure 1.1). From the map of the electric field on the surface one may infer the location of a central generator in the depth, often illustrated as a center of opposing charges, or dipole.[45,72,91,103]

Although the location of the dipole may suggest the approximate location of the anatomical structures generating the electric field of the EP peak, the definite identification of these structures as generators requires further validation by clinical correlations or direct recordings with depth electrodes. So far, the generators of most EP peaks have been determined only approximately. There is much debate about the brain stem generators of each of the peaks of the short-latency AEP; it seems that a precise correlation between each peak and one structure cannot be made because more than one structure may contribute to the production of one peak, and each generator may contribute to more than one peak. Discussions of the generators of the scalp SEP have not yet answered the question of which peak indicates arrival of the afferent volley at the cortical level. Although the generator of the major peak of the VEP must be the visual cortex, it is not known to what extent the primary visual cortex on the mesial surface

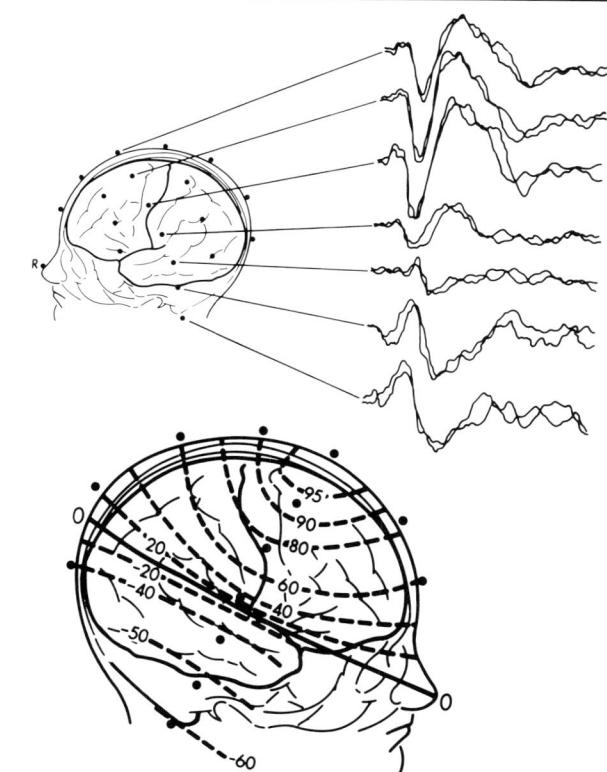

Figure 1.1. Plotting of isopotential lines. Top: A set of AEPs recorded from electrodes along a coronal line from vertex to mastoid with reference to an electrode on the tip of the nose. Bottom: Isopotential lines connecting all points on the scalp that have equal amplitude of the major peak. From Vaughan [91] with permission by the author and Academic Press, Inc.

and the secondary visual cortex on the lateral surface contribute to the VEP.

The general principles of the production of EPs by elements of the nervous system have been the subject of many experimental studies that suggest some general conclusions of practical importance.[26,90,104] Two types of nerve cell activity seem to contribute, in different proportions, to (1) cortical EPs, (2) subcortical EPs, and (3) EPs recorded from sensory nerves.

1. Cortical EPs are due largely to the spatial and temporal summation of excitatory and inhibitory postsynaptic potentials generated at the membrane of nerve cell bodies and dendrites in response to the input produced by the stimulus. The resulting potential differences generate currents which penetrate to the cortical surface and the scalp and produce electric fields that are modified by the electric properties of the intervening structures.[2,13,82]
2. Subcortical EPs are probably a mixture of two components: Postsynaptic potentials generated in groups of neurons of subcortical relay nuclei, and action potentials of the connecting axonal tracts. The first component, consisting of stationary generators of electric fields, is probably responsible for those subcortical potentials that can be recorded with similar latency at various distant electrode sites. The second component, consisting of propagated waves of depolarization, may explain some subcortical EPs that appear with delays of up to a few milliseconds at different recording sites.
3. EPs recorded from sensory nerves are due to a wave of depolarization propagated synchronously along the membrane of the nerve fibers. When passing under a stationary recording electrode on the skin, the wave produces a major surface-negative deflection that may be preceded and followed by minor positive deflections due to the approaching and disappearing wave. The compound action potential may include later deflections generated by fiber groups of lower conduction velocity, but these deflections are of low amplitude due to the greater temporal dispersion of slowly conducted impulses.

For all three kinds of EPs, the shape, size, and timing of an EP recorded from the scalp or skin depend on many factors, including the duration of the potential change, the size and spatial orientation of the generator to the recording electrodes, the distances between generator and electrodes and between the electrodes themselves, and the electric conductivity of the intervening structures. In general, it seems that long-lasting potential changes generated by stationary sources in large structures with uniform geometrical orientation have a better chance of being recorded than brief potentials generated by traveling waves in small structures with disparate geometrical orientation. In both cases, the chances for a potential to be recorded decrease as the distance between the source and the recording electrodes increases, but even relatively long distances can be overcome with far-field recording methods (1.3.4.2).

2

General methods of stimulating and recording

SUMMARY

2.1. Methods of stimulating and recording should be designed to best distinguish between normal and abnormal conduction in sensory pathways. Stimulus type, intensity, duration, rate, and number all affect the EP.

2.2. Recording electrodes are placed using measured coordinates that refer to anatomical landmarks; the electrodes should be applied so that electrode impedances are equal and low.

2.3. The amplifier is coupled to the recording electrodes either directly or through an input board containing selector switches that allow quick selection of recording electrodes as inputs to the amplifier. Calibration may be done either by comparison with standard electric pulses or by the use of cursors giving direct readouts of amplitude and latency. Differential amplification increases any voltage difference between recording electrodes coupled to input 1 and input 2 of the amplifier, but is insensitive to electric changes of the same voltage occurring at both inputs. This method serves to amplify electric potentials that are generated by the brain and affect one electrode more than the other and to reject extraneous potentials, or artifacts, that affect both electrodes equally. The direction of the amplifier output depends on the polarity convention used in the construction of the recording system and on the polarity of the potential changes at the electrodes connected to input 1 and input 2; therefore, for an EP to be interpretable, both the polarity and the electrode combination must be specified. Because in many recording situations, both electrodes contribute to the EP, the distinction between active and inactive, indifferent, or reference electrode is often not justified. Amplification is described by gain, i.e., the ratio of output voltage over input voltage, or sensitivity, or the ratio of input voltage over output deflection.

2.4. Filters are used to eliminate potential changes that are slower or faster than those of the EP. Filtering reduces noise and thereby diminishes the number of responses that need to be averaged. However, filtering with the commonly used analog filters may distort the timing of peaks and yield false latencies; filter settings must therefore be chosen carefully for each EP type.

2.1 STIMULUS METHODS

2.1.1 Stimulus type

In each stimulus modality, EPs differ not only depending on the stimulus type in general, such as visual stimulation to alternating checkerboard patterns versus visual stimulation to checkerboard pattern onset or to diffuse flashes, but also depending on the specific methods of generating the stimulus: VEPs to alternating checkerboard patterns differ depending on whether they are produced on a TV screen, with a projecting mirror system, or by a matrix of light-emitting diodes. AEPs to clicks differ from those to tones, and AEPs to clicks vary depending on whether the clicks move the air in the external auditory canal toward or away from the eardrum. Only SEPs, practically always generated by electric stimuli, are less capricious.

2.1.2 Stimulus intensity

In general, a stimulus must have a minimum intensity to produce an EP; this is the threshold. This minimum may be difficult to determine because EPs to weak stimuli have low amplitude and may require averaging of a great number of responses. An increase of stimulus intensity above threshold usually increases the amplitude and may decrease the latency of the EP; it may also change the shape and add more peaks to the EP. Amplitude does not usually grow linearly with stimulus intensity but may increase in proportion to the power or logarithm of stimulus intensity. Amplitude increases with stimulus intensity to a maximum after which it is usually stable but may decrease slightly if the stimulus is increased further. The latency of an EP peak may saturate at stimulus intensities different from those that give maximum amplitude of the peak. Latency and amplitude of each peak of an EP may vary independently of other peaks.

2.1.3 Stimulus duration

An increase in the duration of short stimuli has an effect similar to that of an increase of stimulus intensity. Such an increase therefore often increases the amplitude and decreases the latency of EP peaks. After reaching a certain duration, a long stimulus produces separate EPs at the onset and the end of the stimulus (1.3.6).

2.1.4 Stimulus rate

Stimulus rate accounts for the qualitative difference between transient and steady-state EPs. Specific stimulus frequencies producing transient and steady-state EPs depend on the EP type (1.3.5).

Stimulus rate is measured in cycles per second or hertz for sine wave stimuli such as sound waves, and in number per second for other stimuli. The denomination hertz is sometimes also used for nonsinusoidal stimuli as in the 40-Hz AEP.

2.1.5 Number of stimuli

The number of stimuli to be used for an average depends on the amplitude of the EP components relative to the amplitude of the ongoing EEG activity (3.3.3).

2.1.6 Size and location of stimulated area

VEPs differ depending on the stimulated part of the visual field. AEPs vary with the frequency of the sound stimulus that excites different parts of the cochlea. SEPs to stimulation of different arm and leg nerves have different latencies, shapes, and distributions.

2.2 RECORDING METHODS

2.2.1 Electrode types and application methods

2.2.1.1 EEG electrodes

EEG electrodes are used most commonly for EP recordings from the scalp and often also for EP recordings from the skin. They consist of a metal cup, often having a central hole and a flat rim with an outer diameter of 4–10 mm. The cup is connected by an insulated lead wire to a connector plug. Before electrodes are applied, the site of electrode application, determined by measurements referring to bony landmarks (2.2.3), is prepared by wiping with alcohol or acetone; a mild abrasive may be used to remove dry superficial skin layers.

Several methods may be used to attach electrodes to the scalp and skin.[80] The collodion technique gives stable recordings with few artifacts. In this technique, a cup electrode with a central hole is placed onto the prepared application site and held in place with a stylus while collodion is applied around the edge of the electrode and spread onto the scalp or skin. Spreading and drying of the collodion are facilitated by a stream of compressed air guided through a tube around the stylus. After the electrode is securely attached, the stylus is removed and the cup is filled with conductive jelly injected through the central hole. Electrodes are removed by dissolving the collodion with acetone. Because of the chemicals used, this method should not be used in areas with limited ventilation or with explosion hazard such as infant isolettes or operating rooms.

Another method of applying electrodes uses a conductive paste that can both hold the electrode in place and provide good electric contact. A dab of paste is placed on the prepared site and a cup electrode with or without a central hole is pressed down until it makes contact with the scalp or skin. The electrode should be covered with a gauze pad or tape to protect it and delay drying of the paste. After the recording, the electrodes can be removed easily and the paste washed off with water. Although this method is faster, it is susceptible to more mechanical and electric problems than the collodion technique.

2.2.1.2 Clip electrodes

An electrode mounted in a clip is sometimes used for recordings from the ear lobe. This electrode should consist of the same kind of metal as that of scalp electrodes so that it can be paired with those electrodes at the inputs of the same amplifier without causing electric problems.

2.2.1.3 Adhesive ECG electrodes

For recordings of the ECG, metal electrodes covered with conductive gel may be attached to the skin with an adhesive patch and should be paired with similar electrodes at the amplifier input.

2.2.1.4 Needle electrodes

Needle electrodes are used for recording in some laboratories. Sharp steel or platinum wires are inserted into the superficial layers of the skin or scalp after disinfection of the insertion site. The lead wires of the needle electrode must be attached to the skin or scalp to avoid pulling the needles out. Although they can be applied rapidly and usually cause no problems, these electrodes carry the risk of discomfort, infection, and electric recording problems and should be used only under restricted conditions. Because they have high impedance, especially for low frequencies, they may be useful for recordings of fast EPs with amplifiers of high input impedance.[29,77] Needle electrodes are occasionally used for electrocochleographic recordings (2.2.1.5), for SEP recordings from the interspinal ligament (19.4.10), and for recordings of sensory nerve action potentials from peripheral nerves (20.1); they may also be used for stimulation of peripheral nerves (15.3.1.1). Because of the risk of infection, needles should not be used during prolonged recordings such as those needed for monitoring in intensive care units. They should be discarded after use in patients who have, or may have, Jakob-Creutzfeldt disease.

2.2.1.5 Electrocochleographic electrodes

For transtympanic recordings, a needle electrode may be placed through the ear drum so that the electrode tip lies against the promontory of the middle ear near the cochlea.[4] These electrodes should be inserted only by otological specialists using general anesthesia in children and local anesthesia in adults. In extratympanic recordings, the electrocochleogram is recorded from a needle electrode inserted into the anterior wall of the external auditory canal[73] or from specially shaped electrodes placed into the lumen of the ear canal.[23,43,78]

2.2.1.6 Electroretinographic electrodes

Recordings of the electroretinogram (ERG) can be made from the eyeball with a contact lens attached to a scleral speculum[54] or with a light-weight corneal contact lens (jet electrode).[41] The lenses contain one or two recording leads and usually degrade visual acuity to some degree but may permit stimulation with patterned light. For unobstructed vision during the recording, a gold foil electrode may be hooked over the lower eyelid[6] or minute silver-impregnated nylon fibers may be placed on the cornea.[28] Periorbital EEG electrodes are less suitable for ERG recordings.[19]

2.2.1.7 Electrocorticographic electrodes
Recordings from the cortical surface may be made during neurosurgical procedures exposing the cerebral cortex. The electrodes consist of spring-mounted metal balls or saline-soaked cotton wicks. A matrix of metal electrodes embedded in a sheet of plastic material may be used for chronic subdural recordings.

2.2.1.8 Intracerebral electrodes
Multicontact wire electrodes may be inserted stereotactically into the brain for acute recordings during an operation or for subsequent chronic recordings.

2.2.2 Electric properties of recording electrodes

Recording electrodes should be capable of conducting, without distortion, the potential changes in the frequency range of the EP to be recorded. To obtain this recording capability, one must (1) choose recording electrodes of suitable material, (2) test electric continuity, if in doubt, by measuring the electrode resistance between the ends of the electrode, (3) evaluate the electric contact between electrodes and scalp or skin by measuring electrode impedance, and (4) avoid electrode polarization and bias potentials.

2.2.2.1 Electrode materials
The materials on the electrode surface should not interact with the electrolytes of the scalp or skin. Electrodes coated with gold, tin, or platinum are satisfactory. Silver electrodes covered with silver chloride are required for recording potentials that are slower than those recorded in routine clinical work (21.2).

2.2.2.2 Electrode resistance
Electric resistance, or opposition to direct current flow, is measured to test the electric continuity of an electrode if a break in continuity is suspected. For this measurement, the two ends of the electrode are connected to an ohmmeter which passes a weak direct current through the electrode and shows a readout of resistance; the resistance of an intact electrode measures no more than a few ohms.

2.2.2.3 Electrode impedance
Electric impedance, or opposition to alternating current flow, is measured to ascertain good electric contact between an electrode and scalp or skin after the electrode

has been applied. This measurement should be made before the start of every recording and should be repeated during the recording if there is reason to suspect bad electric contact. The impedance of an EEG scalp electrode should be between 1,000 and 5,000 Ω.

Electrode impedance is measured with an impedance meter which passes a weak alternating current from the electrode selected for testing to all other electrodes connected to the meter. Alternating current is used for this measurement since it is more representative of the alternating potentials in the EP than direct current and because it avoids electrode polarization caused by direct current.

Both very low and very high impedance are undesirable. Very low impedance short-circuits the amplifier input and is often due to an abnormal conduction pathway between the recording electrodes, for instance, an excess of electrolyte jelly or paste or the presence of saline or sweat; electrodes with impedances of less than 1,000 Ω should be inspected, cleaned, and reapplied. An electrode of very high impedance, paired with an electrode of lower impedance at the input of a differential amplifer, may cause an electric imbalance that favors the recording of interference, especially of 60 Hz artifact from power lines. Only when the electrode impedance is so high that it equals or exceeds the input impedance of the amplifer will it significantly reduce the amplitude of the recording. Electrodes with impedances over 5,000–10,000 Ω should be checked and usually need improvement of their mechanical and electric contact with the scalp or skin, although a break in continuity, such as a break between electrode cup and lead wire or between lead wire and plug, may also cause very high electrode impedance.

2.2.2.4 Electrode polarization and bias potentials

Electrode polarization and bias potentials may distort EP recordings but are easily avoided with modern techniques. Polarization is caused by the flow of electric current through the recording electrode. The current distributes ions at the electrode so that current flows better in one direction than in the other and thereby distorts the recording of alternating potentials. Polarization can be minimized by measures that reduce the flow of current through recording electrodes, namely, by using amplifiers with high input impedance and electrodes with fairly large contact areas, and by avoiding steady current flow, especially that used to measure electrode resistance while the electrode is in contact with scalp or skin. Bias potentials are caused by ion exchanges not due to current flow and can be avoided by using pure metal electrodes with clean surfaces and by pairing similar electrodes at the inputs of each amplifier.

2.2.3 Electrode placement

The placement of recording electrodes depends on the EP type to be recorded and generally aims at obtaining the highest amplitude and clearest definition of the peaks of an EP. Electrodes are placed as closely as possible to the presumed generator in near-field recordings (1.3.4.1); widely spaced electrodes are used in far-field recordings from distant structures (1.3.4.2).

To make electrode placements constant, the site of each electrode must be determined by measuring coordinates with reference to standard landmarks such as bridge of the nose, mastoid process, parts of the ear, cervical or lumbar vertebrae, and clavicle. A tape measure and marker pen should be used in each case. The International 10–20 System,[47] widely used in clinical EEG recordings, may be employed to determine the positions of the vertex and midfrontal electrode locations often used in EP recordings. The 10–20 system may also be used to designate the position of other scalp electrodes by indicating their distances relative to the points defined by the 10–20 system.

2.3 AMPLIFIERS

In many modern averagers, the amplifiers, input selector switches, calibration units, and filters are housed together with the computer in a single unit; only the input board is separated from the other electronic apparatus so that it can be brought near the patient to connect the recording electrodes through a cable to the amplifier. Other averagers, especially those requiring external amplifiers, have no input board so that the amplifiers have to be placed near the recording site.

2.3.1 Input board

The input board consists of a box that connects the recording electrodes on the subject, through a shielded cable, to the amplifier and computer. This box contains receptacles for the electrode connector plugs. The receptacles are usually labeled either in pairs as *input 1* and *input 2* (2.3.4.1), connecting directly to the inputs of the amplifiers, or with numbers or symbols indicating electrode placement. They are connected to selector switches that are used to select two electrodes as inputs to each amplifier. Input selector switches are very useful in recording EPs with averagers having more than one channel because they allow quick and easy input selections.

The input board also provides a receptacle for the ground electrode. A ground electrode on the patient should always be connected to this receptacle unless the subject is grounded through another connection, for instance, through a ground connection to a stimulator or an ECG monitor, in which case the ground of the recording equipment must be connected to the ground of the other equipment. Because grounding is most effective in reducing interference if the ground connection is close to the recording site, it may be necessary to use a ground electrode placed near the recording electrodes for EP recordings and to disconnect the patient from the ground lead of other equipment. For the safety of the patient, the input board should be constructed so that no more than 20 μA of current can flow through any of the electrodes and the patient to ground.[76]

The input board may also contain a meter for measuring electrode impedance (2.2.3.3).

2.3.2 Calibration

To determine the voltage represented by vertical EP deflections, the recording system must be calibrated at the start of each recording session. This is done by feeding square pulses of known voltage to the inputs. These pulses must be amplified with the same gain and filter settings, averaged for the same number of times, and displayed with the same gain settings as the responses recorded from the nervous system. This calibration procedure requires a source of selectable, very precise electric pulses that can be driven to appear at a constant interval after a trigger pulse. The selection should include pulses of 5 and 10 μV and of 10–50 msec for cortical EPs and of 0.5 and 1 μV and of 2–5 msec for subcortical EPs. In averagers with more than one channel, each channel must be calibrated separately. The horizontal deflection of the averager display hardly ever needs to be calibrated because of the great precision of the timing circuits used in digital computers.

Most commerical averagers offer the option of cursors that can be placed on any point of the computer display of an EP and give the readout of the amplitude and latency of that point (3.4.4). In these averagers, the averaged calibration pulse may be used to verify the amplitude readout of the cursors: When one cursor is placed on the calibration pulse and the other one on the baseline, the readout should equal the voltage of the averaged pulse. In averagers not equipped with cursors, a hard copy of the calibration pulse is made with the same gain settings as used for copies of EPs; the height of the calibration pulse is measured and compared with the height of EP deflections. For instance, if a calibration pulse of 10 μV has a height of 5 cm, and an EP peak has a height of 2.5 cm when displayed in the same manner, then the peak has an amplitude of 5 μV.

2.3.3 Input impedance

The electric impedance of the amplifier input, as encountered by signals applied differentially between input 1 and input 2, must be high compared with that of the recording electrodes to avoid loss of amplitude of the signals recorded (2.2.2.3). The input impedance of the amplifiers should be 10 MΩ or more.

2.3.4 Differential amplification and polarity convention

Biological amplifiers serve (1) to discriminate cerebral potential changes, or signals, that appear at one of the two electrodes connected to the amplifer inputs without affecting the other electrode in the same manner, (2) to reject extracerebral potential changes, or artifacts, that affect both electrodes, and (3) to increase the amplitude of the discriminated potentials.

2.3.4.1 Discrimination of cerebral potentials and polarity convention

Electric potential differences are recorded between two electrodes that are electrically coupled to the two inputs of an amplifier, called input 1 and input 2. The basic operation of a differential amplifier is to subtract the voltage at input 2 from that at input 1 and to amplify only the voltage difference. This results in amplifying mainly cerebral potentials, which usually have different voltage and timing at the two electrodes, and in discriminating them from extracerebral potentials, or artifacts, especially 60 Hz interference, which often have the same voltage and timing at the two electrodes and are therefore canceled at the amplifier input.

The difference between the voltages at input 1 and input 2 makes the display of the amplifier output go either up or down. Because a potential change at either input can cause a deflection at the output, the output of a single channel cannot reveal which electrode is active; furthermore, the direction of the deflection at the output indicates polarity of the input activity only if the active input is known: An upward deflection of the output may mean either positivity at one input or negativity at the other input. The direction of the output also depends on the polarity convention used in the construction of the amplifier.

Figure 2.1 illustrates the two polarity conventions used in amplifiers of averagers. (1) Some averagers, especially many of those manufactured in the United States, are wired to give an upward deflection at the output in response to a relative positivity at input 1 and a downward deflection in response to a relative positivity at input 2 (Figure 2.1a). (2) Other averagers, commonly those made in Europe, follow the EEG convention of giving a downward deflection at the output as a result of relative positivity at input 1 and

Figure 2.1. The two polarity conventions (*a, b*) used in differential amplifiers. Amplifiers are represented by triangles with two input leads to the left and an output lead to the right; input 1 is above input 2. The circles at the tip of the amplifiers in *b* indicate that these amplifiers invert the output polarity as compared to the amplifiers in *a*. This figure illustrates only the polarity relationships between input and output; amplification is not shown. In polarity convention *a*, a positive signal at input 1 produces an upward deflection at the output (1); a positive signal at input 2 causes a downward output deflection (2). In polarity convention *b*, these signals cause output deflections opposite to those of polarity convention *a*: A positive potential at input 1 causes a downward deflection at the output (1) whereas a positive potential at input 2 produces an upward output deflection (2).

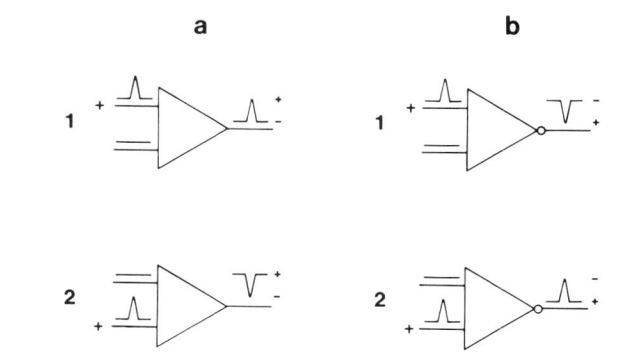

an upward deflection as a result of relative positivity at input 2 (Figure 2.1b). To identify the polarity of EP tracings, the polarity convention used to depict an EP must be indicated (4.4).

The results of differential amplification using either polarity convention are illustrated in Figures 2.2 and 2.3. Figure 2.2 shows the effect of signals of different amplitude and polarity; Figure 2.3 shows the effect of signals of different timing and polarity. Both figures acknowledge the fact that neither electrode is likely to be entirely unaffected by potentials near the other electrode. In summary, the output of the amplifier always indicates the net difference between the inputs; the polarity of the output depends on the polarity convention of the specific amplifier; and the amplitude depends on the gain setting used.

The cerebral potential changes reflected in EPs rarely come from only one electrode, although this could be the case in a recording using an electrode located on the scalp directly over a small area of cortex generating the EP and another electrode located at a great distance from the generator. In these rare instances, the first electrode may deserve the name *active* or *exploring electrode* and the other one, *inactive*, *indifferent*, or *reference electrode*. However, the terms *active* and *reference* electrode are often used less strictly to indicate that one electrode is closer to the generator than the other one. The term *referential recording* is sometimes used to describe this recording situation. The terms *monopolar* or *unipolar recording*, occasionally used to suggest that signals originate only at one electrode, should not be used in any description of differential recordings because all such recordings are made between two points.

In most EP recordings, both electrodes pick up some potential changes so that each electrode adds or subtracts parts of the EP; the contribution by each electrode depends mainly on its distance and spatial orientation with respect to the generator. The distinction between *active* and *reference electrode* becomes unimportant. It loses its meaning entirely in far-field recordings where neither electrode is located much closer to the generator than the other one. The term *bipolar recording* is sometimes used to denote

Figure 2.2. Differential amplification of signals of different amplitude and polarity appearing simultaneously at both inputs of an amplifier using polarity convention *a* of Figure 2.1. (1) Signals of equal amplitude and polarity are rejected. (2) Signals of equal amplitude and opposite polarity are added. (3) A signal of lower amplitude and equal polarity at input 2 is subtracted from the signal at input 1. (4) A signal of lower amplitude and opposite polarity at input 2 is added to the signal at input 1. As Figure 2.1, this figure does not show amplification between inputs and output.

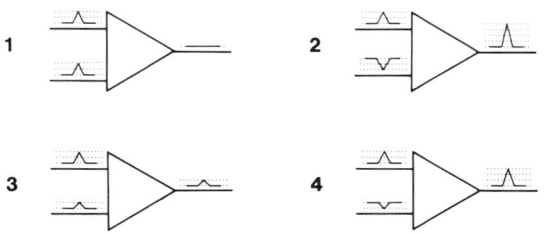

recordings between two electrodes that are both known to contribute to the EP. However, the terms *bipolar* and *referential* are best reserved for multichannel recordings.

The term *average reference electrode* is used to denote recordings attempting to find an average potential level against which to record potential changes at another electrode. This average is produced by using more than one electrode as input 2 of an amplifier so that each electrode selected contributes equally to the average potential level at that input. For instance, electrodes on both ears or both mastoids may be connected to one input and used as an average reference.

Simultaneous recordings of EPs in more than one channel are made by selecting different electrode pairs as inputs to each channel, creating different montages. Two types of montages are used in multichannel recordings. (1) Referential montages use the same electrode as input 2 of all channels; this electrode, usually located at some distance from the areas generating the EP, becomes the reference for the various electrodes connected to input 1 of all channels. (2) Bipolar montages connect different pairs of electrodes to the different amplifier inputs; often the electrode at input 2 of one channel is also used at input 1 of the next channel so that the electrode pairs of successive channels form a chain along or across the head. Both electrodes in a pair contribute to the EP, each to a different degree.

Figure 2.3. Differential amplification of signals of the same amplitude appearing with different timing at the inputs of an amplifier using polarity convention *a* of Figure 2.1. (1) Input signals of opposite polarity that partly overlap in time produce a prolonged output deflection. (2) Signals of opposite polarity that appear successively at inputs 1 and 2 produce a succession of deflections of the same polarity as the potential at input 1. (3) Input signals of the same polarity that partly overlap in time produce a rapidly changing biphasic output deflection. (4) Signals of the same polarity that appear successively at inputs 1 and 2 cause a more slowly changing biphasic output deflection. As Figure 2.1, this figure does not show amplification of signals between input and output.

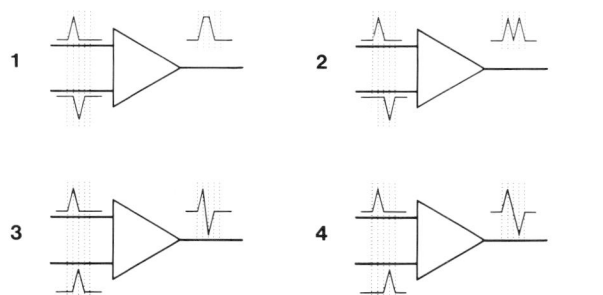

2.3.4.2 Rejection of artifacts

It should be understood that differential amplification rejects only those artifacts that cause identical potential changes at both inputs; such potential changes are said to be in common mode. Other artifacts, greater at one electrode than at the other, are amplified like cerebral activity. By the same rule of differential amplification, cerebral activity, such as the EEG or the responses averaged in EPs, is rejected to the extent that it has the same polarity, amplitude, and timing at both inputs. The ability to reject common mode signals is expressed as the common mode rejection ratio. This is the ratio of the amplifier output produced by a signal applied differentially, i.e., between the two inputs, over the amplifier output produced by the same signal when it is applied in common mode, i.e., between both inputs tied together and the amplifier ground. The common mode rejection ratio of amplifiers used for EP recordings should be 10,000:1, i.e., 80 dB or more.

Effective discrimination, or rejection of artifacts, depends on several factors. A defect in any one of these factors is often revealed by the appearance of power-line artifact in the recording. This artifact is due to interference from power lines. This interference is present to some degree at all EP recording sites; in some areas, interference may be strong enough to require electric shielding of the recording room. The artifact has a frequency of 60 Hz in the United States and of 50 Hz in countries using alternating current of that frequency. Interference is normally kept out of recordings by discrimination: Because interference usually appears with similar amplitude at the inputs of the differential amplifier, it is rejected at this stage. However, various defects of the recording electrodes may make the interference appear with different amplitude at the two inputs. This amplitude difference is amplified and appears at the output.

The most common cause of 60 Hz artifact is unequal electrode impedance. This is often due to partial or complete loss of contact between one electrode and scalp or skin or to a break in the continuity of an electrode, for instance, a break between lead wire and metal cup or connector plug. Infinitely high impedance may result from the inadvertent failure to connect an electrode to an open input and has a similar effect as grounding one input: The amplifier then no longer operates differentially but increases signals between the remaining input and the ground electrode. Problems with the ground electrode are a frequent cause for picking up 60 Hz interference in differential recordings. If the ground is not connected to the subject, the amplifier inputs float without reference to the potential level at the recording site, and the amplifier uses its own internal ground potential level to subtract signals at input 2 from those at input 1. Input potentials that are equal with reference to the subject's head, and would therefore be eliminated by differential recording with a proper ground connection to the head, may now appear different with respect to the amplifier ground and not be cancelled. For instance, 60 Hz or heart-beat artifacts may not be nulled. Cerebral potentials may show little obvious distortion when recorded with floating inputs. However, electric hazards for the subject are generally increased in the absence of a ground connection between subject and recording equipment.

If only one amplifier input is connected to the subject, i.e., if both the other input and the ground connection are open, the output of the amplifier represents the potential difference between the electrode on the subject and the amplifier ground; this kind of recording, although at times resembling a normal recording for brief periods, is meaningless and usually marred by artifacts.

The electronic parts of the amplifier themselves give rise to potential fluctuations that are added to the output generated by the input signals. The noise level of the amplifier, measured when both inputs are closed and the filters wide open, should not exceed a few microvolts.

2.3.4.3 Amplification

2.3.4.3.1 Gain and sensitivity. The increase of the voltage of a signal between the input and the output of an amplifier can be described by gain or sensitivity. Gain is the ratio of signal voltage obtained at the amplifier output to the signal voltage applied at the input: An amplifier set to give an output signal of 1 V for an input signal of 10 μV has a gain of 100,000. This voltage ratio is sometimes expressed in decibels, the number of decibels amounting to 20 times the logarithm 10 of the voltage gain (dB = $20 \cdot \log_{10}$ gain). For instance, a gain of 10 equals that of 20 dB and a gain of 100,000 equals that of 100 dB. In contrast, sensitivity is the ratio of input voltage over the size of the deflection it produces in a tracing of the output. For instance, an amplifier set to produce a vertical deflection of 1 cm for an input of 10 μV has a sensitivity of 10 μV/cm, or of 1 μV/mm.

2.3.4.3.2 Recording and display gain. The differential amplifier at the input of an averager increases the voltage of the signal as it is being recorded. Another amplifier is used to display the output of the computer, namely, the digitized EP, after the completion of averaging. Although both the recording and the display gain affect the same substrate—the vertical size of the tracing—they do so at different stages of the recording and may substitute for each other only within limits. In particular, the recording gain must be set to match the vertical resolution of the analog-to-digital converter for the most effective averaging (3.3.5). The display gain is used to adjust the size of the completed average for viewing and plotting purposes.

2.4 FILTERS

Filters serve to exclude from the recording those potential changes that have frequencies different from the frequencies represented in the response under study. This form of signal conditioning reduces unwanted waveforms before they reach the computer. The noise-reducing, signal-enhancing action of filters resembles the effect of averaging and indeed can reduce the number of responses that need to be collected for the clear definition of an EP. Even though averaging alone can enhance the EP, without filtering the number of responses required for clear definition of the EP may be impractically large (3.3.3). Also, filtering of high-frequency components is needed in conjunction with averaging to eliminate components that exceed the limit imposed by the sampling rate (3.3.4).

Filtering can be accomplished in several ways. The most

commonly used filters reduce or eliminate frequencies higher or lower than a middle range that contains the frequencies of the EP components under study. Most filters operate on the recording before it is digitized and are said to be analog filters; they can be further divided into passive filters made mainly of resistive and capacitive components and active filters containing power-consuming components such as transistors. Analog filters not only reduce the amplitude of unwanted components, but also distort the time relationship between some of the desired signal components passing through the filter, especially those near the cutoff frequencies. This phase shift can be avoided, and a very sharp cutoff can be accomplished, by the use of digital filters, but these filters consist of computer programs and are more costly and difficult to handle.

The effect of high and low frequency filters is usually specified in terms of the frequency of sine waves that are reduced in amplitude by a certain fraction. The frequency range between the low and high cutoff points, which is not significantly affected by the filter settings, is called the *bandwidth,* or *bandpass.* Because most EPs do not have the shape of sine waves, the choice of the most suitable filter settings may be difficult. A very rough estimate of filter action can be made by measuring the lengths of the waves in an EP. For instance, BAEPs contain wavelets lasting about 1–2 msec, corresponding with frequencies of 500–1,000 Hz. However, filter settings excluding all but this narrow range would not be satisfactory because BAEPs also contain slower and faster rising and falling phases that would be deformed by filtering unless much wider frequency settings were used. A good practical estimate of filter action can be obtained by comparing the effects of different filter settings on the same EP; for this purpose, an EP may be recorded simultaneously in different channels with different filter settings, or responses may be stored on magnetic tape and played back repeatedly for averaging with different filter settings. Although such an estimate cannot be made for each recording, estimates for each type of EP have been obtained often enough to provide practical guidelines for suitable filter settings for each EP type. Theoretically, the effect of filters can be predicted by computing the power spectrum of an EP, which shows the spectrum of frequencies that are contained in the EP and thereby indicates the frequencies that may be eliminated by filtering.

2.4.1 Analog filters

2.4.1.1 Low frequency filters or high-pass filters
Low frequency filters reduce the amplitude of slow waves without attenuating faster waves. This action can be described in terms of the low filter frequency setting and the time constant. The low filter frequency indicates the frequency of sine waves that are reduced in amplitude by a specified fraction. The effects of different low frequency filter settings on sine waves of different frequencies are shown in Figure 2.4 with dotted lines. In the commonly used passive filter made of a capacitor (C) in series and a resistor (R) in parallel to the signal path, the cutoff is usu-

Figure 2.4. Effect of low frequency filters (dotted lines) and high frequency filters (dashed lines) on the amplitude of signals of different frequency. Relative amplitude (vertical scale) indicates the amplitude of the amplified signals at different frequencies (horizontal scale). Note that the low frequency filter setting of 10 Hz reduces the amplitude of a signal of 10 Hz by about 30 percent, reduces signals of slightly higher frequency to a lesser degree, and reduces signals below 10 Hz much more without completely abolishing them. The effects of the high frequency filters are mirror images of those of the low frequency filters. The combination of a high and a low filter frequency designates the bandwidth of the frequencies that will be unaffected.

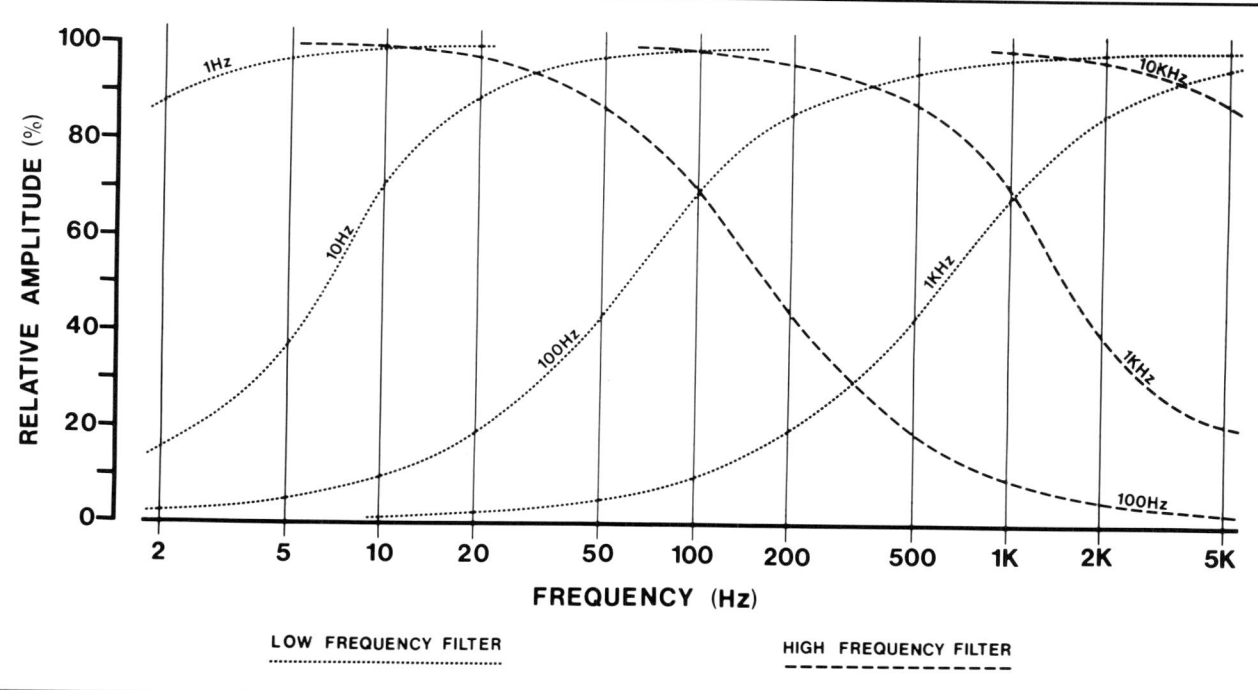

ally defined as the 3-dB point, i.e., the frequency at which a signal at the filter output is reduced by 3 dB, or 30 percent; this frequency (*f*) amounts to

$$f = \frac{1}{2\pi \cdot R \cdot C}$$

Slightly higher frequencies are affected to a minor degree; lower frequencies are progressively more reduced. This progression has a roll-off slope of 6 dB/octave, i.e., the amplitude drops by 50 percent as the frequency is reduced by a factor of 2. Passing the signal through another filter of the same setting, or using the same filter twice, for instance, during tape recording and playback, doubles the steepness of the filter slope. Very steep slopes can be obtained with active filters. For clinical EP recordings, roll-off slopes for low-frequency filters should not exceed 12 dB/octave because steeper slopes may cause too much phase shift (2.4.1.4)

The time constant is the time required by a sudden sustained voltage change to decay to 37 percent of its peak value. The time constant T of a simple RC filter is $T = R \cdot C$; it is related to the cutoff frequency (*f*) by the equation

$$T = \frac{1}{2 \cdot \pi \cdot f}$$

The time constant may be determined by measuring the time interval between the onset of a sustained calibration pulse and the point at which its amplitude has dropped to 37 percent.

The limits of low frequency vary with the design of the coupling of amplifiers. Coupling through low frequency filters limits the recording to alternating signals such as alternating current (AC). To record sustained potential changes such as generated by direct current (DC) requires direct coupling. Amplifiers used in routine clinical EP recordings should have low frequency filter settings that include 0.1 Hz. Lower filters or direct coupling (DC recording) may be needed for very slow potential changes such as those of some of the ERPs (21.2).

2.4.1.2 High frequency filters or low-pass filters
High frequency filters reduce the amplitude of waves of high frequency and let waves of low frequency pass without attenuation. The action of different high frequency filter settings is shown in Figure 2.4 with dashed lines. This action is usually specified by the high filter frequency which is the mirror image of the low filter frequency, having 3-dB cutoff points at

$$f = \frac{1}{2\pi \cdot R \cdot C}$$

and roll-off slopes of 6 dB/octave for filters made of a resistor in series and a capacitor in parallel with the signal path. The time constant, or the time required for a sudden input voltage to rise to 63 percent of its peak value, equals $T = R \cdot C$ in an RC filter.

Amplifiers to be used for clinical EP recordings should have high frequency filter settings including at least 5,000 Hz. High frequency filters should have roll-off slopes not exceeding 24 dB/octave.

2.4.1.3 60-Hz notch filter

Some amplifiers provide a filter that greatly reduces the amplitude of waves of frequencies in a very narrow band centering at 60 Hz, the most common artifact from power-line interference in countries using AC of 60 Hz; a 50-Hz notch filter is used in countries with that power-line frequency. Although this filter affects only a very narrow part of the spectrum of frequencies in an EP, the frequencies of the filtered band may form an important part of the EP, and the EP may be significantly distorted by the use of the filter.[7] The filter should therefore not be used routinely. If it must be used, the same responses should also be averaged without the filter to assess how useful the filter was in reducing the artifact and how harmful it was in distorting parts of the EP.

2.4.1.4 Phase shift

Analog filters alter not only the amplitude but also the timing of signals. Low frequency filters affect mainly slow waveforms and make them appear earlier than fast ones (phase lead); high frequency filters are of importance for fast deflections and make them appear later than slow ones (phase lag) (Figure 2.5). The phase lead and lag induced by a passive RC filter at different low and high cutoff frequencies, respectively, are shown in Figure 2.6. Although these effects are most prominent at frequencies near the cutoff and beyond, they extend into the range of the frequencies that are passed with little attenuation and may significantly distort the timing of EP peaks.[12,27,29,40,52,61] In particular, phase shifts of either kind may falsify the interval between the stimulus and the EP peaks and may distort the time interval between EP peaks, especially between peaks that are composed of different frequencies. Passive filters generally shift phase less than active filters, although the degree of phase shift varies with the type of active filter.[35,61]

2.4.2 Digital filters

Digital filtering consists of the use of computer programs that operate either on the digitized responses before averaging or on the EP after completion of averaging. A relatively simple digital filtering method consists of smoothing (3.4.3). Other methods are more complex and have only rarely been incorporated in routine clinical averaging procedures. They include methods for elimination of low and high frequencies, Wiener filtering, and time-varying filters.

Computer programs can eliminate components of low or high frequency with very sharp cutoff and without phase shift. In contrast to analog filtering, these methods cause no temporal distortion of EPs.[12,27,34,40,88,92,96]

Figure 2.5. Phase shift by low frequency filter (LFF) on a VEP containing relatively slow wave forms (top two tracings), and by high frequency filter (HFF) on a BAEP containing relatively fast wave forms (bottom two tracings); note the different time scales for VEP and BAEP. An increase of the low frequency filter setting from 0.3 Hz to 100 Hz shifts the VEP peaks toward the left, or shorter latency; the gain had to be increased for this demonstration because the filter change also reduced the amplitude (see calibration bars). A decrease of the high frequency filter setting from 30 kHz to 1 kHz shifts the BAEP peaks toward the right, or longer latency.

Figure 2.6. Effect of low frequency filter (top) and high frequency filter (bottom) on the phase of signals of different frequency (horizontal axis). Values above the horizontal zero line indicate the amount of time by which the low frequency filters shift signals of lower frequency ahead of faster signals not affected by the low frequency filter. Values below zero indicate the amount of time by which high frequency filters shift signals of high frequency past those of low frequency. For instance, a low frequency filter setting of 10 Hz makes waves of 20 Hz (i.e., waves commonly represented in the VEP and other cortical potentials) appear almost 4 msec early; a low frequency filter setting of 100 Hz, commonly used in BAEPs and other subcortical EPs, makes waves of 200 Hz (i.e., waves prominently represented in the BAEP) appear nearly 0.5 msec early. On the other hand, a high frequency filter setting of 100 Hz, often used for cortical EPs, may delay even waves of low frequency of less than 50 Hz by 1–2 msec, whereas a high frequency filter of 1 kHz, sometimes used for subcortical EPs, delays waves of high frequencies of 200–500 Hz by about 0.15 msec. These filter curves were derived from measurements of the filters of the same commercial amplifier as used in Figure 2.4. Comparison of the two figures shows that phase shift by filters affects waves of frequencies that are also affected by amplitude changes; for instance, a low frequency filter setting of 100 Hz shifts the phase of a 200 Hz wave by 0.4 msec, or about 10 percent of its period, and reduces its amplitude by less than 20 percent.

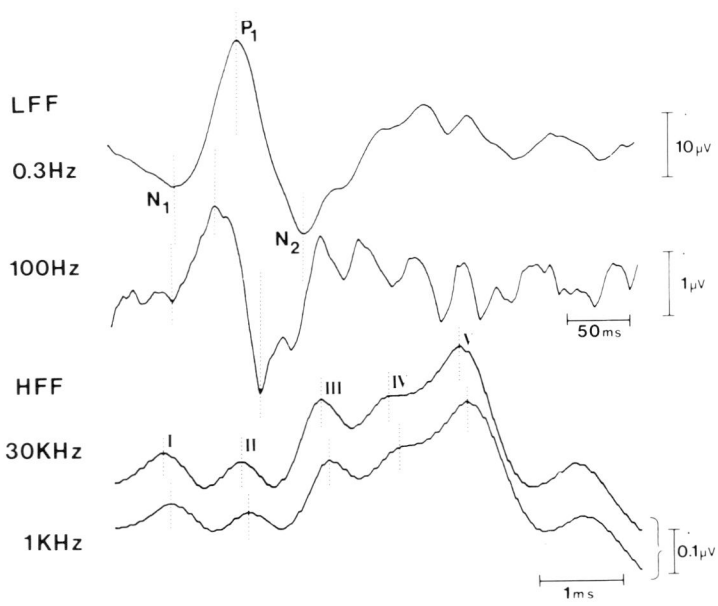

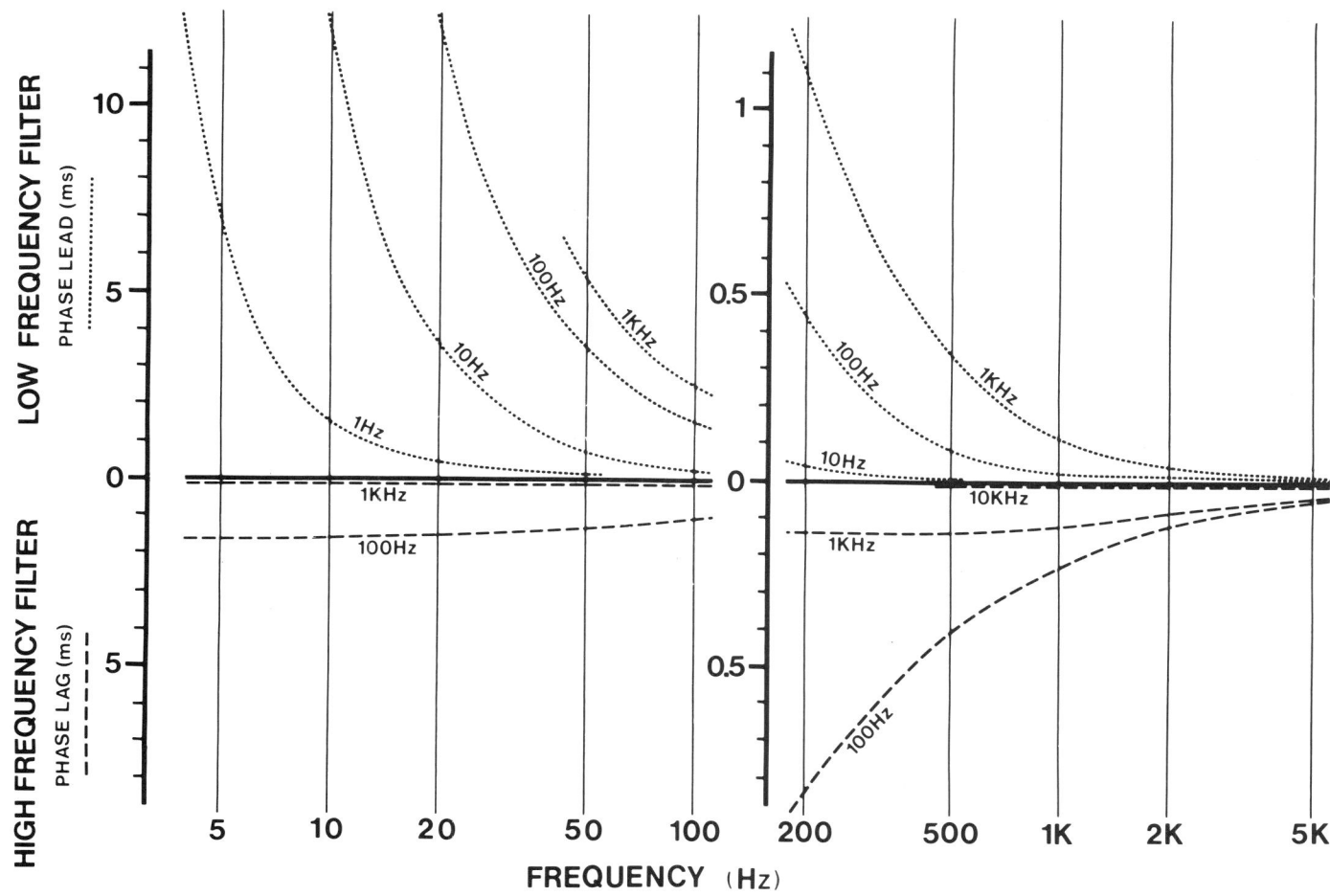

Wiener filtering separates frequency components of the signal from those of noise either when the characteristics of both components are known beforehand[32,94] or after they have been calculated from a small sample of EPs.[44] The value of this method is still in doubt.[18,33,88,97]

Time varying filters are needed if signal or noise is not stationary.[40] Adaptive or learning filters are designed to progressively develop criteria for detection of the signal.[62,63,95,105]

3
Averaging

SUMMARY

3.1. In principle, averaging extracts the time-locked responses to a stimulus from unrelated potential changes such as the EEG, artifacts, and other noise by repeatedly adding the responses following each stimulus and dividing by the number of responses.

3.2. Averaging computers are either special-purpose computers that are permanently programmed or general-purpose computers that may be programmed for averaging and other tasks.

3.3. Several recording parameters must be selected by the operator. The number of channels is usually limited to a maximum of 2, 4, or 8. The averaging period may be started at, or shortly before or after, the stimulus. The number of responses collected ranges from 50 to 200 for cortical EPs and from 1,000 to 4,000 for subcortical EPs. In theory, the number of responses required depends on the relative amplitude of the responses and of the noise. In practice, the number should be large enough that two successively averaged EPs can be superimposed without significant differences. The length of the interval to be averaged (*analysis period, epoch,* or *sweep*) is usually 250–500 msec for VEPs, 10–15 msec for BAEPs, and 40–100 msec for SEPs; the sweep must be long enough to include all important parts of the EP with sufficient resolution of detail. The recording gain should be adjusted so that the response amplitude fills about one-half to two-thirds of the vertical range of the computer input.

3.4. Several features of the computer can facilitate EP recording and evaluation. An artifact rejection option, provided in most averagers, is used to exclude sweeps containing excessively large deflections. Oscilloscope displays are used to monitor the incoming responses and the accumulating average during the recording. The completed EP may be smoothed to eliminate unnecessary high frequency components. Cursors serve to indicate latency and amplitude of any point of the EP. Data manipulations include addition, inversion, and subtraction of EPs. EPs may be stored as hard copies in the form of XY plots, photographs, or photosensitive paper copies of the oscilloscope screen. EPs may be stored in digital form on magnetic discs for later measurements, manipulation, and hard copying.

3.5. Magnetic tape is used occasionally to store continuously recorded responses for later averaging.

3.6. Methods other than averaging that can be used to analyze electric responses of the CNS to sensory stimulation include frequency analysis and power spectral analysis.

3.1 PRINCIPLE OF AVERAGING

The electric responses of the brain, brain stem, or spinal cord to a single stimulus, when recorded at the surface of the body, are small and obscured by EEG, ECG, muscle activity, and other biological and extraneous electric activity so that individual responses cannot be clearly distinguished and seem to fluctuate in repeated recordings. Averaging serves to extract the responses, time-locked to the stimulus and considered signal in this context, from potential changes unrelated to the stimulus and here summarily considered noise. Averaging is done by presenting sensory stimuli repeatedly, collecting and adding each response to the preceding ones, and dividing the sum by the number of responses. This procedure enhances the signal by reducing the noise toward zero.

Averaging is carried out by a digital computer that (1) records electric activity during the selected time period, (2) converts the continuous voltage change of the recording (analog recording) during that period into a sequence of numbers (digital recording), (3) adds the numbers representing recordings after successive stimuli to each other and scales them to the average (Figure 3.1). The period of analysis (epoch or sweep) is started by a trigger pulse from the stimulator and must be long enough to include the response under study. The analysis period is divided into equal segments of time, called *bins* or *points*. The analysis dwells on each point for a dwell time or bin width amounting to a fraction of a second per point. This time represents the reciprocal of the sampling rate which equals the number of points sampled each second. For the analog-to-digital (A/D) conversion, or digitization, the amplitude of the recording segment contained in each bin is measured and converted into a number of a corresponding size. Because averaging computers operate with digital numbers, the numbers representing amplitude consist of a string of binary digits, or bits, of 0 or 1. The result of the A/D conversion of a single response is a sequence of binary numbers which are stored at discrete locations, or addresses, of the computer memory and can be displayed on the oscilloscope as a sequence of illuminated dots approximating the outline of the response. To obtain an average,

Figure 3.1. Averaging of EPs. From the recordings of continuous voltage changes appearing at the output of the amplifier (analog, top left), the computer admits for analysis the recording during the selected poststimulus interval (epoch), divides it into regular intervals (bins) at a sampling rate of many intervals per second, and measures the voltage at each interval. These measurements are converted into a sequence of digital numbers represented by dots on the oscilloscope display (response 1, top right). The next response is digitized in the same manner (response 2) and the numbers representing the voltages for each bin are added, bin for bin, to those of the first response, and the sum in each bin is divided by two: (R1 + R2)/2. Successive responses are added in the same manner and the sums are scaled after each addition of a response to obtain a new running average: (R1 + R2 . . . + Rn)/n. After the designated number of responses have been averaged, the EP may be displayed in digital form (bottom right) or converted back into analog form (bottom left).

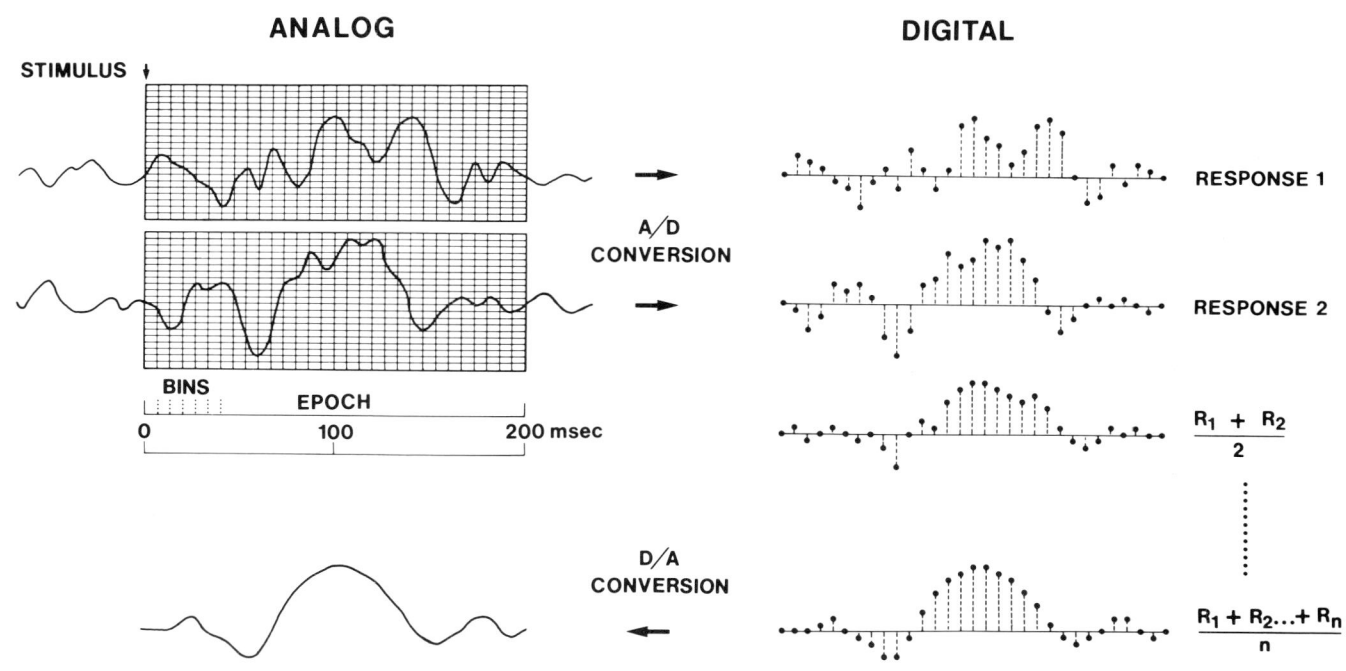

successive responses are digitized and added to the preceding ones; after each addition, the numbers in each bin are divided by the number of responses collected to give the running average until the designated number of responses have been averaged.

3.2 COMPUTER TYPES

Two types of digital computers can be distinguished by the method of programming. Special-purpose, or hard-wired, computers, have one or a few programs permanently stored in their memory. When dedicated to evoked response averaging, these computers are also called averaging computers, or averagers. Averaging computers usually include amplifiers, filters, calibration signal sources, synchronizing inputs and outputs from and to stimulators, and other options for recording cerebral responses; they may include stimulators. Parameters of stimulation, recording, and averaging can be selected with front-panel controls. Special-purpose computers are usually less expensive than general-purpose computers.

In contrast, general-purpose computers are capable of many different tasks and must be programmed for each task by a sequence of instructions given in a specific code, or language. Once a program for averaging has been developed, it can be stored on magnetic tape or disc and loaded into the computer when needed. Parameters of analysis are selected by typing keyboard commands. General-purpose computers are more versatile, but usually require more time for setting up and running a program than special-purpose computers. General-purpose computers are usually sold without matching biological amplifiers, stimulators, and other options that make the use of special-purpose computers for averaging easier.

3.3 ACQUISITION PARAMETERS

3.3.1 Number of channels

Most averagers have 2, 4, or 8 channels from which the operator can select the number of channels needed for a study. In most averagers, the number of addresses per channel is constant, but in some averagers, it decreases as the number of channels selected is increased. To maintain the same sweep length, the sampling rate must be decreased in these averagers.

The multichannel option may be used to record EPs either simultaneously from different areas or successively from the same area; successive recordings are often made to evaluate the reproducibility of EPs or to study EPs to stimulation of either side of the body. Such recordings can be compared directly by superimposing the EPs on the oscilloscope screen and measuring them with cursors; these methods are much more expedient than the comparison of EPs on hard copies.

3.3.2 Triggering

To average responses to successive stimuli, each epoch must be started at the same time with respect to the stimulus. In principle, there are three choices of starting points for the averaging of transient EPs (Figure 3.2). Not all of them are available on every averager.

Figure 3.2. Three options of timing the start of the averager. (1) Averaging starts at the stimulus onset (*a*) and includes the entire response (*b*). (2) Averaging starts before the onset of the stimulus *a* and includes a prestimulus recording period *b*. (3) Averaging starts after the stimulus *a* and excludes the stimulus artifact and the early part of the poststimulus period *b*.

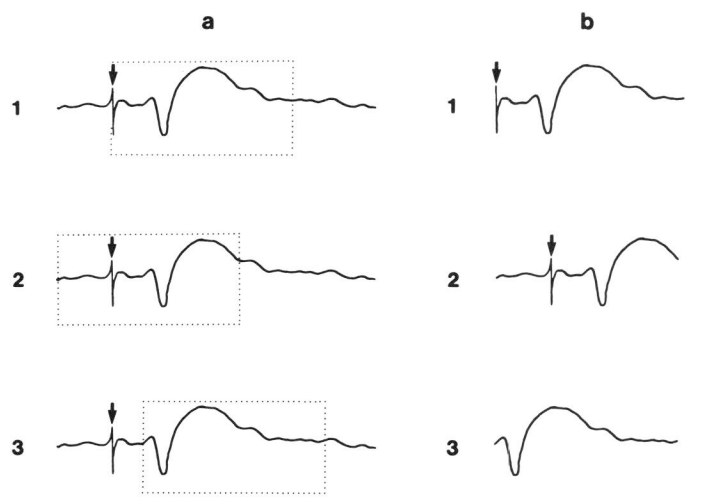

1. Triggering at the onset of the stimulus (1*a* in Figure 3.2) is the most commonly used method. It includes in the average most of the stimulus artifact and the earliest part of the EP (1*b*). In this technique, either the stimulator triggers the computer or vice versa.
2. Prestimulus triggering (2*a* in Figure 3.2) includes a period of recording before the stimulus (2*b*) which can be used both as a baseline against which to measure the amplitude of EP peaks and as an estimate of the amount of residual noise in the average (3.3.3). Prestimulus averaging may be accomplished either by delaying the stimulus against the synchronizing pulse which starts the computer or by using a computer capable of retaining a period of recording before the appearance of the triggering pulse.
3. Poststimulus triggering (3*a* in Figure 3.2) can be used to eliminate from the EP (3*b*) a period between stimulus and onset of the response if this period is very long or contaminated by stimulus artifact. Poststimulus delays are usually generated by the computer, but may be produced by delaying the synchronizing pulse to the computer against the stimulus.

Steady-state EPs are most commonly averaged by triggering the computer at the onset of one of the repetitive stimuli. More than one cycle of the rhythmical steady-state EP is usually included in the epoch. Stimuli occurring during the epoch cannot trigger the computer, and only the first stimulus after the epoch is effective in triggering a new sweep. Because the responses to the first few stimuli may undergo adaptive changes before steady state is reached, collection of steady-state EPs is best started after at least several stimuli have been given.

3.3.3 Number of responses to be averaged: Noise reduction

The number of responses is selected with a sweep counter that stops the acquisition of data after a preset number of sweeps have been collected. The number of sweeps needed for an EP can be determined on the basis of practical and theoretical considerations.

In practice, the number of responses collected should be large enough that successively averaged EPs do not differ from each other. Each EP should be recorded at least twice, and the tracings should be superimposed to ascertain that they resemble each other closely enough in latency, amplitude, and shape. The criteria for replication vary among laboratories but usually amount to a few milliseconds for VEP peaks and to fractions of milliseconds for short-latency AEP and SEP peaks. If two successive EPs do not replicate within these limits, more averages must be obtained until it is clearly established whether an EP of constant features is present. In most cases, satisfactory near-field EPs such as VEPs are obtained by averaging 50–200 epochs, whereas far-field EPs such as the BAEP require averaging 1,000–4,000 or more epochs; SEPs, usually including both near- and far-field recordings, require 500–2,000 epochs. The number of epochs averaged equals the number of stimuli and of responses in the case of transient EPs. Steady-state EPs, usually averaged for more than one cycle of the rhythmical response, require larger numbers of stimuli.

In theory, the number of epochs required depends on the ratio of the amplitude of the stimulus-locked components in the recording, or of the signal, to the amplitude of the unrelated components, or of the noise. Because averaging reduces the noise, this ratio improves with the number of responses averaged. The relationship between signal-to-noise ratio and number of responses is not linear: The number of responses increases with the square of that ratio. Therefore, averaging improves the signal-to-noise ratio by a factor equaling the square root of the number of epochs averaged. This is strictly true only if the signal does not vary and if the noise consists of randomly distributed frequencies. Although these conditions are not met exactly in clinical EP recordings, the rule of the square root gives a rough estimate of the number of responses to be averaged for the various EPs. For instance, a cortical response which has an amplitude of 10 μV and is embedded in EEG activity of 20 μV, has a signal-to-noise ratio of 1:2. Averaging only 4 responses improves this ratio by a factor of 2, increasing the signal to the size of the noise. An average of 64 responses, which increases the signal-to-noise ratio by a factor of 8, makes the signal four times larger than the noise. To make the signal eight times larger than noise, i.e., to improve the original ratio of 1:2 by a factor of 16, would require 256 epochs.

Subcortical EPs have a less favorable signal-to-noise

ratio and therefore need averaging of larger numbers of responses than cortical EPs. These responses often have amplitudes of only 0.5 μV; if embedded in EEG of about 20 μV, their signal-to-noise ratio is 0.025. To bring the EP to the level of the noise alone requires a 40-fold improvement, or 1,600 responses; to make the EP twice as large as the noise requires an 80-fold improvement, or an average of 6,400 responses. Improvement by another factor of 2 would require an average of 25,600 responses. The nonlinear relationship between number of responses and improvement of signal-to-noise ratio puts a practical limitation on the ability to enhance EPs by averaging. This illustrates the importance of using good recording methods to minimize artifacts and of using filters to exclude those portions of the frequency spectrum that do not contain signal components; both these measures reduce the noise component, let the computer begin to work at a more favorable signal-to-noise ratio, and thereby effectively reduce the number of responses required for the definition of an EP. For instance, the filtering of slow EEG waves during the recording of a BAEP may reduce the noise component to 5 μV. A response of 0.5 μV then requires only 100 repetitions for the 10-fold increase to the noise level and will be 4 times larger than the noise after only 1,600 repetitions.

Long periods of recording an EP should be avoided because they may increase the chance that the signal or noise characteristics vary, for instance, as a result of changes in the subject's level of attention or alertness. Variation of the response characteristics during the averaging will reduce those response parts that vary; such variations can be detected by breaking the average of a large number of responses into smaller parts. Variations of the noise component, for instance, the appearance of slow waves of drowsiness in the EEG, may diminish the signal-to-noise ratio and thereby impair the definition of the EP. Of great practical importance are gross deviations from random noise such as those caused by intermittent large transients, for instance, artifacts or K complexes of the sleep EEG. If such a transient enters the average, it is not reduced by the factor predicted for noise reduction because it does not represent randomly distributed noise. This is why it is important to exclude large artifacts with an artifact rejection option (3.4.1).

The degree of noise reduction is manifested by the amplitude of residual noise in the averaged EP. An estimate of residual noise can be obtained by various methods which may be used to answer the question whether an average contains EP peaks or only noise (4.1.3.1).

3.3.4 Horizontal parameters of analysis: Analysis period, number of points, sampling rate, and dwell time

The choice of the analysis period, epoch, or sweep length, depends on the duration of the EP under study. For transient EPs, this period extends only through the fraction of the interstimulus interval that contains the EP peaks of interest. The most common analysis periods are 200–250 msec for VEPs, 10–15 msec for BAEPs, and 50–100 msec for SEPs. Longer periods are needed to capture such long-latency peaks as encountered in EPs of infants, pathologically delayed EPs, and some event-related potentials.[21] For steady-state EPs, the analysis period is made longer than the interstimulus interval so that more than one cycle of the rhythmical response is recorded, usually a few cycles. Because the timing circuits of digital computers operate with binary numbers, the actual durations of analysis periods are 1.024 times longer than the rounded-off numbers indicated on the dials of most averagers (Table 3.1, columns 1 and 2).

The length of the analysis period is related to the number of points available for averaging and the dwell time or bin width which is the reciprocal of the sampling rate (3.1). The analysis period equals the number of points multiplied by the dwell time or divided by the sampling rate. In most commercial averagers, the analysis period is selected directly; this selection, in conjunction with the number of points available, determines the dwell time and sampling rate. However, in a few computers, dwell time or sampling rate is selected directly, and analysis period is an indirect result of this selection and of the number of available points. Most averagers offer 1,024 points per channel. Some averagers with the option of selecting different numbers of channels divide the 1,024 points available for single-channel operation into halves of 512 points, and quarters of 256 points, when using two or four channels, respectively; other multichannel averagers use a full complement of 1,024 points for each channel regardless of how many are in use (3.3.1). Combinations of the most commonly used analysis periods, numbers of points, dwell times, and sampling rates are illustrated in Table 3.1.

In selecting the horizontal parameters of analysis, two requirements must be fulfilled: (1) The analysis period must be long enough to include the important peaks of the EP to be studied, and (2) the sampling rate or dwell time must be sufficient to resolve the highest frequencies of EP peaks. In particular, care must be taken not to choose a combination which spreads the number of points available so widely that they cannot adequately depict the high-frequency components of the response under study. To depict a sine wave digitally, at least two points are needed for each cycle. As a minimum, the sampling rate of an averager therefore must be twice as fast as the fastest sine wave component in the signal to be resolved. This critical sampling rate is sometimes referred to as the *Nyqvist frequency* and is listed in the sixth column of Table 3.1; the seventh column indicates the shortest wavelength corresponding with this frequency. The Nyqvist frequency specifies only the minimum frequency for temporal resolution of components of the EP, but does not ensure the most effective averaging. An increase of sampling rate above this minimum can greatly reduce the number of responses required for a clear definition of the EP; usually at least 10–20 points are used to depict one cycle of the fastest sine wave component of an EP.

The relations between the signal frequency and the sampling rate are illustrated in Figure 3.3, which shows the effect of using the same sampling rate on three sine waves of increasing frequency; whereas waves of relatively low

TABLE 3.1. Relation between parameters of horizontal resolution

Designated epoch (msec)	Analysis period (msec)	Number of points	Dwell time (msec)	Sampling rate (kHz)	Nyqvist frequency (kHz)	Shortest sine wave resolved (msec)
1,000	1,024	1,024	1	1	0.5	2
		512	2	0.5	0.25	4
		256	4	0.25	0.125	8
500	512	1,024	0.5	2	1	1
		512	1	1	0.5	2
		256	2	0.5	0.25	4
250	256	1,024	0.25	4	2	0.5
		512	0.5	2	1	1
		256	1	1	0.5	2
200	204.8	1,024	0.2	5	2.5	0.4
		512	0.4	2.5	1.25	0.8
		256	0.8	1.25	0.625	1.6
100	102.4	1,024	0.1	10	5	0.2
		512	0.2	5	2.5	0.4
		256	0.4	2.5	1.25	0.8
50	51.2	1,024	0.05	20	10	0.1
		512	0.1	10	5	0.2
		256	0.2	5	2.5	0.4
20	20.48	1,024	0.02	50	25	0.04
		512	0.04	25	12.5	0.08
		256	0.08	12.5	6.25	0.16
10	10.24	1,024	0.01	100	50	0.02
		512	0.02	50	25	0.04
		256	0.04	25	12.5	0.08

and medium frequency are well resolved in the digital representation, a sine wave slightly faster than the critical value of one-half the sampling rate is not correctly represented after digitization: The frequency of the digitized wave form is lower than that of the original analog sine wave. This erroneous representation of an analog wave by digital values suggesting a lower frequency, due to a sampling rate of less than twice that of the original waveform, is called

Figure 3.3. The relationship between sampling rate and signal frequency: Effects of different signal frequencies at the same sampling rate. Sine waves at three different frequencies (*a, b, c*) are sampled at the same rate. At each sampling interval, their amplitude is marked by a dot. These dots, connected by straight dashed lines, are shown as a digital display at the bottom (*d, e, f*). The sine wave in *a* is sampled about 14 times per cycle and well depicted in the digital representation in *d*. The sine wave in *b*, being sampled at a rate only 6 times higher than its own frequency, is depicted with less, but still fair, detail after digitization in *e*. In contrast, the sine wave in *c*, falling slightly below the critical value of one-half the sampling rate, is quite distorted in the digital representation in *f*, which shows fewer peaks than the original wave train, thereby falsely suggesting the presence of lower frequencies than contained in the signal wave.

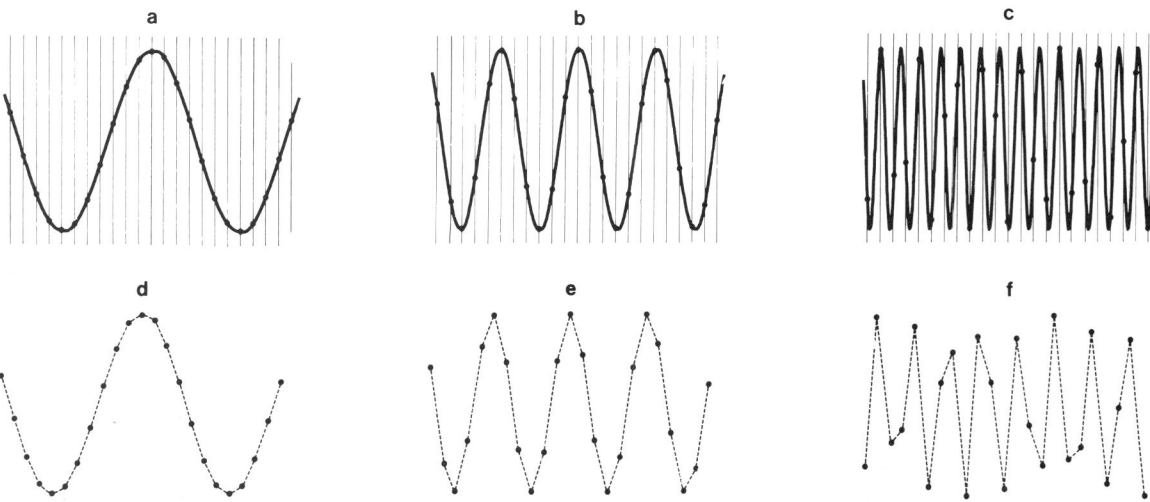

aliasing. Aliasing may be the source of significant errors in EPs and can be prevented by the choice of high frequency filter settings that eliminate components above the critical sampling rate.

Therefore, when setting up facilities to record a specific EP type, one must determine the frequency of the fastest component in the EP to be recorded and then calculate the limits of the high frequency resolution resulting from analysis period and number of points to ensure that the highest frequencies in the EP will be adequately represented. Although manufacturers of averagers generally tailor their equipment so that this requirement is likely to be fulfilled for the most commonly used EPs, there is no guarantee, and the burden of proof is on the operator. As a minimum, averagers used in clinical practice should be capable of a sampling rate of 12.5 kHz, i.e., a dwell time of no more than 80 μsec per address; they should have at least 250 addresses per channel.

3.3.5 Vertical parameters of analysis: Number of bits, resolution, and recording gain

At each sampling interval, the digitizer measures the amplitude of the response and converts it into a binary number that consists of binary digits, or bits. How well the amplitude of the signal is reproduced digitally depends on the signal size and the number of discrete vertical intervals available for the digital representation. The range of vertical measurements possible is given by the number of bits that the digitizer can assign to the amplitude of the signal at each sampling interval. Such a string of bits is called *word*.

Most averaging computers have A/D converters with a word length of 8 bits; this number of binary digits can represent decimal numbers up to 2^8, or 256. The capacity to resolve 256 discrete steps of amplitude can also be expressed as resolution which refers to the smallest change detectable within a total of discrete numbers. An 8-bit A/D converter, having a total capacity of 256 parts, can resolve 1 part in 255, or about 0.4 percent of the amplitude of a signal filling the entire vertical range. Averagers operating with longer words have higher resolutions.

The relationship between signal size and vertical resolution is illustrated in Figure 3.4. Signal parts exceeding the range of the digitizer are distorted; signals occupying about half the digitizer range or more are fairly faithfully depicted in digital form, but smaller signals are inadequately resolved.

For each EP recording, the computer operator must match the signal size with the vertical capacity of the digitizer by adjusting the recording gain. This is easy in averagers that are built so that the vertical range of the digitizer coincides with the height of the oscilloscope display. In these averagers, the gain of the recording amplifier is adjusted so that most responses fill about half, or slightly more, of the oscilloscope face. Although an amplitude of 100 percent of the digitizer would theoretically be better, responses vary so much that selecting a gain that increases the average response amplitude to the maximum capacity of the digitizer results in the distortion of many excessively large responses. Distortions can be avoided if the computer is equipped with an artifact rejection option (3.4.1) that can exclude such excessive responses. Lower gain settings require collection of larger numbers of responses; below a minimum gain, even the largest number of responses cannot compensate for small amplitude.

Addition and storage of the responses in the computer memory require a wider range of amplitudes than those dealt with by the digitizer; the word length of the memory should exceed that of the digitizer by at least 8 bits.

Figure 3.4. The effect of digital conversion on signals of different amplitude. Sine waves of three amplitudes (*a, b, c*) are digitized using vertical intervals of equal size. At each step, the amplitude, marked by a dot, is measured. These measures are shown at the bottom (*d, e, f*) where the dots are connected by straight dashed lines. The signal in *a* exceeds the vertical range of the digitizer and is distorted in the digital representation in *d*. The signal in *b* fills about one-half of the digitizer range and is fairly well resolved in *e*. The signal in *c* fills only a small part of the vertical range and is poorly resolved in *f*.

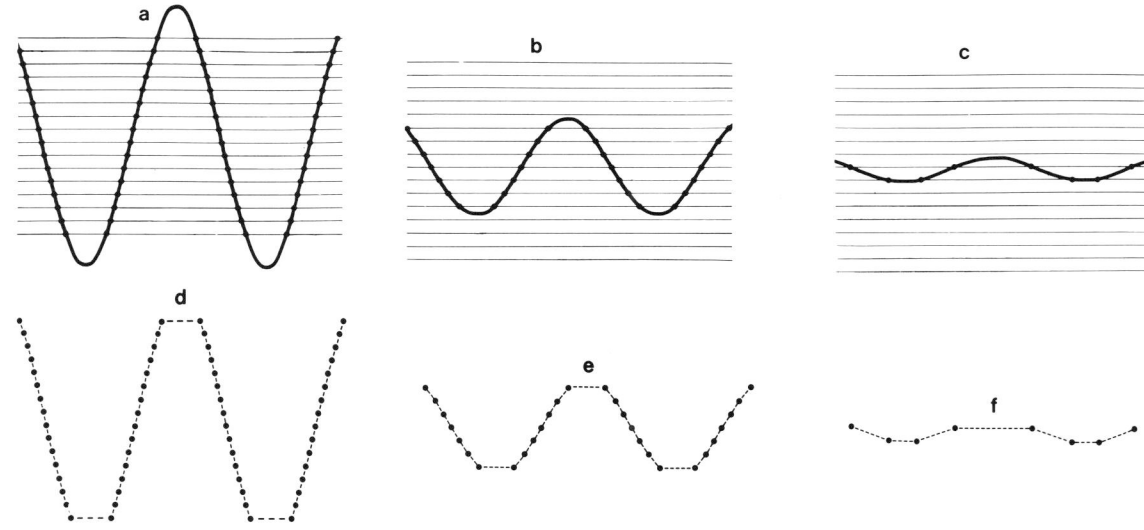

3.4 AVERAGING DEVICES

3.4.1 Artifact rejection option

A relatively simple way to prevent some artifacts from entering the average EP is to exclude responses containing deflections of an amplitude that exceeds a selected voltage range. This procedure is based on the assumption that responses of excessive amplitude are likely to contain artifact. The procedure requires selection of a voltage level likely to include all or most acceptable variations of response amplitude while excluding artifacts that are larger than those variations; obviously, smaller artifacts will not be rejected and unusually large responses may be rejected. In many averaging computers with built-in amplifiers, an artifact rejection option is coupled with the recording gain so that the computer rejects every response that is larger than the voltage range displayed on the oscilloscope. Other computers show the artifact rejection option as horizontal lines that mark the allowable amplitude limits of the response. When triggered by a deflection of excessive amplitude, the artifact rejection option prevents the entire response from entering the average (Figure 3.5). An arti-

Figure 3.5. Artifact rejection. (*a*) A single response falling within the amplitude boundaries selected for artifact rejection is accepted and counted for the average. (*b*) A response partly exceeding the boundaries of the artifact rejection option is rejected in toto and not counted for the average. (*c, d*) Subsequent responses falling into the selected amplitude range are accepted, counted and added to the average.

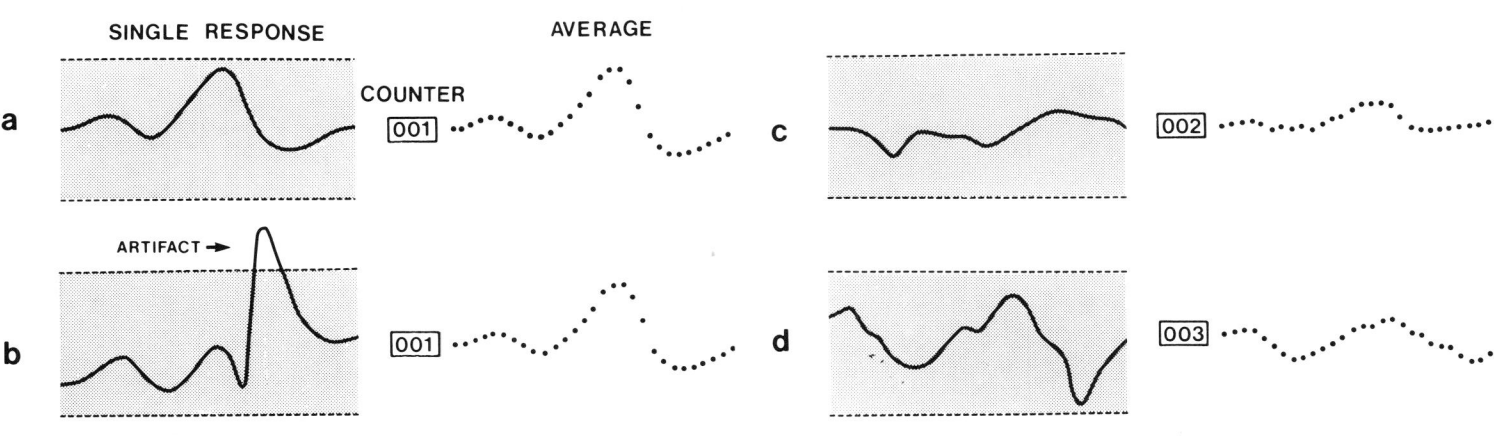

fact rejection option can be added to computers not having one.[66]

3.4.2 Oscilloscope displays

Averagers should have at least one oscilloscope to permit monitoring of the averaging process at two or three stages (Figure 3.6). (1) The average digitized EP is displayed as a sweep which recurs with each response during the averaging and as a steady display after the completion of the averaging (Figure 3.6c). (2) Single digitized responses are observed before they are added to the average (Figure 3.6b). In some averagers, this option is chosen with a selector switch as an alternative to the display of the accumulating average; other averagers use a second oscilloscope to display the digitized response. This display is used to adjust the recording gain for the optimum utilization of the digitizer (3.3.5) and to set the threshold of the artifact rejection option (3.4.1); it is also useful for the identification of artifacts in the recording. (3) Single responses may be observed before digitization and compared with the digitized response to ascertain that the digitized tracing adequately represents the analog response, especially in the range of high frequencies (Figure 3.6a). This option is not a standard feature in most averagers.

Another feature found in a few averagers is the option to let the averager run continuously so that the oscilloscope shows a continuous display of the activity at the electrodes; this display may be used to monitor the EEG and to identify artifacts in the recording.

Figure 3.6. Simultaneous display options during averaging. (*a*) Single response, representing the input to the computer. (*b*) Same response, after digitization. (*c*) Digital average of all responses collected.

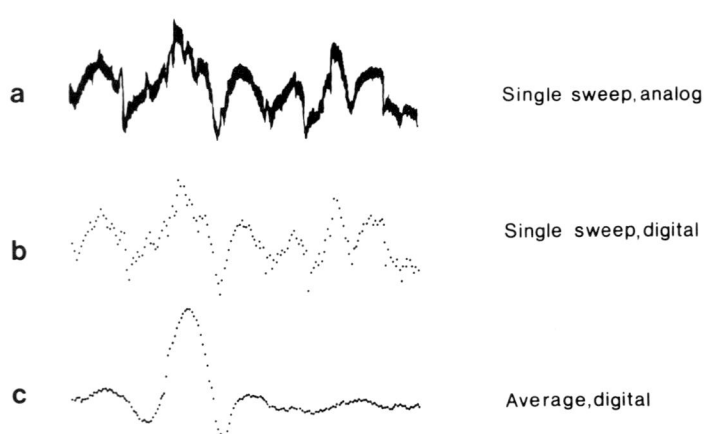

3.4.3 Smoothing

High frequency noise components are often superimposed on EPs and can be reduced by smoothing, a fairly simple digital filtering method (Figure 3.7). This operation replaces each point of a digital tracing by a moving average of the 3, 5, or more neighboring points. The smoothing operation may be used repeatedly but must be used with caution. Even though it does not distort the phase relationship between EP peaks, it may change the EP shape by reducing the amplitude of short waves more than that of long ones.

3.4.4 Cursors

A cursor is a marker represented by an increased brightness of one dot in the line of dots forming a digital display

Figure 3.7. Effect of 3-point smoothing on a BAEP containing high frequency noise. (*a*) BAEP before smoothing. (*b*) Same BAEP after smoothing once. (*c*) Same BAEP after smoothing five times. Note the disappearance of high frequency components including the stimulus artifact and the better definition of the peaks.

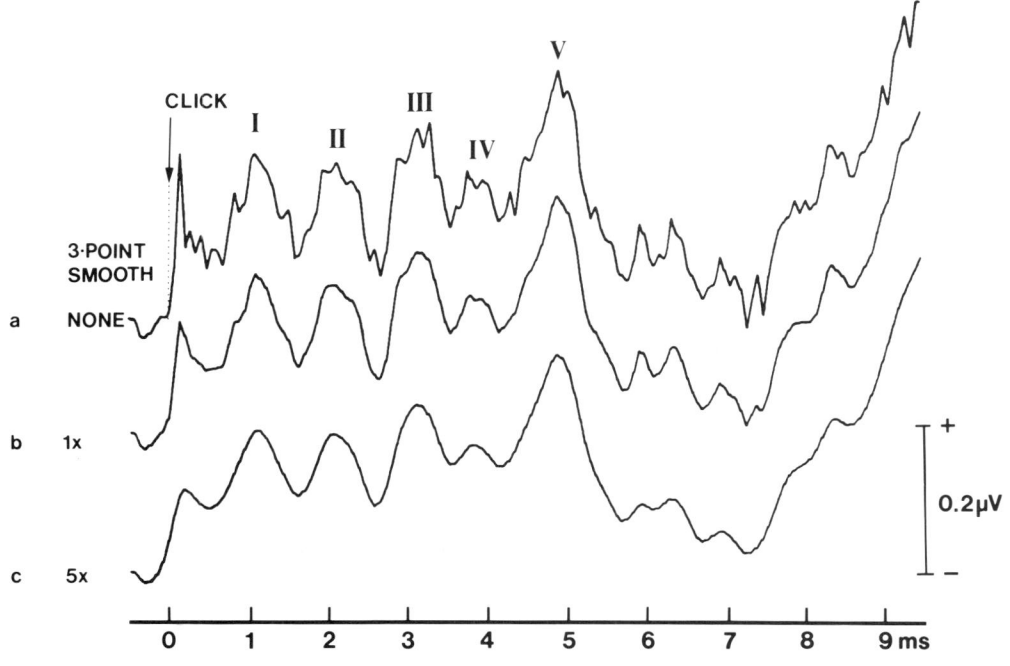

(Figure 3.8) or by a vertical line above or below that display. The cursor can be placed on any point of the digital display. The numerical values of the amplitude and latency of the point indicated by the cursor are shown on the oscilloscope screen. The accuracy of the horizontal readout is limited by the dwell time; the accuracy of the vertical readout depends on the voltage represented by each bit of the digitizer. Most averagers provide a selection of one or two cursors. If one cursor is selected for a measurement, the readout indicates amplitude and latency of the selected point with reference to the first point of the display. If two cursors are selected, the readout indicates the difference in amplitude and time between the two points marked by the cursors (Figure 3.8); this facilitates measurements of interpeak latencies (4.1.3.2.2) and of peak-to-peak amplitude (4.1.4.1.2). In most averagers, changes of the vertical display gain made after completion of the averaging do not alter the voltage readout, but changes of the sweep length do alter the readout and may cause errors in latency measurements by the cursors if latency is read with sweep settings different from those used during averaging.

Figure 3.8. EP with cursors. The tracing consists of a string of points representing average voltage at each point. Two cursors, indicated by brighter points in the tracings and marked by arrows for this illustration, are placed on the major downward and upward deflections. As a result, the readout above the EP indicates the difference of amplitude (21.3 μV) and latency (20.00 msec) between the peaks.

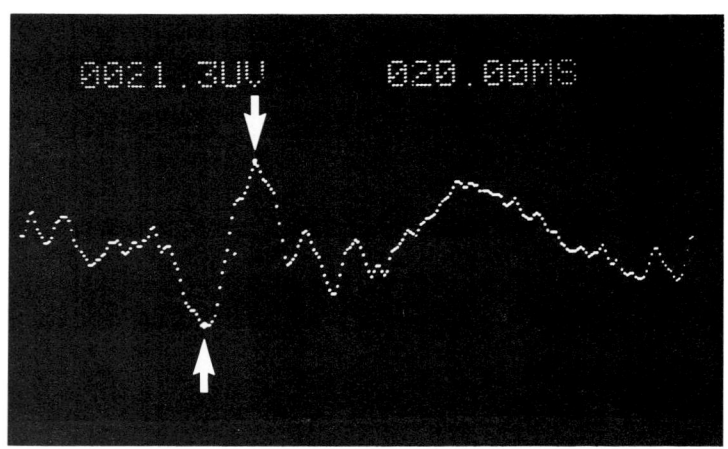

3.4.5 Addition, inversion, and subtraction of EPs

Many averagers offer the ability to add the contents of parts of the memory to each other, to invert them, and to subtract them from each other. The ability to add tracings stored in two halves or two quarters of the averager memory enables the operator to collect averages of smaller numbers of responses in two channels, to compare them with each other for an assessment of constancy or variability of the EP, and to add them for a grand average for final measurement of the EP parameters. The ability to invert the polarity and to subtract tracings stored in different parts of the memory may be used occasionally to determine whether an EP is present in an average (4.1.3.1) or to separate peaks that vary with stimulus polarity from those that do not, for instance, to separate the cochlear microphonic from the action potential of the acoustic nerve (14.6).

3.4.6 Copies of EPs

EPs are copied to produce a permanent record and, in some instances, to obtain measurements for analysis. Several options are available. Although hard copies such as XY plots, photographs, and photosensitive paper copies of the oscilloscope screen are all satisfactory for producing permanent records, they are not very satisfactory for precise measurements, which are best made with computer cursors (3.4.4). EPs can also be stored on magnetic disc or tape for later measurements with cursors and for production of hard copies.

3.4.6.1 Hard copies

3.4.6.1.1 XY plots. EPs may be plotted in analog form, as a continuous line, or digitally, point-for-point. Digital plotting allows inclusion of letters and numbers such as used for cursor readouts and labeling of coordinates.

The XY plot is usually larger than a photograph, giving better resolution, and plotting paper is relatively cheap. However, XY plotting is slower and may be noisier than other copying methods.

3.4.6.1.2 Photographs. Tracings on the face of the oscilloscope may be photographed with a polaroid camera; the photograph may include an illuminated grid and cursor readouts. Measurements from photographs are accurate only if the horizontal and vertical amplifiers of the oscilloscope are perfectly linear; an illuminated grid on the oscilloscope screen can be used as reference only if it is flush with the oscilloscope display and avoids parallactic distortion.

3.4.6.1.3 Photosensitive paper copies. Special copying devices may be used for making copies of the oscilloscope display on photosensitive paper. These copies have a resolution approaching that of XY plots, yet have the advantage of being faster, quieter, and cleaner.

3.4.6.2 Magnetic copies

EPs may be stored on magnetic disc or tape. These records must include information on acquisition parameters, especially sweep length and gain, and may be filed under the patient's name or an identification number. Magnetically stored EPs may later be retrieved and loaded back into the computer memory for digital measurements with cursors, for changes of display gain, smoothing, addition to other averages, and making of hard copies.

3.5 MAGNETIC TAPE RECORDING

Instead of averaging EPs during the recording session, or on-line, one may want to continuously record the amplified

EEG on magnetic tape for averaging of the responses after the recording, or off-line. This method is especially useful when responses are recorded in many channels simultaneously; they can then be averaged successively with a computer having a smaller number of channels. Tape recording also allows repeated averaging of the same responses and may be used to investigate the effects of different instrument settings (2.4). Synchronizing pulses marking each stimulus must be included on one tape channel to serve as a trigger for averaging. The tape recorder must be capable of recording and reproducing the entire range of frequencies that are represented in the EP and included in the instrument settings of the averaging equipment. Because of the low frequency components in most EPs, frequency-modulated (FM) tape recorders must be used. However, tape recording is rarely used in routine clinical EP work because it requires more time and increases the chances of technical errors. The increase in the number of channels afforded by tape recording is hardly needed now that four-channel averagers are easily available.

3.6 OTHER METHODS: FREQUENCY AND POWER SPECTRAL ANALYSIS

Electric responses of the nervous system to sensory stimulation may be analyzed with methods other than averaging. Steady-state responses can be separated from the EEG with a Fourier analyzer or a filter sharply tuned to the stimulus frequency.[37,69,87] These methods measure response components having the same frequency as the stimulus. Another method uses power spectral analysis to search for response components defined by the sine wave frequencies of these components.[20] Both frequency analysis and power spectral analysis yield measures representing the amplitude of the response; the clinically more important response latency is not measured directly (4.1.3.3).

4

General characteristics of evoked potentials

SUMMARY

4.1. EPs consist of a series of upward and downward deflections, here called *peaks* or *waves,* which are characterized by polarity, number in sequence, latency, amplitude, waveshape, and distribution. Different nomenclatures name peaks by either polarity and latency (N20, P100), number in sequence (I, II, III), or polarity and number in sequence (N1, P1, N2, P2).

4.2. The variability of an EP depends on the EP type and is generally greater among subjects than in the same subject at different times.

4.3. For clinical use, those EPs have been selected that vary least among normals and are most sensitive to pathology.

4.4. Labeling of EP records requires identification of several variables of the subject and of the stimulus and recording conditions.

4.1 GENERAL DESCRIPTION OF EPs

EPs consist of a series of peaks or waves, each characterized by (1) positive or negative electric polarity, (2) number of the wave in a sequence, (3) latency from the onset of the stimulus or from a preceding peak, (4) amplitude with respect to the baseline or to the preceding or subsequent peak of opposite polarity, and (5) waveshape.

4.1.1 Polarity

Polarity refers to positivity and negativity between the two electrodes connected to the inputs of the recording system. The relationship between the electric potential changes at the electrodes and the upward and downward deflections of an EP tracing depends on (1) which electrode is connected to input 1 and which to input 2, and (2) which polarity convention is used in the recording system (2.3.4.1). Therefore, upward and downward deflections in an EP must be explained in both terms.

4.1.2 Number in sequence

Most EPs have more than one peak. Peaks are labeled differently for each EP type. In some EP types, the number of a peak in the sequence of peaks is used to name the peak. Other EP types are named by combinations of polarity and number in sequence or of polarity and latency (4.1.7.1).

4.1.3 Latency

4.1.3.1 Peak identification
Peaks are often not clearly outlined but obscured by noise. The problem of having to identify peaks ("peak picking") in a noisy tracing can often be prevented by reducing the sources of artifacts such as interference and muscle contraction by using filters that eliminate high and low frequency components outside the range of the frequencies represented in the EP, and by collecting a large number of responses. However, in many instances, the problem cannot be eliminated by these means. The tracing then may show abundant noise and raise the questions of (1) how to determine the point at which latency and amplitude of a peak should be measured and (2) whether there is any EP peak at all.

If two or more small deflections of equal or similar amplitude are found riding on a larger EP wave, the point to be used for measurement of latency and amplitude may be chosen as the highest or the first of the deflections; instead, latency may be measured to the middle between the smaller deflections or to the point of intersection of two lines drawn through the ascending and descending portions of the wave on which the smaller deflections ride (Figure 4.1). Methods using computer programs to identify peaks and measure latency and amplitude have been described.[9,17,22,38,39,99,100] While any of these methods may be applied, only one should be used as a routine once it has been adopted. All these methods should be applied only to small deflections that are closely spaced, superimposed on a clearly defined single EP peak and obviously due to residual noise. None of the methods should be used if there is a possibility that the deflections represent two separate peaks; this possibility may be suggested by a separation larger than that between noise components in other parts of the same EP. The presence of two such peaks instead of one may represent a true variation of normal EP waveshape. This phenomenon, commonly described as a bifid or bilobed peak, should be characterized by measurement of both tips.

The question whether a noisy average contains any EP peak often arises when stimuli of threshold intensity are used. An answer may be obtained by comparing the EP to weak stimuli with an EP to stronger stimuli that shows a clearly defined peak and thereby indicates where to look for that peak in the EP to weak stimuli. Another method of detecting peaks in noisy recordings compares the EP in question with an average showing only the noise present during the averaging of the response. An estimate of this noise may be obtained (1) by collecting an average of recordings made without stimulation, (2) by including a period before the stimulus in the average of the responses[50] (3.3.2), or (3) by averaging the responses while alternating the polarity of the recording to eliminate stimulus-related components (plus-minus average).[74,75,101] These methods are used in some of the peak-measuring computer programs mentioned above.

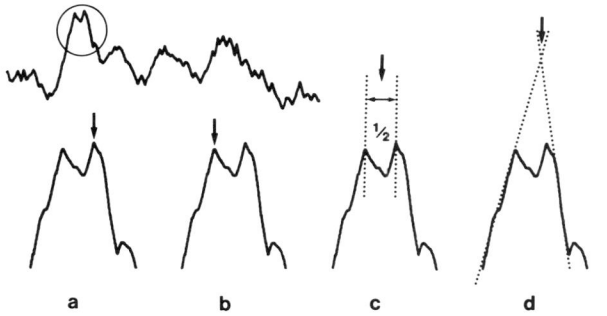

Figure 4.1. Four methods of selecting the point for measuring the latency and amplitude of an EP peak that is split into two tips by superimposed noise: (*a*) Highest tip, (*b*) first tip, (*c*) latency to the middle between the latency of each tip, (*d*) latency to the intersection of lines through rising and falling flanks of the tips.

4.1.3.2 Latency of transient EPs

For transient EPs, latency is usually expressed either as peak latency or as interpeak latency. Latency to the onset of an EP deflection is used only rarely because the onset is usually difficult to identify precisely.

4.1.3.2.1 Peak latency. The latency to a peak is measured from stimulus onset to the point of maximum amplitude of a negative or positive peak (L_{N_1}, L_{P_1}, L_{N_2} in Figure 4.2). The stimulus onset, usually selected as the trigger for the sweep (3.3.2), can be identified easily in most instances because most clinically used stimuli begin suddenly. However, gradually increasing stimuli such as tone bursts of initially increasing intensity require specific definitions of stimulus onset, for instance the 50 percent amplitude point or the hearing threshold.

Figure 4.2. How to measure peak latency, peak amplitude, and peak-to-peak amplitude of a transient EP. Peak latency is measured from the stimulus at the beginning of the tracing to the first negative (L_{N_1}), first positive (L_{P_1}), and second negative (L_{N_2}) peak. Peak amplitude is measured from a baseline drawn through a quiet part of the tracing (A_{1P_1}). Peak-to-peak amplitude is measured with reference to the preceding (A_{2P_1}) or following (A_{3P_1}) peak of opposite polarity.

4.1.3.2.2 Interpeak latency and central conduction time. Interpeak latencies (IPLs) are measured between peaks of the same EP (Figure 4.3) and are commonly used in the evaluation of BAEPs. They represent the time of conduction

Figure 4.3. How to measure interpeak latencies of a transient EP. IPL I–II is the difference between the latency of peak I and that of peak II of the EP; IPL I–III is the interval between peak I and peak III, and so on.

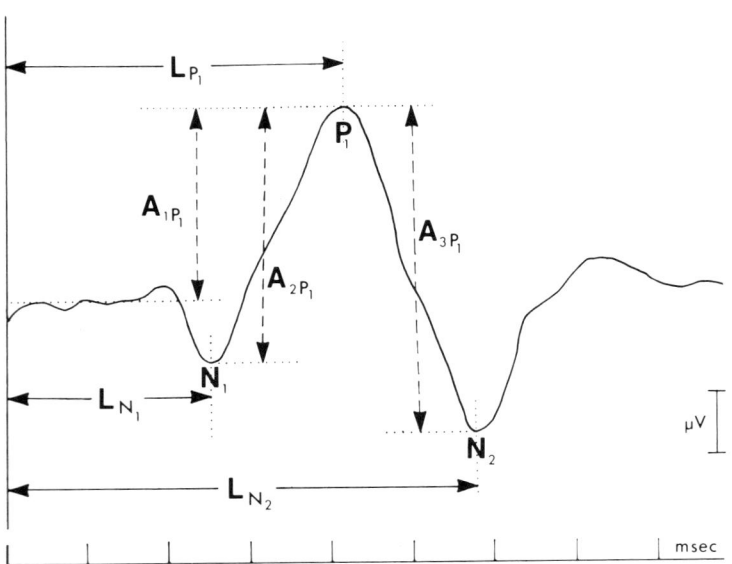

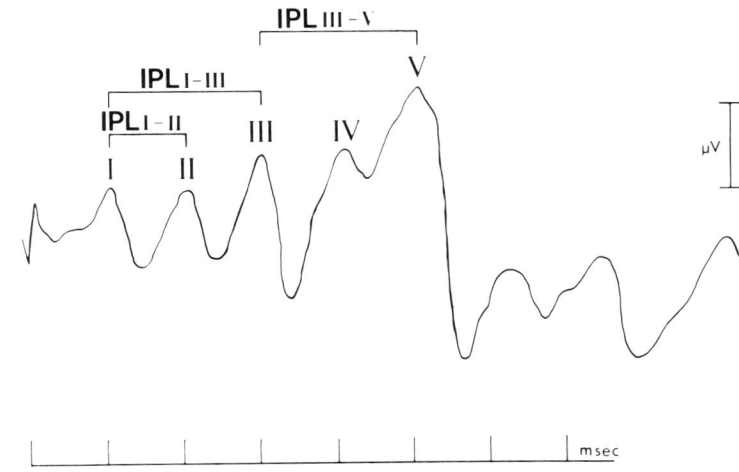

between the structures that generate the peaks. The name *central conduction time* is sometimes used instead of *interpeak latency* and is often used in SEP recordings to denote the difference between latencies of EPs recorded simultaneously at different points in the same sensory pathway. Measurement of interpeak latency is preferred to the measurement of peak latency because it is less variable and rather independent of peripheral factors. Peak latency is increased by lesions in both the peripheral and central parts of the sensory system, whereas interpeak latency generally is increased only by central lesions but not by peripheral lesions which shift peak latencies to a similar degree (5.3.1.2).

4.1.3.3 Latency of steady-state EPs

Peak latency of steady-state EPs cannot be measured reliably because each peak of these rhythmical responses is the composite result of more than one preceding stimulus; furthermore, the time interval between successive peaks and the time interval between each stimulus and the next peak varies with stimulus frequency. Several methods can be used to characterize the time relationship between stimuli and response peaks (Figure 4.4): (1) Latency may be arbitrarily defined as the interval between a stimulus and the next upward (L_1) or downward (L_2) peak; a later peak may be chosen for measurements (L_3), especially if its latency and polarity resemble those of the major peak of the same polarity of the transient EP. (2) The timing may be described by considering the interval between stimuli as a cycle of 360° and calculating the interval between stimulus and response peak in terms of the phase angle between stimulus and peak. For instance, peaks occurring 75 msec after stimuli given at 10/sec, i.e., at 100 msec intervals, may be said to lag behind the stimuli by three-fourths of a full cycle of 360°, or by 270°. From measurements of phase lag, an apparent latency of the EP can be calculated.[70,98] (3) Other methods of describing the latency of steady-state EPs have been used.[30]

4.1.4 Amplitude

The amplitude of EP peaks is most precisely measured with cursors. Various types of measurements may be selected for different purposes, but only one type should be used once the selection has been made.

4.1.4.1 Absolute amplitude measurements

4.1.4.1.1 Peak amplitude. The peak amplitude is the vertical distance, representing the voltage difference, between a peak and a reference level representing zero amplitude. This zero level may be defined as a horizontal line drawn through a brief prestimulus interval of the EP recording (3.3.2), through an early poststimulus period not containing EP peaks (Figure 4.2, A_1P_1), or through the average level of the entire EP. Measurements of peak amplitude are useful in cases of EPs with a stable baseline and with clearly separated peaks, but these measurements present

Figure 4.4. How to measure latency and amplitude of steady-state EPs. The steady-state EP consists of peaks of the same frequency as the stimuli, indicated on the bottom line. Latency may be measured as the time interval from a stimulus to the first peak of one polarity (L_1), to the first peak of opposite polarity (L_2), or to a later peak (L_3). As an alternative, latency may be expressed as the phase angle ϕ which corresponds with the ratio of the interval from stimulus to first peak over the length of the interstimulus interval $\left(\phi = 360°\frac{x}{a}\right)$. Amplitude A is measured as peak-to-peak voltage between lines drawn through the average of peak heights.

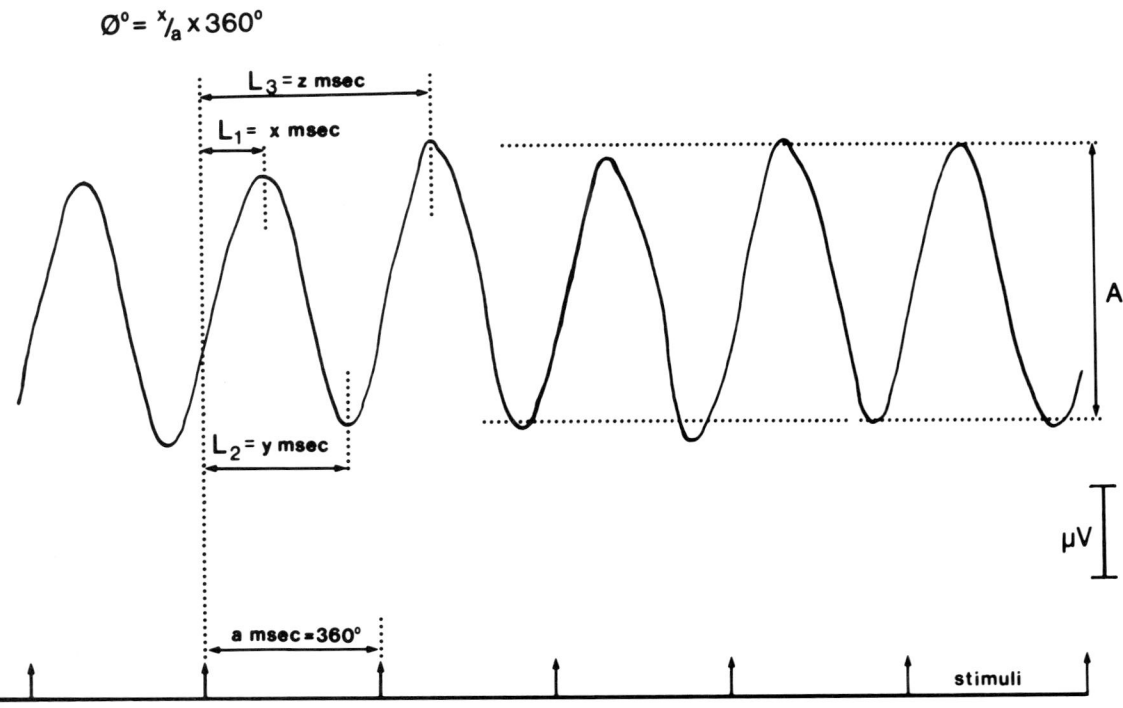

problems if the baseline is noisy or if a peak begins before the end of the preceding one.

4.1.4.1.2 Peak-to-peak amplitude. The peak-to-peak amplitude is the vertical distance between successive peaks of opposite polarity and may be measured as the amplitude of the ascending limb (Figure 4.2, A_2P_1), of the descending limb (Figure 4.2, A_3P_1), or as the mean of the two. Whichever method is chosen, it must be used consistently. Measurements of peak-to-peak amplitude are useful in EPs having stable sequences of peaks of opposite polarity but these measurements present problems when successive peaks vary seemingly independently of each other, or when a peak of high amplitude is preceded or followed by a smaller peak of the same polarity before the tracing returns to baseline. Peak-to-peak amplitude is an important measurement of steady-state EPs (Figure 4.4).

4.1.4.2 Amplitude ratios
The ratio of the amplitude of two peaks may be calculated for peaks in the same EP or for peaks that are recorded simultaneously from different locations such as the two sides of the head or successively from midline and lateral electrode positions.

4.1.5 Waveshape

The form of EPs may be fairly characteristic for an EP type and may be described, for instance, as W-shaped. However, waveshape usually varies more than latency and amplitude and is therefore not usually used in the evaluation of clinical EP recordings.

4.1.6 Distribution

The appearance of EPs over different parts of the head depends not only on the sensory modality but also on the stimulus used. In clinical practice, distribution is not studied extensively, although recordings from both sides of the head are usually compared in evaluating BAEPs and in VEP investigations of retrochiasmal lesions.

The earliest peaks of cortical EPs are restricted to the specific sensory receiving areas of the stimulus modality. Subsequent peaks have a wider distribution. The two types of peaks, although often difficult to clearly distinguish from each other, are sometimes referred to as specific and nonspecific, or primary and secondary.

4.1.7 Naming of peaks

4.1.7.1 Nomenclatures
A uniform nomenclature for all EPs has not yet been established, although suggestions for a standardization of terminology have been made.[3,5,31] In practice, different naming systems are used for different EP types. Even the peaks of the same EP type may be given different names in different laboratories.

4.1.7.1.1 Polarity-latency. Peaks may be named by a combination of the letters P or N, standing for positive or negative polarity of the peak, and a number representing peak latency. This method is now used for most EP types except BAEPs. For instance, the major occipital positive peak of a VEP to checkerboard reversal which occurs at about 100 msec is labeled P100. Terminological problems arise if a peak normally occurring at 100 msec does not occur until 140 msec in a patient (4.1.7.2).

4.1.7.1.2 Number of a peak in sequence. Peaks may be named by numbering them in the order of their occurrence. This method is used mainly for BAEPs whose peaks are labeled I–VII (12.1). Since the polarity of the peak is not indicated in the name, it must be known by convention; thus, only vertex-positive peaks are labeled by the numbers of the BAEP. Terminological problems arise if peaks are missing and the numbering of subsequent peaks is in doubt.

4.1.7.1.3 Polarity-number in sequence. Positive and negative peaks may be named by the letters P and N and by a number indicating their order of appearance among peaks of the same polarity. For example, in some instances the major positive peak of the checkerboard reversal VEP is named P1 and the negative peaks preceding and following it are called N1 and N2. This method may be useful in EPs which show more variation in latency than in number of peaks; it may be more effective than the polarity-latency nomenclature in dealing with the large changes of latency normally seen between premature and adult age. Problems arise if a peak is missing in the EP of a particular subject.

4.1.7.2 Observed versus characteristic peaks
Each nomenclature may be used to describe either observed or characteristic peaks. For instance, the major positive peak in the VEP to checkerboard reversal, occurring at a latency of about 100 msec in normal adults, is usually called P100. The major positive peak in the VEP of a patient may appear at 140 msec and is generally presumed to represent the same peak as that occurring at 100 msec in normal subjects, delayed by 40 msec. This peak may be described as an observed P140, equivalent to the characteristic P100 of normal subjects.

It has been suggested that a horizontal bar be placed over the label of a characteristic peak to distinguish it from the label of an observed peak:[31] The patient may be said to have a $\overline{P100}$ at 140 msec, or a P140 corresponding with the $\overline{P100}$. A characteristic label, such as $\overline{P300}$, may be used to describe peaks whose observed latency varies over a few hundred milliseconds. Because only characteristic peaks and no individual observations are dealt with in this book, the overbar is used to describe all peaks labeled with this nomenclature.

A similar problem arises for the sequential naming systems: A wave V may be the first or second clearly distinguishable peak in a BAEP recording, or a peak at the latency of the characteristic N2 of another EP may be the first negative peak. As a rule, such peaks are still named

V and N2, respectively. For instance, the term P3, or P_3, is used in some laboratories instead of $\overline{P300}$ regardless of the number of the preceding peaks.

4.2 VARIABILITY OF EPs

4.2.1 Intraindividual and interindividual variability

EPs recorded sequentially from the same subject without change in stimulus or recording conditions may differ from each other. EPs recorded with identical methods from different subjects may differ even more. This intraindividual and interindividual variability is very important for clinical diagnostic EP work because the ability to detect abnormalities decreases with increasing variability: The narrower the range of normal EPs, the greater the chance of detecting even slight abnormalities.

4.2.1.1 Intraindividual variability

Several factors may alter the ongoing EEG activity, the responses to sensory stimuli, or both. Variation of background activity, or noise, has been held responsible for most of the intraindividual variability of EPs.[21,51] However, the response, or signal, may also vary during the averaging of an EP.[25,81]

4.2.1.1.1 Changes of attention and alertness. Attention and habituation may alter EPs,[46,58,65] especially the late cortical peaks that are unimportant for most clinical studies.[49,53,64,71] The same is true for sleep. The clinically important BAEPs and SEPs, having relatively short latencies, may be recorded during sleep, which reduces muscle and movement artifacts. Pattern VEPs require focusing on the stimulus and must be recorded awake.

4.2.1.1.2 Background activity. VEPs vary with amplitude[48] and phase[86] of alpha rhythm.

4.2.1.1.3 Heart rate and blood pressure. EPs vary with heart rate and blood pressure[16,93]; this is usually unimportant except in very small EPs. To avoid distortion of the EP by the heart beat, the stimulus may be time-locked to the cardiac cycle by triggering each stimulus at a constant interval after heartbeats.

4.2.1.1.4 Time between averages. EPs recorded from the same subject in different sessions vary more than EPs recorded during the same session.[8,36,57]

4.2.1.1.5 Age. EPs change fairly rapidly during infancy and then gradually mature to adult levels; most EPs show slighter changes during adult age. The variation of EPs between subjects of the same age is greater in infants and children than in adults.[36]

4.2.1.1.6 Difference between laboratories. EPs recorded from the same subject in different laboratories with different equipment vary more than EPs recorded from the same subject at different times in the same laboratory.[8]

4.2.1.1.7 Abnormal cerebral function. EPs vary more over a diseased than over an intact cerebral hemisphere.[102]

4.2.1.2 Interindividual variability
Variations in EPs between individuals are much greater than intraindividual variation because they are a product of all the factors of intrasubject variability and of many other factors which distinguish different persons. These factors include sex, body size, and variations of sensory receptors and the anatomy of the peripheral and central nervous system, especially differences of spatial orientation of the structures generating the electric fields recorded in an EP and differences in the thickness, electric resistance, and capacitance of the structures between the nervous system and the recording electrodes. Studies of monozygotic twins, dizygotic twins, and other relatives suggest that some of these factors are probably inherited.[14,15,55,60]

4.3 SELECTION OF EP TYPE AND OF EP PARAMETERS TO BE MEASURED

4.3.1 Selection of EP type

The most commonly used clinical EPs are VEPs to checkerboard pattern alternation, BAEPs to clicks, and SEPs to electric stimulation of the median and the common peroneal or posterior tibial nerve. The most important parameters are summarized in Tables 7.1, 12.1, 16.1, and 18.1.

Transient EPs are used in all three modalities because the latencies of these EPs can be measured easily and are most suitable for detection of abnormal function.

VEPs to checkerboard reversal are less variable, and therefore diagnostically more useful, than VEPs to diffuse light flashes. Cortical EPs, although more variable than subcortical EPs, must be used for visual studies because subcortical VEPs cannot be recorded reliably with present methods. Cortical VEPs are recorded with near-field recording methods (1.3.4.1). They have a long latency and require longer interstimulus intervals than subcortical EPs, but they have larger amplitude and therefore require collection of fewer responses for an average. Sweep and dwell times are relatively long, and filter settings favor rather low frequencies.

Among AEPs, the BAEP to click stimuli has become the favorite EP for the diagnosis both of brain stem lesions and of hearing defects in infants. The BAEP is recorded with far-field recording methods (1.3.4.2). Because it has short latency and duration, it can be elicited with relatively rapid stimulation and recorded with a short dwell time and with filter settings favoring high frequencies. Because the responses have low amplitude, a few thousand of them must be collected.

SEPs are elicited by stimulation of a sensory or mixed nerve of an arm or leg. SEPs to arm stimulation are simultaneously recorded from scalp, neck, and brachial plexus, while SEPs to leg stimulation are recorded from scalp and thoracolumbar spine; these recordings show the ascension of the volley at different points in the afferent pathway. Because SEPs at the different recording sites have different latencies and durations, the parameters of stimulation, recording, and computing are a compromise between those used for far-field and near-field EPs.

4.3.2 Selection of EP parameters to be measured

Some EP peaks are less susceptible to variation[57,84] and more sensitive to pathology[1,24,83] than others. In general, latency is the most important measurement used in detecting EP abnormalities. The diagnostic usefulness of EPs having more than one peak or being recorded at different levels is enhanced by measuring IPLs rather than peak latencies because IPLs are less variable and more sensitive to pathology than peak latencies; furthermore, they can distinguish central from peripheral lesions (4.1.3.2.2).

Amplitude, showing great normal variability, is much less useful than latency. However, the variability of amplitude measurements may be reduced by measuring peak-to-peak amplitude and calculating amplitude ratios (4.1.4). Peak-to-peak amplitude must be used in the evaluation of steady-state EPs.

4.4 LABELING EP RECORDS

Because EPs depend on a great number of variables, the specific conditions for each EP must be documented on the EP record. The guidelines for clinical evoked potential studies of the American EEG Society[3] suggest that the following items be included:

1. The subject's name, age, sex, and identification number.
2. The date of the examination and the number of the procedure.
3. Identification of the technologist.
4. Identification of the electrodes connected to inputs 1 and 2, in that order, for each channel.
5. The type, intensity, rate, and, when relevant, polarity of the stimulus, and the side and site of stimulation.
6. Description of other conditions important for the results such as masking noise applied to the nonstimulated ear, state of retinal adaptation, level of alertness, degree of cooperation, and apparent fixation on visual stimuli.
7. The bandpass, specified in -3-dB cutoff points for high and low filter frequencies in hertz, and the filter roll-off slopes, in decibels per octave, for the entire recording system including the amplifier, averager, magnetic tape recorder, and any other equipment used in the recording.
8. The number of responses averaged in each EP.
9. The horizontal resolution, as dwell time in milliseconds per data point or as sampling rate in kilohertz, for each channel.
10. A time calibration in the form of a horizontal line of the length of the analysis period, with divisions appropriate for the timing of the recorded EP peaks. The

stimulus onset should be indicated by a mark such as an arrow. When a prestimulus interval is included in the average, the zero point of the time calibration should correspond with the onset of the stimulus.
11. A voltage calibration in the form of a vertical line and a label indicating the voltage represented by that line.
12. The polarity convention indicated by plus and minus signs at the upper and lower ends of the voltage calibration line. These signs should indicate the polarity of the electrode connected to input 1 of the amplifier relative to that of the electrode connected to input 2.
13. Marks that indicate which peaks were identified and at what points latency and amplitude were measured.

5

Clinical Interpretation

SUMMARY

5.1. Lesions in the three major sensory systems can be detected by delays, decreases of amplitude, or abolition of the characteristic peaks of VEPs, AEPs, and SEPs.

5.2. EP abnormalities are detected by comparison with control values recorded from normal subjects; separate control groups must be established for subjects differing in age and other variables that cause differences in EP values. Normal values must be defined statistically; normal ranges should be set to yield an acceptable proportion of false positive and false negative readings.

5.3. Different types of EP abnormalities have different implications. Most important are increases in peak latency of VEPs and in interpeak latency of BAEPs and SEPs; these increases usually indicate a conduction defect in the sensory pathway. Low amplitude is a less reliable indicator of lesions and is best evaluated by calculating amplitude ratios. Complete absence of an EP, although the severest degree of abnormality, may be caused by problems outside the sensory system, especially technical problems with stimulating and recording equipment.

5.4. Strategy of localization and lateralization uses correlations between EP peaks and anatomical structures for BAEPs and SEPs, and combinations of monocular or half-field stimulation with midline or lateral occipital recordings for VEPs.

5.5. A report of the EP findings and their interpretation should be provided for each clinical EP study.

5.1 PRINCIPLES OF CLINICAL INTERPRETATION

In clinical practice, EPs are used to evaluate conduction in the three major sensory systems. Lesions involving these systems often delay, reduce, or abolish EPs; EPs may be abnormal even when clinical examination and other tests are normal. In many instances, the EP abnormalities indicate the location of lesions. Localization to certain parts of the sensory system may be made either by correlating abnormalities of specific EP peaks with anatomical structures known to generate these peaks or by strategies of unilateral stimulation and recording which take advantage of the anatomical characteristics of crossed and uncrossed portions of the sensory systems.

EPs are less often used to study the peripheral portion of sensory systems, although they may be used to evaluate hearing or visual ability in infants and other noncommunicating subjects.

5.2 DEFINITION OF NORMAL: CONTROL GROUPS

Because normal EPs vary in and between subjects (4.2), normal values are established by studying normal subjects in several age groups. In general, age-specific normal values should be obtained for each week of age in the perinatal period, for each month in young infants, for quarters in older infants, and for decades in children and older adults. The specific requirements for age groups vary with EP types. The two sexes should be represented equally in each group of normals. A few laboratories use normative values matched both for age and for sex for EP types showing slight differences between the sexes, especially for the BAEP (12.2.2). The normal range of EP characteristics is defined statistically for each subject group, EP type, and EP characteristic. Each group should comprise at least 20 subjects. Measurements of EPs to stimulation of the right and left eye, ear, and peripheral nerves should not be lumped together unless they show no or negative correlation; however, paired values are useful for the determination of the normal range of interocular, interaural, and lateral somatosensory differences of latency and amplitude.

Because EPs depend so much on the details of stimulating and recording conditions, control values for each response type to be used in a clinical laboratory must be obtained in that laboratory and cannot be adopted from other laboratories. The guidelines of the American EEG Society[3] state that in setting up a new laboratory it is acceptable to use the detailed normative values published by some large laboratories, provided two requirements are fulfilled. (1) Stimulus, recording, and other conditions in the new laboratory are identical or fully compatible with those of the reference laboratory. (2) A study in the new laboratory of at least 20 normal subjects spanning the age range of the patients to be examined in this laboratory shows that a specified portion, such as 95 percent or 99 percent, of these normal values falls within the limits derived from the subset studied in the reference laboratory.

5.2.1 Characteristics of normal subjects

Normative values are obtained from subjects who are in good general health, free of CNS disease, not under the influence of drugs affecting the CNS, and without history or findings of neurological diseases and of disorders of the specific sensory modality under study. Subjects serving as normal controls in VEP studies should have an ophthalmological examination including (1) testing of visual acuity showing a corrected refractive error of no more than -5 diopters for myopes, (2) studies of visual fields and color sensitivity, and (3) fundoscopic examination. Normal controls for AEPs should have an audiometric examination including pure tone audiometry with threshold measurements for air and bone conduction and tests of discrimination, acoustic impedance, and acoustic reflex. Controls for SEPs should be evaluated with special care if they have a history of sensory defects, trauma, or fractures. For studies making comparisons between scalp recordings from the two sides, handedness and eye or ear dominance as perceived by the subject should be specified.[3]

5.2.2 Statistical definition of normal ranges

The normal range of each EP measurement is defined depending on the distribution of the values of this measurement in normal subjects. Because a statistically normal distribution allows convenient definition of the normal range in terms of mean and standard deviations, an attempt should be made to determine whether the control values for the various measurements are distributed normally, i.e., whether a plot of the incidence of each value versus its distance from the mean approximates a bell-shaped curve. Normal distribution may be assessed further by plotting the control measurements on probability paper or, more stringently, by evaluating their fit with the normal distribution curve and by calculating their deviation from symmetry, or skewness, and their degree of peaking, or kurtosis. If the distribution of the measurements is not normal, an attempt should be made to transfer them to a normal distribution by such means as calculating their logarithms, square roots, or reciprocals.

After the necessary number of normally distributed groups for each measurement have been determined and a sufficient number of normal values in each group have been collected, the normal range for this measurement may be defined in each group as the mean ± 2, 2.5, or 3 standard deviations. This is the normal range for the control sample which represents only a very small part of the normal population. The statistical significance of a measurement that is abnormal with reference to this range can be illustrated by calculation of the tolerance limits which indicate the chance that a certain percentage of the normal population will show measurements falling into the specified normal range. Because in clinical EP studies only excessively large latency values and excessively small amplitude values are considered abnormal, one-tailed tolerance limits may be used.

If the control values for a measurement are not distributed normally, one may plot their cumulative frequency in the control group and use the percentile method to indicate the portion of the normal group having measurements similar to that of an observed EP. To obtain reliable results with this method, the control group should comprise at least 100 subjects. One may use the percentile method to set tolerance limits which exclude 95 percent or 99 percent of the normal population, meaning that an observation outside these limits has a 5 percent or 1 percent chance to be normal.

Every laboratory must decide on how to set the limits for its normal ranges. These limits represent a compromise between the numbers of false positive and false negative readings that can be accepted. Choosing a wider normal range has the advantage of decreasing the number of false positive readings but also has the disadvantage of increasing the number of false negative readings. Where best to draw the line between normal and abnormal values is a practical matter that is ultimately determined by the ability of the test results to discriminate healthy from diseased subjects. Most laboratories use a normal range of 2.5 or 3 standard deviations or tolerance limits of 99 percent.

A simple graphic method may be used to compare latency and amplitude measurements of an EP peak with normal values. The normal values for this peak may be plotted against each other, showing latency on the horizontal axis and amplitude on the vertical axis. The normal range can then be drawn in the form of an ellipse containing a defined percentage of all normal values; abnormal values fall outside this probability ellipse.[1,10,22,85]

5.3 TYPES AND CLINICAL IMPLICATIONS OF EP ABNORMALITIES

5.3.1 Abnormally long latency

Increased latency is much more reliable than decreased amplitude in signaling clinical abnormalities in a sensory system. Increased latency generally indicates decreased conduction velocity and is often due to lesions which cause slowing of axonal conduction in the central part of the system. However, peripheral lesions may also increase the conduction time. The latency of VEPs to patterned stimuli may be increased as a result of ocular conditions that blur the retinal image; BAEPs may be delayed by cochlear lesions, while SEPs may be delayed due to impaired peripheral nerve conduction. Therefore, before an increased latency can be interpreted as indicating a lesion in the central part of a sensory pathway, peripheral lesions must be excluded. Measurements of interpeak latencies may help to distinguish central from peripheral lesions (4.1.3.2.2).

A very common cause of slowed conduction is the demyelination of nerve fibers which reduces conduction velocity both by delaying saltatory conduction[68] and by replacing it with very slow continuous conduction.[11] These factors can account for the delays of EP peaks sometimes seen in traumatic, ischemic, degenerative, and other demyelinating lesions. Reduced conduction velocity contributes to the very long delays characteristic of multiple sclerosis.[42,56,98] The resulting dispersion of impulses probably reduces the efficiency of temporal and spatial summation of postsynaptic potentials and may explain the greater length of the delays in multiple sclerosis.[59] Restoration of conduction by remyelination is likely to play an important part in the recovery from acute and chronic compressive lesions of the CNS but not in the remissions of multiple sclerosis.[67,79] This lack of effective remyelination probably explains the rather high incidence of abnormal EPs in patients with stable multiple sclerosis. Abnormal synaptic transmission may also increase the latency of EPs.[10]

5.3.1.1 Abnormally long peak latency
Increased peak latency indicates a defect between the point of stimulation and the structure generating the EP peak, i.e., the stretch from retina to occipital cortex in VEPs, from cochlea to pons and midbrain in BAEPs, and from peripheral nerve to spinal cord or somatosensory cortex in SEPs. Peak latency is the major parameter of VEPs; it is less important than IPLs in BAEPs and SEPs.

5.3.1.2 Abnormally long interpeak latency
Interpeak latencies can be measured in BAEPs and SEPs (4.1.3.2.2) and are preferred to peak latency for three reasons: (1) The peak representing acoustic nerve excitation in BAEPs and the peak indicating brachial plexus or cauda equina excitation in SEPs serve as benchmarks that help to distinguish peripheral lesions, which cause a delay of these peaks and all subsequent ones, from central lesions, which leave the latency of the benchmark peaks intact but delay the subsequent ones and thereby increase the IPL or central conduction time. (2) Latencies measured between peaks may help to localize lesions to the structure in the central sensory path assumed to generate these peaks. (3) Interpeak latencies vary less than peak latencies and are more sensitive to pathology; they are therefore better indicators of abnormal function.

5.3.1.3 Abnormal differences of peak or interpeak latencies
Differences of peak or interpeak latency between EP peaks recorded from the midline in response to successive stimulations of the eye or ear on either side (interocular or interaural latency difference) or between EPs recorded simultaneously on both sides of the head (left-right latency difference) may suggest an abnormality of the EP having longer latency. Not all EP measurements identified as abnormal by comparison with the normal range of latency differences fall outside the normal range of absolute peak or interpeak latency values.

5.3.2 Abnormal amplitude

A decrease of amplitude often accompanies the increase of latency seen in cases of reduced conduction velocity and is probably explained by dispersion due to general slowing of conduction, increased scatter of conduction velocities, and complete dropout of part of the fibers. Primary axonal defects produce reductions of amplitude rather than increases of latency.[56]

However, low amplitude has limited value as an indicator of impaired conduction because amplitude varies much more than latency in normal subjects and because amplitude reduction does not necessarily indicate conduction defects but may be due to other defects, including technical problems.

5.3.2.1 Reduced amplitude
Amplitude reduction can suggest a conduction defect but usually cannot prove it.

5.3.2.2 Abnormal amplitude ratios
The amplitude ratio of EP peaks is a somewhat better indicator of abnormality than is absolute amplitude because it is less variable; also, an abnormal amplitude ratio is less likely to be due to technical problems. Several kinds of amplitude ratios may be used. (1) Amplitude ratios of peaks in the same EP are used in the evaluation of BAEPs; a commonly used measurement is the amplitude ratio of waves V/I, or its reciprocal, which may indicate brain stem lesions. (2) Amplitude ratios of corresponding peaks of EPs recorded simultaneously from the left and right side of the head may be used to search for unilateral hemispheric lesions, such as retrochiasmal lesions of the visual pathway. (3) Amplitude ratios of peaks of EPs recorded successively from the midline are used to evaluate VEPs to stimulation of either eye in the search for a prechiasmal abnormality. (4) Amplitude ratios of peaks of EPs recorded successively from the two sides of the head may be used in studies of SEPs to stimulation of peripheral nerves on either side of the body and may reveal unilateral lesions in the somatosensory system.

5.3.2.3 Complete absence of an entire EP
Absence of an EP may indicate complete interruption of conduction at a point before that generating the first EP peak. This severest degree of EP abnormality may represent the maximum effect of a lesion which in milder form produces only a decrease of amplitude, an increase of latency, or both. However, complete absence of an EP does not necessarily prove a conduction defect because it may be due to other factors including peripheral problems such as eye or ear disorders, technical problems such as failure of stimulation or of synchronization between stimulator and averager, faulty electrode connections, or defects of electrodes, input boards, selector switches, amplifiers, averager, and oscilloscope. Only after these problems have been excluded can the absence of an EP be attributed to a CNS lesion.

5.3.3 Abnormal shape

The waveshape of EPs cannot be easily defined and is not a reliable indicator of abnormality.

5.4 STRATEGY OF LOCALIZATION AND LATERALIZATION

Chiefly two principles are used to determine the side and level of a lesion.

5.4.1 Correlation between EP peaks and anatomical structures

Certain peaks in the BAEP and the SEP correspond with excitation of certain structures in the auditory and somatosensory pathways so that lesions of these structures can be detected by abnormalities of the corresponding peaks; subsequent peaks, representing excitation of more proximal structures which receive abnormal input from the damaged structure, are usually also abnormal. Correlations of EP peaks with anatomical structures have been made empirically and are useful for an approximate localization even though it is now agreed that one-to-one correlations between a peak and a specific point in the sensory pathway cannot be made because most peaks are due to activity in more than one structure and any structure may contribute to more than one peak.

5.4.2 Localization by combinations of unilateral stimulation and recording in partially or completely crossed pathways

Each of the major sensory pathways may be divided into an uncrossed, crossing, and crossed part, as shown in Figure 5.1. The crossing is partial in the visual and auditory systems, connecting each side of the body to both hemispheres, and complete in the somatosensory system, connecting one side of the body to the opposite hemisphere only. Lesions in these three parts can be detected by combinations of unilateral stimulation and recording. A lesion in the uncrossed part causes abnormalities in EPs to stimulation of only that side. A lesion in the crossing part produces abnormalities of EPs to stimulation of either side. These abnormalities often differ from each other. A lesion in the crossed part of the somatosensory system causes abnormalities of the SEP to contralateral stimulation. A lesion in the crossed part of the visual system is best searched for by selective excitation of each hemisphere with visual half-field stimuli and by recording bilateral EPs. A lesion in the crossed part of the auditory system is difficult to detect with any strategy.

5.5 REPORT

A report should be provided for every clinical EP study. The report should give a description of the findings, including measurements of peak latency, interpeak latency, amplitude ratio, and other criteria used for the definition of normal responses. The report should indicate which measurements are abnormal and what methods were used to define the normal range, for instance, the number of standard deviations beyond the normal mean or the tolerance limits. Relevant details such as visual acuity, hearing threshold, treatment with drugs having possible effects on EPs, level of alertness, and cooperation during the test should be included.

The report should give a clinical interpretation, especially in cases of abnormal findings. The interpretation should point out the implications of the abnormal findings in general pathophysiological terms and with respect to the clinical data of the specific case under study.

Figure 5.1. Strategy of localizing lesions in different parts of the central sensory pathwways. The pathways of VEP, AEP, and SEP consist of an uncrossed, distal part, a crossing part, and a crossed, proximal part. Lesions of the uncrossed part cause abnormalities of EPs to stimulation on the side of the defect. Lesions of the crossing part cause abnormalities of EPs to stimulation of either side. Lesions of the crossed part cause abnormalities that differ for each modality.

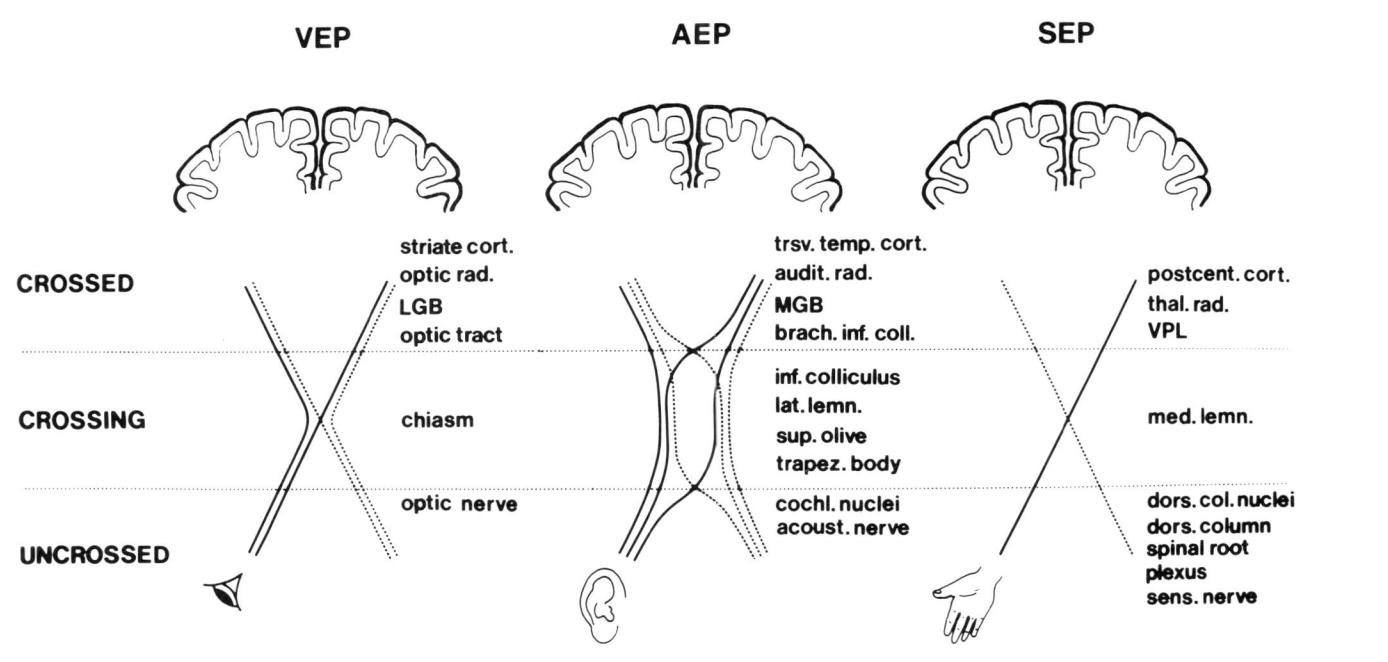

References for part A

1. Abbruzzese, G., Cocito, L., Ratto, S., Abbruzzese, M., Leandri, M., and Favale, E. 1981. A reassessment of sensory evoked potential parameters in multiple sclerosis: A discriminant analysis approach. J. Neurol. Neurosurg. Psychiatry 44:133–139.
2. Allison, T., Goff, W.R., Williamson, P.D., and Van-Gilder, J.C. 1980. On the neural origin of early components of the human somatosensory evoked potential. Prog. Clin. Neurophysiol. 7:51–68.
3. American Electroencephalographic Society. 1984. Guidelines for clinical evoked potential studies. J. Clin. Neurophysiol. 1:3–53.
4. Aran, J.M. 1971. The electro-cochleogram. Recent results in children and in some pathological cases. Arch. Klin. Exp. Ohren-, Nasen- Kehlkopfheilkd. 198:128–141.
5. Arden, G.B., Bodis-Wollner, I., Halliday, A.M., Jeffreys, A., Kulikowski, J.J., Spekreijse, H., and Regan, D. 1977. Methodology of patterned visual stimulation (report of the Brussels symposium ad-hoc committee). In: Visual evoked potentials in man: New developments, ed. J.E. Desmedt, pp. 3–15. Oxford: Clarendon Press.
6. Arden, G.B., Carter, R.M., Hogg, C., Siegel, I.M., and Margolis, S. 1979. A gold foil electrode: Extending the horizons for clinical electroretinography. Invest. Ophthalmol. Visual Sci. 18:421–426.
7. Arlinger, S. 1981. Technical aspects on stimulation, recording and signal processing. Scand. Audiol. Suppl. 13:41–53.
8. Aunon, J.I., and Cantor, F.K. 1977. VEP and AEP variability: Interlaboratory vs. intralaboratory and intersession vs. intrasession variability. Electroencephalogr. Clin. Neurophysiol. 42:705–708.
9. Billings, R.J. 1981. Automatic detection, measurement and documentation of the visual evoked potential using a commercial microprocessor-equipped averager. Electroencephalogr. Clin. Neurophysiol. 52:214–217.
10. Bodis-Wollner, I., Yahr, M.D., Mylin, L., and Thornton, J. 1982. Dopaminergic deficiency and delayed visual evoked potentials in humans. Ann. Neurol. 11:478–483.
11. Bostock, H., and Sears, T.A. 1978. The internodal axon membrane: Electrical excitability and continuous conduction in segmental demyelination. J. Physiol. 280:273–301.
12. Boston, J.R., and Ainslee, P.J. 1980. Effects of analog and digital filtering on brain stem auditory evoked potentials. Electroencephalogr. Clin. Neurophysiol. 48:361–364.
13. Broughton, R., Rasmussen, T., and Branch, C. 1981. Scalp and direct cortical recordings of somatosensory evoked potentials in man (circa 1967). Can. J. Psychol. 35:136–158.
14. Buchsbaum, M.S. 1974. Average evoked response and

stimulus intensity in identical and fraternal twins. Physiol. Psychol. 2:365–370.
15. Buffington, V., Martin, D.C., and Becker, J. 1981. VER similarity between alcoholic probands and their first-degree relatives. Psychophysiology 18:529–533.
16. Callaway, E., and Buchsbaum, M. 1965. Effects of cardiac and respiratory cycles on averaged visual evoked responses. Electroencephalogr. Clin. Neurophysiol. 19:476–480.
17. Callaway, E., Halliday, R., and Herning, R.I. 1983. A comparison of methods for measuring event-related potentials. Electroencephalogr. Clin. Neurophysiol. 55:227–232.
18. Carlton, E.H., and Katz, S. 1980. Is Wiener filtering an effective method of improving evoked potential estimation? IEEE Trans. Biomed. Eng. 27:187–192.
19. Celesia, G.G. 1978. Visual evoked potentials in neurological disorders. Am. J. EEG Technol. 18:47–59.
20. Celesia, G.G., Soni, V.K., and Rhode, W.S. 1978. Visual evoked spectrum array and interhemispheric variations. Arch. Neurol. (Chicago) 35:678–682.
21. Cigánek, L. 1975. Section V. Visual evoked responses. In: Handbook of electroencephalography and clinical neurophysiology, vol. 8A, ed. A. Rémond, pp. 33–59. Amsterdam: Elsevier.
22. Cigánek, L., Smiešková, A., Hrubý, M., and Mladonický, P. 1984. Processing and analysis techniques for brainstem auditory evoked potentials with localization of brainstem lesions. Electroencephalogr. Clin. Neurophysiol. 57:92–96.
23. Coats, A.C. 1974. On electrocochleographic electrode design. J. Acoust. Soc. Am. 56:708–711.
24. Collins, D.W.K., Black, J.L., and Mastaglia, F.L. 1978. Pattern-reversal visual evoked potential. Method of analysis and results in multiple sclerosis. J. Neurol. Sci. 36:83–95.
25. Coppola, R., Tabor, R., and Buchsbaum, M. 1978. Signal to noise ratio and response variability measurements in single trial evoked potentials. Electroencephalogr. Clin. Neurophysiol. 44:214–222.
26. Creutzfeldt, O.D., Watanabe, S., and Lux, H.D. 1966. Relations between EEG phenomena and potentials of single cortical cells. I. Evoked responses after thalamic and epicortical stimulation. Electroencephalogr. Clin. Neurophysiol. 20:1–18.
27. Dawson, W.W., and Doddington, H.W. 1973. Phase distortion of biological signals: Extraction of signal from noise without phase error. Electroencephalogr. Clin. Neurophysiol. 34:207–211.
28. Dawson, W.W., Trick, G.L., and Litzkow, C.A. 1979. Improved electrode for electroretinography. Invest. Ophthalmol. Visual Sci. 18:988–991.
29. Desmedt, J.E. 1977. Some observations on the methodology of cerebral evoked potentials in man. Prog. Clin. Neurophysiol. 1:12–29.
30. Diamond, A.L. 1977. Latency of the steady state visual evoked potential. Electroencephalogr. Clin. Neurophysiol. 42:125–127.
31. Donchin, E., Callaway, E., Cooper, R., Desmedt, J.E., Goff, W.R., Hillyard, S.A., and Sutton, S. 1977. Publication criteria for studies of evoked potentials (EP) in man. Report of a committee. Prog. Clin. Neurophysiol. 1:1–11.
32. Doyle, D.J. 1975. Some comments on the use of Wiener filtering for the estimation of evoked potentials. Electroencephalogr. Clin. Neurophysiol. 38:533–534.
33. Doyle, D.J. 1977. A proposed methodology for evaluation of the Wiener filtering method of evoked potential esti-

mation. Electroencephalogr. Clin. Neurophysiol. 43:749–751.
34. Doyle, D.J., and Hyde, M.L. 1981. Analogue and digital filtering of auditory brainstem responses. Scand. Audiol. 10:81–89.
35. Doyle, D.J., and Hyde, M.L. 1981. Bessel filtering of brain stem auditory evoked potentials. Electroencephalogr. Clin. Neurophysiol. 51:446–448.
36. Ellingson, R.J., Lathrop, G.H., Danahy, T., and Nelson, B. 1973. Variability of visual evoked potentials in human infants and adults. Electroencephalogr. Clin. Neurophysiol. 34:113–124.
37. Freeman, R.D., and Thibos, L.N. 1975. Visual evoked responses in humans with abnormal visual experience. J. Physiol. (Lond.) 247:711–724.
38. Fridman, J., John, E.R., Bergelson, M., Kaiser, J.B., and Baird, H.W. 1982. Application of digital filtering and automatic peak detection to brain stem auditory evoked potential. Electroencephalogr. Clin. Neurophysiol. 53:405–416.
39. Gabriel, S., Durrant, J.D., Dickter, A.E., and Kephart, J.E. 1980. Computer identification of waves in the auditory brain stem evoked potentials. Electroencephalogr. Clin. Neurophysiol. 49:421–423.
40. Glaser, E.M., and Ruchkin, D.S. 1976. Principles of neurobiological signal analysis. New York: Academic Press.
41. Grounauer, P.A. 1982. The new single use ERG corneal electrode contact lens electrode and its clinical application. Doc. Ophthalmol. Proc. Ser. 31:89–93.
42. Halliday, A.M., and McDonald, W.I. 1977. Pathophysiology of demyelinating disease. Br. Med. Bull. 33:21–27.
43. Harder, H., and Arlinger, S. 1981. Ear-canal compared to mastoid electrode placement in BRA. Scand. Audiol. Suppl. 13:55–57.
44. Hartwell, J.W., and Erwin, C.W. 1976. Evoked potential analysis: On-line signal optimization using a mini-computer. Electroencephalogr. Clin. Neurophysiol. 41:416–421.
45. Henderson, C.J., Butler, S.R., and Glass, A. 1975. The localization of equivalent dipoles of EEG sources by the application of electrical field theory. Electroencephalogr. Clin. Neurophysiol. 39:117–130.
46. Hillyard, S.A., Picton, T.W., and Regan, D.M. 1978. Sensation, perception and attention: Analysis using ERPs. In: Event-related brain potentials in man, ed. E. Callaway, pp. 273–347. New York: Academic Press.
47. Jasper, H.H. 1958. Report of the committee on methods of clinical examination in electroencephalography. Electroencephalogr. Clin. Neurophysiol. 10:370–375.
48. Jones, K.G., and Armington, J.C. 1977. The removal of alpha from the VECP by means of selective averaging. Vision Res. 17:949–956.
49. Josiassen, R.C., Shagass, C., Roemer, R.A., Ercegovac, D.V., and Straumanis, J.J. 1982. Somatosensory evoked potential changes with a selective attention task. Psychophysiology 19:146–159.
50. Kobayashi, H., and Yaguchi, K. 1981. A statistical method of component identification of average evoked potentials. Electroencephalogr. Clin. Neurophysiol. 151:213–214.
51. Kooi, K.A., and Marshall, R.E. 1979. Visual evoked potentials in the central disorders of the visual system. Philadelphia: Harper and Row.
52. Lane, R.H., Mendel, M.I., Kupperman, G.L., Vivion, M.C., Buchanan, L.H., and Goldstein, R. 1974. Phase distortion of averaged electroencephalic response. Arch. Otolaryngol. 99:428–432.

53. Lavine, R.A., Buchsbaum, M.S., and Schecter, G. 1980. Human somatosensory evoked responses: Effects of attention and distraction on early components. Physiol. Psychol. 8:405–508.
54. Lawwill, T., and Burian, H.M. 1966. A modification of the Burian-Allen contact-lens electrode for human electroretinography. Am. J. Ophthalmol. 61:1506–1509.
55. Lewis, E.G., Dustman, R.E., and Beck, E.C. 1972. Evoked response similarity in monozygotic, dizygotic and unrelated individuals: A comparative study. Electroencephalogr. Clin. Neurophysiol. 23:309–316.
56. McDonald, W.I. 1977. Pathophysiology of conduction in central nerve fibres. In: Visual evoked potentials in man: New developments, ed. J.E. Desmedt, pp. 427–437. Oxford: Clarendon Press.
57. Meienberg, O., Kutak, L., Smolenski, C., and Ludin, H.P. 1979. Pattern reversal evoked cortical responses in normals: A study of different methods of stimulation and potential reproducibility. J. Neurol. 222:81–93.
58. Näätänen, R. 1975. Selective attention and evoked potentials in humans: A critical review. Biol. Psychol. 2:237–307.
59. Noël, P., and Desmedt, J.E. 1980. Cerebral and far-field somatosensory evoked potentials in neurological disorders involving the cervical spinal cord, brainstem, thalamus and cortex. Prog. Clin. Neurophysiol. 7:205–230.
60. Osborne, R.T. 1970. Heritability estimates for the visual evoked response. Life Sci. 9:481–490.
61. Osterhammel, P. 1981. The unsolved problems in analog filtering on the auditory brain stem responses. Scand. Audiol. Suppl. 13:69–74.
62. Peregrin, J., and Valach, M. 1981. Averaging, selective averaging and latency-corrected averaging. Pfluegers Arch. 391:154–158.
63. Pfurtscheller, G., and Cooper, R. 1975. Selective averaging of the intracerebral click evoked responses in man: An improved method of measuring latencies and amplitudes. Electroencephalogr. Clin. Neurophysiol. 38:187–190.
64. Picton, T.W., and Hillyard, S.A. 1974. Human auditory evoked potentials. II. Effects of attention. Electroencephalogr. Clin. Neurophysiol. 36:191–199.
65. Picton, T.W., Hillyard, S.A., and Galambos, R. 1976. Habituation and attention in the auditory system. In: Handbook of sensory physiology, vol. V/3, Auditory system, part 3 (Clinical and special topics), ed. W.D. Keidel and W.D. Neff, pp. 343–389. Berlin: Springer.
66. Pratt, H., Rogowski, Z., and Bental, E. 1982. Analog delay line artifact rejector for evoked potential studies. Electroencephalogr. Clin. Neurophysiol. 53:565–567.
67. Prineas, J.W., and Connell, F. 1979. Remyelination in multiple sclerosis. Ann. Neurol. 5:22–31.
68. Rasminsky, M., and Sears, T.A. 1972. Internodal conduction in undissected demyelinated nerve fibres. J. Physiol. (Lond.) 227:323–350.
69. Regan, D. 1966. Some characteristics of average steady-state and transient responses evoked by modulated light. Electroencephalogr. Clin. Neurophysiol. 20:238–248.
70. Regan, D. 1972. Evoked potentials in psychology, sensory physiology and clinical medicine. London: Chapman and Hall.
71. Roger, M., and Galand, G. 1981. Operant conditioning of visual evoked potentials in man. Psychophysiology 18:477–482.
72. Ryding, E. 1980. A mathematical model for localization

72. ...of the source of cortical evoked potentials. Electroencephalogr. Clin. Neurophysiol. 48:312–317.
73. Salomon, G., and Elberling, C. 1971. Cochlear nerve potentials recorded from the ear canal in man. Acta Oto-Laryngol. 71:319–325.
74. Schimmel, H. 1967. The (±) reference: Accuracy of estimated mean components in average response studies. Science 157:92–94.
75. Schimmel, H., Rapin, I., and Cohen, M.M. 1975. Improving evoked response audiometry. Results of normative studies for machine scoring. Audiology 14:466–479.
76. Seaba, P. 1980. Electrical safety. Am. J. EEG Technol. 20:1–13.
77. Siivola, J., and Järvilehto, M. 1982. Spinal evoked potentials evaluated with two relevant electrode types. Acta Physiol. Scand. 115:103–107.
78. Singh, C.B., Mason, S.M., and Brown, P.M. 1980. Extratympanic electrocochleography in clinical use. In: Evoked potentials: Proceedings of an international evoked potentials symposium, held in Nottingham, England, ed. C. Barber, pp. 357–366. Lancaster: MTP Press.
79. Smith, K.J., Blakemore, W.F., and McDonald, W.I. 1981. The restoration of conduction by central remyelination. Brain 104:383–404.
80. Spehlmann, R. 1981. EEG Primer. Amsterdam: Elsevier Biomedical.
81. Spreng, M. 1981. Variability of the early (0–10 msec) auditory evoked extracranial components. Scand. Audiol. Suppl. 11:79–89.
82. Stohr, P.E., and Goldring, S. 1969. Origin of somatosensory evoked scalp responses in man. J. Neurosurg. 31:117–127.
83. Tackmann, W., Strenge, H., Barth, R., and Sojka-Raytscheff, A. 1979. Diagnostic validity for different components of pattern shift visual evoked potentials in multiple sclerosis. Eur. Neurol. 18:243–248.
84. Thornton, A.R.D. 1975. The measurement of surface-recorded electrocochleographic responses. Scand. Audiol. 4:51–58.
85. Thornton, A.R.D. 1975. Statistical properties of surface-recorded electrocochleographic responses. Scand. Audiol. 4:91–102.
86. Trimble, J.L., and Potts, A.M. 1975. Ongoing occipital rhythms and the VER. I. Stimulation at peaks of the alpha-rhythm. Invest. Ophthalmol. Visual Sci. 14:537–546.
87. Tyler, C.W., Apkarian, P., Levi, D.M., and Nakayama, K. 1979. Rapid assessment of visual function: An electronic sweep technique for the pattern visual evoked potential. Invest. Ophthalmol. Visual Sci. 18:703–713.
88. Van der Tweel, L.H., Estévez, O., and Strackee, J. 1980. Measurements of evoked potentials. In: Evoked potentials: Proceedings of an international evoked potentials symposium held in Nottingham, England, ed. C. Barber, pp. 19–41. Lancaster: MTP Press.
89. Van der Tweel, L.H., and Verduyn Lunel, H.F.E. 1965. Human visual responses to sinusoidally modulated light. Electroencephalogr. Clin. Neurophysiol. 18:587–598.
90. Vaughan, H.G. 1968. The relationship of brain activity to scalp recordings of event-related potentials. In: Average evoked potentials, ed. E. Donchin and D.B. Lindsley, pp. 45–94. NASA SP-191.
91. Vaughan, H.G. 1974. The analysis of scalp-recorded brain potentials. In: Bioelectric recording techniques, Part B, Electroencephalography and human brain potentials, ed.

R.F. Thompson and M.M. Patterson, pp. 157–207. New York: Academic Press.

92. Vickery, J.C. 1979. Digital signals and techniques. In: Auditory investigation: The scientific and technological basis, ed. H.E. Beagley, pp. 32–53. London: Clarendon Press.
93. Walker, B.B., and Sandman, C.A. 1982. Visual evoked potentials change as heart rate and carotid pressure change. Psychophysiology 19:520-527.
94. Walter, D.O. 1969. A posteriori "Wiener filtering" of average evoked responses. Electroencephalogr. Clin. Neurophysiol. Suppl. 27:61–70.
95. Wastell, D.G. 1977. Statistical detection of individual evoked responses: An evaluation of Woody's adaptive filter. Electroencephalogr. Clin. Neurophysiol. 42:835–839.
96. Wastell, D.G. 1979. The application of low-pass linear filters to evoked potential data: Filtering without phase distortion. Electroencephalogr. Clin. Neurophysiol. 46:355–356.
97. Wastell, D.G. 1981. When Wiener filtering is less than optimal: An illustrative application to the brain stem evoked potential. Electroencephalogr. Clin. Neurophysiol. 51:678–682.
98. Waxman, S.G. 1980. The structural basis for axonal conduction abnormalities in demyelinating diseases. Prog. Clin. Neurophysiol. 7:170–189.
99. Wicke, J.D., Goff, W.R., Wallace, J.D., and Allison, T. 1978. On-line statistical detection of average evoked potentials: Application to evoked response audiometry. Electroencephalogr. Clin. Neurophysiol. 44:328–343.
100. Wilkison, D.M. 1983. Estimation of thresholds for evoked potentials using a laboratory computer. J. Neurosci. Methods 7:253–260.
101. Wong, P.K.H., and Bickford, R.G. 1980. Brain stem auditory evoked potentials: The use of noise estimate. Electroencephalogr. Clin. Neurophysiol. 50:25–34.
102. Wong, P.K.H., Lombrosco, C.T., and Matsumiya, Y. 1982. Somatosensory evoked potentials: Variability analysis in unilateral hemispheric disease. Electroencephalogr. Clin. Neurophysiol. 54:266–274.
103. Wood, C.C. 1982. Application of dipole localization methods to source identification of human evoked potentials. Ann. N.Y. Acad. Sci. 388:139–155.
104. Wood, C.C., and Allison, T. 1981. Interpretation of evoked potentials: A neurophysiological perspective. Can. J. Psychol. 35:113–135.
105. Woody, C.D. 1967. Characterization of an adaptive filter for the analysis of variable latency neuroelectric signals. Med. Biol. Eng. 5:539–553.

B

Visual evoked potentials

6. VEP types, principles, and general methods of stimulating and recording
    6.1 VEP types
        6.1.1 Visual content: Patterned and diffuse light
        6.1.2 Stimulus rate: Transient versus steady-state VEPs
        6.1.3 Presentation mode: Pattern reversal and pattern shift, pattern appearance and disappearance, patterned and diffuse light flashes
        6.1.4 VEPs most commonly used in clinical practice
    6.2 The subject
    6.3 General description of stimulus parameters
        6.3.1 Stimulus rate
        6.3.2 Intensity
        6.3.3 Brightness contrast of pattern stimuli
            6.3.3.1 Contrast borders: Checkerboard and sine wave grating patterns
            6.3.3.2 Contrast depth: Counterphase and depth modulation stimuli
        6.3.4 Size of the elements of pattern stimuli
            6.3.4.1 Visual angle
            6.3.4.2 Spatial frequency
        6.3.5 Size of the stimulus field
        6.3.6 Full-field versus half-field stimulation, upper versus lower retinal stimulation
        6.3.7 Pattern orientation
    6.4 General description of recording parameters

7. **The normal transient VEP to checkerboard pattern reversal**
   7.1 Description of the normal transient pattern reversal VEP
   7.2 Subject variables
      7.2.1 Age
         7.2.1.1 Before adulthood
         7.2.1.2 Adults
      7.2.2 Sex
      7.2.3 Visual acuity
      7.2.4 Pupillary size
      7.2.5 Ocular dominance
   7.3 Effect of stimulus characteristics
      7.3.1 Stimulators
         7.3.1.1 Television screens
         7.3.1.2 Projection with a pivoting mirror
         7.3.1.3 Light-emitting diodes
         7.3.1.4 Custom-made stimulators
            7.3.1.4.1 Tachistoscopic displays
            7.3.1.4.2 Two projectors with synchronized shutters
            7.3.1.4.3 Patterned mirrors
            7.3.1.4.4 Polarized light
            7.3.1.4.5 Maxwellian systems
      7.3.2 Stimulus rate and phase
      7.3.3 Contrast
         7.3.3.1 Sharpness of contrast borders
         7.3.3.2 Effect of contrast depth
      7.3.4 Luminance
      7.3.5 Check size and field size
      7.3.6 Location of the stimulus in the visual field
         7.3.6.1 VEPs to full-field and half-field stimulation
         7.3.6.2 VEPs to upper and lower retinal stimulation
      7.3.7 Monocular versus binocular stimulation
      7.3.8 Effect of pattern orientation
   7.4 Effect of recording parameters
      7.4.1 Electrode placements and combinations for full-field and half-field VEPs
         7.4.1.1 VEPs to full-field stimulation
         7.4.1.2 VEPs to half-field stimulation
      7.4.2 Other recording parameters
   7.5 Prechiasmal, chiasmal, retrochiasmal, and ophthalmological strategies
      7.5.1 Prechiasmal strategy
      7.5.2 Chiasmal and retrochiasmal strategies
      7.5.3 Choice of strategies
      7.5.4 Ophthalmological strategies

8. **The abnormal transient VEP to checkerboard pattern reversal**
   8.1 Criteria distinguishing abnormal pattern reversal VEPs
      8.1.1 Absence of VEPs
      8.1.2 Abnormally long $\overline{P100}$ latency
      8.1.3 Abnormally long interocular $\overline{P100}$ latency difference
      8.1.4 Abnormally long $\overline{N75}$ latency
      8.1.5 Abnormally large $\overline{P100}$ amplitude ratio
   8.2 General clinical interpretation of abnormal VEPs to checkerboard pattern reversal

8.2.1 Technical and ocular problems and lack of focusing
8.2.2 Abnormal VEPs to monocular full-field stimulation
    8.2.2.1 Increase of latency
    8.2.2.2 Absence or reduced amplitude of VEPs
8.2.3 Abnormal VEPs to half-field stimulation
    8.2.3.1 Abnormal VEPs to stimulation of the temporal half-fields
    8.2.3.2 Abnormal VEPs to stimulation of corresponding half-fields
    8.2.3.3 Abnormal VEPs to stimulation of a monocular half-field
8.2.4 Ophthalmological problems

8.3 Neurological disorders that cause abnormal pattern reversal VEPs
  8.3.1 Prechiasmal lesions
    8.3.1.1 Retrobulbar neuritis
    8.3.1.2 Multiple sclerosis
    8.3.1.3 Tumors
    8.3.1.4 Ischemic optic neuropathy
    8.3.1.5 Friedreich's ataxia, hereditary cerebellar ataxia, and hereditary spastic ataxia
    8.3.1.6 Charcot-Marie-Tooth disease
    8.3.1.7 Neurosyphilis
    8.3.1.8 Hereditary spastic paraplegia
    8.3.1.9 Leber's hereditary optic neuropathy
    8.3.1.10 Traumatic optic nerve lesions
    8.3.1.11 Leukodystrophies
    8.3.1.12 Chronic renal failure
    8.3.1.13 Endocrine orbitopathy
    8.3.1.14 Pernicious anemia
    8.3.1.15 Sarcoidosis
    8.3.1.16 Diabetes
    8.3.1.17 Alcoholism
    8.3.1.18 Phenylketonuria
    8.3.1.19 Parainfectious optic neuritis
    8.3.1.20 Postconcussion syndrome
    8.3.1.21 Hysterical blindness
  8.3.2 Chiasmal lesions
  8.3.3 Retrochiasmal lesions

8.4 Ophthalmological disorders that cause abnormal pattern reversal VEPs
  8.4.1 Visual acuity
  8.4.2 Refractive errors
  8.4.3 Amblyopia ex anopia
  8.4.4 Nutritional and toxic amblyopia
  8.4.5 Retinopathy
  8.4.6 Glaucoma
  8.4.7 Corneal opacity, miosis, and cataract
  8.4.8 Congenital oculomotor apraxia

9. THE TRANSIENT VEP TO DIFFUSE LIGHT STIMULI
  9.1 Clinical usefulness of the flash VEP
  9.2 The normal transient flash VEP
  9.3 Subject variables
    9.3.1 Age
      9.3.1.1 Premature infants

- 9.3.1.2 From full term to adult age
- 9.3.1.3 Old age
- 9.3.2 Pupillary size
- 9.4 Stimulus characteristics
  - 9.4.1 Stimulators
    - 9.4.1.1 Xenon flash tubes
    - 9.4.1.2 Electromechanical stimulators
    - 9.4.1.3 Fluorescent lamps
    - 9.4.1.4 Light-emitting diodes
    - 9.4.1.5 Full-field (Ganzfeld) flash stimulators
  - 9.4.2 Stimulus rate
  - 9.4.3 Stimulus intensity
- 9.5 Recording parameters: Electrode placements and combinations
- 9.6 Criteria distinguishing abnormal flash VEPs
- 9.7 Neurological disorders that cause abnormal flash VEPs
  - 9.7.1 Prechiasmal lesions
    - 9.7.1.1 Multiple sclerosis
    - 9.7.1.2 Surgical monitoring of optic nerve compression
    - 9.7.1.3 Other prechiasmal lesions
  - 9.7.2 Retrochiasmal lesions
    - 9.7.2.1 Head injury, increased intracranial pressure, and hydrocephalus
    - 9.7.2.2 Anoxic encephalopathy and brain death
    - 9.7.2.3 Cortical blindness
    - 9.7.2.4 Seizures
    - 9.7.2.5 Other retrochiasmal disorders
- 9.8 Ophthalmological disorders that cause abnormal flash VEPs
  - 9.8.1 Increased intraocular pressure and glaucoma
  - 9.8.2 Injury to the eye or optic nerve
  - 9.8.3 Retinal detachment
  - 9.8.4 Retinitis pigmentosa
  - 9.8.5 Amblyopia
  - 9.8.6 Cataract
  - 9.8.7 Other ophthalmological disorders

10. VEPs to other stimuli
- 10.1 Transient VEPs to checkerboard appearance, disappearance, and flash
- 10.2 Steady-state VEPs to checkerboard pattern reversal, appearance, disappearance, and flash
  - 10.2.1 The normal steady-state pattern VEP
  - 10.2.2 Maturation
  - 10.2.3 Neurological disorders that cause abnormal steady-state pattern VEPs
  - 10.2.4 Ophthalmological disorders
    - 10.2.4.1 Refraction and accommodation
    - 10.2.4.2 Amblyopia
    - 10.2.4.3 Macular lesions
    - 10.2.4.4 Glaucoma
- 10.3 Steady-state VEPs to diffuse light stimuli
  - 10.3.1 The normal steady-state flash VEP
  - 10.3.2 Prechiasmal lesions that cause abnormal steady-state flash VEPs

10.3.3 Retrochiasmal lesions that cause abnormal steady-state flash VEPs
10.4 VEPs to sine wave gratings
    10.4.1 The principle of sine wave grating stimuli
    10.4.2 Stimulators for sine wave gratings
    10.4.3 Transient VEPs to sine wave grating stimuli
        10.4.3.1 The normal transient sine grating VEP
        10.4.3.2 Neurological disorders that cause abnormal transient sine grating VEPs
            10.4.3.2.1 Multiple sclerosis
            10.4.3.2.2 Parkinson's disease
        10.4.3.3 Ophthalmological disorders that cause abnormal transient sine grating VEPs
    10.4.4 Steady-state VEPs to sine wave gratings
        10.4.4.1 The normal steady-state sine grating VEP
        10.4.4.2 Neurological disorders that cause abnormal steady-state sine grating VEPs
        10.4.4.3 Ophthalmological disorders that cause abnormal steady-state sine grating VEPs
10.5 VEPs to bar gratings
10.6 VEPs to small macular light spots
10.7 VEPs to moving and stereoscopic random dot patterns
10.8 VEPs to other patterned stimuli
10.9 Photopic versus scotopic VEPs
10.10 Color VEPs
10.11 VEPs to blinking and shift of gaze
10.12 Electric stimulation of the eye
10.13 The ERG to diffuse and patterned light

References for part B

6

VEP types, principles, and general methods of stimulating and recording

SUMMARY

6.1. VEP types vary with the type of stimulus. Different types of VEPs are produced by stimuli of different visual content such as checkerboard and other patterns or diffuse light. VEPs are classified by stimulus rate as transient and steady-state VEPs. VEPs can be further divided by presentation mode into VEPs to alternating patterns (pattern reversal and pattern shift), VEPs to pattern appearance and disappearance, and VEPs to patterned or diffuse light flashes. The VEP most commonly used in clinical practice is the transient VEP to checkerboard pattern reversal or shift.

6.2. During the recording, the subject should sit comfortably in a dark quiet room and look at the stimulus. Pupils should be dilated for stimulation with diffuse light. Refractory errors should be corrected for recordings of pattern VEPs.

6.3. Stimulus parameters differ for each response type. The stimulus rate is usually 1/sec or 2/sec for transient VEPs and about 10/sec for large steady-state VEPs; higher rates are used to determine the critical frequency of photic driving. Stimulus intensity is most important for VEPs to diffuse light stimuli. Most important for patterned stimuli are the brightness contrast between light and dark pattern elements and the size and location of the pattern elements in the visual field.

6.4. Recording parameters are similar for most VEPs: Filters are set to a bandwidth of about 1–200 Hz, sweep length is usually 250 msec, and 100–200 responses are collected for each VEP. Recording electrode placements differ slightly for some VEPs.

6.1 VEP TYPES

VEPs are distinguished by (1) the visual content, (2) the rate, and (3) the mode of presentation of the stimulus; these characteristics are used to classify VEPs as outlined in Table 6.1. VEPs further differ depending on (4) the type of stimulator used and (5) stimulus parameters such as stimulus size and brightness; these aspects are discussed in the sections on the specific types of VEPs. The classification of VEPs, which depends primarily on stimulus characteristics, differs from the classification of AEPs, which depends on recording methods, and of SEPs, which depends on the location of stimulus and recording electrodes.

6.1.1 Visual content: Patterned and diffuse light

The visual content of a stimulus can be divided into two types: patterned and diffuse. The most important patterned

TABLE 6.1. VEP types

A. VEPs to patterned light
  1. VEPs to checkerboard patterns
     a. Transient VEPs
        (1) VEP to checkerboard pattern reversal and shift*
        (2) VEP to checkerboard pattern appearance
        (3) VEP to checkerboard pattern disappearance
        (4) VEP to checkerboard pattern flashes
     b. Steady-state VEPs
        (1) VEP to checkerboard pattern reversal and shift
        (2) VEP to checkerboard pattern appearance-disappearance
        (3) VEP to checkerboard pattern flashes
  2. VEPs to sine wave gratings
     a. Transient VEPS
        (1) VEP to sine wave grating reversals
        (2) VEP to sine wave grating appearance
        (3) VEP to sine wave grating flashes
        (4) VEP to sine wave grating contrast-depth modulation
     b. Steady-state VEPs
        (1) VEP to sine wave grating reversals
        (2) VEP to sine wave flashes
  3. VEPs to other patterns
     VEPs to bar gratings, small light spots, random dots
B. VEPs to diffuse light
  1. Transient VEPs
     a. VEP to brief flashes of diffuse light*
     b. VEP to onset of diffuse light
     c. VEP to end of diffuse light
  2. Steady-state VEPs
     a. VEP to repetitive flashes*
     b. VEP to sine wave modulation of light intensity
C. Other VEPs
  Scotopic VEP, VEP to colored light stimuli, VEPs to blinking, eye movements, pattern motion, electric stimulation of the eye

*VEPs most commonly used in clinical practice.

stimuli are (1) checkerboard patterns consisting of light and dark squares with sharp borders and (2) sine wave grating patterns consisting of light and dark stripes with a gradual transition of brightness between them. Other patterns such as bars with sharp borders, small light spots, and random dots are used occasionally. Whereas VEPs to patterned stimuli (pattern VEPs) are due mainly to the visual content of the pattern, especially the density of light and dark contrast borders, VEPs to diffuse light flashes (flash VEPs) are due to changes of luminance only.

6.1.2 Stimulus rate: Transient versus steady-state VEPs

Transient VEPs consist of a sequence of different peaks which occur at a constant latency after each stimulus. Steady-state VEPs consist of a rhythm of uniform peaks occurring at the same frequency as the repetitive stimulus or at harmonic frequencies. In clinical studies, transient VEPs are used much more often than steady-state VEPs.

6.1.3 Presentation mode: Pattern reversal and pattern shift, pattern appearance and disappearance, patterned and diffuse light flashes

Presentation modes differ for patterned and diffuse light stimuli. In the commonly used method of alternating patterns, the stimulus consists of a sudden change of all light pattern elements into dark ones, and vice versa. This alternation may be accomplished either by replacing each stationary pattern element by an element of the opposite phase (pattern reversal) or by displacing the pattern by the width of one pattern element (pattern shift); the term *pattern reversal* is often used to denote both stimulus mechanisms. Pattern appearance and disappearance are produced by presenting a pattern alternating with a diffusely illuminated or dark field; either the onset or the end of the pattern presentation is used as a stimulus. Patterned light flashes are pattern presentations lasting so briefly that they elicit a VEP combining the responses to onset and end of the pattern presentation. Diffuse light stimuli are usually presented as brief flashes producing combinations of responses to the onset and end of illumination. Longer light pulses elicit VEPs to either the onset or the end of diffuse light and are not used for routine clinical studies. Only a few of the many possible combinations of patterns and presentation modes are used clinically.

6.1.4 VEPs most commonly used in clinical practice

In general, transient VEPs are preferred to steady-state VEPs because their normal range is more concisely defined. Pattern VEPs, especially transient VEPs to alternating checkerboard patterns, are generally preferred to flash VEPs for several reasons: (1) Pattern VEPs vary less between subjects and are more sensitive to lesions that impair conduction through the visual pathway;

(2) patterned light can be used more conveniently for separate stimulation of the left and right visual half-fields which is needed for the investigation of chiasmal and retrochiasmal lesions; (3) patterned stimuli can be used for ophthalmological testing of visual acuity and refractory errors.

Transient flash VEPs have been used extensively in the past but are now largely abandoned because they vary so much in and between subjects that they are rather insensitive indicators of abnormalities. However, they are still used in patients who cannot focus on a patterned stimulus, especially infants and small children, comatose patients, and subjects with very poor visual acuity. Steady-state VEPs to diffuse light flashes are used in a few laboratories to determine the highest flash rate capable of eliciting a steady-state VEP, or the critical frequency of photic driving.

VEPs to sine wave gratings have not yet been used much in clinical studies even though their diagnostic value may equal or exceed that of checkerboard patterns. Sine wave grating stimuli are more difficult to handle, and less information on normal and abnormal VEPs has as yet been accumulated.

There is no ideal method of eliciting VEPs. The type of VEP to be used depends on the requirements of the case. Although transient VEPs to checkerboard reversal generally have the greatest ability to detect abnormalities, in some patients abnormalities can be detected only with other stimuli. Therefore, a few laboratories now routinely use more than one type of VEP, for instance, a combination of transient VEPs to checkerboard pattern reversal and steady-state VEPs to diffuse light. In some cases, stimulus and recording methods may need to be changed and other VEPs may need to be added for a complete examination.

6.2 THE SUBJECT

Subjects should be seated in a comfortable chair in a quiet room and should remain alert during the recording. Use of a chin or neck rest can reduce muscle artifact in the recording. Visual stimuli may be presented on a TV screen, or on a transparent or reflecting screen at a fixed distance from the subject's eyes. Stroboscopic flashes may be applied from a lamp directed at the subject's eyes. The stimulator should not produce any noise that could interfere with the VEP or elicit an AEP.

A steady, dim background illumination is usually maintained to reduce the importance of stray light from various sources and to stabilize dark adaptation during long intervals between stimuli. Pupillary dilatation or an artificial pupil should be used for flash stimulation but is not needed for routine patterned light stimulation (6.3.2), although extreme miosis, ptosis, and cataracts may reduce the intensity of pattern stimuli enough to affect the VEP (7.3.4). More important for pattern VEPs is visual acuity: Subjects requiring corrective lenses for the stimulus distance should wear their glasses and should acknowledge that they can see the stimulus pattern sharply because blurring of small pattern elements may alter the VEP. Furthermore, without their glasses, some myopic subjects squint to see the target more clearly, and this pinhole effect may reduce the stimulus luminance and thereby change the VEP.

Monocular stimulation is best carried out by covering the nonstimulated eye with an eyepatch. Having the patient close one eye or hold an ocular occluder is less satisfactory because it is likely to introduce muscle and movement artifacts into the recording. A fixation point is often used with full-field stimulation and must be used with half-field stimulation studies (6.3.6). Responses to the first few stimuli are often not included in the average because they may be affected by startle or head and eye movements. The beginning of averaging should be delayed, especially for steady-state VEPs, which often develop only after an initial transition period of at least a few stimuli.

6.3 GENERAL DESCRIPTION OF STIMULUS PARAMETERS

A change of a stimulus parameter can change VEP amplitude and latency. This effect is not easily predictable because it occurs only within a certain range between a minimum, or threshold, and a maximum, or saturation point; even within this range, the relation between a change of stimulus parameter and changes of VEPs may be nonlinear. Furthermore, changes of one parameter interact strongly with other parameters and may shift the entire range of effectiveness of another parameter.[85,145,186,419,430,444,450] As a practical consequence, laboratories doing clinical VEP studies must select a specific type of stimulus defined by visual content, presentation mode, stimulator, and stimulus parameters and strictly adhere to that selection in order to avoid the variations of VEPs that occur with the slightest deviation from routine (4.3).

6.3.1 Stimulus rate

For transient VEPs, stimulus rates of 1–2/sec are usually used. Slower stimulus rates unnecessarily prolong the recording session and introduce the risk of increasing the VEP variability due to changes in attention. At stimulus rates of about 4–6/sec, individual VEPs begin to interact with each other. A steady-state response of uniform, sinusoidal shape is usually obtained at rates over 6–8/sec and may be largest at one or more specific stimulus rates. Flash stimuli over 50/sec may be needed to determine the critical frequency of photic driving (10.3.1).

6.3.2 Intensity

Luminance is the most important parameter for diffuse light stimuli because they act only by virtue of a change of luminance. In contrast, patterned light stimuli are used with the aim of avoiding overall luminance changes; however, changes of luminance affect the VEPs to both types of stimuli.

Luminance refers to the amount of light coming from a surface and can be measured by comparing the brightness of the surface with that of a standard light source.[70,395] The luminance of a patterned light stimulus is defined by indicating the luminance of the light and dark pattern elements, which may be measured with a precision spot photometer. The mean luminance of a pattern equals the mean of its light and dark elements (Figure 6.1). The luminance of short flashes can be measured with integrating flash meters or can be defined by the luminance of the steady light produced by flashes given at a rate above fusion frequency; however, in most stroboscopic lamps, the light output per flash decreases with increasing flash rate. Because it is difficult to measure the intensity of brief flashes, intensity is often characterized by specifying the control settings and the distance of the lamp from the subject's eye for equipment made by a particular manufacturer.

Figure 6.1. Stimulus patterns of dark and light elements with gradual (*a*) and sudden (*c*) transitions between the dark and light phases. Graphs of the luminance represent the gradual transition by a spatial sine wave (*b*) and the sudden transition by a spatial square wave (*d*). Either stimulus pattern is characterized by minimum (L_{min}), maximum (L_{max}) and mean (L_{mean}) luminance.

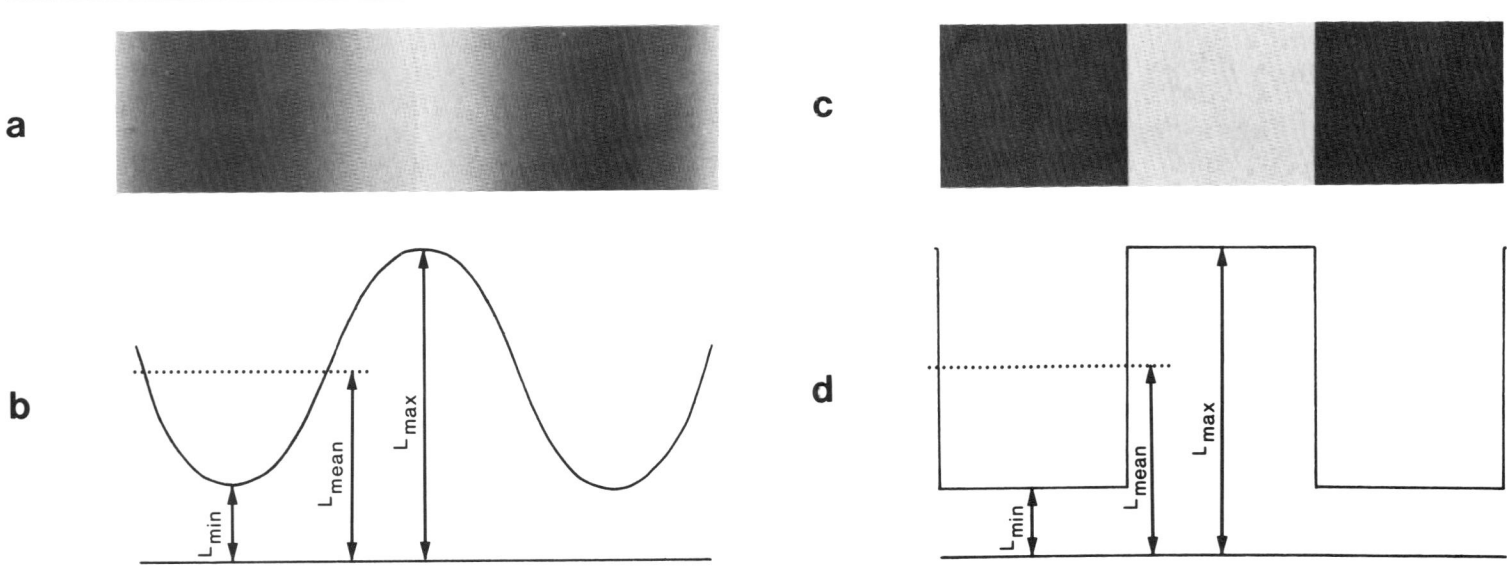

Luminance is measured as candela per square meter (cd/m^2); older measures, which can be converted into candela per square meter are stilb (cd/cm^2), footlambert ($cd/\pi ft^2$), lambert ($cd/\pi cm^2$), and apostilb ($cd/\pi m^2$). Retinal illumination is measured in trolards, one trolard being the illumination that results when a surface with a luminance of 1 cd/cm^2 is viewed through an aperture of 1 mm^2.

The effect of a light stimulus varies with pupillary size to a degree that depends on the stimulus type. VEPs to patterned stimuli that are not associated with significant luminance changes are not affected by changes in pupillary size unless they are extreme; pupillary size is therefore usually not controlled in recordings of pattern VEPs. In contrast, VEPs to diffuse light, depending entirely on the change of retinal luminance, vary greatly with pupillary size, which therefore must be controlled in recordings of these VEPs to reduce the variability of the responses. The effect of pupillary size can be controlled with cycloplegics, an artificial pupil, or a Maxwellian system.

The effectiveness of a light stimulus depends to some extent on the amount of ambient light present during the recording. Although ambient light generally reduces the effect of a visual stimulus, keeping the background of the recording room very dimly lit helps to make stray light from oscilloscope sweeps, illuminated dials, control lamps, and other light sources insignificant and to prevent dark adaptation, which may develop between stimulations. The ambient luminance can be determined by measuring luminance at several sites near the stimulus field and by calculating the mean of these measurements.

6.3.3 Brightness contrast of pattern stimuli

6.3.3.1 Contrast borders: Checkerboard and sine wave grating patterns

Transitions between light and dark pattern elements have sharp borders in checkerboard patterns and in bar gratings. These transitions can be represented by a spatial square wave (Figure 6.1*a*). In sine wave gratings with sinusoidally modulated brightness contrast, the change of luminance between light and dark stripes is gradual and has the form of a sine wave (Figure 6.1*b*).

6.3.3.2 Contrast depth: Counterphase and depth modulation stimuli

The depth of contrast is defined as the difference between the luminances of a light (Lmax) and a dark (Lmin) pattern element divided by their sum: Contrast = (Lmax − Lmin)/(Lmax + Lmin). By this definition, maximum contrast has a value of 1, minimum contrast a value of 0. In clinical practice, contrast of stimulus patterns is usually set to at least 0.5, i.e., to a ratio of at least 3:1 between maximum and minimum luminances.

Clinical studies use such extreme changes of contrast as produced by pattern reversal, appearance, and disappearance. Research studies have investigated the effect of partial changes of contrast produced by intermittent increase and decrease of the brightness of the light and dark pattern elements. This contrast depth modulation is a less effective stimulus than the complete pattern reversal, or counterphase contrast modulation, used in clinical studies (10.4.4.1). The timing of the contrast reversal is usually made as sudden as possible (7.3.1), but a gradual alternation is produced by stimulators using rotating polaroids (7.3.1.4.4).

6.3.4 Size of the elements of pattern stimuli

The bright and dark elements of checkerboard and grating patterns most commonly used in clinical practice have equal size. Figure 6.2 illustrates measurements of visual angle and spatial frequency commonly used to describe this size.

6.3.4.1 Visual angle
The visual angle β describes the size of the image of one light or dark element at the retina and is derived from $\tan \beta = a/b$, where a is the side length of the element and b its distance from the eye (Figure 6.2a). Visual angles of less than 1° can be approximately calculated by $\beta = 3,450\ a/b$ in minutes of arc; angles of 1° or more can be calculated by $\beta = 57.3\ a/b$ in degrees of arc. Visual angle is commonly used to indicate the size of pattern elements with sharp borders such as the squares in checkerboard patterns.

6.3.4.2 Spatial frequency
The spatial frequency equals the number of repetitions of one light plus one dark element, or of one cycle of the stimulus pattern, per degree of visual angle (Figure 6.2b).

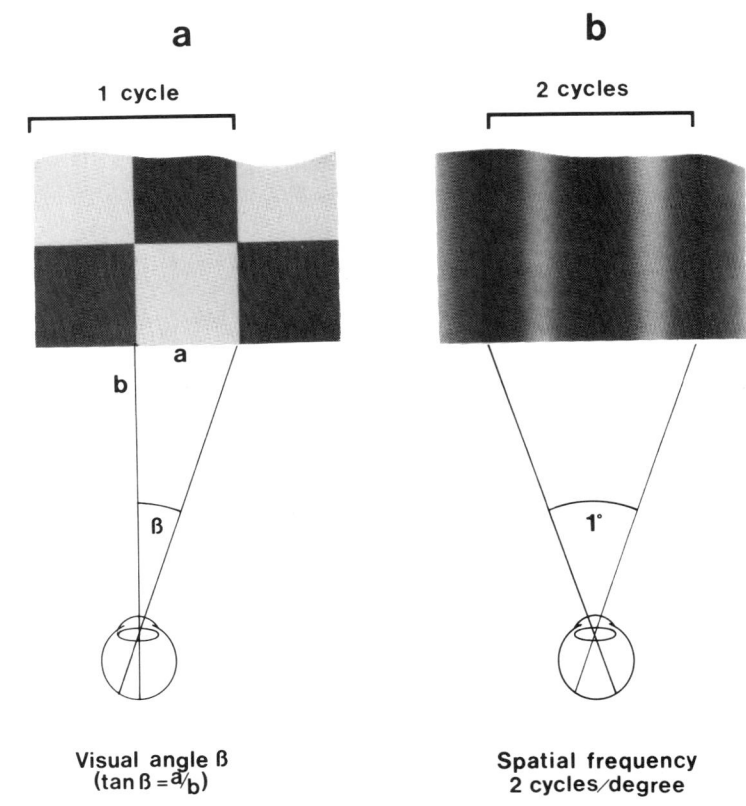

Figure 6.2. Two methods of defining the size of pattern elements in checkerboard and grating patterns. (a) Visual angle β ($\tan \beta = a/b$) is defined as the subtense of a single light or dark element at the retina. (b) Spatial frequency is the number of recurring pattern elements, or cycles, per one degree of visual angle.

The elements defined by spatial frequency and visual angle are different: A light or dark element of a visual angle of 15′ is only half the element described by the corresponding spatial frequency; this element, consisting of a light and a dark unit of 15′ each, has a visual angle of 30′.

Spatial frequency, rather than visual angle, is commonly used to define the size of elements in sine wave gratings that represent a single spatial frequency. Spatial frequency is used less often to indicate the size of checkerboard squares. The spatial square waves representing the light and dark squares of a checkerboard pattern consist of a mixture of frequencies. If the light and dark elements of a checkerboard pattern were to be represented by alternating light and dark stripes, the direction of the stripes would run at angles not of 90° but of 45° to the squares. The major Fourier component of the spatial frequencies contained in checkerboard patterns is oriented diagonally to the squares and has a wavelength of 1.4 times the side-length of the squares.

6.3.5 Size of the stimulus field

Field size is best described in terms of visual angle (6.3.4.1).

6.3.6 Full-field versus half-field stimulation, upper versus lower retinal stimulation

For full-field stimulation, the stimulus pattern extends equally to both sides of a fixation point, such as a small light-emitting diode, placed in the middle of the pattern. Half-field stimulation may be accomplished in two ways: (1) One half of the stimulus pattern is presented to the left or right of the central fixation point; (2) the entire stimulus pattern is used by positioning the fixation point at the right or left margin of the pattern. With either method, the fixation point should be about 0.5–1° outside the stimulated half-field to avoid stimulation of the opposite half-field due to small involuntary shifts of gaze. In method (1), a dark strip, about 1° wide, may be placed vertically across the middle of the stimulus pattern, and the fixation point may be positioned in the middle of the strip in the upper or middle part of the pattern.

Stimulation of the upper and lower hemiretina may be accomplished either by presenting only the part of the stimulus pattern above or below the fixation point or by moving the fixation point to the upper or lower border of the entire pattern.

6.3.7 Pattern orientation

Patterns whose components are oriented vertically are more effective than obliquely oriented patterns. The major spatial Fourier component of checkerboard patterns is diagonal to the checks.[244]

6.4 GENERAL DESCRIPTION OF RECORDING PARAMETERS

Whereas stimulus characteristics differ for each type of VEP, most recording parameters are similar for all VEP types.

The number of channels should be four for a complete examination including the retrochiasmal part of the visual pathway, but fewer channels are often used for testing of the prechiasmal part.

Recording electrodes are placed on the occipital scalp in the midline and laterally; reference electrodes are positioned on the frontal or central scalp or on the earlobes. The precise location of these electrodes and the electrode combinations used for recording differ widely in different laboratories and vary for different types of VEPs. Specific recommendations have been made[8] for VEPs to checkerboard pattern stimulation (7.4.1) and to diffuse light flashes (9.5).

The polarity convention is not standardized; major positive or negative peaks are displayed upward in some laboratories and downward in others.

Filters are set to a bandwidth between 0.2–1 Hz and 200–300 Hz. A decrease of the high frequency filter to 100 Hz may cause an apparent increase of peak latency[86] (2.4.1.4).

The analysis period for transient VEPs is about 250 msec in normal adults and up to 500 msec in infants and for abnormally delayed VEPs at any age. The requirements for the temporal resolution of VEP waves during these periods are easily met by conventional averagers: Even in an averager with only 250 points per channel, a sweep length of 500 msec results in a dwell time of 2 msec, or a sampling rate of 500 Hz, which can resolve waves up to 250 Hz (3.3.4), i.e., waves much faster than those contained in VEPs. Usually 100–250 responses are averaged and at least two averages are superimposed to ascertain replication.

7

The normal transient VEP to checkerboard pattern reversal

SUMMARY

7.1. The normal transient VEP to checkerboard pattern reversal or shift has a major positive occipital peak at about 100 msec (P100), often preceded by negative peaks (N75, N145). The precise latency and amplitude of the P100 depend on subject variables such as age, sex, and visual acuity, on stimulus characteristics such as type of stimulator, full-field or half-field stimulation, check size, contrast, and luminance, and on recording parameters such as placement and combination of recording electrodes.

7.2. The most important subject variable affecting the pattern VEP is age.

7.3. Stimulus characteristics that affect the pattern VEP include the type of stimulator, stimulus rate, brightness contrast, luminance, check and field size, location of the stimulus in the visual field, monocular or binocular stimulation, and pattern orientation.

7.4. Important recording parameters are the electrode placements and montages for full-field and half-field stimulation.

7.5. The general strategy of stimulating and recording depends on the portion of the visual pathway to be studied: Prechiasmal studies commonly use monocular full-field stimulation and rely mainly on midline occipital recordings; chiasmal and retrochiasmal investigations require monocular half-field stimulation and both midline and lateral occipital recordings. Although VEPs are used mainly to study conduction through the central visual pathway, they are occasionally used to evaluate vision and to investigate ocular problems.

7.1 DESCRIPTION OF THE NORMAL TRANSIENT PATTERN REVERSAL VEP

The most commonly used methods of producing and analyzing VEPs are summarized in Table 7.1. Figure 7.1 shows a normal midline occipital VEP to checkerboard pattern reversals at 2/sec. It is characterized by a positive occipital peak at a latency of 90–110 msec and an amplitude of about 10 μV. This peak is often referred to as $\overline{P100}$ or P1 (4.1.7.2).

TABLE 7.1. VEPs to checkerboard pattern reversal

A. Subject variables
 1. Age: Separate normal control groups are needed for subjects up to about 5 years and over 60 years.
 2. Sex: Separate controls for males and females are used in some laboratories.
 3. Visual acuity: Visual acuity should be better than 20/200 for squares of about 1° or less. The subject should see the stimulus pattern clearly. Refractory errors should be corrected.
 4. Pupillary size: Pupils should neither be altered by medication nor fixed at extremely large or small size by disease.

B. Stimulus characteristics
 1. Type: Checkerboard reversal on TV screen, checkerboard pattern shift by pivoting mirror projection.
 2. Rate: 1–2/sec.
 3. Phase: Responses to reversal into one phase are averaged with responses to reversal into the opposite phase.
 4. Brightness contrast: Greater than 0.5 for bright versus dark squares.

TABLE 7.1. (*continued*)

 5. Intensity: Luminance of bright and dark squares and ambient luminance must be kept constant.
 6. Monocular stimulation: Monocular stimuli should be used for full-field and half-field stimulation.
 7. Full-field versus half-field stimulation: The stimulus pattern extends equally to the sides of the fixation point for full-field stimulation, but only to one or the other side for half-field stimulation.
 8. Check size: 28–31′ for full-field stimulation of the central retina; 50–90′ for half-field stimulation of the peripheral retina.
 9. Field size: Over 8° for full-field stimulation; over 10–16° for half-field stimulation.
 10. Color: The color of the pattern elements should be specified.

C. Recording parameters
 1. Number of channels: 3–4.
 2. Electrode placements: Midoccipital (MO), 5 cm above the inion; left and right occipital (LO, RO), 5 cm lateral to MO; left and right posterior temporal (LT, RT), 10 cm lateral to MO; midfrontal (MF), 12 cm behind the nasion; ground electrode at the vertex or elsewhere on the head.
 3. Montages:
  a. Full-field stimulation:
  Channel 1: RO–MF
  Channel 2: MO–MF
  Channel 3: LO–MF
  A fourth channel may be used to record the pattern ERG between an infraorbital and a lateral orbital electrode.

TABLE 7.1. *(continued)*

    b. Half-field stimulation
       (1) Right half-field stimulation:
          Channel 1: RO–MF
          Channel 2: MO–MF
          Channel 3: LO–MF
          Channel 4: LT–MF
       (2) Left half-field stimulation:
          Channel 1: RT–MF
          Channel 2: RO–MF
          Channel 3: MO–MF
          Channel 4: LO–MF
4. Filter settings: Low frequency filters: 0.2–1 Hz; high frequency filters: 200–300 Hz.
5. Number of responses averaged: 100–200 for full-field stimulation; 200 for half-field stimulation.
6. Sweep length: 250 msec; up to 500 msec in infants and for abnormally delayed VEPs.

D. Analysis
  1. Normal peaks
    a. Full-field VEPs: $\overline{N75}$, $\overline{P100}$, and $\overline{N145}$ in midline and lateral VEPs.
    b. Half-field VEPs: $\overline{N75}$, $\overline{P100}$, and $\overline{N145}$ over midline and ipsilateral to the stimulated half-field; $\overline{P75}$, $\overline{N105}$, and $\overline{P135}$ contralateral to the stimulated half-field.
  2. Criteria for abnormal VEPs
    a. Full-field VEPs
      (1) Absence of any peaks.
      (2) Abnormally long peak latency of $\overline{P100}$.
      (3) Abnormally long $\overline{P100}$ interocular latency difference.
      (4) Questionable criteria: Abnormally long peak latency and interocular latency difference of $\overline{N75}$, abnormally large lateral occipital amplitude ratio of full-field VEPs.
    b. Half-field VEPs
      (1) Absence of any peaks ipsilateral and contralateral to the stimulated half-field.
      (2) Questionable criteria: Abnormal latency and waveform.

In many cases, this major positive peak is preceded by a smaller negative peak at about 60–80 msec, sometimes called $\overline{N75}$ or N1, and followed by a second negative peak, $\overline{N145}$ or N2. A second positive peak and later peaks vary considerably in and between subjects and depend on the location of the reference electrode. In some instances, the first positive peak is so closely followed by a second positive peak that the two peaks may seem to be part of a single deflection having two smaller peaks. Such split peaks, also called bifid or bilobed peaks, differ from the splitting of a single peak produced by superimposed residual noise (4.1.3.1) in that the separation between the split peaks is deeper and wider than that caused by noise in other parts of the same VEP. Bifid positive peaks should be measured separately, and only the first one considered equivalent to the $\overline{P100}$. A very small positive peak occasionally precedes the $\overline{P100}$ (Figure 7.1) and is of no clinical importance, but has led some investigators to use the terms P1 for this peak and P2 for the major positive peak.

The precise values of the mean latencies and amplitudes depend on many factors described below. Most of these factors have similar effects on transient and steady-state VEPs, regardless of whether they are produced by pattern reversal, appearance, disappearance, or flash. The VEPs to checkerboard pattern appearance, disappearance, and flashes and the steady-state VEPs to checkerboard stimuli are presented elsewhere (10).

Figure 7.1. Normal occipital midline VEP to monocular full-field pattern reversal stimulation showing $\overline{N75}$, $\overline{P100}$, and $\overline{N145}$ peaks. Stimulation with squares of 30′ sidelength in a 14°- (horizontal) by-11° (vertical) field of squares reversing at 2/sec. Recording between midoccipital (MO) and midfrontal (MF) electrodes. Positivity at the occiput is plotted upward. Two averages of 100 responses each are superimposed.

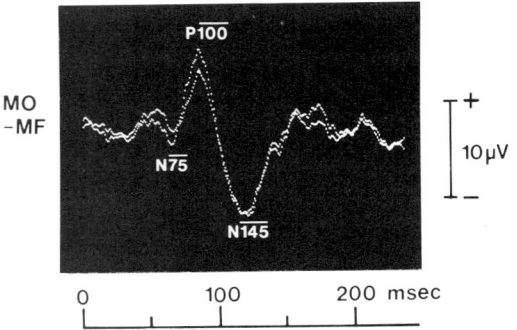

7.2 SUBJECT VARIABLES

7.2.1 Age

7.2.1.1 Before adulthood

Pattern VEPs can usually not be recorded in young children because they do not fixate well. However, careful investigations of pattern VEPs starting at preterm age have shown that the latency decreases rapidly during the first year of life and reaches adult values at the end of the first year for check sizes of 50–60′ and at over 5 years for checks of 12–15′.[336] Accommodation in infants is better for larger than for smaller checks.[437] VEP amplitude and latency have been reported to remain stable in older children,[154] although a few studies report that latency[149,251] and amplitude[215,426] decrease during adolescence, or that amplitude decreases to the fourth decade.[418]

7.2.1.2 Adults

The latency of the $\overline{P100}$ increases after the age of 60 years so that age-dependent normal controls are needed to correctly distinguish normal from abnormal in the elderly.[7,149,186] Several studies suggest that the $\overline{P100}$ latency increases through the entire adult life.[88,251,417,419,438] Age effects on latency have been reported to be more prominent for low luminance levels[417] and for small check sizes.[438]

Studies of the effect of age on amplitude report a decrease with old age,[426,418] no change,[88] or an increase.[215]

7.2.2 Sex

Women were found to have shorter $\overline{P100}$ latency than men in some studies.[7,145,251,450]

7.2.3 Visual acuity

A visual acuity of less than 20/200[190] is likely to reduce VEP amplitude[193] and to increase latency, especially for small check sizes[102] and low contrast.[471]

7.2.4 Pupillary size

Normal variations of pupillary size do not affect the VEP to patterned stimuli, but extreme miosis and mydriasis may alter stimulus luminance and thereby change VEP amplitude and latency (7.3.4).

7.2.5 Ocular dominance

Although the VEP to stimulation of the dominant eye has been reported to have slightly shorter latency and higher amplitude,[412] clinical studies do not use different normative values for dominant and nondominant eye stimulation.

7.3 EFFECT OF STIMULUS CHARACTERISTICS

7.3.1 Stimulators

7.3.1.1 Television screens
TV picture tubes under control of pattern-generating circuitry are commonly used. These stimulators may be bought commercially or built from published circuits.[16,237,446] TV pattern stimulators have several advantages: They are sufficiently large and bright, can generate checks and horizontal or vertical bars of different sizes and present them at different rates, and may be used to stimulate visual half-fields or quadrants without shift of the fixation point. Patterns can be made to reverse or to appear and disappear in alternation with a diffusely illuminated or a dark background. The stimulus pattern may be mixed with a TV program, such as a cartoon, to keep young children looking at the screen[16] without significant degradation of the VEP.[283,442,475]

The onset of the TV pattern stimulus is slower than that obtained with other pattern generators. This accounts for some latency differences between laboratories, but does not reduce the value of TV stimulation for clinical VEP testing. The slow, gradual onset of the stimulus is due to the way in which the stimulus image is generated on TV. The change of the TV picture progresses in horizontal lines from the top to the bottom of the screen and is completed in the time required for the presentation of a TV picture frame, namely in 16.7 msec at the power-line frequency of 60 Hz used in the United States and in 20 msec at the line frequency of 50 Hz used in many other countries. Although the frame rate can be increased,[237] this reduces the sharpness of the image.[214] The gradual appearance of the pattern change on the TV screen leads to a gradual projection of the stimulus pattern on the retina; the commonly used patterns with relatively small check sizes become effective as a visual stimulus when they reach the central retina. If the pattern change always begins at the top of the screen, the time to its appearance at the fixation point is added to the VEP latency measured from the beginning of the reversal.[473] While this makes the latency of pattern reversal VEPs longer than that obtained with fast pattern shifts,[450] the difference seems to become insignificant if no fixation point is used,[298] i.e., if the interval between pattern change and its appearance at the central retina is allowed to be irregular. However, it is not desirable to lock the pattern change to the onset of the TV frame because this timing also locks the onset of the averaging epoch to the power-line frequency and favors the buildup of power-line interference artifact in the average. Therefore, the change in pattern is usually unlocked from the start of the TV frame and begins in different parts of the screen with each stimulus.[214,473] This gives the responses a slightly more variable latency, resulting in a broader VEP peak.[16]

TV pattern generators have minor disadvantages: Contrast borders between dark and light are not absolutely sharp, and the control of contrast and luminance is limited and not linear.

7.3.1.2 Projection with a pivoting mirror

The image of a checkerboard pattern is projected onto a screen by reflecting it off a small mirror mounted on a pen motor or a fast-moving galvanometer. The mirror is rotated intermittently to shift the pattern by exactly the width of one square.[191] This results in a pattern change which appears on all parts of the retina at the same time and with a speed depending on the duration of the mirror movement.[192,463] Movements of 5–10 msec give sharply peaked VEPs with latencies of sufficiently small variation for good clinical diagnostic use.[192] This method is quite effective and widely used even though it is less versatile than the TV technique: The choices of square sizes, half-field and quadrantic presentations, and stimulus rates are limited; contrast is difficult to control.

7.3.1.3 Light-emitting diodes

Light-emitting diodes are mounted in a square matrix and connected so that half of them, forming the light squares of a checkerboard pattern, are turned on while the other half, forming the dark squares, are turned off. The electronic switching between the two phases gives the advantage of very fast pattern reversals.[143,147] Disadvantages are invariant element size, small stimulus field size, colored light, low luminance, and contrast borders that are often not sharp.

7.3.1.4 Custom-made stimulators

7.3.1.4.1 Tachistoscopic displays. Tachistoscopes can present patterns with switching times of less than 1 msec.[15,234]

7.3.1.4.2 Two projectors with synchronized shutters. Slides showing patterns may be projected alternately on a screen from two projectors with electronically controlled shutters having switching times of a few milliseconds. Pattern reversal is produced by projecting complementary checkerboard patterns; pattern appearance and disappearance are generated by alternately projecting a pattern and a uniformly gray field.[9]

7.3.1.4.3 Patterned mirrors. Mirrors with alternating reflecting and transparent elements have been used in conjunction with independently switchable light sources to project reversing or appearing and disappearing checkerboard patterns.[444]

7.3.1.4.4 Polarized light. Pattern reversal stimuli may be generated by passing light through a pattern of polaroid squares with alternating horizontal and vertical polarizing axes and by viewing the slide through a rotating polaroid disc[44]; this generates pattern reversals with a sinusoidally changing time course, suitable for steady-state VEPs.

7.3.1.4.5 Maxwellian systems. The stimulus may be viewed through an eyepiece which focuses the light from a pattern so that it forms an image at the level of the pupil and becomes independent of pupillary size.[379,491]

7.3.2 Stimulus rate and phase

Transient checkerboard pattern reversal VEPs are usually elicited at 2/sec. Slower rates of stimulation produce no change of the VEP. An increase of stimulus rate to 4/sec may significantly increase the latency of the transient VEP.[450]

The stimulus rate of pattern reversal stimuli equals twice the alternation or stimulus pulse rate because each complete alternation or stimulus pulse generates two stimuli, namely, the transition of the pattern from one phase into the other and the return to the original phase. Responses to transitions into both phases are usually averaged together because these responses are similar.

7.3.3 Contrast

7.3.3.1 Sharpness of contrast borders

Blurring of contrast borders degrades the pattern VEP.[440] Reduction of amplitude and increase of latency are especially prominent with small check sizes.[211,388,436] Since blurring may be caused by refractory errors, these should be corrected for VEP studies so that they will not interfere with the evaluation of conduction through the central visual pathway.[102] As a rule, a decrease of visual acuity does not alter VEPs to checkerboard pattern stimulation unless it reaches 20/200 or unless the smallest normally effective check sizes are used (7.2.3).

7.3.3.2 Effect of contrast depth

A decrease of the steady contrast between light and dark squares, usually set at 0.5 or higher (6.3.3.2), may increase the latency and decrease the amplitude of the VEP, but the range and the degree of this effect depends on many other variables including luminance and size of the squares.[388,444,468] An increase of contrast above the fairly high levels usually used in clinical VEP work does not change the VEP.

7.3.4 Luminance

An increase of the mean luminance of a checkerboard pattern stimulus decreases the latency and increases the amplitude of the VEP to a degree that depends on other variables, including check size and contrast.[25,44,419,450,468,475] Under commonly used conditions, decreasing the mean luminance of a reversing checkerboard pattern by a factor of 10 increases the latency of the major positive peak of the VEP by 10–15 msec[79,122,187,193,450] and decreases its amplitude by about 15 percent.[187] The luminance of light and dark pattern elements used in different laboratories varies widely: The light squares used in some studies are darker than the dark squares in others.[122,187]

7.3.5 Check size and field size

A decrease of check size increases VEP amplitude[440] to a maximum at a visual angle of 10–15′[212,387] and increases latency,[396,434] especially at the smallest effective check sizes. The effect of check size varies in different parts of the retina. The central 4–5° of the retina, or fovea, generates the largest part of the VEP.[230,287,444,505] The fovea is most sensitive to small check sizes of about 10′ whereas the peripheral parts of the retina are more sensitive to larger check sizes; locations up to 7.5° from the fovea are susceptible to progressively increasing sizes. The check size most effective at a given eccentricity depends on the stimulus field size. Check sizes exceeding those that are effective as pattern stimuli in a particular part of the retina produce responses mainly due to local retinal luminance changes.[13,24,44,206,288,327,388] The crossover between pattern and luminance effects depends on many variables,[352] such as location of the stimulus on the retina, brightness contrast, orientation of checks, luminance, and level of light adaptation.[24,468] Because of the different effect of small and large check sizes on different parts of the retina, small checks need to be presented only in a small central field, whereas large checks require larger fields to cause their pattern effect in the peripheral retina; while they act as pattern stimuli at their respective effective sites, small checks have little effect in the peripheral retina, but large checks cause luminance effects in the central retina. The smallest effective check size of about 10′, although producing the largest VEP under ideal conditions, is not used in routine clinical studies because blurring, often caused by refractory errors, affects latency and amplitude of the VEP to the smallest size much more than those of VEPs to larger sizes.

7.3.6 Location of the stimulus in the visual field

7.3.6.1 VEPs to full-field and half-field stimulation

Stimulation of the left and right half-fields of the retina (6.3.6) produces VEPs different from those to full-field stimulation. This difference is small if only the central part of the retina is stimulated with checks of small size presented in a small field. It becomes larger if the periphery of the retina is stimulated with checks of 50–90' in a field of over 16°. The difference between full-field and half-field VEPs is greatest in recordings that include lateral occipital or posterior temporal electrode placements (7.4.1).

7.3.6.2 VEPS to upper and lower retinal stimulation

Stimulation of the upper hemiretina seems to produce VEPs of shorter latency and more anterior distribution than does stimulation of the lower hemiretina.[2,243,286] Clinical studies usually do not use this variable but place the fixation point in the middle of the vertical extent of the stimulus pattern.

7.3.7 Monocular versus binocular stimulation

VEPs to stimulation of either eye are normally very similar to each other. The VEP to stimulation of both eyes may have slightly larger amplitude but normally has the same latency as VEPs to stimulation of each eye. In patients with an abnormal VEP to stimulation of one eye, the VEP to binocular stimulation is usually normal.[223]

7.3.8 Effect of pattern orientation

Because the major spatial Fourier component of checkerboard patterns is oriented diagonally to the checks (6.3.7), checkerboards are more effective when presented as diamonds than as squares.[323] However, squares are used in clinical testing because they are easier to generate.

7.4 EFFECT OF RECORDING PARAMETERS

7.4.1 Electrode placements and combinations for full-field and half-field VEPs

7.4.1.1 VEPs to full-field stimulation

Normal full-field pattern VEPs have a maximum at the midline of the head and are usually fairly symmetrical on the two sides (Figures 7.2 and 7.3), but may be much larger on one side than on the other (8.1.5). The lateral extension of the VEP varies considerably in normal subjects: The amplitude of the $\overline{P100}$ drops off to the sides much more rapidly in some subjects than in others. The amplitude of VEP recordings further depends on the choice of the reference electrode and may be low with ear or vertex references because similar potential changes may appear both at the occipital and the reference electrode and lead to partial cancellation of the VEP due to differential amplification (2.3.4). Some normal subjects show no VEPs in recordings between occipital and vertex electrodes because of complete cancellation.

For full-field studies of pattern VEPs, recordings are made in three or four channels. To conform with place-

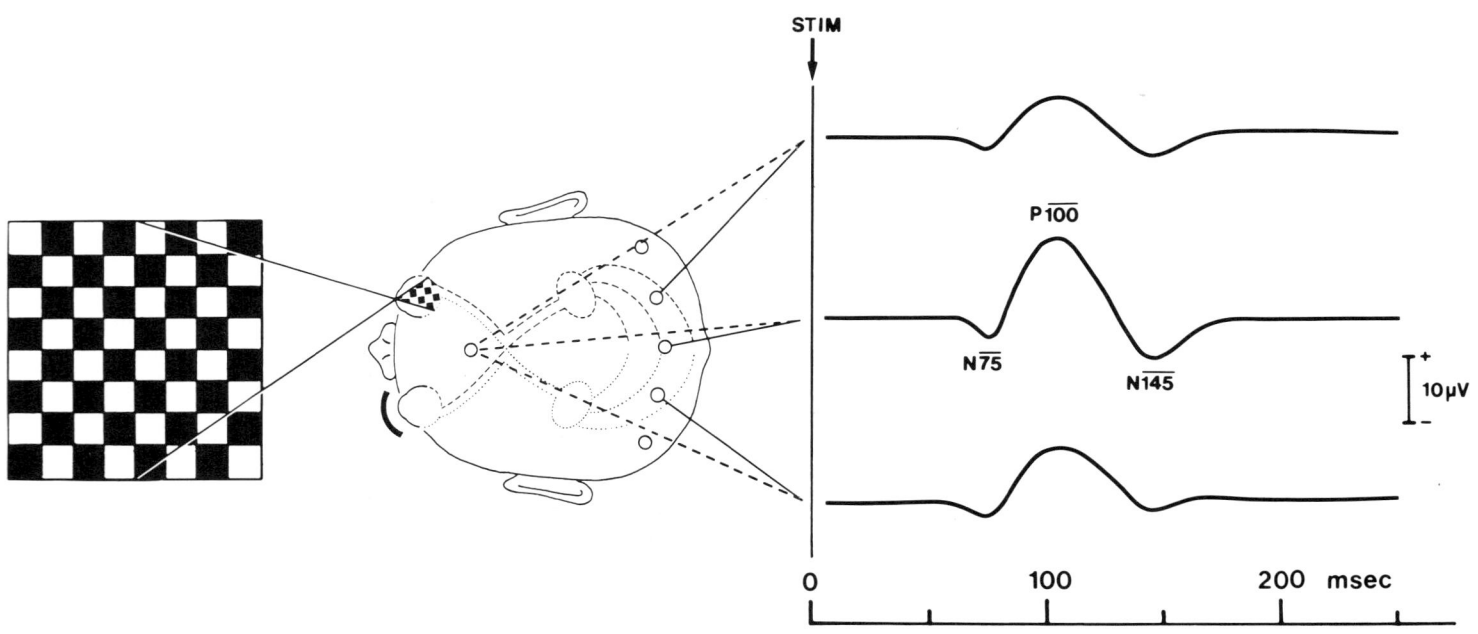

Figure 7.2. Schematic diagram of normal transient pattern VEPs to monocular full-field stimulation. Stimulation of one eye produces VEPs that are distributed approximately symmetrically over both occipital areas with a maximum at the midline. They have a major positive peak ($\overline{P100}$), preceded and followed by smaller negative peaks ($\overline{N75}$, $\overline{N145}$).) Positivity at the occiput is plotted upward.

Figure 7.3. Normal midline and lateral occipital VEPs to stimulation of the left eye (top) and right eye (bottom) with full-field (*a, d*), left half-field (*b, e*) and right half-field (*c, f*) checkerboard pattern reversals. The latencies measured to P$\overline{100}$ of the midline VEP (L1, L2) are normal. The amplitudes of the half-field VEPs ipsilateral to the stimulated half-field are higher than those on the other side. A P$\overline{105}$ is not seen in this case in which posterior temporal electrodes were not used. Stimulation with checkerboard patterns of squares of 30' (*a, d*) and 60' (*b, c, e, f*) side-length. Recordings between left occipital (LO), midoccipital (MO), right occipital (RO), and midfrontal (MF) electrodes. Positivity at the occiput is plotted upward.

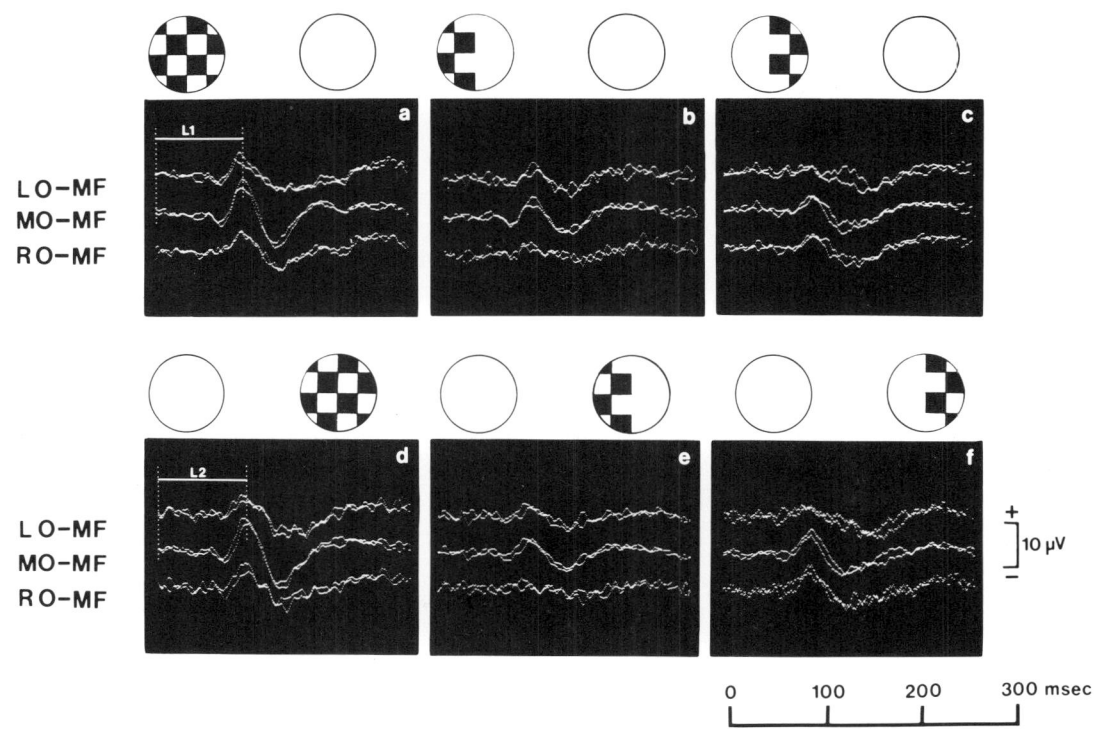

ments commonly used for half-field studies, recording electrodes are placed on a coronal line 5 cm above the inion in the midline and 5 cm to the left and right. A midfrontal electrode is positioned 12 cm above the bridge of the nose. A ground electrode is placed somewhere on the head, for instance at the vertex. Four-channel recordings should use three channels to record between each of the three occipital electrodes and the midfrontal electrode (Table 7.1). The fourth channel may be used to record the VEP between the midoccipital and a vertex electrode or to record the ERG (2.2.1.6, 10.13).

7.4.1.2 VEPs to half-field stimulation
Normal half-field VEPs to checkerboard stimulation have a paradoxical distribution that is best seen with a technique that uses (1) stimulation with fairly large check sizes of about 1°, presented in a field sufficiently large to affect the retinal periphery, and (2) recordings from midline and lateral occipital, or posterior temporal, electrodes in reference to a midfrontal electrode. With this technique, stimulation of one visual half-field, which excites the opposite hemisphere, produces a VEP that is largest in scalp recordings from the midline and the side of the stimulated half-field; these recordings show $\overline{N75}$, $\overline{P100}$, and $\overline{N145}$ peaks resembling those of the full-field VEP (Figure 7.3 and 7.4). On the opposite side, i.e., over the excited hemisphere, the VEP usually has lower amplitude (Figure 7.3, *b–f*). Recordings from lateral occipital or posterior temporal electrode positions may show a VEP with a shape resembling a mirror image of the contralateral VEP, having $\overline{P75}$, $\overline{N105}$, and $\overline{P135}$ peaks (Figure 7.4). Some subjects have symmetrical VEPs to half-field stimulation. Like full-field VEPs, half-field VEPs extend laterally to a degree that varies between different subjects.

The full-field VEP consists of the sum of the two asymmetrical half-field VEPs (Figure 7.5). The predominance of the half-field VEP on the side of the stimulated field is thought to be due to the spatial orientation of the visual cortex in the depth of the occipital lobe: Excitation of the visual cortex on one side seems to generate an electric field of higher amplitude on the other side of the head.[28,52,54,183,425] However, this paradoxical distribution is not seen with all stimulating and recording methods. VEPs with a maximum over the excited hemisphere may be recorded with different recording electrode placements[118,287] and with stimulation using small field sizes,[225,285] pattern onset rather than reversal,[268] diffuse light,[340] or repetition rates producing steady-state VEPs.[279,350]

Most half-field studies now use four recording channels. Three channels are connected to the same electrode combinations as used for full-field studies. An additional left posterior temporal electrode (LT) and right posterior temporal electrode (RT) are placed 10 cm lateral to the midline on the coronal line of occipital electrodes 5 cm above the inion. The fourth channel is used to record between the posterior temporal electrode contralateral to the stimulated half-field and the midfrontal electrode[8] (Table 7.1, C.3). If more than four channels are available, recordings may be made between all five posterior electrodes and

Figure 7.4. Schematic diagram of normal transient pattern VEPs to monocular half-field stimulation. Stimulation of each half-field produces occipital VEPs that have maximum amplitude, and peaks similar to those of the full-field VEP, at the midline and ipsilateral to the stimulated half-field, i.e., opposite the excited hemisphere; VEPs opposite the stimulated half-field, i.e., over the excited hemisphere, usually have lower amplitude and may show $P\overline{75}$, $N\overline{105}$, and $P\overline{135}$ peaks. Because most averagers have no more than four channels and cannot record from all five posterior electrodes simultaneously, routine studies may exclude recordings from the posterior temporal electrode ipsilateral to the stimulated half-field. Positivity at the occiput is plotted upward.

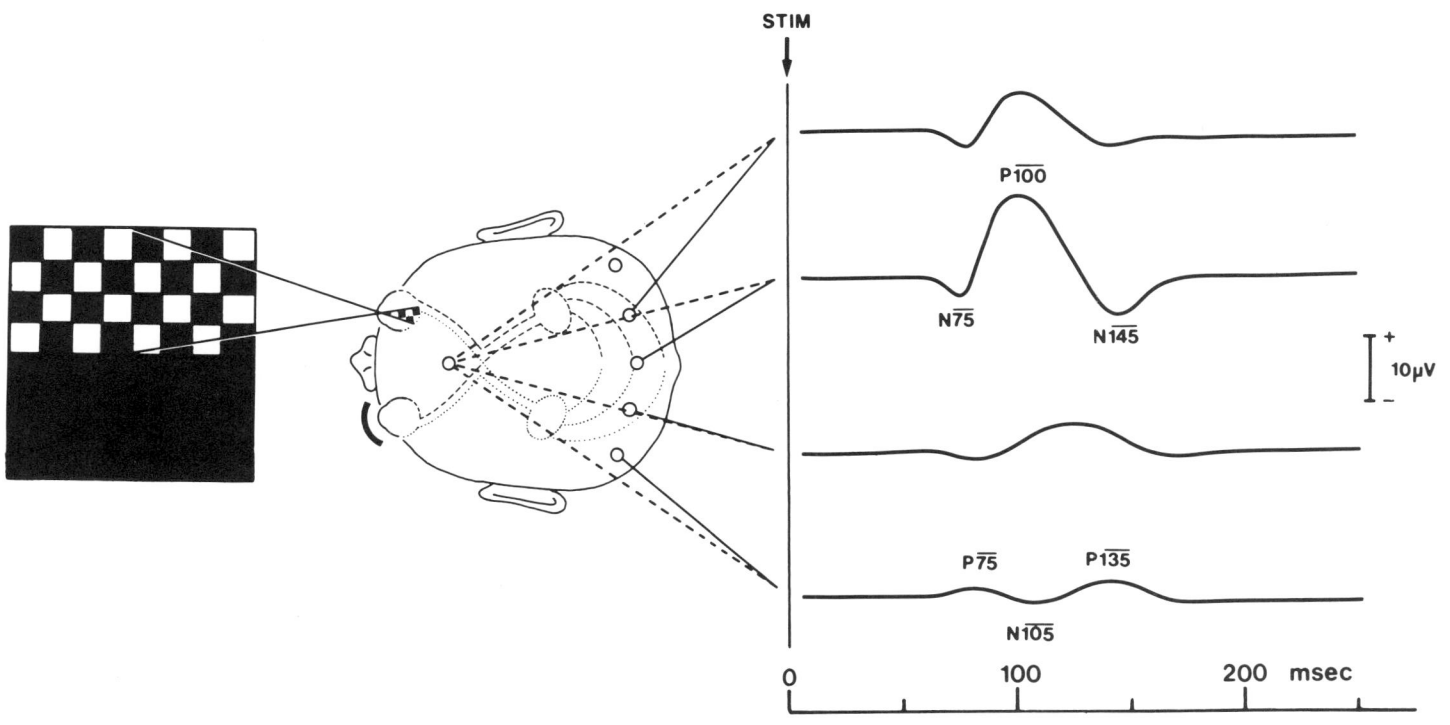

Figure 7.5. Distribution of transient pattern shift VEPs to half-field and full-field stimulation in midline and lateral occipital recordings with a frontal reference electrode. Stimulation of the left half-field (left column) and right half-field (middle column) of the left eye with alternating checkerboard patterns of large check and field sizes produces asymmetrical VEPs with a maximum in the midline and ipsilateral to the stimulated half-field. Full-field stimulation (right column, solid lines) produces fairly symmetrical VEPs with a maximum at the midline; these full-field VEPs resemble the algebraic summation of the asymmetric half-field VEPs (right column, dashed lines). Electrode placements as indicated in the head diagram; midfrontal reference electrode. Negativity at the occiput is plotted upward; the P$\overline{100}$ points down. From Blumhardt et al.[50] with permission by the authors and British Medical Journal.

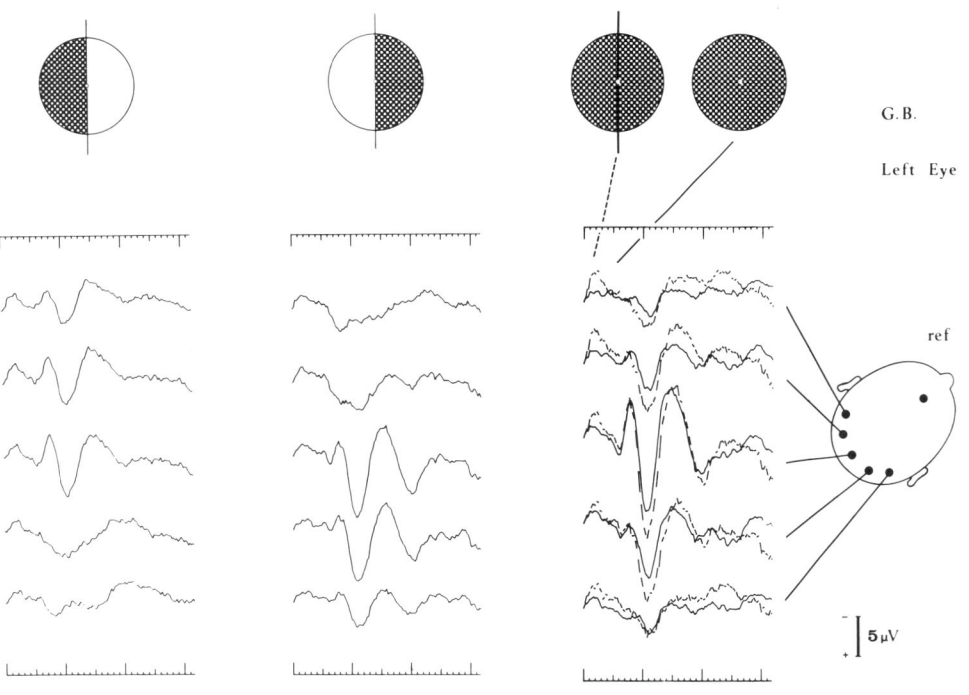

the midfrontal electrode. Other electrode placements such as those of the International 10–20 System are used occasionally.[183] Two-channel recordings are barely sufficient for half-field studies. Bipolar montages of electrode chains across the occipital areas produce gross distortions of VEPs.

7.4.2 Other recording parameters

Most other recording parameters are indicated in the general section (6.4). The number of recording channels should be at least three for full-field stimulation and at least four for half-field stimulation. The number of responses in each average is about 100 for full-field VEPs and about 200 for half-field VEPs.

7.5 PRECHIASMAL, CHIASMAL, RETROCHIASMAL, AND OPHTHALMOLOGICAL STRATEGIES

Different combinations of monocular and half-field stimulation and of recording from midline and lateral occipital electrodes are used to detect lesions in the three major portions of the visual pathway, namely (1) the prechiasmal part, consisting of retina and optic nerve, (2) the optic chiasm, and (3) the retrochiasmal portion, consisting of optic tract, lateral geniculate body, optic radiation, and visual cortex.

7.5.1 Prechiasmal strategy

To test the prechiasmal parts of the visual pathway, one must stimulate each eye separately and record from the midline of the occiput; lateral occipital recordings are often added (Figure 7.6). Because prechiasmal studies aim at large full-field VEPs to stimulation of the central retina, they require fields of no more than about 10° in combination with centrally more effective check sizes of about 30'.

With this strategy, a prechiasmal lesion on one side is detected by an abnormality of the VEP to stimulation of the eye on that side (Figure 7.6). Whereas complete interruption of an optic nerve abolishes the monocular VEP, partial lesions usually increase the latency and decrease the amplitude of the VEP. A prechiasmal lesion affecting only the nasal or temporal half of the retina or optic nerve of one eye may cause an abnormal VEP only on stimulation of the opposite half-field of that eye (7.5.3).

Lesions of both optic nerves or eyes may produce VEP abnormalities to stimulation of each eye. Such bilateral monocular abnormalities often differ from each other. If they are similar, they must be distinguished from those produced by chiasmal and retrochiasmal lesions; testing with half-field stimulation is needed in these cases (7.5.2). A reduction of VEP amplitude is not always due to lesions: Symmetrically low amplitude, or even a slight difference of amplitude between monocular midline VEPs to stimulation of each eye, may be seen in some normal subjects and can therefore not be used as a reliable indicator of a prechiasmal lesion.

Figure 7.6. Strategy for detecting prechiasmal conduction defects by monocular full-field stimulation and midline recording. Top: Stimulation of the right eye and occlusion of the left eye produces a normal VEP. Bottom: Stimulation of the left eye and occlusion of the right eye produces no VEP because of a lesion completely interrupting the left optic nerve. Incomplete lesions may increase the latency and reduce the amplitude. Positivity at the occiput is indicated by an upward deflection.

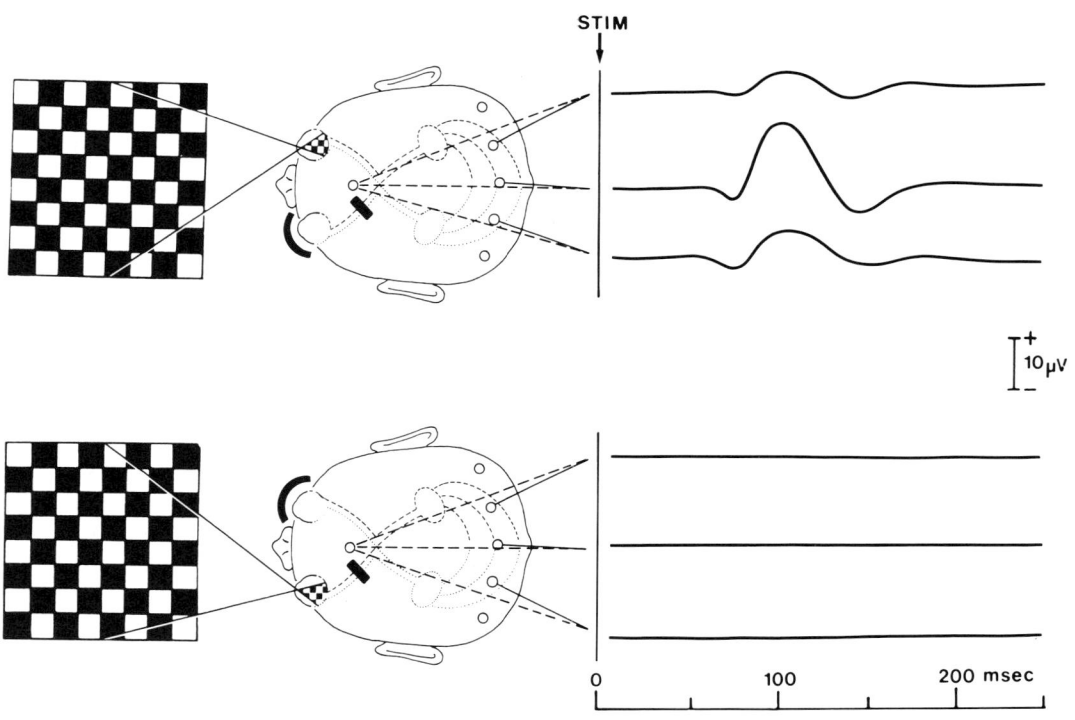

The prechiasmal strategy yields abnormal VEPs with lesions both of the eye and of the optic nerve. Although VEPs cannot always distinguish between these two types of lesions, ocular lesions generally reduce VEP amplitude without affecting latency, whereas marked increases of latency are caused only by optic nerve lesions, especially by demyelination. However, a slight increase of latency may have ocular causes, whereas some optic nerve lesions, such as tumors and ischemia, may be manifested mainly by a decrease of VEP amplitude; complete absence of an EP may be due to either ocular or nerve lesions. Several other characteristics may help to distinguish ocular from optic nerve lesions (8.2.1).

7.5.2 Chiasmal and retrochiasmal strategies

To detect a lesion at or behind the chiasm, one must stimulate the right and left visual half-fields separately and record VEPs from the occipital areas of each hemisphere. Half-field stimulation is necessary because, contrary to expectations, a lesion of the chiasm or of one side of the retrochiasmal pathway may produce inconclusive or normal VEPs even in recordings from each side as long as full-field stimuli are used. The reason for this is probably that the normal VEP generated by the intact hemisphere, distributed rather widely over both sides of the head in most persons, may overshadow the unilateral abnormality; furthermore, normal full-field VEPs may be fairly asymmetrical. Therefore, to detect more reliably chiasmal and retrochiasmal abnormalities, each hemisphere must be excited separately by stimulating either the right or the left visual half-field while recording from each side of the occiput. Because retrochiasmal studies attempt to excite the peripheral halves of the retina separately, they require larger stimulus fields and use the peripherally more effective larger checks sizes of 50–90′.

When studied with this method, chiasmal lesions, which characteristically interrupt fibers from the nasal half of each eye, cause abnormalities of the VEPs to stimulation of the temporal half-fields of each eye. Fibers from the nasal half-fields remain intact and produce VEPs which, although generated on the side of the stimulated eye, have a maximum over the hemisphere opposite that eye (Figure 7.7). Retrochiasmal lesions affect the VEP to stimulation of the contralateral visual half-field of each eye (Figure 7.8). Thus, a left-sided retrochiasmal lesion may abolish the VEPs to stimulation of the right visual half-field, which is normally represented maximally over the right, uninvolved, side; this lesion leaves intact the VEP to stimulation of the left visual half-field, which has a maximum over the left, involved hemisphere.

Half-field stimulation can detect only massive lesions that completely abolish the VEP to stimulation of the opposite half-fields; such lesions cause large field defects, usually fairly complete hemianopsias. Less extensive lesions, for instance those causing only quadrantic field defects, are not detected reliably, and lesions of the posterior hemispheres not causing any field defects do not produce VEP abnormalities.[53]

Figure 7.7. Strategy for detecting chiasmal lesions by monocular half-field stimulation and bilateral occipital recording. Top: Stimulation of the nasal half-field of the right eye, exciting the temporal half of the eye and the right hemisphere, produces normal VEPs over both hemispheres, larger on the left side. Bottom: Stimulation of the temporal half-field of the right eye, exciting the nasal half of the eye, produces no VEPs on either side because of a lesion completely interrupting the optic nerve fibers that cross in the chiasm. Study of the other eye would show no VEPs to temporal half-field stimulation and normal VEPs to nasal half-field stimulation. Incomplete lesions may reduce the amplitude, distort the shape and, occasionally, increase the latency of the involved VEPs.

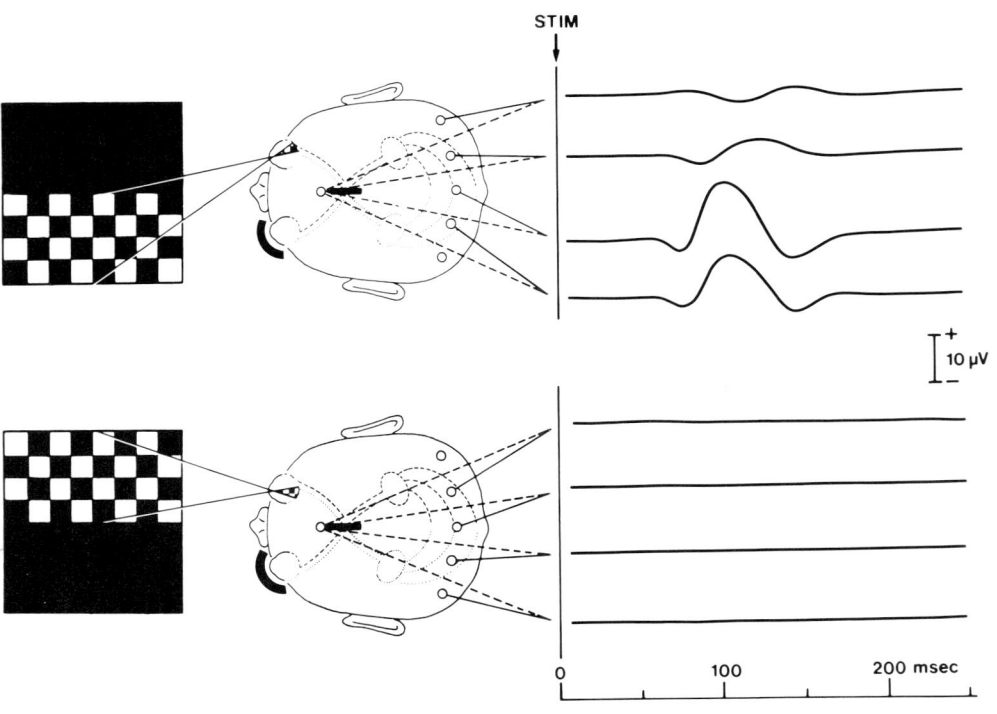

Figure 7.8. Strategy for detecting retrochiasmal lesions by monocular half-field stimulation and bilateral occipital recording. Top: Stimulation of the nasal half-field of the right eye, exciting the temporal half of the eye and the right hemisphere, produces normal VEPs over both hemispheres, larger on the left side. Bottom: Stimulation of the temporal half-field of the right eye, exciting the nasal half of the eye, produces no VEPs on either side because of a lesion completely interrupting the fibers of the left optic radiation. Study of the other eye would show normal VEPs to temporal half-field stimulation and no VEPs to nasal half-field stimulation. Incomplete lesions may reduce the amplitude, distort the shape and, occasionally, increase the latency of the involved VEPs.

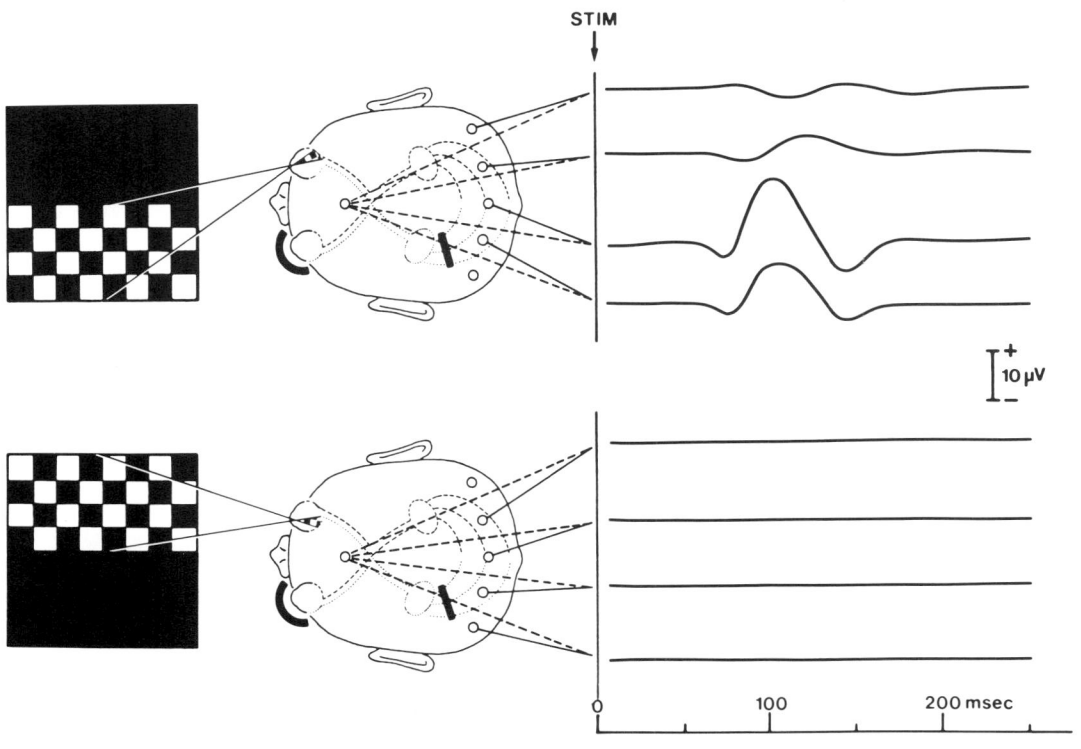

Full-field stimulation in combination with lateral occipital recordings may reveal VEP asymmetries similar to those obtained by stimulation of the intact preserved half-field. Retrochiasmal lesions leave intact VEPs with a maximum on the side of the lesion and therefore cause an abnormality that does not change sides with the stimulated eye, or an uncrossed asymmetry. Chiasmal lesions leave intact VEPs opposite the stimulated eye and thereby produce a crossed asymmetry[50] (8.3.2).

Although chiasmal and retrochiasmal lesions mainly alter VEP amplitude, they occasionally produce an increase of latency which may be recorded with the prechiasmal strategy and could be confused with prechiasmal lesions if not evaluated with the retrochiasmal strategy.[279]

7.5.3 Choice of strategies

Ideally, all three portions of the central visual pathway should be examined if an abnormality is suspected in any one of them. Clinically most useful is the prechiasmal strategy: It can detect lesions not found with other methods of examination; the chiasmal and retrochiasmal strategies are much less sensitive, giving abnormal findings usually only in cases that also show large visual field defects. Although full-field studies are most rewarding and usually provide sufficient evidence for a prechiasmal lesion if the VEP abnormality is clearly monocular, lesions of the chiasmal and retrochiasmal portions must be considered in many cases of abnormal full-field VEPs, especially those showing bilateral monocular abnormalities; these cases require further investigation with half-field studies. Furthermore, half-field stimulation may increase the sensitivity of prechiasmal studies: Abnormal VEPs may be found on stimulation of only one or the other half-field of one eye in cases with normal full-field VEPs and may suggest prechiasmal lesions restricted to the nasal or temporal side of the optic nerve or eye.[404] Some laboratories therefore use both full-field and half-field studies in every patient.

7.5.4 Ophthalmological strategies

VEPs are sometimes used to study ocular lesions, especially those producing visual defects. VEPs have been used to answer the question whether a person can see. These studies are based on the general rule that VEPs are present in subjects who can see and absent in subjects who cannot see. However, VEPs may be present and vision absent in patients with cortical blindness, whereas VEPs may be absent and vision present in several conditions including optic neuropathies, equipment failure, and unintentional or intentional lack of focusing on a patterned stimulus (8.2.1). VEPs can therefore not be reliably used to determine whether a subject can see. Nevertheless, the presence of a normal VEP to pattern stimuli in the absence of bilateral occipital lesions strongly suggests that a subject is able to see, whereas the absence of VEPs, characteristic of all kinds of blindness except cortical blindness, is by no means diagnostic of blindness.

VEPs have also been used to study ocular lesions, visual acuity, and refraction, especially in infants and other noncommunicating subjects. Visual acuity can be estimated by determining the size of pattern elements giving VEPs of maximum amplitude or by determining the smallest size of pattern elements capable of producing a VEP; refractive errors and astigmatism have been estimated by selecting corrective lenses that give the largest VEP. Other ophthalmological applications include studies of brightness contrast sensitivity, color vision, adaptation, and binocularity.

8

The abnormal transient VEP to checkerboard pattern reversal

SUMMARY

8.1. Major criteria distinguishing abnormal from normal VEPs are the complete absence of a VEP, an abnormally increased latency of the $\overline{P100}$, and an abnormally prolonged interocular latency difference.

8.2. Clinical interpretation uses abnormal VEP findings to detect and localize prechiasmal, chiasmal, and retrochiasmal lesions.

8.3. Clinical disorders most effectively studied with VEPs are multiple sclerosis and other conditions affecting the prechiasmal portion of the visual pathway. VEPs are used occasionally in neurological studies of chiasmal and retrochiasmal lesions.

8.4. Ophthalmological disorders are studied rarely with VEP methods and include refractory errors, amblyopia, and retinopathies.

8.1 CRITERIA DISTINGUISHING ABNORMAL PATTERN REVERSAL VEPs

8.1.1 Absence of VEPs

The complete absence of any peak larger than residual noise (4.1.3.1) is abnormal and indicates a clinical defect if lack of focusing and technical problems are excluded (8.2.1). To demonstrate the absence of even a very delayed VEP, an analysis time of 500 msec should be used if no peak appears in recordings with shorter analysis times.

8.1.2 Abnormally long $\overline{P100}$ latency

The latency of the $\overline{P100}$ is measured between the pattern reversal, usually coinciding with the beginning of the analysis period, and the peak of the first major occipital positive peak.

8.1.3 Abnormally long interocular $\overline{P100}$ latency difference

The interocular latency difference is determined by subtracting the $\overline{P100}$ latency of the midline VEP to stimulation of one eye from the $\overline{P100}$ latency of the midline VEP to stimulation of the other eye. This value may exceed the limit of normal interocular latency differences even if both absolute latencies remain within the normal range.

8.1.4 Abnormally long $\overline{N75}$ latency

The latency of the $\overline{N75}$ is often increased in cases also showing a delayed $\overline{P100}$, but an abnormal delay of the negative peak alone can merely raise a suspicion of abnormal conduction.

8.1.5 Abnormally large $\overline{P100}$ amplitude ratio

A lateral occipital amplitude ratio may be calculated by measuring $\overline{P100}$ amplitudes of VEPs on the two sides of the head and dividing the larger $\overline{P100}$ amplitude by the smaller one. The normal amplitude ratio for full-field VEPs varies considerably in and between subjects. Amplitude ratios for half-field VEPs indicate higher amplitude over the hemisphere on the side of the stimulated half-field in most subjects.

8.2 GENERAL CLINICAL INTERPRETATION OF ABNORMAL VEPs TO CHECKERBOARD PATTERN REVERSAL

Abnormal VEP findings must be interpreted with regard to the possible location and type of the underlying defect. Before VEP abnormalities can be accepted as indicators of lesions in the visual pathway, several technical problems must be excluded. Lesions affecting the prechiasmal, chiasmal, and retrochiasmal parts of the visual pathway can be recognized by fairly characteristic constellations of VEP abnormalities. Abnormal VEP findings and their clinical implications are outlined in Table 8.1.

8.2.1 Technical and ocular problems and lack of focusing

Absence of VEPs may be due to various technical problems. Such problems must be suspected especially in cases where no VEP can be elicited by full-field and half-field stimulation of either eye with check sizes up to 1°, even though the subject can see the stimulus pattern. Technical problems are less likely to explain the absence of a monocular full-field or half-field VEP if stimulation of the other eye or half-field produces a VEP. Common technical problems are failure to synchronize the stimulator with the sweep of the averager, and faulty recording electrodes, amplifiers, and averaging channels.

VEP abnormalities may be the result of various extraocular and ocular lesions: Ptosis, cataracts, and extreme miosis may reduce illumination of the retina; corneal opacities, cataracts, retinal lesions, and refractory errors may interfere with the sharp projection of small pattern elements onto the retina. Although reduced luminance and blurring of a patterned stimulus usually reduce VEP amplitude, they may also slightly increase VEP latency. However, in contrast to optic nerve lesions, ocular lesions produce no more than a slight increase of latency (8.4.6). As a rule, ocular defects can be distinguished from optic nerve lesions in that they alter the VEP only if they also interfere with vision, i.e., if they reduce visual acuity or cause central scotomata; the VEP may remain normal with scotomata that spare the central 3°. Ocular lesions causing abnormal VEPs can usually be detected by ophthalmological examination. A few ophthalmological disorders have been made the subject of VEP studies (8.4). ERGs to diffuse and patterned light may be of help (10.13).

Another possible cause of reduced VEP amplitude or slightly prolonged latency is unintentional or intentional lack of focusing on the stimulus pattern.[41]

8.2.2 Abnormal VEPs to monocular full-field stimulation

8.2.2.1 Increase of latency
A significant increase of $\overline{P100}$ latency to stimulation of one eye or an increased $\overline{P100}$ latency difference between VEPs to stimulation of each eye practically always indicates an optic nerve lesion on the side of the longer latency, especially if stimulation of the other eye produces a normal VEP. A prolonged latency of VEPs to stimulation of either eye usually indicates a lesion of both optic nerves; rarely is it caused by chiasmal or retrochiasmal lesions. The magnitude of the latency increase may differ for both eyes in prechiasmal and chiasmal lesions, but is similar for both eyes in retrochiasmal lesions. Bilateral optic nerve lesions caused by degenerative and metabolic disorders cause bilateral delays that often, but not always, have the same

TABLE 8.1. Clinical interpretation of abnormal VEPs

Abnormal VEP findings	Interpretation
A. Technical problems occurring with any strategy	
1. Absent VEPs	Lack of stimulus; lack of synchronization between stimulator and averager; faulty recording electrodes and equipment; lack of focusing on pattern stimulus
2. Decreased amplitude, with or without slightly increased latency	Decreased retinal illumination; lack of focusing
B. Monocular full-field stimulation: Prechiasmal strategy	
1. Increased latency	
a. Increased latency to stimulation of one eye with	
(1) normal VEP to stimulation of the other eye	Ipsilateral optic nerve lesion, especially demyelination
(2) decreased or absent VEP to stimulation of the other eye	Bilateral optic nerve lesions, chiasmal lesion
b. Increased latency to stimulation of either eye with	
(1) different increase for stimulation of each eye	Bilateral optic nerve lesions, chiasmal lesion
(2) similar increase for stimulation of each eye	Bilateral optic nerve lesions, chiasmal lesion, or bilateral retrochiasmal lesion
c. Increased interocular latency difference, both latencies within the normal range	Optic nerve lesion on side of longer latency
2. Absence or decreased amplitude of VEP	
a. Complete absence of VEP to stimulation of one eye with	
(1) normal VEP to stimulation of the other eye	Lesion of optic nerve or eye
(2) increased latency to stimulation of other eye	Bilateral optic nerve lesions, chiasmal lesion
(3) complete absence of VEP to stimulation of the other eye	Rule out technical problems; bilateral prechiasmal lesions, chiasmal lesion, bilateral retrochiasmal lesions
(4) very low amplitude of VEP to stimulation of the other eye	Suspect bilateral prechiasmal or chiasmal lesion

TABLE 8.1. *(continued)*

Abnormal VEP findings	Interpretation
b. Very low amplitude of VEP to stimulation of one eye with	
(1) normal VEP to stimulation of the other eye	Suspect lesion of optic nerve or eye
(2) increased latency to stimulation of the other eye	Suspect bilateral optic nerve lesions, chiasmal lesion
(3) similar amplitude of VEP to stimulation of the other eye	Possible bilateral prechiasmal lesions, chiasmal lesion, or bilateral retrochiasmal lesions; may be normal
C. Half-field stimulation: Chiasmal and retrochiasmal strategies	
1. Abnormal VEPs to stimulation of both temporal half-fields	
a. complete absence of VEPs	Chiasmal lesion
b. increased latency, decreased amplitude	Suspect chiasmal lesion
2. Abnormal VEPs to stimulation of both left or right half-fields	
a. complete absence of VEPs	Retrochiasmal lesion
b. increased latency, decreased amplitude	Suspect retrochiasmal lesion
3. Abnormal VEP to stimulation of one half-field of one eye	
a. increased latency of VEP	Lesion of optic nerve or eye; rule out chiasmal or retrochiasmal lesion
b. complete absence of VEP	Suspect lesions as in a.
D. Ophthalmological strategy	
1. Decreased amplitude, with or without slightly increased latency	Rule out technical problems; lesions of optic media causing blurring (corneal opacity, cataract), retinopathy
2. Increase of effective check size, maximum VEP with corrective lenses	Reduced visual acuity, refractive error

magnitude on the two sides. Demyelinating lesions are more likely than other lesions to produce significant unilateral or bilateral increases of VEP latency.

8.2.2.2 *Absence or reduced amplitude of VEPs*
Absence of a VEP to monocular stimulation is usually abnormal and due to an ipsilateral prechiasmal lesion involving either the eye or the optic nerve; if the VEP to stimulation of the other eye is also abnormal, bilateral prechiasmal lesions, a chiasmal lesion, and bilateral retrochiasmal lesions must be considered. The possibility of technical problems has to be carefully eliminated, especially in cases of bilateral absence of VEPs. Very low amplitude of a VEP is a less reliable indicator of lesions than is increased latency, but may occur in ocular lesions and optic nerve compression, ischemia, or injury; low amplitude may be combined with increased latency. A very marked reduction of a VEP recorded from a lateral occipital electrode, causing an abnormal lateral occipital amplitude ratio, suggests the possibility of a retrochiasmal lesion.

8.2.3 Abnormal VEPs to half-field stimulation

8.2.3.1 *Abnormal VEPs to stimulation of the temporal half-fields*
Absence of VEPs to stimulation of the temporal half-fields suggests a chiasmal lesion and is usually associated with bitemporal visual field defects. Reduced amplitude, increased latency, and abnormal distribution of these VEPs are less reliable indicators of clinical lesions.

8.2.3.2 *Abnormal VEPs to stimulation of corresponding half-fields*
Absence of VEPs to stimulation of both left or right half-fields strongly suggests a retrochiasmal lesion opposite the stimulated half-field and is usually associated with homonymous field defects. Reduced amplitude, increased latency, and abnormal distribution of these VEPs suggest the possibility of a retrochiasmal lesion.

8.2.3.3 *Abnormal VEPs to stimulation of a monocular half-field*
Absence or increased latency of a VEP to stimulation of a half-field of only one eye may suggest a partial lesion of the eye or optic nerve on the side opposite the stimulated half-field.

8.2.4 Ophthalmological problems

Reduced amplitude with or without a slight increase of latency may be seen in ocular disorders such as corneal opacities, cataracts, and glaucoma. Stimulation with patterns of different check size may reveal an increase in the check size causing the largest VEP; this is usually associated with reduced visual acuity. In patients with poor visual acuity due to refractive errors, the amplitude of VEPs to smaller pattern elements may be raised with corrective lenses and give an estimate of the magnitude of the refractive error.

8.3 NEUROLOGICAL DISORDERS THAT CAUSE ABNORMAL PATTERN REVERSAL VEPs

Most disorders can be easily divided into prechiasmal, chiasmal, and retrochiasmal. Some disorders, such as multiple sclerosis, leukodystrophies, and renal encephalopathies, may involve more than one segment of the central visual pathway and are here arbitrarily classified according to the site of their lesions most often studied with VEPs.

8.3.1 Prechiasmal lesions

VEPs can be very useful in the diagnosis of multiple sclerosis because they can detect clinically silent conduction defects in the optic nerve and thereby indicate the presence of multiple lesions. VEPs may also help in the diagnosis of tumors compressing the optic nerve and in the diagnosis of ischemic optic neuropathies. VEPs are occasionally used in patients with other diseases that sometimes involve the optic nerve; in these instances, VEPs are usually used not to make the diagnosis of the disease but to determine whether the optic nerve is involved by it. In hereditary or familial conditions involving the optic nerve, an abnormal VEP may be the only or the earliest evidence that a person at risk is affected. In some of the diffuse disorders discussed here, the distinction between prechiasmal and chiasmal or retrochiasmal involvement is difficult to make; chiasmal and retrochiasmal lesions rather than prechiasmal lesions may be partly or entirely responsible for VEP abnormalities, especially those appearing on stimulation of either eye.

8.3.1.1 Retrobulbar neuritis

Acute retrobulbar neuritis, usually occuring in patients with multiple sclerosis, markedly reduces or abolishes VEPs to stimulation of the eye on the affected side[63,189,191,194] (Figure 8.1). On recovery, the amplitude increases as visual acuity improves and may eventually become normal. However, the latency remains abnormally prolonged in most cases of retrobulbar neuritis and multiple sclerosis,[194,343,416] although it may return to the normal range in some.[123,189,318,319,400]

8.3.1.2 Multiple sclerosis

Transient VEPs to alternating checkerboard stimulation are abnormal in a high percentage of patients with multiple sclerosis. The rate of abnormality is high in patients who have a history of retrobulbar neuritis or findings of optic pallor, reduced visual acuity, and visual field defects, but abnormal VEPs are also found in patients without any history or clinical findings suggesting involvement of the visual

Figure 8.1. Transient pattern shift VEPs recorded 3 and 10 days after the onset of acute bilateral optic neuritis in a 14-year-old girl. The first recording (left) shows no peaks to stimulation of the left eye and a questionable peak to stimulation of the right eye; visual acuity of the left and right eye (V.A.L., V.A.R.) was reduced to finger counting. In the second recording (right), both VEPs had returned but were abnormally delayed. Visual acuity had greatly improved. Stimulation of the left eye (upper tracings) and right eye (lower tracings). Recording between midoccipital and midfrontal electrodes. Negativity at the occiput is plotted upward; the $\overline{P100}$ points down. Time scale 10, 50, and 100 msec. From Halliday and McDonald[189] with permission by the authors and Churchill Livingstone Medical Division of Longman Group Ltd.

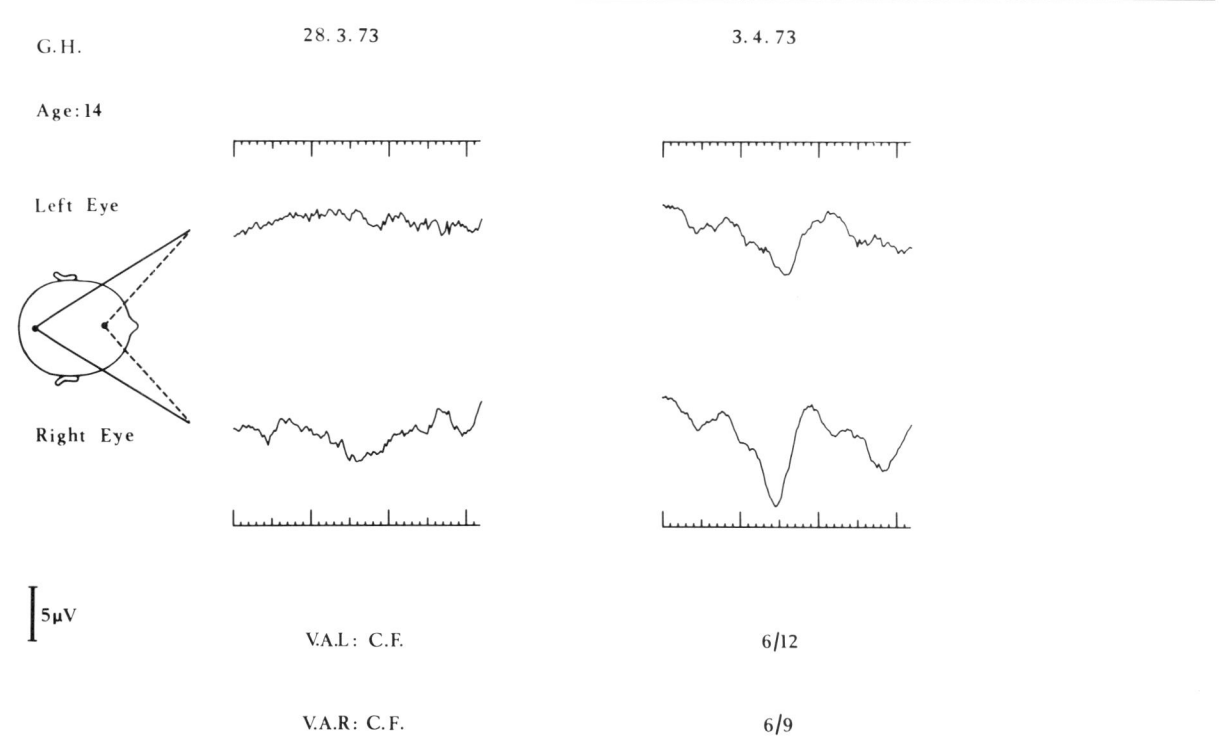

system. Even though abnormal VEPs are most common in patients with definite multiple sclerosis, less common in probable, and least common in possible multiple sclerosis,[30,38,64,187,192,251,253,274,320,321,343,416,421] they are clinically most useful when the diagnosis is in doubt. The VEP may help to distinguish the spinal form of multiple sclerosis from transverse myelitis of other causes.[30,51,71,171,216,356,399,449] Follow-up studies have confirmed the diagnostic value of VEPs.[64,123,321,484]

The most common and diagnostically most important VEP abnormality in multiple sclerosis is an increase of $\overline{P100}$ latency (Figure 8.2) which in many cases is much longer than the delays found in other diseases. This delay may be associated with decreased amplitude and increased duration of the $\overline{P100}$ (Figure 8.2), suggesting dispersion of impulses traveling in abnormal nerve fibers (5.3.1). Complete absence of a VEP is rare except in patients with acute retrobulbar neuritis or severely impaired vision. VEP abnormalities are often monocular but may be binocular. If binocular, they usually are different on the two sides but sometimes are similar or identical (Figure 8.3). Although binocular abnormalities are most commonly due to bilateral optic nerve lesions of multiple sclerosis, they may also be due to chiasmal and retrochiasmal lesions, which should be excluded by appropriate studies (7.5.3). The use of stimuli of lower luminance has been reported to increase the detection rate of abnormal VEPs in one study,[79] but to decrease it in another.[122] Also, testing with checkerboard patterns of different orientation (7.3.8) has been said to increase the incidence of abnormal findings.[109]

The degree of the prolongation of latency may vary over time with some correspondence to visual function.[43,103,318] As in patients having retrobulbar neuritis without evidence of multiple sclerosis, latency may normalize after months or years in some cases of multiple sclerosis. The amplitude of the VEP may transiently decrease in patients experiencing deterioration of vision as a result of physical exercise.[363] An increase of body temperature alone may also reduce the VEP amplitude in some patients to a greater degree[33] than in normal subjects[317]; heating increases the incidence of abnormal VEP findings in patients with multiple sclerosis.[366]

The diagnostic value of checkerboard reversal stimuli has been compared with that of other VEP types. It is generally agreed that alternating checkerboard stimuli are much more effective than stimulation with diffuse light flashes.[191,298,498] Stimulation with large check sizes may be less effective than stimulation with a small foveal light spot (10.6). Measurement of the critical frequency of photic driving to diffuse light flashes (10.3) may complement the recording of the checkerboard reversal VEP[98] and, in one study, was even slightly more sensitive.[101] Another study found VEPs to a reversing sine wave grating pattern more effective than VEPs to checkerboard pattern shift.[74] VEPs to pattern reversal are as effective as those to pattern shift.[298]

Comparisons of alternating checkerboard VEPs with other EPs have given diverse results: VEPs were more effective than SEPs in all cases,[328,373] in definite[177,484] and probable[97] cases, or in early cases[176] of multiple sclerosis; SEPs were more effective than alternating checkerboard

Figure 8.2. Midline and lateral occipital VEPs to monocular full-field and half-field stimulation with reversing checkerboards in a patient with probable multiple sclerosis. Stimulation of the right eye (bottom) produces normal VEPs. Stimulation of the left eye (top) produces full-field VEPs (*a*) with abnormally prolonged latency and with slightly lower amplitude and longer duration than seen in the full-field VEPs to stimulation of the right eye (*d*). Stimulation of the left half-field of the left eye produces no peaks that clearly exceed the noise level (*b*), and stimulation of the right half-field of the left eye produces only low amplitude peaks of prolonged latency (*c*). This indicates a conduction defect in the left optic nerve. Stimulation with checkerboard patterns of squares of 30' (*a, d*) and 60' (*b, c, e, f*) sidelength. Recording between left occipital (LO), midoccipital (MO), right occipital (RO), and midfrontal (MF) electrodes. Positivity at the occiput is plotted upward. This 38-year-old man has a history of two bouts of transverse myelitis and residual spastic paraparesis. Normal myelogram; no history, symptoms, or signs suggesting involvement of the visual system. The finding of abnormal VEPs in this patient makes the diagnosis of multiple sclerosis likely.

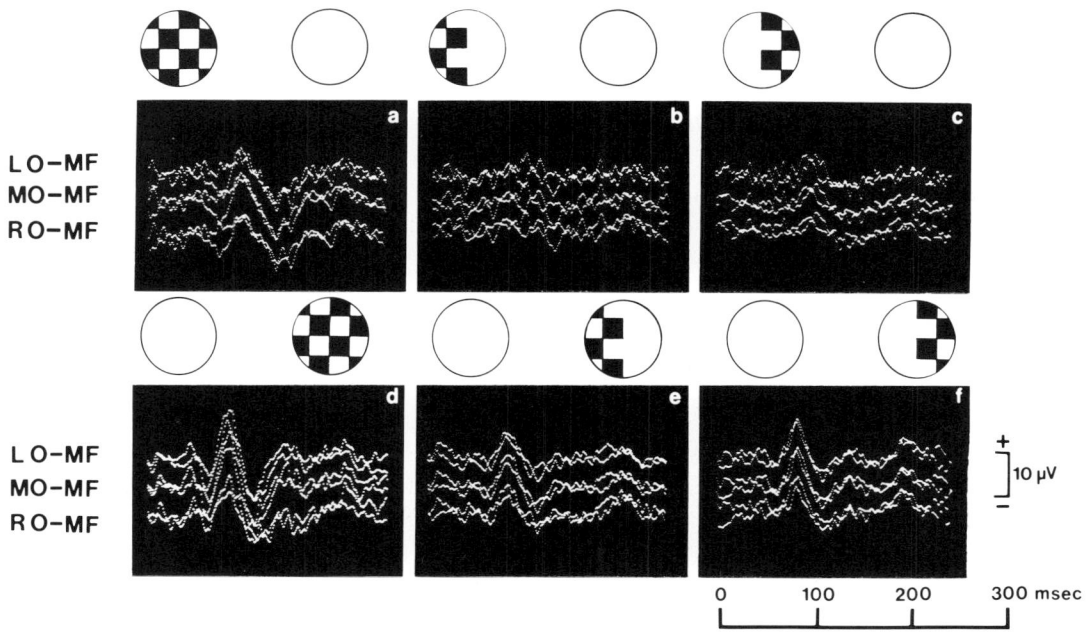

Figure 8.3. Transient midline occipital VEPs to full-field checkerboard pattern shift stimulation of the left (L) and right (R) eye in four patients (*A, B, C, D*) with multiple sclerosis. (*A*) The latency of the P100 is abnormally increased to a similar degree on stimulation of either eye. (*B*) The P100 is delayed on stimulation of the left eye only. (*C*) The P100 is delayed on stimulation of either eye, much more on stimulation of the right eye. (*D*) The P100 and earlier peaks show similar delays on stimulation of either eye. Recording between midoccipital and vertex electrodes. Negativity at the occiput is plotted upward; the P100 points down. From Asselman et al.[30] with permission by the authors and Oxford University Press.

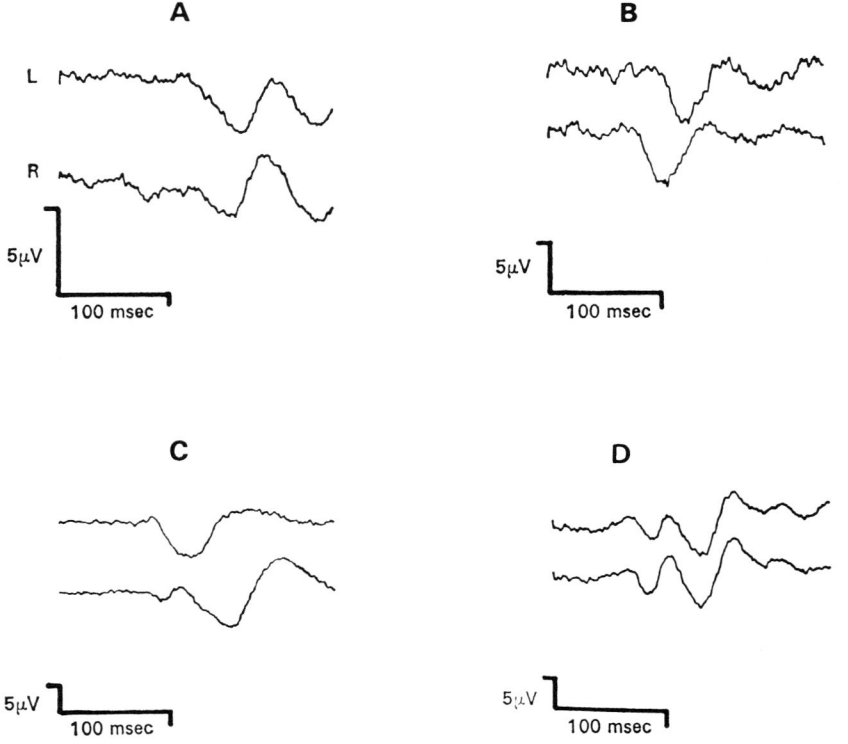

VEPs in other studies,[38,265,316] especially in the important diagnostic categories of probable and possible multiple sclerosis[315,461,463] (17.5.1); cervical SEPs remain more stable than checkerboard VEPs between relapses.[318] These comparisons depend on whether SEP studies use only arm stimulation or also leg stimulation (17.5.1, 18.4.1). BAEPs are generally less effective than checkerboard VEPs in detecting lesions in most diagnostic categories of multiple sclerosis.[252,265,297,328,373]

8.3.1.3 Tumors

Tumors compressing the optic nerve may reduce the amplitude and distort the shape of the VEP (Figure 8.4); in the extreme, they may completely abolish the VEP.[188,231] Although latency may also be increased,[30,188,231,343,474] delays of over 30 msec, such as can be seen in multiple sclerosis, are rare in tumors.[186] VEP testing is more often used to confirm than to detect optic nerve involvement by tumors: Patients with abnormal VEPs due to tumors usually have abnormal vision, optic atrophy, or visual field defects indicating an optic nerve lesion.[188] Papilledema due to tumors not directly compressing the optic nerve usually does not cause abnormal VEPs.[30]

8.3.1.4 Ischemic optic neuropathy

Ischemic neuropathy of the optic nerve often reduces VEP amplitude.[202,231,497] As in tumors, latency may also be increased,[30,216] but this VEP abnormality is not as characteristic and as pronounced as in multiple sclerosis. Patients with carotid occlusion have been reported to show blurring of vision and reduction of VEP amplitude on exposure to bright light.[125]

8.3.1.5 Friedreich's ataxia, hereditary cerebellar ataxia, and hereditary spastic ataxia

The majority of patients with Friedreich's ataxia show a bilateral and fairly symmetrical increase of P100 la-

Figure 8.4. Transient pattern shift VEPs in a patient with right optic nerve compression due to a sphenoid wing meningioma. Stimulation of the left eye with alternating checkerboard patterns produces a normal midline VEP (top). The VEP to stimulation of the right eye is distorted and delayed (bottom). Visual acuity was 6/9 in each eye. Time scale 10, 50, and 100 msec. Recording between midoccipital and midfrontal electrodes. Occipital negativity is plotted upward; the P100 points down. From Halliday et al.[188] with permission by the authors and Oxford University Press.

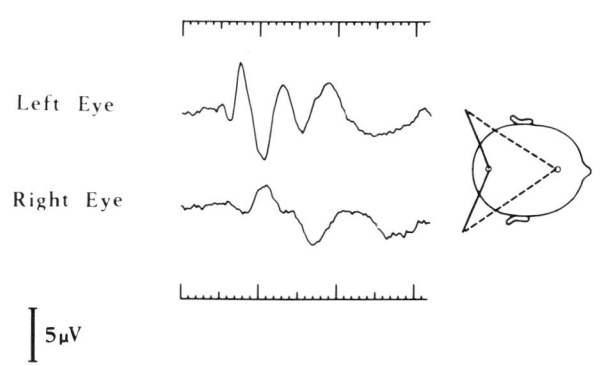

tency[47,84,249,293,346,361,489]; amplitude may be reduced more often than in multiple sclerosis,[84] especially in patients with severe visual impairment[293] (Figure 8.5). Studies of other kinds of ataxia showed abnormal VEPs in several patients with hereditary cerebellar ataxia[361] and in one of two patients with olivopontocerebellar degeneration.[197] Only a small portion of patients with hereditary spastic ataxia had abnormal VEPs,[293] whereas patients with other hereditary ataxias had normal VEPs.[47] However, a high incidence of abnormal VEPs in many forms of hereditary ataxias has also been reported.[346]

8.3.1.6 Charcot-Marie-Tooth disease
The VEP was delayed in most[455] or some[48] patients, even in the absence of clinically apparent optic nerve lesions.

8.3.1.7 Neurosyphilis
VEP latency was found to be increased in one-fifth of patients with neurosyphilis, most often in those with tabes dorsalis; the VEP was not more useful for the diagnosis than other neuroophthalmological tests.[105]

8.3.1.8 Hereditary spastic paraplegia
VEP latency may be increased.[361]

8.3.1.9 Leber's hereditary optic neuropathy
Leber's disease increases the latency, decreases the amplitude, and eventually abolishes VEPs, in keeping with optic

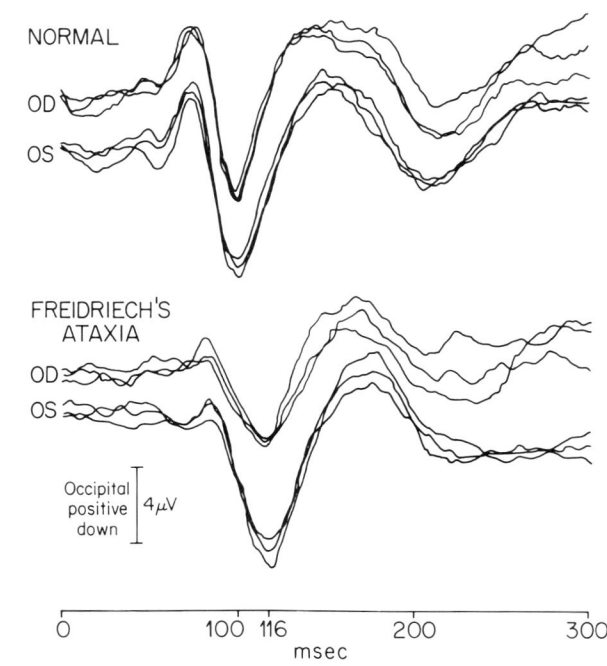

Figure 8.5. Checkerboard pattern shift VEPs in a normal subject (top two tracings) and in a patient with Friedreich's ataxia (bottom two tracings). The patient's VEPs to stimulation of either eye show abnormally delayed $\overline{P100}$ peaks. Monocular full-field stimulation of the right (OD) and left (OS) eye. Recording between midoccipital and vertex electrodes. Negativity at the occiput is plotted upward; the $\overline{P100}$ points down. Three tracings are superimposed for each condition. From Nuwer et al.[346] with permission by the authors and Little, Brown and Company.

nerve involvement and decreasing vision; changes in VEP shape have also been reported. Asymptomatic family members show little or no VEP abnormalities.[83,126,294,303]

8.3.1.10 Traumatic optic nerve lesions
Head injuries may acutely reduce the VEP amplitude suggests indirect optic nerve damage.[202]

8.3.1.11 Leukodystrophies
Pelizaeus-Merzbacher disease, adrenoleukodystrophy, and metachromatic leukodystrophy abolish or delay VEPs.[314] The latency may be increased in subclinical cases of adrenoleukodystrophy.[309]

8.3.1.12 Chronic renal failure
The VEP may be delayed, especially in patients on chronic hemodialysis.[100,272,296,398]

8.3.1.13 Endocrine orbitopathy
The VEP may be delayed even in patients with preserved visual acuity, presumably due to demyelination of the optic nerve.[494]

8.3.1.14 Pernicious anemia
Subacute combined degeneration of the spinal cord due to pernicious anemia was associated with slightly delayed VEPs in a few cases.[155,271,464]

8.3.1.15 Sarcoidosis
VEP latencies may be abnormal even in patients without clinically evident brain or eye involvement.[453]

8.3.1.16 Diabetes
Abnormally delayed VEPs have been reported in visually unimpaired patients with diabetes.[374]

8.3.1.17 Alcoholism
VEP latency may be increased in a small number of chronic alcoholics and in many patients with Korsakoff's psychosis.[370]

8.3.1.18 Phenylketonuria
A small portion of patients were found to have increased VEP latency.[112]

8.3.1.19 Parainfectious optic neuritis
Optic neuritis after viral diseases may initially abolish VEPs and later be associated with delayed VEPs.[410]

8.3.1.20 Postconcussion syndrome
The $\overline{P100}$ latency may be increased after head injuries, especially those leading to loss of consciousness.[397]

8.3.1.21 Hysterical blindness
Normal VEPs were found in patients with hysterical blindness.[190]

8.3.2 Chiasmal lesions

Pituitary tumors, craniopharyngiomas, and other lesions near the sella turcica compressing the optic chiasm produce abnormalities of the monocular full-field VEP that often differ for stimulation of each eye. The abnormalities include reductions of amplitude and gross distortions of waveform,[174,188,277] but increased latency may also occur[224,342]; lateral occipital recordings may show marked asymmetries[188] (Figure 8.6) suggesting the site of the abnormality (7.5.2).
   Monocular half-field stimulation in conjunction with midline occipital recordings may produce abnormal VEPs on stimulation of the temporal visual half-field.[73,359] Even

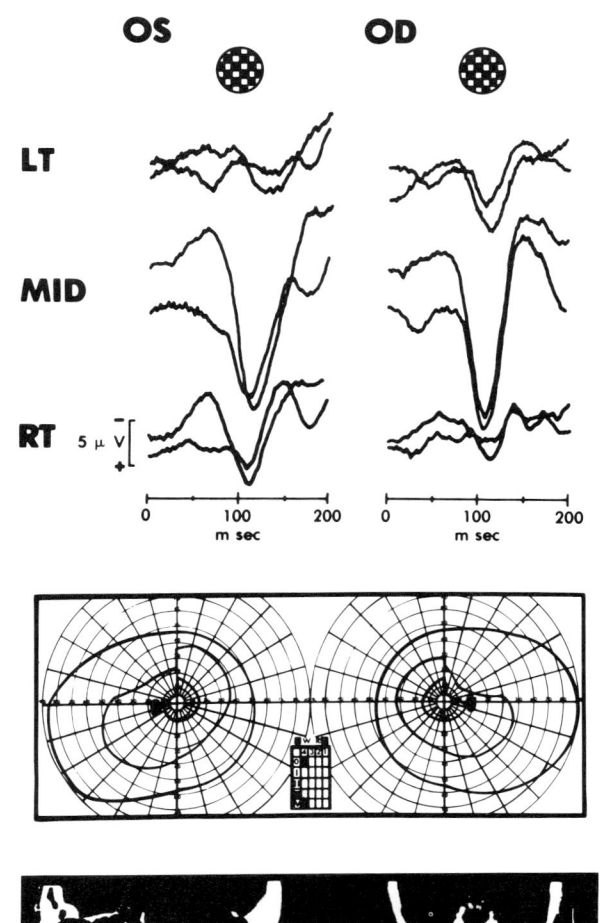

Figure 8.6. Transient checkerboard reversal VEPs (top) in a patient with bitemporal hemianopsia (middle) due to a large intrasellar and suprasellar pituitary tumor (bottom). Full-field stimulation of the left eye (OS) and right eye (OD) and recording with midline (MID) and left (LT) and right (RT) occipital electrodes referred to a midfrontal electrode produce transient VEPs with maximum P100 amplitude opposite to the stimulated eye; this crossed asymmetry results from stimulation of the intact nasal half-field of each eye. Occipital negativity is plotted upward; the P100 points down. From Maitland et al.[308] with permission by the authors and Modern Medicine Publications, Inc.

monocular half-field stimulation combined with lateral occipital recordings (Figure 8.7), although the most powerful chiasmal strategy, shows that VEPs are only of limited value in detecting chiasmal lesions even in cases that show visual field defects.[184,308,350] The abnormal crossing of optic nerve fibers in albinism produces abnormal asymmetries of the full-field and half-field VEPs[82] (10.1).

8.3.3 Retrochiasmal lesions

Tumors, infarcts, and other lesions of the occipital and parietal lobes are difficult to detect with VEPs to full-field stimulation even when they can be detected by visual field defects and computerized tomography.[452] As expected, bilateral hemispheric lesions may cause abnormal VEPs to stimulation of either eye.[29] Diffuse cerebral involvement in Alzheimer's disease may increase latency and amplitude of

Figure 8.7. Transient checkerboard reversal VEPs in two patients (*A* and *B*) with bitemporal hemianopsia due to chiasmal lesions. Stimulation of the left and right half-fields of the left eye (OS) and right eye (OD) produces VEPs when the nasal half-fields are stimulated. Stimulation of the temporal half-fields produces no VEPs in patient *A*; in patient *B* stimulation of the temporal half-field of the right eye produces questionable VEPs, and stimulation of the temporal half-field of the left eye produces no VEPs. Recording between occipital, temporal, and midfrontal electrodes as indicated by abbreviations of the International 10–20 System for EEG electrode placement. Negativity at the occiput is plotted upward; the $P\overline{100}$ points down. From Haimovic and Pedley[184] with permission by the authors and Elsevier Scientific Publishers Ireland Ltd.

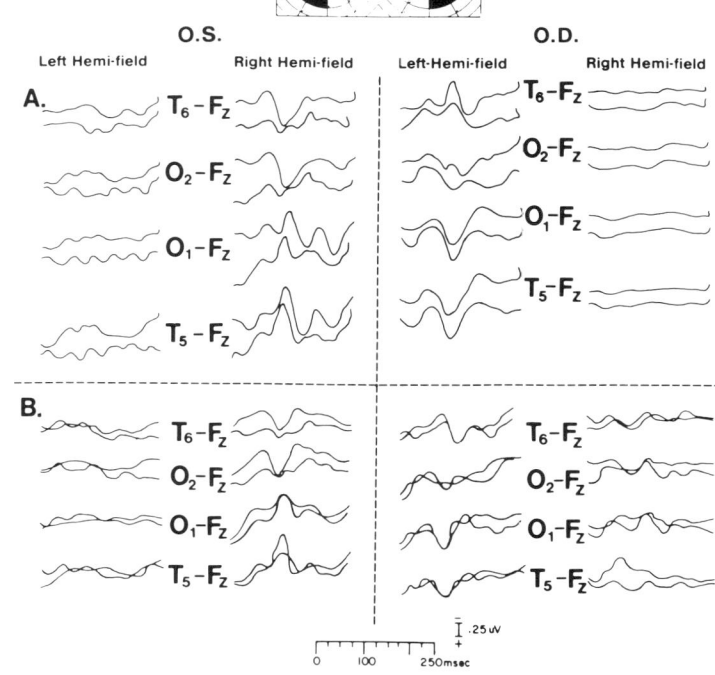

late peaks.[99] Huntington's chorea was found to reduce the amplitude of monocular VEPs in patients and some offspring in one study,[349] but to produce no VEP changes in another investigation.[137] In Parkinson's disease, the VEP amplitude has been reported to be decreased[198] or slightly increased[136]; an increase of latency was found in only one study[170] and was thought to depend on low contrast stimuli.[136] In myotonic dystrophy, abnormalities of latency and amplitude were found in many patients without significant ocular involvement and were thought to be due to prechiasmal and retrochiasmal lesions.[173] VEPs were normal in Tourette's syndrome.[270] Infantile neuraxonal dystrophy with myoclonus epilepsy was associated with increased VEP amplitude.[127] VEP latency was shortened in some cases of photosensitive epilepsy.[148] A binocular increase of VEP latency was found in a patient with palinopsia[116] and in a patient with pseudotumor cerebri.[449] However, bilateral occipital infarcts leading to cortical blindness (9.7.2.3), although not usually tested with patterned light stimuli, do not necessarily eliminate the VEP to checkerboard patterns.[87,91]

The use of binocular visual half-field stimulation and midline occipital recording has shown abnormal VEPs in cases of occipitoparietal tumors and infarcts.[73] The potentially more powerful technique of half-field stimulation in conjunction with bilateral occipital recordings (7.5.2) has so far demonstrated its validity mainly in cases of occipital lobectomies[50,54,92] and in patients with lesions of the occipital cortex or optic radiations who also had massive visual field defects (Figure 8.8); even here, the reliability and sensitivity of this method needs further investigation.[4,53,90,114,184,279,308,350,404]

8.4 OPHTHALMOLOGICAL DISORDERS THAT CAUSE ABNORMAL PATTERN REVERSAL VEPs

Because VEPs depend on sharpness of focus and on check size, they may be used to investigate conditions interfering with the sharp projection of small checks onto the retina. In general, ocular disorders may reduce VEP amplitude, but they increase VEP latency only little or not at all.[195,429]

8.4.1 Visual acuity

The smallest check size giving a VEP may be used to measure the limits of resolution.[424]

8.4.2 Refractive errors

At normal visual acuity, maximum VEP amplitude is obtained with checks of 10–15′ (7.3.5). Refractive errors increase the size of the most effective check size. The magnitude of the error is indicated by the corrective lenses required to reverse this effect.[195]

8.4.3 Amblyopia ex anopia

The amplitude of transient VEPs to small check sizes is reduced[429]; the latency may be increased.[325,431] Although this finding suggests a prechiasmal conduction defect, it is not easily mistaken for an optic nerve lesion because it is characterized by a history dating from childhood. Amblyopia has been studied more extensively with steady-state VEPs to alternating checkerboard patterns (10.2.4.2) and to sine wave gratings (10.4.4.3).

8.4.4 Nutritional and toxic amblyopia

Most patients with nutritional amblyopia who showed defective color vision and central or cecocentral scotoma had

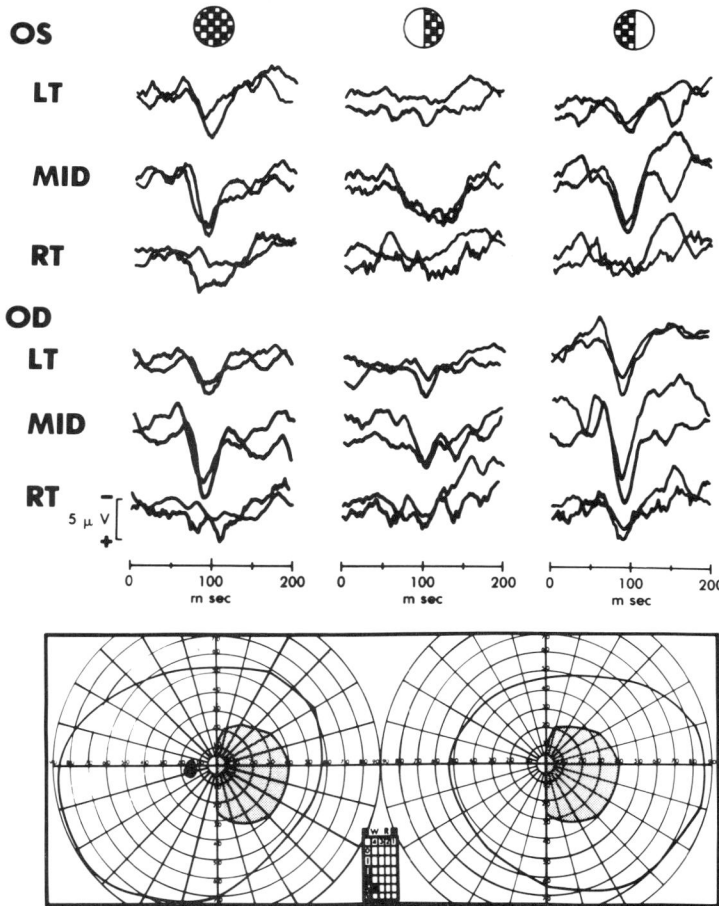

Figure 8.8. Full-field and half-field VEPs to checkerboard pattern reversal (top) in a patient with right homonymous scotomas for form and color (bottom) due to a left occipital defect. Stimulation of the left eye (upper 3 rows of tracings) and of the right eye (lower 3 rows of tracings) with full-field (left column), right half-field (middle column), and left half-field (right column) stimuli produces transient VEPs at midline (MID), left (LT), and right (RT) occipital electrodes referred to a midfrontal electrode. Full-field and left half-field VEPs are similar and slightly larger ipsilateral to the intact visual half-field. Stimulation of the defective right half-field elicits hardly any VEPs. Two samples are superimposed for each VEP. Occipital negativity is plotted upward; the $\overline{P100}$ points down. From Maitland et al.[308] with permission by the authors and Modern Medicine Publications, Inc.

delayed VEPs in one study,[267] but not in another.[278] VEP amplitude may be reduced symmetrically.[231] Quinine amblyopia may cause asymmetrical VEP abnormalities.[168] Increased latency may reveal subclinical optic neuropathy during ethambutol treatment.[506]

8.4.5 Retinopathy

Idiopathic central serous retinopathy may increase the latency and decrease the amplitude of the VEP of the involved eye.[353]

8.4.6 Glaucoma

The latency of the VEP may be slightly increased as the result of glaucomatous optic nerve damage associated with field defects.[230,433]

8.4.7 Corneal opacity, miosis, and cataract

A significant reduction of the stimulus luminance reduces VEP amplitude[195,429] and may increase VEP latency slightly.[213]

8.4.8 Congenital oculomotor apraxia

Normal visual acuity has been demonstrated with VEPs.[172]

9

The transient VEP to diffuse light stimuli

SUMMARY

9.1. The clinical usefulness of transient VEPs to diffuse light flashes is limited mainly to conditions in which pattern VEPs cannot be obtained.

9.2. Normal VEPs to flashes have several peaks in most subjects but vary in and between subjects so that a clinically useful standard VEP cannot be easily defined. Flash VEPs depend on many subject variables, stimulus characteristics, and recording parameters.

9.3. The most important subject variables affecting the flash VEP are age and pupillary size.

9.4. Stimulus characteristics that affect the flash VEP include the type of stimulator and the stimulus rate and intensity.

9.5. Important recording parameters are electrode placements and combinations.

9.6. The most reliable criterion of abnormality is the complete absence of a flash VEP to monocular stimulation. Deviations of latency and amplitude must be interpreted with extreme caution. Various abnormalities have been described in a great variety of prechiasmal and retrochiasmal disorders, but, because the flash VEP is so variable, its sensitivity to these disorders is rather low.

9.7. Neurological disorders that have been studied with flash VEPs are multiple sclerosis and many other prechiasmal disorders. Retrochiasmal problems include coma due to head injury or anoxia and cortical blindness.

9.8. Ophthalmological disorders that have been investigated with flash VEPs include glaucoma and injury to the eye and optic nerve.

9.1 CLINICAL USEFULNESS OF THE FLASH VEP

VEPs to diffuse light flashes, although widely used in early clinical VEP studies, have now been largely replaced by VEPs to patterned light except (1) in testing infants, children, and other subjects who do not focus reliably on pattern stimuli or who cannot clearly see the stimulus pattern even if it consists of large pattern elements; (2) for the study of patients who show no responses to patterned light stimuli; (3) for determining the critical frequency of photic driving (10.3). Because reports of abnormal flash VEPs in other disorders have lost much of their clinical importance, they are here discussed only briefly.

9.2 THE NORMAL TRANSIENT FLASH VEP

Reports on normal VEPs to diffuse light flashes usually describe up to seven peaks (Figure 9.1) but differ widely with regard to peak polarity, latency, and amplitude.[94,95,98,106,169,259,261,362,391] These differences are partly due to different stimulating and recording methods[95] which alone would not necessarily reduce the value of the flash VEP as a diagnostic tool as long as stimulus and recording conditions were maintained rigidly constant in the testing laboratory. However, even VEPs recorded in the same laboratory vary so much between subjects, and in the same subject with time, that it is very difficult to define a standard normal VEP as having a certain number of peaks of constant polarity, latency, and amplitude[27,351,362,490] unless VEPs from many subjects are averaged together. For diagnostic purposes, each laboratory may select one or a few of the least variable VEP parameters for use in clinical testing. Usually, these are the latency of the major positive peak, occurring some time between 50 and 100 msec after the flash, and the latency of the subsequent, large negative peak, occurring between 100 and 250 msec after the flash; a negative wave preceding the major positive wave may also be suitable.

Peaks of shorter and longer latency than 50–250 msec are found even less constantly. Early peaks of low amplitude[98] may be recorded in some normal subjects with the usual recording methods. Recordings with narrow filter settings may pick up short-latency wavelets, some of which probably represent excitation of subcortical segments of the visual pathway,[110,371,405,478,493] but these potentials have not so far been put to clinical use in the same manner as the far-field recordings commonly used in AEP and SEP recordings. Late peaks of the cortical VEP to diffuse light flashes are seen often and may have rather large amplitude but are more variable than those between 50 and 250 msec. The early and late peaks of the VEP may be followed by a train of rhythmical waves at or near the frequency of the alpha rhythm[34] (Figure 9.1). This rhythmical afterdischarge varies enormously and has no known clinical importance.

9.3 SUBJECT VARIABLES

9.3.1 Age

9.3.1.1 Premature infants
The flash VEP after 24 weeks of gestation consists of an occipitally negative peak at 200–300 msec. At 32 to 35 weeks, a positive peak of less than 200 msec appears before the negative peak. With further development, the amplitude of the positive peak increases, that of the negative peak decreases, and the latencies of both peaks decrease. The features of these VEPs differ in the different sleep stages of the premature infant.[141,229,467,487] With increasing

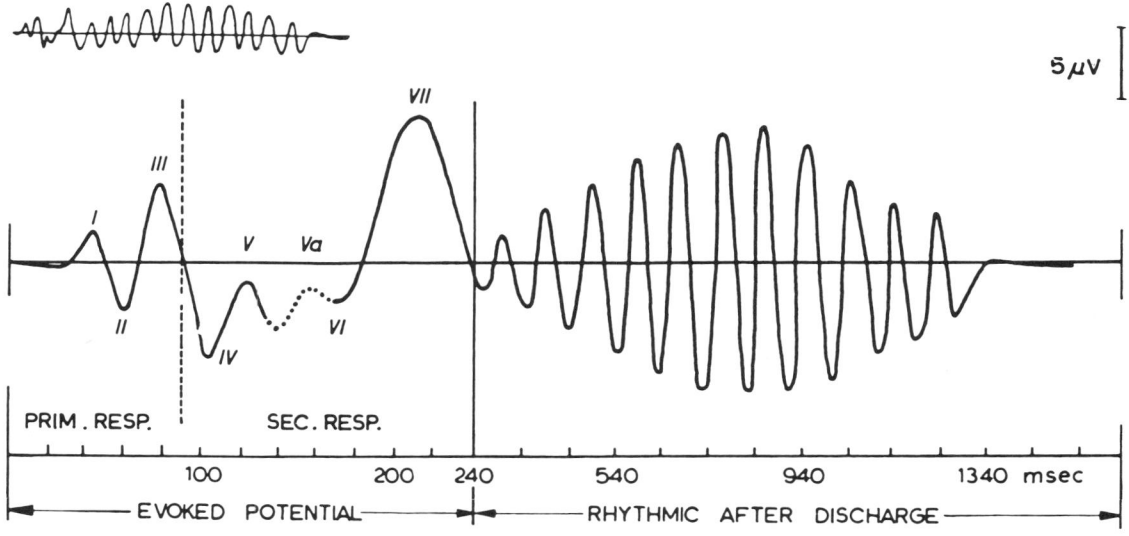

Figure 9.1. Diagram of the normal VEP to diffuse light flashes. This tracing is the grand average of VEPs from 75 subjects. The VEP is here divided into a primary response (prim. resp.) and a secondary response (sec. resp.) and is followed by a rhythmical afterdischarge. Note the change in time scale between VEP and afterdischarge. The insert in the left upper corner shows VEP and afterdischarge on the same time scale. Recording between midoccipital and midparietal electrodes. Negativity at the occipital electrode is plotted upward. From Cigánek[94] with permission by the author and Elsevier Scientific Publishers Ireland Ltd.

gestational age, the distribution of the VEP expands from the occipital to the frontocentral areas.[479]

9.3.1.2 *From full term to adult age*
At term, the flash VEP often consists of a sequence of negative-positive-negative peaks. Later, additional peaks appear and peak latencies decrease. The latencies of the early peaks reach adult values in early childhood while later peaks do not completely mature until puberty.[49,130,141,281] Latency and amplitude of VEPs vary even more in children than in adults.[72,142,258] Flash VEPs in children therefore do not reliably indicate maturation and are even less sensitive in detecting lesions than flash VEPs in adults. Longitudinal studies, if practical, may be more rewarding.[36]

9.3.1.3 *Old age*
The latency of VEP peaks increases after the age of 65.[130,451] The amplitude may increase[451] or decrease.[131]

9.3.2 Pupillary size

An increase of pupillary size acts like an increase of stimulus intensity (9.4.3). An artificial pupil or other devices are needed to keep retinal illuminance by stimulus and background light constant throughout the recording (6.3.2). The subject's eyes should be open during stimulation.

9.4 STIMULUS CHARACTERISTICS

9.4.1 Stimulators

9.4.1.1 *Xenon flash tubes*
Stroboscopic flash stimulators, found in most EEG laboratories, are the most commonly used source of diffuse light flashes in EP laboratories. They give brief and very intense flashes at rates up to more than 100/sec. The lamp may be directed at the eyes of the subject, or at a reflecting or transparent screen in front of the subject's eyes; light coming from a screen is more likely to illuminate larger parts of the retina evenly and to avoid stimulation by the pattern represented by the image of the lamp.

Two features of flash tubes may have clinical importance: (1) the light intensity may decrease at higher stimulus rates so that VEPs at different rates may not be directly comparable; (2) the flashes may be accompanied by a click that may act as an auditory stimulus and should be eliminated by acoustic shielding.

9.4.1.2 *Electromechanical stimulators*
The beam of a constant light source is brought into a narrow focus where it can be interrupted by an electronically controlled shutter producing brief flashes or sustained light pulses.

9.4.1.3 *Fluorescent lamps*

Fluorescent tubes that can be driven by direct current may be used to give light pulses of different length and shape; these lamps may be placed behind a diffuser to give uniform illumination.

9.4.1.4 *Light-emitting diodes*

Light-emitting diodes may be used for flash stimulation[447] and may be mounted in goggles or contact lenses for VEP studies during surgery[152] (9.7.1.2) or for testing of infants, small children, and patients in coma.

9.4.1.5 *Full-field (Ganzfeld) flash stimulators*

Brief flashes of specified luminance are delivered into a reflecting sphere into which the subject looks through a round cutout. This arrangement provides for illumination of the entire visual field independent of the direction of gaze; distance and intensity of the stimulus and background light can be easily controlled. Ganzfeld stimulation is used for quantitative studies, especially for the ERG.

9.4.2 Stimulus rate

Because VEPs to diffuse light stimuli may last 500 msec, stimulus rates should not exceed 2/sec for transient VEPs.

9.4.3 Stimulus intensity

Most VEP peaks vary with the stimulus intensity: Latency decreases and amplitude increases with increasing intensity to a saturation level that depends on many other parameters such as monocular versus binocular stimulation.[98,104,106,422,478] The distance between the flash lamp or the screen used to present flash stimuli and the subject's eyes should be 30–45 cm and must be kept constant in each laboratory. Moderate ambient light is usually present and should be rigidly controlled because it alters the luminance change produced by the stimulus.

9.5 RECORDING PARAMETERS: ELECTRODE PLACEMENTS AND COMBINATIONS

Routine recording electrodes may be placed in midoccipital and left and right occipital positions as for recordings of checkerboard pattern reversal VEPs (7.4.1). Each of these electrodes may be referred to interconnected ear electrodes for three channels of recording. A fourth channel may be added to record VEPs between a vertex electrode and the interconnected ear electrodes or to record the ERG (10.13).

The distribution of the flash VEP has been studied extensively.[6,65,261,391] A few points are of practical importance. Early peaks have a maximum over occipital areas, whereas some later peaks are most prominent over the vertex.[106,169,259,391] Latency shows little asymmetry, but amplitude may differ markedly in normal subjects, and a VEP may be completely absent over one side of the head in up to 5 percent of normal subjects.[204] Occipital VEPs recorded from the scalp are similar to VEPs recorded from the calcarine cortex.[107,478] Artifacts from ERG and muscle contractions may contaminate scalp recordings in some areas.[6]

The VEP is due largely to stimulation of the fovea.[121] It is difficult to evaluate the relationship between other retinal stimulus sites and VEP distribution precisely: Restricted flash stimuli scatter in the optic media of the eye and reach wide parts of the retina. Diffuse light stimuli are therefore not of much value for half-field stimulation. Very small light spots act as a patterned stimulus (10.6).

9.6 CRITERIA DISTINGUISHING ABNORMAL FLASH VEPs

Because of the great variability of flash VEPs, the only reliable criterion of abnormality is the complete absence of the monocular flash VEP unexplained by technical problems. Some other criteria may suggest possible abnormalities. In most laboratories, the latencies to one or more of the major occipital positive or negative peaks are chosen. Different normative values should be determined for different ages, especially for the young age groups in which patterned stimuli cannot be used.

9.7 NEUROLOGICAL DISORDERS THAT CAUSE ABNORMAL FLASH VEPs

The division into prechiasmal, chiasmal, and retrochiasmal is satisfactory except for disorders that involve the brain widely; these are arbitrarily classified depending on the site of involvement most frequently studied with VEPs.

9.7.1 Prechiasmal lesions

9.7.1.1 Multiple sclerosis
VEPs to flash stimuli may be used in patients with impaired vision who cannot see patterned stimuli. The preservation of flash VEPs in the absence of pattern VEPs indicates that at least some fiber connections between retina and occiput are preserved. The flash VEP has been described to have abnormally increased latency, decreased amplitude, and deformed shape in many patients with multiple sclerosis.[31,140,150,341,358,393] In patients who can see checkerboard patterns, VEPs to this stimulus have much greater diagnostic power than VEPs to flash stimuli.[191,498]

9.7.1.2 Surgical monitoring of optic nerve compression
The flash VEP has been used to monitor compression and decompression of the optic nerve during orbital[502] and intracranial[5,153,499] surgery. However, the VEP varies with

the level of general anesthesia[375,466] and has been found to be an unreliable indicator of optic nerve function during surgery.[377]

9.7.1.3 Other prechiasmal lesions

Abnormal flash VEPs have been described in many conditions, some of which may also cause retrochiasmal lesions:

1. Tumors compressing the optic nerve or chiasm[31]
2. Ischemic optic neuropathy[497]
3. Friedreich's ataxia[249]
4. Hereditary spastic paraplegia[199]
5. Lipid storage diseases[117,200,201]
6. Leukodystrophies[81,314]
7. Leber's hereditary optic atrophy[31,202]
8. Toxic optic neuropathy[267,278]

9.7.2 Retrochiasmal lesions

9.7.2.1 Head injury, increased intracranial pressure, and hydrocephalus

Abnormal VEPs, AEPs, and SEPs, seen in many patients in coma after severe head injuries,[179] have been found helpful in diagnosing focal deficits[178] and in predicting general outcome.[10,180,292,376] Head trauma may cause abnormal VEPs due to optic nerve injury (9.8.2). In postconcussion syndrome, VEP abnormalities have been reported to identify patients with organic symptoms.[151]

Increased intracranial pressure due to severe head injury or to hydrocephalus with shunt malfunction may increase VEP latency.[508] Removal of spinal fluid in hydrocephalus may decrease VEP latency.[304]

9.7.2.2 Anoxic encephalopathy and brain death

In acute cerebral anoxia, the VEP generally deteriorates and disappears earlier than the EEG and ERG,[390,485] but may be preserved in the absence of the EEG.[462] Although VEPs are absent in brain death,[462] their absence does not prove brain death[67]: The VEP seems to reflect the condition of the hemispheres more than that of the brain stem centers essential for survival.

9.7.2.3 Cortical blindness

The bilateral occipital lesions causing cortical blindness abolish occipital VEPs, as expected, in many cases, but peaks of normal[87,218,357] or prolonged latency may persist at the occiput[441] or at the vertex[262,359] in some cases. Children with cerebral blindness often have preserved VEPs showing some abnormalities.[37,57,161] The preservation of VEPs in cortical blindness reduces the ability of the VEP to diagnose blindness, especially in children.

The prognostic value of a preserved VEP to diffuse flashes in cortical blindness is also limited. Even though

VEPs often improve with return of visual function,[128,262] preservation of VEPs and degree of VEP abnormality do not necessarily indicate chances for recovery.[161] Stimulation with diffuse light has been used much more often for the study of cerebral blindness than stimulation with patterned light (8.3.3, 10.2.3, 10.4.4.2).

9.7.2.4 Seizures
Patients with seizures triggered by flash stimuli (photosensitive, photic, or photogenic seizures),[32,69,203,221,299] children with generalized seizures,[42] and patients with myoclonus epilepsy[127] have been found to have abnormally large VEPs. Other interictal VEP abnormalities are inconsistent.

VEPs may be preserved during 3/sec spike-and-wave discharges.[334] Seizures induced by unilateral electroconvulsive treatment depress VEPs on the treated side for 15 minutes postictally.[269]

9.7.2.5 Other retrochiasmal disorders
Abnormal VEPs have been studied in many conditions, some of which also cause prechiasmal problems:

1. Tumors and strokes[260,261,351,480]
2. Senile and presenile dementia (Alzheimer's disease)[451,482]
3. Jakob-Creutzfeldt disease[282,357,401]
4. Parkinson's disease[333]
5. Wilson's disease[300]
6. Huntington's chorea[138,408,457]
7. Renal failure and hemodialysis[196,291,403]
8. Down's syndrome[131]
9. Phenylketonuria[407]
10. Hyperglycinemia[313]
11. Menke's "kinky hair disease"[164]
12. Hyperthyroidism[456]
13. Hypothyroidism[227,344]
14. Alcohol[290,370,392,411,423]
15. Smoking[500]
16. Developmental and perinatal disorders[167,185,228,481]
17. Learning disorders and dyslexia[345,427]
18. Hyperactive children on treatment with methylphenidate[257]
19. Psychiatric disorders[124,241,413,476]

9.8 OPHTHALMOLOGICAL DISORDERS THAT CAUSE ABNORMAL FLASH VEPs

Flash VEPs may be of some value in ocular disorders that cause impaired vision precluding the use of patterned stimuli. Some retinal lesions cause ERG abnormalities in addition to VEP abnormalities and may be investigated by combined ERG and VEP recordings (10.13).

9.8.1 Increased intraocular pressure and glaucoma

Flash stimulation of glaucomatous eyes elicits VEPs of abnormally low amplitude, especially at low stimulus luminances. These VEPs are very sensitive to induced increases of intraocular pressure.[39,40] Experiments raising the intraocular pressure of normal eyes show that the VEP disappears at the point of pressure blinding, presumably because of retinal ischemia.[160] The VEP is more sensitive than the ERG, which persists even when the intraocular pressure briefly rises above the blood pressure of the ophthalmic artery, suggesting that the retina is less sensitive to pressure than the prelaminar part of the optic nerve.[39]

9.8.2 Injury to the eye or optic nerve

Eye injury[115] and head trauma with indirect optic nerve injury[31,202] may decrease the amplitude and increase the latency of the flash VEP.

9.8.3 Retinal detachment

The flash VEP may be preserved even in cases of severely abnormal ERG and completely detached retina.[472]

9.8.4 Retinitis pigmentosa

The flash VEP is usually normal or less reduced than the ERG.[135,232,338]

9.8.5 Amblyopia

The VEP to diffuse light is usually normal in amblyopia ex anopia.[295] Toxic amblyopia due to alcohol and tobacco[31] or to quinine[168] has been reported to cause VEP abnormalities.

9.8.6 Cataract

The flash VEP has been said to be of some help in predicting visual function after cataract surgery.[26,458,483]

9.8.7 Other ophthalmological disorders

Abnormal flash VEPs have been described in the following conditions:

1. Central retinal artery occlusion[232]
2. Optic nerve hypoplasia[445]
3. Optic nerve head drusen[448]
4. Congenital nystagmus[326]

10

VEPs to other stimuli

SUMMARY

10.1. Transient VEPs to checkerboard pattern appearance, disappearance, and flash differ from each other and from transient VEPs to checkerboard pattern reversal or shift even though they probably test similar mechanisms. They are used clinically in only a few laboratories.

10.2. Steady-state VEPs to checkerboard patterns are usually produced at rates of 6 to 10/sec. They have been used occasionally in neurological studies and somewhat more often in ophthalmological investigations.

10.3. Steady-state VEPs to diffuse light can be elicited with stimulus rates ranging usually from at least 5/sec to 50/sec. Abnormal steady-state VEPs have been reported in some prechiasmal and retrochiasmal lesions. The highest rate eliciting a steady-state VEP, or the critical frequency of photic driving (CFPD), has been found to be reduced in optic nerve demyelination and some retrochiasmal lesions.

10.4. VEPs to sine wave gratings, i.e., to light and dark stripes having sinusoidal transitions rather than sharp borders between bright and dark, may be as effective in detecting abnormalities as are VEPs to checkerboard patterns. Sine wave gratings may be used to study the effects of specific spatial frequencies on transient and steady-state VEPs.

10.5. VEPs to bar gratings, i.e., to light and dark stripes with sharp borders, have been used only rarely for clinical diagnosis.

10.6. VEPs to small macular light spots are of limited clinical use.

10.7. VEPs to moving and stereoscopic random dot patterns may help in the investigation of binocular vision.

10.8. VEPs to other patterned stimuli may be used to study visual acuity and cognitive functions.

10.9. Scotopic VEPs have been used experimentally.

10.10. VEPs to colored light may be of interest in the study of color vision.

10.11. VEPs to blinking and shift of gaze are naturally evoked visual potentials.

10.12. VEPs to electric stimulation of the eye have not been thoroughly explored.

10.13. ERGs to diffuse and patterned light probably have different generators.

10.1 TRANSIENT VEPS TO CHECKERBOARD APPEARANCE, DISAPPEARANCE, AND FLASH

The VEP to pattern appearance, or onset, differs from that to pattern disappearance, or offset; both differ from the VEP to check reversal (Figure 10.1). The VEP to pattern offset is usually more similar to the VEP to pattern reversal[146,268] than is the VEP to pattern onset,[414] although this may vary with stimulus conditions.[287,347] VEPs to pattern onset differ depending on whether the onset is asso-

Figure 10.1. Normal transient VEPs to onset, offset, and reversal of a checkerboard pattern. Each tracing is the grand average of recordings from 10 subjects. Right monocular stimulation with 50′ checks in a 32° round field. Recording between midoccipital and midfrontal electrodes. Occipital negativity is plotted upward; the P100 points down. From Kriss and Halliday[268] with permission by the authors and MTP Press Lancaster.

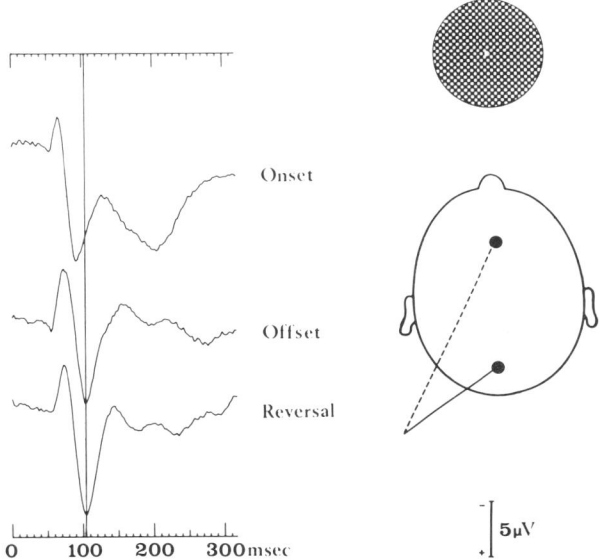

ciated with an increase, a decrease, or no change of overall luminance.[352]

When pattern presentation is shortened to about 30 msec, separate VEPs to onset and offset can no longer be distinguished; VEPs to such brief flashes of patterned light are the compound effect of onset and end of pattern presentation. These VEPs differ in shape and distribution from VEPs to other checkerboard presentations.[118,233,234,235]

Maturation of VEPs to checkerboard onset and flash resembles that of VEPs to checkerboard alternation in that both the peak latency and the effective check size decrease during infancy and childhood[208,242,442] in a manner suggesting that spatial resolution improves rapidly to the age of six months[209] and more slowly to puberty.[443] Amplitude is said to be higher at young and old than at middle age[133] and to depend on luminance in a manner changing with age.[132]

Neurological studies have shown VEPs to pattern onset or flash to be abnormal in multiple sclerosis,[406] hemispherectomy,[92] phenylketonuria,[407] and schizophrenia.[415] Occasional comparisons with VEPs to pattern reversal have suggested that VEPs to pattern onset may be more effective in the diagnosis of optic nerve lesions in multiple sclerosis[9,394] and in the study of abnormal fiber crossing in human albinism.[11,111,113]

Ophthalmological investigations have used these VEPs to study refraction,[129,301] amblyopia ex anopia,[295] strabismic amblyopia,[507] cataracts,[471] and binocular vision.[139]

10.2 STEADY-STATE VEPs TO CHECKERBOARD PATTERN REVERSAL, APPEARANCE, DISAPPEARANCE, AND FLASH

10.2.1 The normal steady-state pattern VEP

Rhythmical VEPs to checkerboard pattern stimulation may be produced at stimulus rates beginning at about 4/sec; these steady-state VEPs are often greatest at 6–8/sec and smaller at higher rates[3] (Figure 10.2). Like transient VEPs, steady-state VEPs are largest with small check sizes.[387,388] Larger check sizes are more effective at higher stimulus rates.[384]

Steady-state VEPs to patterned stimuli, like those to diffuse light, are largely explained by superimposition of transient VEPs,[246] but have been thought to reflect different aspects of visual function than do transient VEPs.[382] Amplitude is the most commonly measured parameter of steady-state VEPs. Because the latency of steady-state VEPs cannot be measured directly, other methods must be used to characterize their timing (4.1.3.3).

10.2.2 Maturation

The check size producing the largest steady-state VEP to patterned flashes decreases during the first few months of life.[242] By six months, reversing checks of 7.5' and 15' pro-

Figure 10.2. Development of the steady-state VEP to checkerboard pattern reversal with increasing stimulus rate; the rate, in pattern reversals per second, is indicated at the left of the tracings. At stimulus rates of 2/sec and 4/sec, transient VEPs with peaks N_1, P_1, N_2, and P_2 can be distinguished. At a rate of 6/sec, VEPs overlap. At 8/sec and above, individual VEP features give way to the rhythmical steady-state VEP consisting of waves at the rate of the stimulus. Occipital negativity is plotted down; the $\overline{P100}$, here named P_1, points up. From Sokol[430] with permission by the author and Churchill Livingstone Medical Division of Longman Group Ltd.

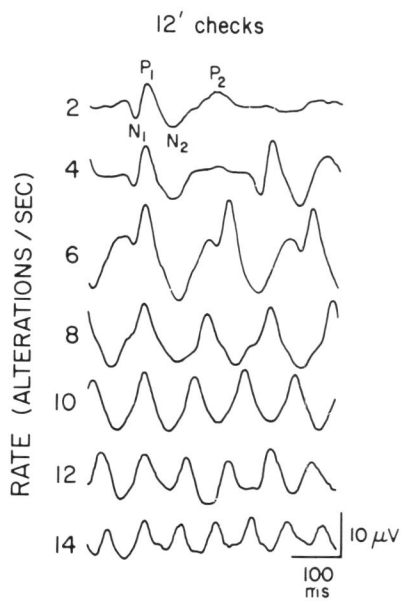

duce the largest response, as they do in adults with 20/20 visual acuity.[432] VEPs to faster stimulation with small sizes mature later than VEPs to slower stimulation.[335]

10.2.3 Neurological disorders that cause abnormal steady-state pattern VEPs

Steady-state checkerboard VEPs are used only rarely in clinical diagnosis. Abnormalities of amplitude and timing have been reported in retrobulbar neuritis[13,63,495] (Figure 10.3), multiple sclerosis,[134] traumatic neuritis,[13] and chiasmal and retrochiasmal tumors and infarcts presenting with visual field defects.[226,496] Study of steady-state VEPs in addition to transient VEPs may increase the diagnostic yield of retrochiasmal VEP studies.[279] Steady-state checkerboard VEPs may be preserved in cortical blindness due to bilateral occipital lesions[57] (9.7.2.3).

10.2.4 Ophthalmological disorders

10.2.4.1 Refraction and accommodation
The amplitude of steady-state VEPs to patterned stimuli may be used to determine the corrective lenses that result in the largest VEP amplitude, indicating optimum visual acuity.[329,331] The same method has been applied to measure accommodation.[330] Estimates of visual acuity have been made with a method using narrow-band filtering of steady-state responses to changing checkerboard sizes[383] and with

Figure 10.3. Steady-state checkerboard reversal VEPs in a normal subject and in a patient with right optic neuritis. In the normal subject, the latency from the stimulus at the beginning of each tracing to the first upward deflection (horizontal arrows) is equal for the right (OD) and left (OS) eye. In the patient with optic neuritis, the time interval from the stimulus to the VEP peak is changed, suggesting a marked prolongation for the clinically affected eye and a moderate prolongation for the clinically normal eye. These phase shifts indicate bilateral optic nerve involvement. Stimulation with 8/sec checkerboard pattern reversals. Recording between midoccipital and earlobe electrodes. Occipital positivity is plotted upward. From Wildberger et al.[495] with permission by the authors and S. Karger AG.

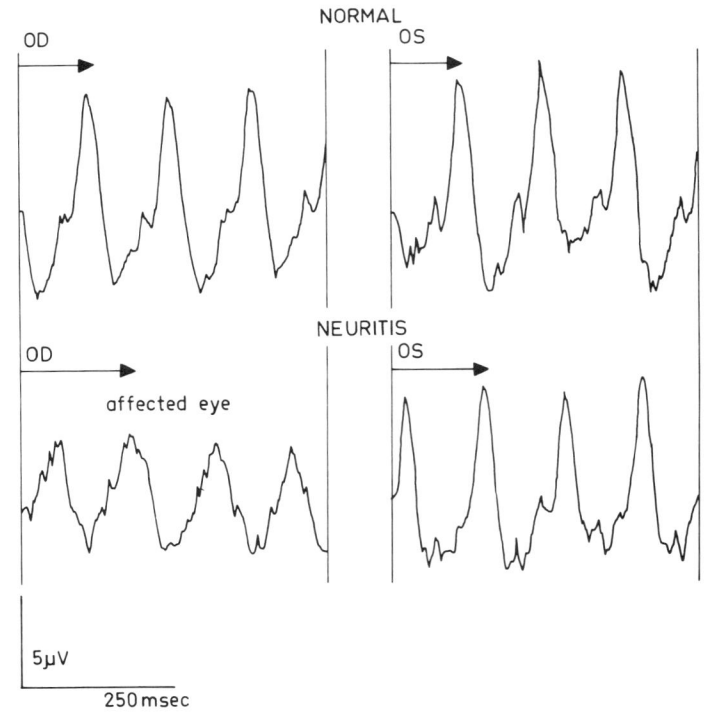

a method using spectral analysis of occipital responses[144] instead of averaging (3.6).

10.2.4.2 Amblyopia

The decreased visual acuity in amblyopia ex anopia is manifested by an increase of the check size producing VEPs of the largest amplitude, suggesting that the VEP of the amblyopic eye is due mainly to parafoveal elements of the retina[14,428]; the phase relation between peaks of VEPs to stimulation of each eye may be changed if binocular vision is not preserved.[14] Toxic amblyopia may produce interocular amplitude differences.[278]

10.2.4.3 Macular lesions

Macular degeneration may reduce the VEP without altering the ERG.[13] A macular cyst was found to increase the effective check size to that which normally stimulates paramacular regions.[44]

10.2.4.4 Glaucoma

The VEPs to stimulation of retinal quadrants in glaucomatous eyes with field defects have been found to have abnormal timing and amplitude; VEPs from eyes with chronic ocular hypertension alone showed no such changes.[80,460] VEPs of glaucomatous eyes had reduced amplitude if the central 10° of the field were defective[40]; they were found to be very sensitive to acute changes of intraocular pressure.[39,40]

10.3 STEADY-STATE VEPs TO DIFFUSE LIGHT STIMULI

10.3.1 The normal steady-state flash VEP

Repetitive stimulation with light flashes or with sinusoidal modulation of luminance at rates over 5/sec produces a rhythmical wave of the frequency of the repetitive stimulus; harmonic components may be superimposed or predominate[470] (Figure 10.4). The timing of each peak with respect to the preceding stimulus shows a rapid phase shift at a stimulus rate of about 10/sec and a phase lag proportional to the stimulus frequency at higher stimulus rates; it is independent of the frequency of the alpha rhythm of the subject under study.[378,470] Like the steady-state VEP to checkerboard pattern stimulation (10.2), the steady-state VEP to diffuse light is largely due to superimposition of transient VEPs.[246]

The steady-state VEP amplitude varies with the stimulus frequency, luminance, size, and many other parameters.[469] Under some laboratory conditions, maximum amplitudes were found at a low frequency of about 10/sec, at a medium frequency of 13–25/sec, and at a high frequency of about 40–60/sec.[382] The distribution of the steady-state VEP varies depending on stimulus frequency, retinal site of stimulation, and other factors.

Steady-state VEPs can be elicited at high rates depending on stimulus intensity and other factors. The highest rate at which a steady-state VEP can be distinguished is called the critical frequency of photic driving (CFPD) and varies inversely with age. It was found to have mean values of 72/sec between the ages of 20 and 30 years, 68/sec between 30 and 60, and 60/sec above 60 years in one laboratory.[88] Steady-state VEPs may be recorded at rates above its subjective counterpart, the critical fusion frequency.[470]

Two problems may arise when steady-state VEPs are used in neurological diagnosis; both pertain to measuring the response. First, VEP amplitude varies so much between subjects that abnormalities can be detected only by comparing VEPs in the same subject, especially VEPs recorded simultaneously on the two sides of the head or VEPs elicited successively by stimulation of either eye. Second, strict criteria for determining the presence of steady-state VEPs at slightly different frequencies have not yet been developed. Nonetheless, it seems that even rather gross estimates of the presence of VEPs at frequencies differing by as much as 10/sec or more are sufficiently sensitive to indicate abnormalities.[89,101]

10.3.2 Prechiasmal lesions that cause abnormal steady-state flash VEPs

Optic neuritis and multiple sclerosis may reduce the amplitude of steady-state VEPs and may change the time relation between stimuli and peaks; however, steady-state VEPs to diffuse light have been found to be diagnostically less useful than VEPs to patterned light stimuli.[134,495] Studies using Fourier analysis of the steady-state VEP instead of averaging showed that flicker rates of 13–25/sec were

Figure 10.4. Steady-state flash VEPs and cortical frequency of photic driving (CFPD) in a normal 21-year-old subject. Each tracing represents a steady-state VEP to diffuse light flashes. Stimulus rate (in flashes per second) is indicated to the left of the tracings. Dots below the tracings mark flash stimuli. Steady-state VEPs are elicited with flashes up to 62/sec but not with flashes at 76/sec. Monocular flash stimulation after pupillary dilatation. Midoccipital recording. Each tracing represents 204.8 msec. From Celesia and Daly,[88] Arch. Neurol. (Chicago) 34:403–407, 1977, Copyright 1977. American Medical Association, with permission by the authors and the American Medical Association.

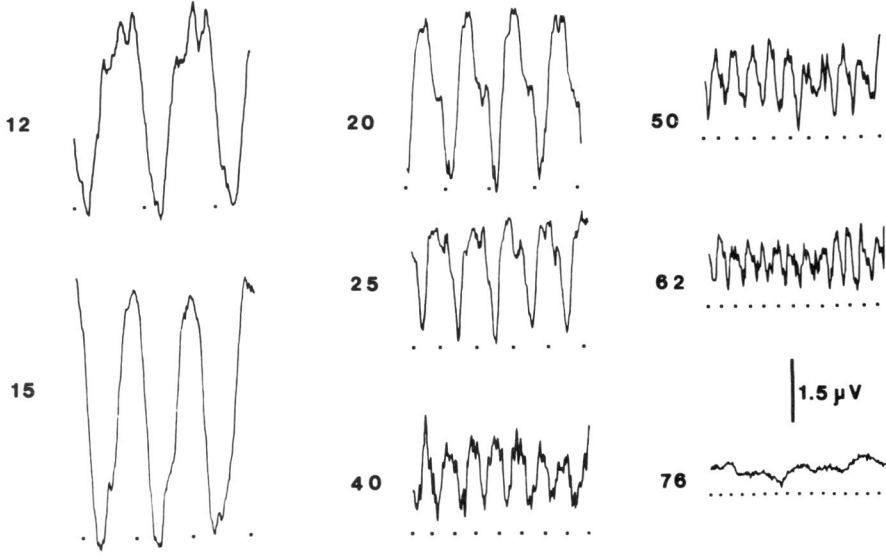

most effective in detecting abnormalities in multiple sclerosis.[332]

Determination of the CFPD may increase the diagnostic yield of transient VEPs to alternating checkerboard patterns in multiple sclerosis.[89] It has been claimed that the CFPD may even be more effective in this regard than the transient pattern VEP.[101] The steady-state VEP has also been reported to be abnormal in patients with chronic papilledema.[248]

10.3.3 Retrochiasmal lesions that cause abnormal steady-state flash VEPs

Steady-state VEPs to diffuse light are not commonly used for the study of retrochiasmal lesions. Power spectral analysis of steady-state VEPs may show abnormal VEP distribution in some cases of tumors and infarcts causing homonymous hemianopsia.[88] Patients with photosensitive seizures have been found to show changes of VEPs elicited by the beginning of repetitive trains of flashes.[283]

10.4 VEPs TO SINE WAVE GRATINGS

10.4.1 The principle of sine wave grating stimuli

Sine wave grating stimuli consist of alternating light and dark stripes with midlines of maximum and minimum luminance and a gradual transition of luminance between them; precisely, the change of luminance has the form of a sine wave (6.3.3.1). The size of the pattern elements is measured between two light or dark stripes and expressed as cycles per degree (c/deg) of visual angle (6.3.4.2). These stripes are used because they contain only a single spatial frequency, namely, that of the sine wave representing the brightness modulation. This differs from patterns with sharp borders between light and dark elements which represent sudden transitions of brightness that can be equated with a spatial square wave (6.3.3.1) and represent a mixture of spatial sine waves of different frequencies. Stimulation with sine wave gratings gives the investigator the opportunity to select one spatial frequency to study the ability of the visual system to transmit information in discrete channels of spatial frequency; another slight advantage of sine wave grating stimuli is that, because they do not contain sharp borders, they are less affected than checkerboard patterns by blurring of up to 2 diopters.

10.4.2 Stimulators for sine wave gratings

Grating stimuli are usually produced with signal generators on an oscilloscope screen,[75] although a computer driving an oscilloscope or video monitor[165] and optical techniques using polarized light[182] have also been described. Laser beams have been used to form sinusoidal grating patterns directly on the retina.[18,75]

10.4.3 Transient VEPs to sine wave grating stimuli

10.4.3.1 The normal transient sine grating VEP
The transient VEP to the reversal of a sine wave grating pattern and the transient VEP to the appearance of a grat-

ing pattern of less than 1 c/deg are similar to the transient checkerboard reversal VEP: They consist mainly of an occipitally positive peak of a latency of about 100 msec; this peak is preceded by a smaller negative peak.[273,369] Transient VEPs to the appearance of a grating pattern of over 1 c/deg consist of two early and two late negative-positive peaks[273] whose latency and amplitude varies with spatial frequency and contrast depth[238,354,369,477] (Figure 10.5).

10.4.3.2 Neurological disorders that cause abnormal transient sine grating VEPs

10.4.3.2.1 Multiple sclerosis. The VEP latency is increased in the majority of patients.[61] Abnormalities are more common with high spatial frequencies.[368] Sine wave gratings may be more effective than checkerboard patterns in detecting optic nerve lesions in multiple sclerosis, and testing with gratings of different spatial orientation may increase the diagnostic yield.[74]

10.4.3.2.2 Parkinson's disease. The latency may be increased.[62,275]

10.4.3.3 Ophthalmological disorders that cause abnormal transient sine grating VEPs

Increased VEP latencies have been reported for ocular hypertension and open-angle glaucoma.[21]

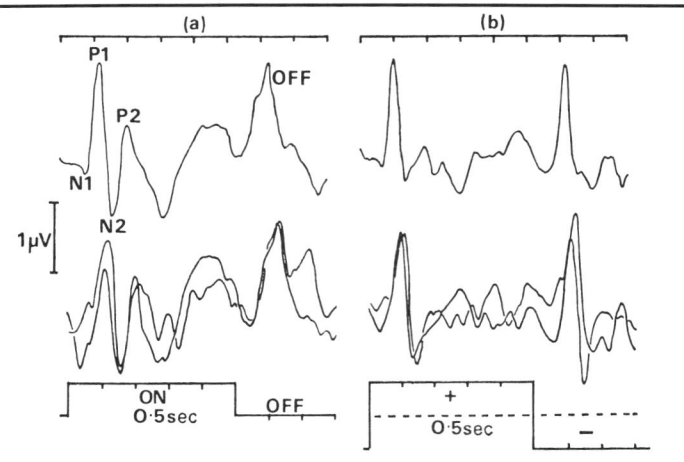

Figure 10.5. Normal transient VEPs to the onset, end, and reversal of a sine wave grating stimulus. Pattern onset, at the upstroke of the marker tracing at the bottom in *a*, produces a VEP with two negative and two positive peaks (N1, P1, N2, P2). Pattern offset, at the downstroke of the marker tracing in *a*, produces a simpler VEP consisting mainly of a positive deflection. Pattern reversals, at the upstrokes and downstrokes of the marker channel in *b*, also produce a VEP consisting mainly of a positive deflection. The lower tracings were obtained after cycloplegia and paralysis of the extraocular muscles which caused no important changes. Monocular stimulation with a grating pattern of 3 c/deg. Recording between midoccipital and temporal electrodes. Occipital positivity is plotted upward. Reprinted from Kulikowski and Leisman,[273] The effect of nitrous oxide on the relation between the evoked potential and contrast threshold, Vision Research 13, Copyright 1973, Pergamon Press, Ltd, with permission by the authors and Pergamon Press, Ltd.

10.4.4 Steady-state VEPs to sine wave gratings

10.4.4.1 The normal steady-state sine grating VEP

Steady-state VEPs to sine wave gratings consist of a rhythm whose amplitude varies with spatial frequency and reaches maximum at a value corresponding with that of the most effective checkerboard pattern (Figure 10.6); the amplitude also increases with contrast within limits.[76,503] The curve relating VEP amplitude to spatial frequency has a similar shape and peak as that obtained from measuring subjective contrast sensitivity, i.e., the threshold for detecting sine wave gratings of different spatial frequencies at different levels of contrast[76,77,78] (Figure 10.6). The measurement of subjective contrast thresholds at different spatial frequencies, or visuogram, gives a more complete description of visual ability than Snellen's method of measuring visual acuity[488] and has been used in some clinical investigations of multiple sclerosis[17,61,274,385,386] and retrobulbar neuritis,[220] other prechiasmal[276] and retrochiasmal[59] lesions, anisometropic and strabismic amblyopia,[219] and glaucoma[222]; the measurement of dynamic contrast sensitivity, using flickering sine wave grating patterns, seems more effective in detecting abnormalities than the measurement of static contrast sensitivity, using stationary gratings. Although visual impairment usually involves discrimination of high spatial frequencies, or small sizes, before affecting medium and low spatial frequencies, or medium and large sizes, some patients show selective involvement of discrimination of medium frequencies; this impairment, although causing complaints of blurred vision, escapes detection by conventional testing with letters that have sharp black and white interfaces representing mixed spacial frequencies. Another approach to the study of contrast effects uses intermittent changes of contrast depth short of reversal as a VEP stimulus (6.3.2.2) to determine the contrast modulation threshold.[60]

Maturation of contrast sensitivity as studied with sine wave grating VEPs, reaches near adult levels for low and medium spatial frequencies at six months of age.[205] However, maximum contrast sensitivity continues to increase, especially for higher spatial frequencies, to the end of the second or third year.[159,367]

10.4.4.2 Neurological disorders that cause abnormal steady-state sine grating VEPs

The steady-state VEP has been used to detect optic nerve lesions in multiple sclerosis.[394] In cortical blindness, the steady-state VEP may persist[57] (9.7.2.3).

10.4.4.3 Ophthalmological disorders that cause abnormal steady-state sine grating VEPs

Amblyopia reduces the amplitude and the effective spatial frequency of steady-state VEPs to sine wave gratings[311] (Figure 10.6). Astigmatic amblyopia is characterized by a reduction of VEP amplitude to gratings oriented in the direction of the reduced visual acuity.[157] Glaucoma preferentially affects VEPs to gratings of low spatial fre-

Figure 10.6. Spatial tuning curves for steady-state VEPs and psychophysical contrast thresholds to sine wave gratings, recorded in a subject with a normal right eye (circles) and an amblyopic left eye (squares). Each curve shows the magnitude of a response (vertical axis) plotted against the spatial frequency of the stimulus in cycles per degree (horizontal axis). The curves in A show the magnitude of the steady-state VEP measured by computing power spectra of the VEPs to sine grating stimuli (signal) and to blank stimulation (noise) and by calculating the ratio of the power at the stimulus frequency under both conditions (S/N ratio). Stimulus contrast was 0.44. Recording between midoccipital and ear electrodes. The curves in b show the contrast threshold which was determined by having the subject view the same grating stimuli, by increasing or decreasing the contrast at each spatial frequency, by recording the contrast level at which the subject indicates appearance or disappearance of the pattern, and by plotting the reciprocal of that contrast level as contrast sensitivity. Both types of tuning curves show a peak at 3–4 c/deg for the normal eye. For the amblyopic eye, the peaks of both curves lie at a lower frequency, and the VEP tuning curve is less sharply peaked. From Levi and Harwerth[289] with permission by the authors and Lippincott/Harper and Row.

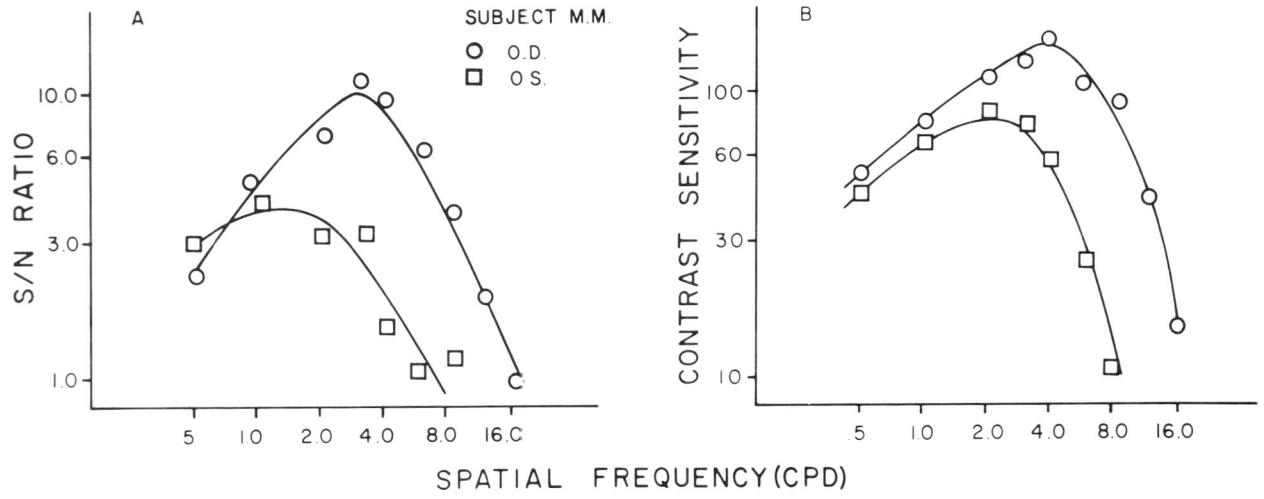

quency.[56] Patients with ocular opacities have been studied with interference fringes produced with a laser interferometer to assess their visual function.[18]

Sine wave grating patterns have been used to determine visual acuity with a method giving faster results than the averaging of VEPs. Steady-state responses are elicited at a fairly high stimulus rate with a grating pattern that gradually changes from low to high spatial frequency. The occipital responses are extracted from the background activity with a filter sharply tuned to the stimulus frequency or with power spectral analysis which locks in on the stimulus frequency and gives a write-out of the response amplitude for the different spatial frequencies[465] (3.6), similar to that obtained with steady-state VEPs to checkerboard reversal (10.2.4.1). The slope of the curve at high spatial frequencies allows extrapolation of a spatial frequency at zero amplitude which gives an electrophysiological measure of visual acuity that corresponds with behavioral acuity (Figure 10.7). The method can be used to study meridional,[163] strabismic and anisometropic[289] amblyopia, ocular rivalry, fusion, and stereopsis.[12]

10.5 VEPs TO BAR GRATINGS

Bar gratings consist of light and dark stripes of equal width with sharp borders between them. They produce VEPs less effectively than do checkerboard patterns because they present contrast borders in only one dimension. They are not used as often as are grating patterns with sinusoidal transitions because, although they represent mainly the spatial frequency represented by the grating itself, the sharp borders contain additional, higher, spatial frequencies that are probably processed by different channels of the visual system. Bar gratings are therefore much less specific for spatial frequency than are sinusoidal gratings.

Like VEPs to checkerboards, bar grating VEPs depend on the size of the stimulus element, the contrast between light and dark elements, the stimulus rate, and other parameters.[322] However, VEPs to bar gratings and to checkerboard patterns have different sizes of effective stimulus elements.[396] Study of maturation of visual acuity, as tested with VEPs to bar grating patterns, indicates that adult acuity is reached between 4 and 6 months,[312] i.e., at a time similar to that determined by checkerboard (7.2.1.1, 10.2.2) and sine grating (10.4.4.1) patterns.

Clinical studies have used bar gratings only rarely.[337] In ophthalmology, bar gratings have been used for refraction,[301] for study of binocularity,[156] and for a demonstration that astigmatism reduces VEPs to stimulation with stripes in the direction of the lower visual acuity[157,162] to an extent unexplained by the normal difference between the effects of oblique and horizontal or vertical stripes.[306,509]

10.6 VEPs TO SMALL MACULAR LIGHT SPOTS

VEPs to stimulation with small light spots projected into the center of fixation have been compared with VEPs to check reversal and to diffuse light in patients with multiple

Figure 10.7. Spatial tuning curves obtained with a rapid linear sweep method. Each tracing represents occipital electric responses to fast repetitive stimulation with reversing sine wave grating patterns of continuously increasing spatial frequency. The recordings are not averaged but filtered with a narrowband filter set to the value of the stimulus rate and reflect response amplitude at the stimulus rate. Three tracings are superimposed for each condition. The intersection of the descending slope of the tracing with the horizontal axis represents VEP acuity; the psychophysically measured visual acuity is indicated by arrows. The effect of defocusing (numbers at the right of the tracings) shows the sensitivity of the method to induced refractive errors and its potential as a means of measuring refractive errors. Recording between electrodes 3 cm above the inion and 3 cm above and lateral. From Tyler et al.[465] with permission by the authors and Lippincott/Harper and Row.

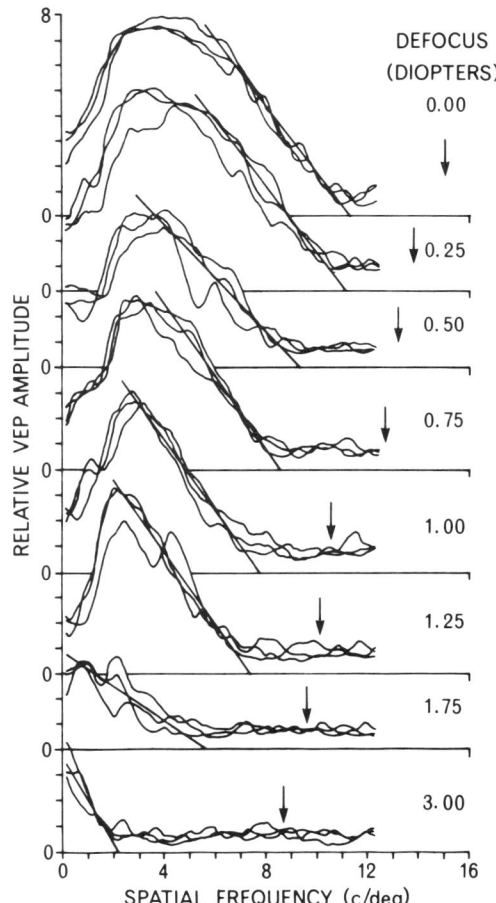

sclerosis and retrobulbar neuritis. Macular stimuli were found to be more effective than diffuse light stimuli,[402,495] but the outcome of comparisons with checkerboard patterns seems to depend on the size of the macular flash stimulus relative to that of the pattern element: Local flashes were inferior to checkerboard patterns when they were much larger than the checks used for comparison,[495] equally effective when of the same size as the checks,[123,348] and more effective when smaller than the checks[216]; this suggests that the small light spot acts as a single element of a stimulus pattern. In one study, stimuli of less than 4° were found to be less useful clinically than larger ones because latencies became longer and more variable and fixation became a problem.[505]

10.7 VEPs TO MOVING AND STEREOSCOPIC RANDOM DOT PATTERNS

VEPs may be induced by moving patterns.[96] Especially effective are patterns of random dots which continue to appear and disappear and are aligned stereoscopically for the two eyes to give the illusion of dots moving through space (dynamic random dot stereogram). The stimulus consists either of a sudden apparent displacement of the dots toward the observer (moving random dot stereogram)[217,284] or of the sudden alternation of the stereogram with a similar pattern without stereoscopic alignment of the images for each eye (dynamic random dot correlogram)[58,240] This stimulus, sometimes called cyclopean, is so obtrusive that it does not require much effort of fixation by the subject and should be useful for the study of children and uncooperative adults; it has been used mainly for study of binocularity or stereopsis, a cortical function.[66,239,364]

10.8 VEPs TO OTHER PATTERNED STIMULI

Printed letters and words have been used as stimuli for VEPs to distinguish between shape and perceptual content of the stimulus.[236,310] Abnormal VEPs were found in reading-disabled children[454] and adults.[372] Similar stimuli have been used to study language and cognitive functions by recording late components such as the $\overline{P300}$ (21.1).

VEPs to stationary dots of various sizes were used to study visual acuity[459] and refractory errors.[207] Grid patterns of different sizes were presented to the two eyes to test stereoacuity.[210] VEPs to horizontal lines were used to study cerebral asthenopia.[181] Patterns of texture contrast elicit VEPs different from similar black-and-white patterns.[302] VEPs to laser speckle patterns[119] may become helpful in cases of cataracts and other opacities of ocular media.[166]

10.9 PHOTOPIC VERSUS SCOTOPIC VEPs

The VEPs used in clinical practice are almost entirely due to the excitation of the photopic retinal cone system, located mainly in the fovea, which operates under conditions of light adaptation and is most sensitive to red.[121] Dark adaptation, stimulation of the peripheral retina, dim stimuli, and blue color all favor excitation of the scotopic rod system which contributes little to VEPs produced with conventional methods[1,108,255,264,501] and has only rarely been tested in neurological diseases such as multiple sclerosis.[358]

10.10 COLOR VEPs

The effect of different wavelengths on transient and steady-state VEPs has been studied with diffuse and patterned stimuli.[254,256,266,355,365,380,381,492] Colored stimuli have occasionally been used in the evaluation of neurological problems, especially of multiple sclerosis[358] and in ophthalmological investigations of abnormal color vision.[245,389]

10.11 VEPs TO BLINKING AND SHIFT OF GAZE

VEPs produced by interruption of visual input due to blinking differ from VEPs produced by interruption of input by other means.[23]

Saccadic shifts of gaze, causing lambda waves in the occipital EEG of some subjects, produce VEPs which are partly, but not fully, explained by the change of the visual input. These VEPs are not identical to VEPs produced by other presentation modes of the same input,[22,35,68,280,409,510] although the onset of the new input following the change of gaze, rather than the termination of the old input from the preceding fixation, appears to be the effective stimulus.[504]

10.12 ELECTRIC STIMULATION OF THE EYE

Electric stimulation of the eye[46] can be used to evoke occipital potentials.[243]

10.13 THE ERG TO DIFFUSE AND PATTERNED LIGHT

The ERG may be elicited by diffuse flashes or by patterned light stimuli. Diffuse illumination of the entire retina with Ganzfeld stimuli (9.4.1.4) produces larger VEPs than restricted, focal retinal stimulation. Recordings are made with electrodes on cornea, sclera, or near the eye (2.2.1.6). Care must be taken to avoid pickup of the ERG from the opposite eye.[360]

The ERG to diffuse light flashes consists of a cornea-negative a-wave and a cornea-positive b-wave which are probably generated by photoreceptors and glial Muller cells, respectively. The scotopic rod system and the photopic cone system contribute to different degrees to these ERG waves. The photopic component can be separated from the scotopic component by stimulation with flashes of different wavelength under light-adapted and dark-adapted conditions.[45,93,175] The ganglion cells of the optic nerve do not participate in the production of the flash ERG. The ERG may be recorded simultaneously with the VEP and may aid in the distinction between ocular and central visual problems.[85,231,232]

The ERG to patterned light differs from that to diffuse light (Figure 10.8) and varies with the size of the pattern elements.[19] The effective size of the pattern elements for the ERG differs from that for the VEP.[24,263,435] In contrast to the flash ERG, the pattern ERG is probably partly or entirely produced by ganglion cells[20,307] or some pregan-

Figure 10.8. Electroretinograms to patterned and diffuse light. The pattern ERG (PERG) and focal ERG to diffuse light (FERG) were produced by presenting a checkerboard pattern or a diffuse light flash in a round 16° field. Both ERGs show mainly a b-wave. This is preceded by a small a-wave of opposite polarity, visible at the higher contrast level, mainly in the FERG. Contrast levels refer to simultaneous brightness contrast between pattern elements for the PERG and to successive brightness contrast between light and dark phase for the FERG. The tracings are averages of recordings from three normal subjects. Recording between a gold foil electrode on the sclera and an electrode on the eyebrow. Positivity at the sclera is plotted upward. From Arden et al.[20] with permission by the authors and the New York Academy of Sciences.

glionic elements.[420] ERGs to pattern stimulation have been found to be abnormal in amblyopia,[439] optic neuritis,[305] some cases of multiple sclerosis,[55,250] glaucoma,[55,486] and various other disorders involving the optic nerve in or near the eye.[120,158,249,324] Because the pattern ERG indicates the entry of the afferent volley into the central visual pathway, it may be used to calculate retinocortical conduction time by subtracting the latency of the pattern ERG from the latency of the simultaneously recorded pattern VEP. This central conduction time may help to distinguish retinal lesions, such as maculopathy, which increase both retinal and cortical latency, from central lesions, such as optic nerve demyelination, which increase only the cortical latency.

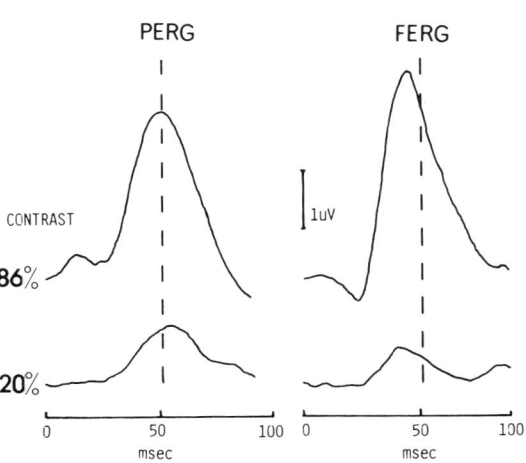

References for part B

1. Adachi-Usami, E. 1978. Scotopic retinal sensitivity in man as determined with visually evoked cortical potentials. Jpn. J. Physiol. 28:171–180.
2. Adachi-Usami, E. 1981. Visual checkerboard-evoked potentials from upper and lower retinal halves, and variation of check size. Neurosci. Lett. 22:245–250.
3. Adachi-Usami, E. 1981. Human visual system modulation transfer function measured by evoked potentials. Neurosci. Lett. 23:43–47.
4. Adachi-Usami, E. 1984. Equipotential maps of pattern-evoked potentials in man. Ophthal. Res. 16:73–79.
5. Allen, A., Starr, A., and Nudleman, K. 1981. Assessment of sensory function in the operating room utilizing cerebral evoked potentials: A study of fifty-six surgically anesthetized patients. Clin. Neurosurg. 28:457–481.
6. Allison, T., Matsumiya, Y., Goff, G.D., and Goff, W.R. 1977. The scalp topography of human visual evoked potentials. Electroencephalogr. Clin. Neurophysiol. 42:185–197.
7. Allison, T., Wood, C.C., and Goff, W.R. 1983. Brain stem auditory, pattern-reversal visual, and short-latency somatosensory evoked potentials: Latencies in relation to age, sex and brain and body size. Electroencephalogr. Clin. Neurophysiol. 55:619–636.
8. American Electroencephalographic Society. 1984. Guidelines for clinical evoked potential studies. J. Clin. Neurophysiol. 1:3–53.
9. Aminoff, M.J., and Ochs, A.L. 1981. Pattern-onset visual evoked potentials in suspected multiple sclerosis. J. Neurol. Neurosurg. Psychiatry 44:608–614.
10. Anderson, D.C., Bundlie, S., and Rockswold, G.L. 1984. Multimodality evoked potentials in closed head trauma. Arch. Neurol. (Chicago) 41:369–374.
11. Apkarian, P.A., Nakayama, K., and Tyler, C.W. 1981. Binocularity in the human evoked potential: Facilitation, summation and suppression. Electroencephalogr. Clin. Neurophysiol. 51:32–48.
12. Apkarian, P., Reits, D., Spekreijse, H., and Van Dorp, D. 1983. A decisive electrophysiological test for human albinism. Electroencephalogr. Clin. Neurophysiol. 55:513–531.
13. Arden, G.B. 1973. The visual response in ophthalmology. Proc. R. Soc. Med. 66:1037–1043.
14. Arden, G.B., Barnard, W.M., and Mushin, A.S. 1974. Visually evoked responses in amblyopia. Br. J. Ophthalmol. 58:183–192.
15. Arden, G.B., Bodis-Wollner, I., Halliday, A.M., Jeffreys, A., Kulikowski, J.J., Spekreijse, H., and Regan, D. 1977.

Methodology of patterned visual stimulation (report of the Brussels symposium ad-hoc committee). In: Visual evoked potentials in man: New developments, ed. J.E. Desmedt, pp. 3–15. Oxford: Clarendon Press.
16. Arden, G.B., Faulkner, D.J., and Mair, C. 1977. A versatile television pattern generator for visual evoked potentials. In: Visual evoked potentials in Man: New developments, ed. J.E. Desmedt, pp. 90–109. Oxford: Clarendon Press.
17. Arden, G.B., and Gucukoglu, A.G. 1978. Grating test of contrast sensitivity in patients with retrobulbar neuritis. Arch. Ophthalmol. 96:1626–1629.
18. Arden, G.B., and Sheorey, U.B. 1977. The assessment of visual function in patients with opacities: A new evoked-potential method using a laser interferometer. In: Visual evoked potentials in man: New developments, ed. J.E. Desmedt, pp. 381–394. Oxford: Clarendon Press.
19. Arden, G.B., and Vaegan. 1983. Electroretinograms evoked in man by local uniform or patterned stimulation. J. Physiol. (Lond.) 341:85–104.
20. Arden, G.B., Vaegan, and Hogg, C.R. 1982. Clinical and experimental evidence that the pattern electroretinogram (PERG) is generated in more proximal retinal layers than the focal electroretinogram (FERG). Ann. N.Y. Acad. Sci. 388:580–601.
21. Arkin, A., Bodis-Wollner, I., Podos, S.M., Wolkstein, M., Mulin, L., and Nitzberg, S. 1983. Flicker threshold and pattern VEP latency in ocular hypertension and glaucoma. Invest. Ophthalmol. Visual Sci. 24:1524–1528.
22. Armington, J.C. 1977. Visual cortical potentials and electroretinograms triggered by saccadic eye movements. In: Visual evoked potentials in man: New developments, ed. J.E. Desmedt, pp. 286–300. Oxford: Clarendon Press.
23. Armington, J.C. 1981. Visually evoked cortical potentials accompanying blinks. Invest. Ophthalmol. Visual Sci. 20:691–695.
24. Armington, J.C., and Brigell, M. 1981. Effects of stimulus location and pattern upon the visually evoked cortical potential and the electroretinogram. Int. J. Neurosci. 14:169–178.
25. Armington, J.C., Corwin, T.R., and Marsetta, R. 1971. Simultaneously recorded retinal and cortical responses to patterned stimuli. J. Opt. Soc. Am. 61:1514–1521.
26. Arnal, D., Gerin, P., Salmon, D., Ravault, M.P., Magnard, P., and Huggonnier, R. 1972. Intérêt des potentiels évoqués moyens visuels en ophtalmologie. Electroencephalogr. Clin. Neurophysiol. 32:615–621.
27. Arnal, D., Gerin, P., Salmon, D., Ravault, M.P., Nakache, J.P., and Peronnet, F. 1972. Les diverses composantes des potentiels évoqués moyens visuels chez l'homme. Electroencephalogr. Clin. Neurophysiol. 32:499–511.
28. Arruga, J., Feldon, S.E., Hoyt, W.F., and Aminoff, M.J. 1980. Monocularly and binocularly evoked visual responses to patterned half-field stimulation. J. Neurol. Sci. 46:281–290.
29. Ashworth, B., Maloney, A.F.J., and Townsend, H.R.A. 1978. Delayed visual evoked potentials with bilateral disease of the posterior visual pathway. J. Neurol. Neurosurg. Psychiatry 41:449–451.
30. Asselman, P., Chadwick, D.W. and Marsden, C.D. 1975. Visual evoked responses in the diagnosis and management of patients suspected of multiple sclerosis. Brain 98:261–282.

31. Babel, J., Stangos, N., Korol, S., and Spiritus, M. 1977. Ocular electrophysiology: A clinical and experimental study of electroretinogram, electro-oculogram and visual evoked response. Stuttgart: Thieme.
32. Bablouzian, B.L., Neurath, P.W., Sament, S., and Watson, C.W. 1969. Detection of photogenic epilepsy in man by summation of evoked scalp potentials. Electroencephalogr. Clin. Neurophysiol. 26:93–95.
33. Bajada, S., Mastaglia, F.L., Black, J.L., and Collins, D.W.K. 1980. Effects of induced hyperthermia on visual evoked potentials and saccade parameters in normal subjects and multiple sclerosis patients. J. Neurol. Neurosurg. Psychiatry 43:849–852.
34. Barlow, S.J. 1960. Rhythmic activity induced by photic stimulation in relation to intrinsic alpha activity of the brain in man. Electroencephalogr. Clin. Neurophysiol. 12:317–326.
35. Barlow, J.S., and Cigánek, L. 1969. Lambda responses in relation to visual evoked responses in man. Electroencephalogr. Clin. Neurophysiol. 26:183–192.
36. Barnet, A.B., Friedman, S.L., Weiss, I.P., Ohlrich, E.S., Shanks, B., and Lodge, A. 1980. VEP development in infancy and early childhood. A longitudinal study. Electroencephalogr. Clin. Neurophysiol. 49:476–489.
37. Barnet, A.B., Manson, J.I., and Wilner, E. 1970. Acute cerebral blindness in childhood. Six cases studied clinically and electrophysiologically. Neurology 20:1147–1156.
38. Bartel, D.R., Markand, O.N., and Kolar, O.J. 1983. The diagnosis and classification of multiple sclerosis: Evoked responses and spinal fluid electrophoresis. Neurology 33:611–617.
39. Bartl, G. 1978. The electroretinogram and the visual evoked potential in normal and glaucomatous eyes. Albrecht von Graefes Arch. Klin. Exp. Ophthalmol. 207:243–269.
40. Bartl, G. 1982. The effects of visual field changes and ocular hypertension on the visual evoked potential. Ann. N.Y. Acad. Sci. 388:227–242.
41. Baumgartner, J., and Epstein, C.M. 1982. Voluntary alteration of visual evoked potentials. Ann. Neurol. 12:475–478.
42. Beck, E.C., Dustman, R.E., and Lewis, E.G. 1975. The use of the averaged evoked potential in the evaluation of central nervous system disorders. Int. J. Neurol. 9:211–232.
43. Becker, W.J. and Richards, I.M. 1984. Serial pattern shift visual evoked potentials in multiple sclerosis. Can. J. Neurol. Sci. 11:53–59.
44. Behrman, J., Nissim, S., and Arden, G.B. 1972. A clinical method for obtaining pattern evoked responses. In: Advances in biology and medicine, vol. 24, ed. G. Arden, pp. 199–206. London: Plenum Press.
45. Berson, E.L. 1981. Electrical phenomena in the retina. In: Adler's physiology of the eye, ed. R.A. Moses, pp. 466–529. St. Louis: Mosby.
46. Bijl, G.K., Melchior, H.J., and Veringa, F. 1980. Recording the visual electrically evoked potential (VEEP). Electroencephalogr. Clin. Neurophysiol. 49:655–656.
47. Bird, T.D., and Crill, W.E. 1981. Pattern-reversal visual evoked potentials in the hereditary ataxias and spinal degenerations. Ann. Neurol. 9:243–250
48. Bird, T.D., and Griep, E. 1981. Pattern reversal visual evoked potentials. Studies of Charcot-Marie-Tooth hereditary neuropathy. Arch. Neurol. 38:739–741.
49. Blom, J.L., Barth, P.G., and Visser, S.L. 1980. The visual

evoked potential in the first six years of life. Electroencephalogr. Clin. Neurophysiol. 48:395–405.
50. Blumhardt, L.D., Barrett, G., and Halliday, A.M. 1977. The asymmetrical visual evoked potential to pattern reversal in one half field and its significance for the analysis of visual field defects. Br. J. Ophthalmol. 61:454–461.
51. Blumhardt, L.D., Barrett, G., and Halliday, A.M. 1982. The pattern visual evoked potential in the clinical assessment of undiagnosed spinal cord disease. Adv. Neurol. 32:463–471.
52. Blumhardt, L.D., Barrett, G., Halliday, A.M., and Kriss, A. 1978. The effect of experimental 'scotomata' on the ipsilateral and contralateral responses to pattern-reversal in one half-field. Electroencephalogr. Clin. Neurophysiol. 45:376–392.
53. Blumhardt, L.D., Barrett, G., Kriss, A., and Halliday, A.M. 1982. The pattern-evoked potential in lesions of the posterior visual pathways. Ann. N.Y. Acad. Sci. 388:264–289.
54. Blumhardt, L.D., and Halliday, A.M. 1979. Hemisphere contributions to the composition of the pattern-evoked potential waveform. Exp. Brain Res. 36:53–69.
55. Bobak, P., Bodis-Wollner, I., Harnois, C., Maffei, L., Mylin, L., Podos, S., and Thornton, J. 1983. Pattern electroretinograms and visual-evoked potentials in glaucoma and multiple sclerosis. Am. J. Ophthalmol. 96:72–83.
56. Bodis-Wollner, I. 1981. Differences in low and high spatial frequency vulnerabilities in ocular and cerebral lesions. Doc. Ophthalmol. Proc. Ser. 30:195–204.
57. Bodis-Wollner, I., Atkin, A., Raab, E., and Wolkstein, M. 1977. Visual association cortex and vision in man: Pattern-evoked occipital potentials in a blind boy. Science 198:629–631.
58. Bodis-Wollner, I., Barris, M.C., Mylin, L.H., Julesz, B., and Kropfl, W. 1981. Binocular stimulation reveals cortical components of the human visual evoked potential. Electroencephalogr. Clin. Neurophysiol. 52:298–305.
59. Bodis-Wollner, I., and Diamond, S.P. 1976. The measurement of spatial contrast sensitivity in cases of blurred vision associated with cerebral lesions. Brain 99:695–710.
60. Bodis-Wollner, I., and Hendley, C.D. 1977. Relation of evoked potentials to pattern and local luminance detectors in the human visual system. In: Visual evoked potentials in man: New developments, ed. J.E. Desmedt, pp. 197-207. Oxford: Clarendon Press.
61. Bodis-Wollner, I., Hendley, C.D., Mylin, L.H., and Thornton, J. 1979. Visual evoked potentials and the visuogram in multiple sclerosis. Ann. Neurol. 5:40–47.
62. Bodis-Wollner, I., and Yahr, M.D. 1978. Measurements of visual evoked potentials in Parkinson's disease. Brain 101:661–671.
63. Bornstein, Y. 1975. The pattern evoked responses (VER) in optic neuritis. Albrecht von Graefes Arch. Klin. Exp. Ophthalmol. 197:101–106.
64. Bøttcher, J., and Trojaborg, W. 1982. Follow-up of patients with suspected multiple sclerosis: A clinical and electrophysiological study. J. Neurol. Neurosurg. Psychiatry 45:809–814.
65. Bourne, J.R., Childers, D.G., and Perry, N.W. 1971. Topological characteristics of the visual evoked response in man. Electroencephalogr. Clin. Neurophysiol. 30:423–436.
66. Braddick, O., Atkinson, J., Julesz, B., Kropfl, W., Bodis-

Wollner, I., and Raab, E. 1980. Cortical binocularity in infants. Nature 288:363–365.
67. Brierley, J.B., Adams, J.H., Graham, D.I., and Simpson, J.A. 1971. Neocortical death after cardiac arrest. Lancet 2:560–565.
68. Brooks, B.A. 1977. Vision and visual evoked potentials during saccadic eye movements. In: Visual evoked potentials in man: New developments, ed. J.E. Desmedt, pp. 301–313. Oxford: Clarendon Press.
69. Broughton, R., Meier-Ewert, K.H., and Ebe, M. 1969. Evoked visual, somato-sensory and retinal potentials in photosensitive epilepsy. Electroencephalogr. Clin. Neurophysiol. 27:373–386.
70. Brown, J.L. 1979. Light and photometry. In Physiology of the human eye and visual system, ed. R.E. Records, pp. 328–350. Philadelphia: Harper and Row.
71. Bynke, H., Olsson, J.E., and Rosen, I. 1977. Diagnostic value of visual evoked response, clinical eye examination and CSF analysis in chronic myelopathy. Acta Neurol. Scand. 56:55–69.
72. Callaway, E., and Halliday, R.A. 1973. Evoked potential variability: Effects of age, amplitude and methods of measurement. Electroencephalogr. Clin. Neurophysiol. 34:125–133.
73. Camacho, L.M., Wenzel, W., and Aschoff, J. 1981. Clinical applications of visual evoked potentials for detection of chiasmal and postchiasmal lesions. Arch. Psychiatr. Nervenkr. 230:243–256.
74. Camisa, J., Mylin, L.H., and Bodis-Wollner, I. 1981. The effect of stimulus orientation on the visual evoked potential in multiple sclerosis. Ann. Neurol. 10:532–539.
75. Campbell, F.W., and Green, D.G. 1965. Optical and retinal factors affecting visual resolution. J. Physiol. (Lond.) 181:576–593.
76. Campbell, F.W., and Maffei, L. 1970. Electrophysiological evidence for the existence of orientation and size detectors in the human visual system. J. Physiol. (Lond.) 207:635–652.
77. Cannon, M.W. 1983. Contrast sensitivity: Psychophysical and evoked potential methods compared. Vision Res. 23:87–95.
78. Cannon, M.W. 1983. Evoked potential contrast sensitivity in the parafovea: Spatial organization. Vision Res. 23:1414–1449.
79. Cant, B.R., Hume, A.L., and Shaw, N.A. 1978. Effects of luminance on the pattern visual evoked potential in multiple sclerosis. Electroencephalogr. Clin. Neurophysiol. 45:496–504.
80. Cappin, J.M., and Nissim, S. 1975. Visual evoked responses in the assessment of field defects in glaucoma. Arch. Ophthalmol. 93:9–18.
81. Carlin, L., Roach, E.S., Riela, A., Spudis, E., and McLean, W.T. 1983. Juvenile metachromatic leukodystrophy: Evoked potentials and computed tomography. Ann. Neurol. 13:105–106.
82. Carroll, W.M., Jay, B.S., McDonald, W.I., and Halliday, A.M. 1980. Pattern evoked potentials in human albinism: Evidence of two different topographical asymmetries reflecting abnormal retino-cortical projections. J. Neurol. Sci. 48:265–286.
83. Carroll, W.M., and Mastaglia, F.L. 1979. Leber's optic neuropathy: A clinical and visual evoked potential study of affected and asymptomatic members of a six generation family. Brain 102:559–580.

84. Carroll, W.M., Kriss, A., Baraitser, M., Barrett, G., and Halliday, A.M. 1980. The incidence and nature of visual pathway involvement in Friedreich's ataxia: A clinical and visual evoked potential study of 22 patients. Brain 103:413–434.
85. Celesia, G.G. 1978. Visual evoked potentials in neurological disorders. Am. J. EEG Tech. 18:47–59.
86. Celesia, G.G. 1982. Clinical applications of evoked potentials. In: Electroencephalography: Basic principles, clinical applications and related fields, ed. E. Niedermeyer and F. Lopes da Silva, pp. 665–684. Baltimore: Urban and Schwarzenberg.
87. Celesia, G.A., Archer, C.R., Kuroiwa, Y., and Goldfader, P.R. 1980. Visual function of the extrageniculo-calcarine system in man: Relationship to cortical blindness. Arch. Neurol. 37:704–706.
88. Celesia, G.G., and Daly, R.F. 1977. Effects of aging on visual evoked responses. Arch. Neurol. 34:403–407.
89. Celesia, G.G., and Daly. R.F. 1977. Visual electroencephalographic computer analysis (VECA): A new electrophysiologic test for the diagnosis of optic nerve lesions. Neurology 27:637–641.
90. Celesia, G.G., Meredith, J.T., and Pluff, K. 1983. Perimetry, visual evoked potentials and visual evoked spectrum array in homonymous hemianopsia. Electroencephalogr. Clin. Neurophysiol. 56:16–30.
91. Celesia, G.G., Polcyn, R.D., Holden, J.E., Nickles, R.J., Gatley, J.S., and Koeppe, R.A. 1982. Visual evoked potentials and positron emission tomographic mapping of regional cerebral blood flow and cerebral metabolism: Can the neuronal potential generators be visualized? Electroencephalogr. Clin. Neurophysiol. 54:243–256.
92. Chain, F., Lesèvre, N., Leblanc, M., Rémond, A., and Lhermitte, F. 1972. Étude topographique des réponses évoquées visuelles dans un cas the lobectomie occipitale. Rev. Neurol. 126:372–378.
93. Chatrian, G.E., Lettich, E., Nelson, P.L., Miller, R.C., McKenzie, R.J., and Mills, R.P. 1980. Computer assisted quantitative electroretinography I: A standardized method. Am. J. EEG Tech. 20:57–77.
94. Cigánek, L. 1961. The EEG response (evoked potential) to light stimulus in man. Electroencephalogr. Clin. Neurophysiol. 13:165–172.
95. Cigánek, L. 1975. Section V. Visual evoked responses. In: Handbook of electroencephalography and clinical neurophysiology, vol. 8A, ed. A. Rémond, pp. 33–59. Amsterdam: Elsevier.
96. Clarke, P.G.H. 1973. Visual evoked potentials to changes in the motion of a patterned field. Exp. Brain Res. 18:145–155.
97. Clifford-Jones, R.E., Clarke, G.P., and Mayles, P. 1979. Crossed acoustic response combined with visual and somatosensory evoked responses in the diagnosis of multiple sclerosis. J. Neurol. Neurosurg. Psychiatry 42:749–752.
98. Cobb, W.A., and Dawson, G.D. 1960. The latency and form in man of the occipital potentials evoked by bright flashes. J. Physiol. (Lond.) 152:108–121.
99. Coben, L.A., Danziger, W.L., and Hughes, C.P. 1983. Visual evoked potentials in mild senile dementia of Alzheimer type. Electroencephalogr. Clin. Neurophysiol. 55:121–130.
100. Cohen, S.N., Syndulko, K., Rever, B., Kraut, J., Coburn, J., and Tourtellotte, W.W. 1983. Visual evoked potentials

and long latency event-related potentials in chronic renal failure. Neurology 33:1219–1222.

101. Cohen, S.N., Syndulko, K., Tourtellotte, W.W., and Potvin, A.R. 1980. Critical frequency of photic driving in the diagnosis of multiple sclerosis. A comparison to pattern evoked responses. Arch. Neurol. 37:80–83.

102. Collins, D.W.K., Carroll, W.M., Black, J.L., and Walsh, M. 1979. Effect of refractive errors on the visual evoked response. Br. Med. J. 1:231–232.

103. Confavreux, C., Mauguière, F., Courjon, G., Aimard, G., and Devic, M. 1981. Course of visual evoked potentials in multiple sclerosis: Electro-clinical correlations and pathophysiological considerations in 25 patients. Rev. Neurol. 137:121–132.

104. Connolly, J.F., and Gruzelier, J.H. 1982. Amplitude and latency changes in the visual evoked potential to different stimulus intensities. Psychophysiology 19:599–608.

105. Conrad, B., Benecke, R., Musers, H., Prange, H., and Behrens-Baumann, W. 1983. Visual evoked potentials in neurosyphilis. J. Neurol. Neurosurg. Psychiatry 46:23–27.

106. Contamin, F., and Cathala, H.P. 1961. Réponses électrocorticales de l'homme normal éveillé à des éclairs lumineux: Résultats obtenues à partir d'enregistrements sur le cuir chevelu, à l'aide d'un dispositif d'intégration. Electroencephalogr. Clin. Neurophysiol. 13:674–694.

107. Corletto, F., Gentilomo, A., Rosadini, G., Rossi, G.F., and Zattoni, J. 1967. Visual evoked potentials as recorded from the scalp and from the visual cortex before and after surgical removal of the occipital pole in man. Electroencephalogr. Clin. Neurophysiol. 22:378–380.

108. Cornu, L., Blanc-Garin, J. 1980. Visual evoked potentials and sensory dimensions. Electroencephalogr. Clin. Neurophysiol. 48:43–51.

109. Coupland, S.G., and Kirkham, T.K. 1982. Orientation-specific visual evoked potential deficits in multiple sclerosis. Can. J. Neurol. Sci. 9:331–337.

110. Cracco, R.Q., and Cracco, J.B. 1978. Visual evoked potential in man: Early oscillatory potentials. Electroencephalogr. Clin. Neurophysiol. 45:731–739.

111. Creel, D., Boxer, L.A., and Fauci, A.S. 1983. Visual and auditory anomalies in Chediak-Higashi syndrome. Electroencephalogr. Clin. Neurophysiol. 55:252–257.

112. Creel, D., and Buehler, B.A. 1982. Pattern evoked potentials in phenylketonuria. Electroencephalogr. Clin. Neurophysiol. 53:220–223.

113. Creel, D., Spekreijse, H., and Reits, D. 1981. Evoked potentials in albinos: Efficacy of pattern stimuli in detecting misrouted optic fibers. Electroencephalogr. Clin. Neurophysiol. 52:595–603.

114. Crevits, L., and van Lith, G.H.M. 1983. Component analysis of pattern evoked occipital potentials in hemianopic patients. Doc. Ophthalmol. 55:295–305.

115. Crews, S.J., Thompson, C.R., and Harding, G.F. 1978. The ERG and VEP in patients with severe eye injury. Doc. Ophthalmol. Proc. Ser. 15:203–209.

116. Cummings, J.L., Syndulko, K., Goldberg, Z., and Trieman, D.M. 1982. Palinopsia reconsidered. Neurology 32:444–447.

117. D'Allest, A.M., Laget, P., and Raimbault, J. 1982. Visual and somesthetic potentials in neurolipidosis. Adv. Neurol. 32:397–407.

118. Darcey, T.M., Ary, J.P., and Fender, D.H. 1980. Spatio-

temporal visually evoked scalp potentials in response to partial-field patterned stimulation. Electroencephalogr. Clin. Neurophysiol. 50:348–355.
119. Dawson, W.W., and Barris, M.C. 1978. Cortical responses evoked by laser speckle. Invest. Ophthalmol. Visual Sci. 17:1209–1212.
120. Dawson, W.W., Maida, T.M., and Rubin, M.L. 1982. Human pattern-evoked retinal responses are altered by optic atrophy. Invest. Ophthalmol. Visual Sci. 22:796–803.
121. DeVoe, R.G., Ripps, H., and Vaughan, H.G. 1968. Cortical responses to stimulation of the human fovea. Vision Res. 8:135–147.
122. Diener, H.C., Koch, W., and Dichgans, J. 1982. The significance of luminance on visual evoked potentials in diagnosis of MS. Arch. Psychiatr. Nervenkr. 231:149–154.
123. Diener, H.C., and Scheibler, H. 1980. Follow-up studies of visual potentials in multiple sclerosis evoked by checkerboard and foveal stimulation. Electroencephalogr. Clin. Neurophysiol. 49:490–496.
124. Domino, E.F., Demetriou, S., Tuttle, T., and Klinge, V. 1979. Comparison of the visually evoked response in drug-free chronic schizophrenic patients and normal controls. Electroencephalogr. Clin. Neurophysiol. 46:123–127.
125. Donnan, G.A., Sharbrough, F.W., and Whisnant, J.P. 1982. Carotid occlusive disease. Effect of bright light on visual evoked response. Arch. Neurol. 39:687–689.
126. Dorfman, L.J., Nikoskelainen, E., Rosenthal, A.R., and Sogg, R.L. 1977. Visual evoked potentials in Leber's hereditary optic neuropathy. Ann. Neurol. 1:565–568.
127. Dorfman, L.J., Pedley, T.A., Tharp, B.A., and Scheithauer, B.W. 1978. Juvenile neuraxonal dystrophy: Clinical, electrophysiological, and neuropathological features. Ann. Neurol. 3:419–428.
128. Duchowny, M.S., Weiss, I.P., Majlessi, H., and Barnet, A.B. 1974. Visual evoked responses in childhood cortical blindness after head trauma and meningitis. Neurology 24:933–940.
129. Duffy, F.H., and Rengstorff, R.H. 1971. Ametropia measurements from the visual evoked response. Am. J. Optom. Arch. Am. Acad. Optom. 48:717–728.
130. Dustman, R.E., and Beck, E.C. 1969. The effects of maturation and aging on the wave form of visually evoked potentials. Electroencephalogr. Clin. Neurophysiol. 26:2–11.
131. Dustman, R.E., Schenkenberg, T., and Beck, E.C. 1976. The development of the evoked response as a diagnostic and evaluative procedure. In: Developmental psychophysiology of mental retardation, ed. R. Karrer, pp. 247–310. Springfield, Ill.: Thomas.
132. Dustman, R.E., Shearer, D.E., and Snyder, E.W. 1982. Age differences in augmenting/reducing of occipital visually evoked potentials. Electroencephalogr. Clin. Neurophysiol. 54:99–110.
133. Dustman, R.E., and Snyder, E.W. 1981. Life-span changes in visually evoked potentials at central scalp. Neurobiol. Aging 2:303–308.
134. Duwaer, A.L., and Spekreijse, H. 1978. Latency of luminance and contrast evoked potentials in multiple sclerosis patients. Electroencephalogr. Clin. Neurophysiol. 45:244–258.
135. Ebe, M., Mikami, T., and Ito, H. 1964. Clinical evaluation of electrical responses of retina and visual cortex in photic

stimulation in ophthalmic diseases. Tohoku J. Exp. Med. 84:92–103.
136. Ehle, A.L., Steward, R.M., Lellelid, N.E., and Leventhal, N.A. 1982. Normal checkerboard pattern reversal evoked potentials in Parkinsonism. Electroencephalogr. Clin. Neurophysiol. 54:336–338.
137. Ehle, A.L., Steward, R.M., Lellelid, N.A., and Leventhal, N.A. 1984. Evoked potentials in Huntington's disease: A comparative and longitudinal study. Arch. Neurol. 41:379–382.
138. Ellenberger, C., Petro, D.J., and Ziegler, S.B. 1978. The visually evoked potential in Huntington disease. Neurology 28:95–97.
139. Ellenberger, C., and Shuttlesworth, D.E. 1978. Electrical correlates of normal binocular vision. Arch Neurol. 35:834–837.
140. Ellenberger, C., and Ziegler, S.B. 1977. Visual evoked potentials and quantitative perimetry in multiple sclerosis. Ann. Neurol. 1:561–564.
141. Ellingson, R.J. 1960. Cortical electrical responses to visual stimulation in the human infant. Electroencephalogr. Clin. Neurophysiol. 12:663–677.
142. Ellingson, R.J., Lathrop, G.H., Danahy, T., and Nelson, B. 1973. Variability of visual evoked potentials in human infants and adults. Electroencephalogr. Clin. Neurophysiol. 34:113–124.
143. Epstein, C.M. 1979. True checkerboard pattern reversal with light-emitting diodes. Electroencephalogr. Clin. Neurophysiol. 47:611–613.
144. Epstein, C.M., Gammon, J.A., Gemmill, M., and Till, J. 1983. Visual evoked potential pattern generation, recording, and data analysis with a single microcomputer. Electroencephalogr. Clin. Neurophysiol. 56:691–693.
145. Erwin, C.W. 1980. Pattern reversal evoked potentials. Am. J. EEG Tech. 20:161–184.
146. Estévez, O., and Spekreijse, H. 1974. Relationship between pattern appearance-disappearance and pattern reversal responses. Brain Res. 19:233–238.
147. Evans, B.T. Binnie, C.D., and Lloyd, D.S.L. 1974. A simple visual pattern stimulator. Electroencephalogr. Clin. Neurophysiol. 37:403–406.
148. Faught, E., and Lee, S.I. 1984. Pattern-reversal visual evoked potentials in photosensitive epilepsy. Electroencephalogr. Clin. Neurophysiol. 59:125–133.
149. Faust, U., Heintel, H., and Hoek, R. 1978. Age dependence of the P2 latency of visually evoked cortical responses to checkerboard pattern reversal. Z. EEG-EMG 9:219–221.
150. Feinsod, M., and Hoyt, W.F. 1975. Subclinical optic neuropathy in multiple sclerosis. J. Neurol. Neurosurg. Psychiatry 38:1109–1114.
151. Feinsod, M., Hoyt, W.F., Wilson, W.B., and Spire, J.P. 1976. Visually evoked response. Use in neurologic evaluation of posttraumatic subjective visual complaints. Arch. Ophthalmol. 94:237–240.
152. Feinsod, M., Madey, J.M.J., and Susal, A.L. 1975. A new photostimulator for continuous recording of the visual evoked potential. Electroencephalogr. Clin. Neurophysiol. 38:641–642.
153. Feinsod, M., Selhorst, J.B., Hoyt, W.F., and Wilson, C.B. 1976. Monitoring optic nerve function during craniotomy. J. Neurosurg. 44:29–31.
154. Fenwick, P.B.C., Brown, D., and Hennesey, J. 1981. The

visual evoked response to pattern reversal in normal 6–11-year-old children. Electroencephalogr. Clin. Neurophysiol. 51:49–62.
155. Fine, E.J. and Hallett, M. 1980. Neurophysiological study of subacute combined degeneration. J. Neurol. Sci. 45:331–336.
156. Fiorentini, A., and Maffei, L. 1970. Electrophysiological evidence for binocular disparity detectors in human visual system. Science 169:208–209.
157. Fiorentini, A., and Maffei, L. 1973. Evoked potentials in astigmatic subjects. Vision Res. 13:1781–1783.
158. Fiorentini, A., Maffei, L., Pirchio, M., Spinelli, D., and Porciatti, V. 1981. The ERG in response to alternating gratings in patients with diseases of the peripheral visual pathway. Invest. Ophthalmol. Visual Sci. 21:490–493.
159. Fiorentini, A., Pirchio, M., and Spinelli, D. 1983. Development of retinal and cortical responses to pattern reversal in infants: A selective review. Behav. Brain Res. 10:99–106.
160. Fox, R., Blake, R., and Bourne, J.R. 1973. Visual evoked cortical potentials during pressure-blinding. Vision Res. 13:501–503.
161. Frank, Y., and Torres, F. 1979. Visual evoked potentials in the evaluation of "cortical blindness" in children. Ann. Neurol. 6:126–129.
162. Freeman, R.D., and Thibos, L.N. 1973. Electrophysiological evidence that abnormally early visual experience can modify the human brain. Science 180:876–878.
163. Freeman, R.D., and Thibos, L.N. 1975. Visual evoked responses in humans with abnormal visual experience. J. Physiol. (Lond.) 247:711–724.
164. Friedman, E., Harden, A., Koivikko, M., and Pampiglione, G. 1978. Menke's disease: Neurophysiological aspects. J. Neurol. Neurosurg. Psychiatry 41:505–510.
165. Fritsch, K., and Keck, M.J. 1978. Grating generation by microcomputer. Vision Res. 18:1083–1085.
166. Fukuhara, J., Uozato, H., Nojima, S., Saishin, M., and Nakao, S. 1983. Visual-evoked potentials elicited by laser speckle patterns: Studies on the characteristics and assessment of visual function in patients with ocular media opacities. Invest. Ophthalmol. Visual Sci. 24:1400–1407.
167. Gambi, D., Rossini, P.M., Albertini, G., Sollazzo, D., Torrioli, M.G., and Polidori, G.C. 1980. Follow-up of visual evoked potential in full-term and pre-term control newborns and in subjects who suffered from perinatal respiratory distress. Electroencephalogr. Clin. Neurophysiol. 48:509–516.
168. Gangitano, J.L., and Keltner, J.L. 1980. Abnormalities of the pupil and visual-evoked potential in quinine amblyopia. Am. J. Ophthalmol. 89:425–430.
169. Gastaut, H., Beek, E., Faidherbe, J., Franck, G., Fressy, J., Rémond, A., Smith, C., and Werre, P. 1963. A transcranial chronographic and topographic study of cerebral potentials evoked by photic stimulation in man. Prog. Brain Res. 1:374–394.
170. Gawel, M.J., Das, P., Vincent, S., and Rose, F.C. 1981. Visual and auditory evoked responses in patients with Parkinson's disease. J. Neurol. Neurosurg. Psychiatry 44:227–232.
171. Gibson, J.M., and Kennedy, P. 1984. Does testing of brainstem auditory response aid diagnosis in progressive spastic paraparesis? Acta Neurol. Scand. 69:107–111.
172. Gittinger, J.W., and Sokol, S. 1982. The visual-evoked

potential in the diagnosis of congenital ocular motor apraxia. Am. J. Ophthalmol. 93:700–703.
173. Gott, P.S., Karnaze, D.S., and Keane, J.R. 1983. Abnormal visual evoked potentials in myotonic dystrophy. Neurology 33:1622–1625.
174. Gott, P.S., Weiss, M.H., Apuzzo, M., and Van der Meulen, J.P. 1979. Checkerboard visual evoked response in evaluation and management of pituitary tumors. Neurosurgery 5:553–558.
175. Gouras, P. 1970. Electroretinography: Some basic principles. Invest. Ophthalmol. 9:557–569.
176. Green, J.B., Price, R., and Woodbury, S.G. 1980. Short-latency somatosensory evoked potentials in multiple sclerosis: Comparison with auditory and visual evoked potentials. Arch. Neurol. 37:630–633.
177. Green, J.B., and Walcoff, M.R. 1982. Evoked potentials in multiple sclerosis. Arch. Neurol. 39:696–697.
178. Greenberg, R.P., Becker, D.P., Miller, J.D., Mayer, D.J. 1977. Evaluation of brain function in severe human head trauma with multimodality evoked potentials. Part 2. Localization of brain dysfunction and correlation with post-traumatic neurological conditions. J. Neurosurg. 47:163–177.
179. Greenberg, R.P., Mayer, D.J., Becker, D.P., and Miller, J.D. 1977. Evaluation of brain function in severe human head trauma with multimodality evoked potentials. Part 1. Evoked brain-injury potentials, methods, and analysis. J. Neurosurg. 47:150–162.
180. Greenberg, R.P., Newlon, P.G., Hyatt, M.S., Narayan, R.K., and Becker, D.P. 1981. Prognostic implications of early multimodality evoked potentials in severely head-injured patients: A prospective study. J. Neurosurg. 55:227–236.
181. Gulmann, N.C., Hammerberg, P.E., Busch Jensen, L., Sommerbeck, K.W., and Orbaek, K. 1979. Visual evoked potential in patients with cerebral asthenopia. Acta Neurol. Scand. 59:324–330.
182. Guzman, O., and Steinbach, M.J. 1981. Manipulating the contrast of sine wave gratings and other light distributions using polaroid. Vision Res. 21:1025–1028.
183. Haimovic, I.C., and Pedley, T.A. 1982. Hemi-field pattern reversal visual evoked potentials. I. Normal subjects. Electroencephalogr. Clin. Neurophysiol. 54:111–120.
184. Haimovic, I.C., and Pedley, T.A. 1982. Hemi-field pattern reversal visual evoked potentials. II. Lesions of the chiasm and posterior visual pathways. Electroencephalogr. Clin. Neurophysiol. 54:121–131.
185. Hakamada, S., Watanabe, K., Hara, K., and Miyazaki, S. 1981. The evolution of visual and auditory evoked potentials in infants with perinatal disorder. Brain Dev. 3:339–344.
186. Halliday, A.M. 1978. Clinical applications of evoked potentials. In: Recent advances in clinical neurology, vol. 2. ed. W.B. Matthews and G.H. Glaser, pp. 47–73. New York: Churchill Livingstone.
187. Halliday, A.M. 1980. Event-related potentials and their diagnostic usefulness. Prog. Brain Res. 54:469–485.
188. Halliday, A.M., Halliday, E., Kriss, A., McDonald, W.I., and Mushin, J. 1976. The pattern-evoked potential in compression of the anterior visual pathways. Brain 99:357–374.
189. Halliday, A.M., and McDonald, W.I. 1977. Pathophysiology of demyelinating disease. Br. Med. Bull. 33:21–27.

190. Halliday, A.M. and McDonald, W.I. 1981. Visual evoked potentials. In: Clinical neurophysiology, neurology 1, Butterworths International Medical Reviews, ed. E. Stalberg and R.R. Young, pp. 228–258. London: Butterworths.
191. Halliday, A.M., McDonald, W.I., and Mushin, J. 1972. Delayed visual evoked response in optic neuritis. Lancet 1:982–985.
192. Halliday, A.M., McDonald, W.I., and Mushin, J. 1973. Visual evoked response in diagnosis of multiple sclerosis. Br. Med. J. 4:661–664.
193. Halliday, A.M., McDonald, W.I., and Mushin, J. 1973. Delayed pattern-evoked responses in optic neuritis in relation to visual acuity. Trans. Ophthalmol. Soc. U.K. 93:315–324.
194. Halliday, A.M., McDonald, W.I., and Mushin, J. 1977. Visual evoked potentials in patients with demyelinating disease. In: Visual evoked potentials in man: New developments, ed. J.E. Desmedt, pp. 439–449. Oxford: Clarendon Press.
195. Halliday, A.M., and Mushin, J. 1980. The visual evoked potential in neuroophthalmology. Int. Ophthalmol. Clin. 20:155–183.
196. Hamel, B., Bourne, J.R., Ward, J.W., and Teschan, P.E. 1978. Visually evoked cortical potentials in renal failure: Transient potentials. Electroencephalogr. Clin. Neurophysiol. 44:606–616.
197. Hammond, E.J., and Wilder, B.J. 1983. Evoked potentials in olivopontocerebellar atrophy. Arch. Neurol. 40:366–369.
198. Hansch, E.C., Syndulko, K., Cohen, S.N., Goldberg, Z.I., Potvin, A.R., and Tourtellotte, W.W. 1982. Cognition in Parkinson disease: An event-related potential perspective. Ann. Neurol. 11:599–607.
199. Happel, L.T., Rothschild, H., and Garcia, C. 1980. Visual evoked potentials in two forms of hereditary spastic paraplegia. Electroencephalogr. Clin. Neurophysiol. 48:233–236.
200. Harden, A., and Pampiglione, P. 1977. Visual evoked potential, electroretinogram, and electroencephalogram studies in progressive neurometabolic "storage" diseases of childhood. In: Visual evoked potentials in man: New developments, ed. J.E. Desmedt, pp. 470–480. Oxford: Clarendon Press.
201. Harden, A., Picton-Robinson, N., Bradshaw, K., and Pampiglione, G. 1980. Ten years' experience of ERG/VEP/EEG studies on visual disorders in paediatrics. In: Evoked potentials: Proceedings of an international evoked potentials symposium held in Nottingham, England, ed. C. Barber, pp. 257–266. Lancaster: MTP Press.
202. Harding, G.F.A., Crews, S.J., and Good, P.A. 1980. VEP in neuro-ophthalmic disease. In: Evoked potentials: Proceedings of an international evoked potentials symposium held in Nottingham, England, ed. C. Barber, pp. 235–241. Lancaster, MTP Press.
203. Harding, G.F.A., and Dimitrakoudi, M. 1977. The visual evoked potential in photosensitive epilepsy. In: Visual evoked potentials in man: New developments, ed. J.E. Desmedt, pp. 509–513. Oxford: Clarendon Press.
204. Harmony, T., Ricardo, J., Otero, G., Fernandez, G., Llorente, S., and Valdes, P. 1973. Symmetry of the visual evoked potential in normal subjects. Electroencephalogr. Clin. Neurophysiol. 35:237–240.
205. Harris, L., Atkinson, J., and Braddick, O. 1976. Visual contrast sensitivity of a 6-month-old infant measured by the evoked potential. Nature 264:570–571.

206. Harter, M.R. 1970. Evoked cortical responses to checkerboard patterns: Effect of check-size as a function of retinal eccentricity. Vision Res. 10:1365–1376.
207. Harter, M.R. 1971. Visually evoked cortical responses to the on- and off-set of patterned light in humans. Vision Res. 11:685–695.
208. Harter, M.R., Deaton, F.K., and Odom, J.V. 1977. Maturation of evoked potentials and visual preference in 6–45 day-old infants: Effects of check size, visual acuity, and refractive error. Electroencephalogr. Clin. Neurophysiol. 42:595–607.
209. Harter, M.R., Deaton, F.K., and Odom, J.V. 1977. Pattern visual evoked potentials in infants. In: Visual evoked potentials in man: New developments, ed. J.E. Desmedt, pp. 332–352. Oxford: Clarendon Press.
210. Harter, M.R., Towle, V.L., Zakrzewski, M., and Moyer, S.M. 1977. An objective indicant of binocular vision in humans: Size-specific interocular suppression of visual evoked potentials. Electroencephalogr. Clin. Neurophysiol. 43:825–836.
211. Harter, M.R., and White, C.T. 1968. Effects of contour sharpness and check-size on visually evoked cortical potentials. Vision Res. 8:701–711.
212. Harter, M.R., and White, C.T. 1970. Evoked cortical responses to checkerboard patterns: Effect of check-size as a function of visual acuity. Electroencephalogr. Clin. Neurophysiol. 28:48–54.
213. Hawkes, C.H., and Stow, B. 1981. Pupil size and the pattern evoked visual response. J. Neurol. Neurosurg. Psychiatry 44:90–91.
214. Hayward, M., and Mills, I.M. 1980. Design effects of video pattern generators on the visual evoked potential. In: Evoked Potentials: Proceedings of an international evoked potentials symposium held in Nottingham, England, ed. C. Barber, pp. 87–92. Lancaster: MTP Press.
215. Heintel, H., Faust, U., and Faust, C. 1979. Age dependence of the P2 amplitudes of visually evoked cortical responses to checkerboard pattern reversal. Z. EEG-EMG 10:194–196.
216. Hennerici, M., Wenzel, D., and Freund, H.J. 1977. The comparison of small-size rectangle and checkerboard stimulation for the evaluation of delayed visual evoked responses in patients suspected of multiple sclerosis. Brain 100:119–136.
217. Herpers, M.J., Caberg, H.B., and Mol, J.M.F. 1981. Human cerebral potentials evoked by moving dynamic random dot stereograms. Electroencephalogr. Clin. Neurophysiol. 52:50–56.
218. Hess, C.W., Meienberg, O., and Ludin, H.P. 1982. Visual evoked potentials in acute occipital blindness: Diagnostic and prognostic value. J. Neurol. 227:193–200.
219. Hess, R.F. 1980. A preliminary investigation of neural function and dysfunction in amblyopia. I. Size-selective channels. Vision Res. 20:749–754.
220. Hess, R.F., and Plant, G.T. 1983. The effect of temporal frequency variation on threshold contrast sensitivity deficits in optic neuritis. J. Neurol. Neurosurg. Psychiatry 46:322–330.
221. Hishikawa, Y., Yamamato, J., Furuya, E., Yamada, Y., Miyazaki, K., and Kaneko, Z. 1967. Photosensitive epilepsy: Relationships between the visual evoked responses and the epileptiform discharges induced by intermittent photic stimulation. Electroencephalogr. Clin. Neurophysiol. 23:320–334.

222. Hitchings, R.A., Powell, D.J., Arden, G.B., and Carter, R.M. 1981. Contrast sensitivity gratings in glaucoma family screening. Br. J. Ophthalmol. 65:515–517.
223. Hoeppner, T.J. 1980. Binocular interaction in the visual evoked response: Temporal factors. J. Neurol. Sci. 47:49–58.
224. Holder, G.E. 1978. The effects of chiasmal compression on the pattern visual evoked potential. Electroencephalogr. Clin. Neurophysiol. 45:278–280.
225. Holder, G.E. 1980. Abnormalities of the pattern visual evoked potential in patients with homonymous visual field defects. In: Evoked potentials: Proceedings of an international evoked potentials symposium held in Nottingham, England, ed. C. Barber, pp. 285–291. Lancaster: MTP Press.
226. Howe, J.W., and Mitchell, K.W. 1980. Visual evoked potentials from quadrantic field stimulation in the investigation of homonymous field defects. In: Evoked potentials: Proceedings of an international evoked potentials symposium held in Nottingham, England, ed. C. Barber, pp. 279–283. Lancaster: MTP Press.
227. Hrbek, A., Fällström, S.P., Karlberg, P., and Olsson, T. 1982. Clinical application of evoked EEG responses in infants. III. Congenital hypothroidism. Dev. Med. Child Neurol. 24:264–172.
228. Hrbek, A., Karlberg, P., Kjellmer, I., Olsson, T., and Riha, M. 1977. Clinical application of evoked electroencephalographic responses in newborn infants. I. Perinatal asphyxia. Dev. Med. Child Neurol. 19:34–44.
229. Hrbek, A., Karlberg, P., and Olsson, T. 1973. Development of visual and somatosensory evoked responses in preterm newborn infants. Electroencephalogr. Clin. Neurophysiol. 34:225–232.
230. Huber, C. 1981. Pattern evoked cortical potentials and automated perimetry in chronic glaucoma. Doc. Ophthalmol. Proc. Ser. 27:87–94.
231. Ikeda, H., Tremain, K.E., and Sanders, M.D. 1978. Neurophysiological investigation in optic nerve disease: Combined assessment of the visual evoked response and electroretinogram. Br. J. Ophthalmol. 62:227–239.
232. Jacobson, J.H., Hirose, T., and Suzuki, T.A. 1968. Simultaneous ERG and VER in lesions of the optic pathway. Invest. Ophthalmol. 6:279–292.
233. Jeffreys, D.A. 1980. The nature of pattern VEPs. In: Evoked potentials: Proceedings of an international evoked potentials symposium held in Nottingham, England, ed. C. Barber, pp. 149–157. Lancaster: MTP Press.
234. Jeffreys, D.A., and Axford, J.G. 1972. Source locations of pattern-specific components of human visual evoked potentials. I. Component of striate cortical origin. Exp. Brain Res. 16:1–21.
235. Jeffreys, D.A., and Axford, J.G. 1972. Source locations of pattern-specific components of human visual evoked potentials. II. Component of extrastriate cortical origin. Exp. Brain Res. 16:22–40.
236. John, E.R., Herrington, R.N., and Sutton, S. 1967. Effects of visual form on the evoked response. Science 155:1439–1442.
237. Johnson, F., Hayward, M., and Walsh, S. 1976. Chequerboard pattern reversal using a modified television display to measure visual evoked responses. Biomed. Eng. 11:57–59.
238. Jones, R., and Keck, M.J. 1978. Visual evoked response

as a function of grating spatial frequency. Invest. Ophthalmol. Visual Sci. 17:652–659.
239. Julesz, B., and Kropfl, W. 1982. Binocular neurons and cyclopean visually evoked potentials in monkey and man. Ann. N.Y. Acad. Sci. 388:37–44.
240. Julesz, B., Kropfl, W., and Petrig, B. 1980. Large evoked potentials to dynamic random-dot correlograms and stereograms permit quick determination of stereopsis. Proc. Natl. Acad. Sci. USA 77:2348–2351.
241. Kadobayashi, I., Nakamura, M., and Kato, N. 1977. Changes in visual evoked potentials of schizophrenics after addition test. Electroencephalogr. Clin. Neurophysiol. 43:837–845.
242. Karmel, B.Z., Hoffmann, R.F., and Fegy, M.J. 1974. Processing of contour information by human infants evidenced by pattern-dependent evoked potentials. Child Dev. 45:39–48.
243. Kato, S., Saito, M., and Tanino, T. 1983. Response of the visual system evoked by an alternating current. Med. Biol. Eng. Comput. 21:47–50.
244. Kelly, D.H. 1976. Pattern detection and the two-dimensional Fourier transform: Flickering checkerboards and chromatic mechanisms. Vision Res. 16:277–287.
245. Kinney, J.A.S., and McKay, C.L. 1974. Test of color-defective vision using the visual evoked response. J. Opt. Soc. Am. 64:1244–1250.
246. Kinney, J.A.S., McKay, C.L., Mensch, A.J., and Luria, S.M. 1973. Visual evoked responses elicited by rapid stimulation. Electroencephalogr. Clin. Neurophysiol. 34:7–13.
247. Kirkham, T.H., and Coupland, S.G. 1981. Abnormal pattern electroretinograms with macular cherry-red spots: Evidence for selective ganglion cell damage. Curr. Eye Res. 1:367–372.
248. Kirkham, T.H., and Coupland, S.G. 1981. Abnormal electroretinograms and visual evoked potentials in chronic papilledema using time-difference analysis. Can. J. Neurol. Sci. 8:243–248.
249. Kirkham, T.H., and Coupland, S.G. 1981. An electroretinal and visual evoked potential study in Friedreich's ataxia. Can. J. Neurol. Sci. 8:289–294.
250. Kirkham, T.H., and Coupland, S.G. 1983. The pattern electroretinogram in optic nerve demyelination. Can. J. Neurol. Sci. 10:256–260.
251. Kjaer, M. 1980. Visual evoked potentials in normal subjects and patients with multiple sclerosis. Acta Neurol. Scand. 62:1–13.
252. Kjaer, M. 1980. Brain stem auditory and visual evoked potentials in multiple sclerosis. Acta Neurol. Scand. 62:14–19.
253. Kjaer, M. 1983. Evoked potentials: With special reference to the diagnostic value in multiple sclerosis. Acta Neurol. Scand. 67:67–89.
254. Klemm, W.R., Goodson, R.A., and Allen, R.G. 1983. Contrast effects of the three primary colors on human visual evoked potentials. Electroencephalogr. Clin. Neurophysiol. 55:557–566.
255. Klingaman, R.L. 1976. The human visual evoked cortical potential and dark adaptation. Vision Res. 16:1471–1477.
256. Klingaman, R.L., and Moskowitz-Cook, A. 1979. Assessment of the visual acuity of human color mechanisms with the visually evoked cortical potential. Invest. Ophthalmol. Visual Sci. 18:1273–1277.
257. Klorman, R., Salzman, L.F., Bauer, L.O., Coons, H.W.,

Borgstedt, A.D., and Halpern, W.I. 1983. Effects of two doses of methylphenidate on cross-situational and borderline hyperactive children's evoked potentials. Electroencephalogr. Clin. Neurophysiol. 56:169–185.

258. Kobayashi, H., and Toyomura, K. 1981. Regional relationship of visual evoked potentials in infants and school-aged children. Electroencephalogr. Clin. Neurophysiol. 52:36–41.

259. Kooi, K.A., and Bagchi, B.K. 1964. Visual evoked responses in man: Normative data. Ann. N.Y. Acad. Sci. 112:254–269.

260. Kooi, K.A., Güvener, A.M., and Bagchi, B.K. 1965. Visual evoked responses in lesions of the higher optic pathways. Neurology 15:841–854.

261. Kooi, K.A., and Marshall, R.E. 1979. Visual evoked potentials in the central disorders of the visual system. Philadelphia: Harper and Row.

262. Kooi, K.A., and Sharbrough, F.W. 1966. Electrophysiological findings in cortical blindness: Report of a case. Electroencephalogr. Clin. Neurophysiol. 20:260–263.

263. Korth, M. 1983. Pattern-evoked responses and luminance-evoked responses in the human electroretinogram. J. Physiol. (Lond.) 337:451–469.

264. Korth, M., and Armington, J.C. 1976. Stimulus alternation and the Purkinje shift. Vision Res. 16:703–711.

265. Koshbin, S., and Hallett, M. 1981. Multimodality evoked potentials and blink reflex in multiple sclerosis. Neurology 31:138–144.

266. Krauskopf, J. 1973. Contributions of the primary chromatic mechanisms to the generation of visual evoked potentials. Vision Res. 13:2289–2298.

267. Kriss, A., Carroll, W.M., Blumhardt, L.D., and Halliday, A.M. 1982. Pattern- and flash-evoked potential changes in toxic (nutritional) optic neuropathy. Adv. Neurol. 32:11–19.

268. Kriss, A., and Halliday, A.M. 1980. A comparison of occipital potentials evoked by pattern onset, offset and reversal by movement. In: Evoked potentials: Proceedings of an international evoked potentials symposium held in Nottingham, England, ed. C. Barber, pp. 205–212. Lancaster: MTP Press.

269. Kriss, A., Halliday, A.M., Halliday, E., and Pratt, R.T.C. 1980. Evoked potentials following unilateral ECT. II. The flash evoked potential. Electroencephalogr. Clin. Neurophysiol. 48:490–501.

270. Krumholz, A., Singer, H.S., Niedermeyer, E., Burnite, R., and Harris, K. 1983. Electrophysiological studies in Tourette's syndrome. Ann. Neurol. 14:638–641.

271. Krumholz, A., Weiss, H.D., Goldstein, P.J., and Harris, K.C. 1981. Evoked responses in vitamin B12 deficiency. Ann. Neurol. 9:407–409.

272. Kuba, M., Peregrin, J., Vít, F., Hanušová, I., and Erben, J. 1983. Pattern-reversal visual evoked potentials in patients with chronic renal insufficiency. Electroencephalogr. Clin. Neurophysiol. 56:438–442.

273. Kulikowski, J.J., and Leisman, G. 1973. The effect of nitrous oxide on the relation between the evoked potential and contrast threshold. Vision Res. 13:2079–2086.

274. Kupersmith, M.J., Seiple, W.H., Carr, R.E., and Weiss, P.A. 1983. The 20/20 eye in multiple sclerosis. Neurology 33:1015–1020.

275. Kupersmith, M.J., Shakin, E., Siegel, I.M., and Lieberman, A. 1982. Visual system abnormalities in patients with Parkinson's disease. Arch. Neurol. 39:284–286.

276. Kupersmith, M.J., Siegel, I.M., and Carr, R.E. 1981. Reduced contrast sensitivity in compressive lesions of the anterior visual pathway. Neurology 31:550–554.
277. Kupersmith, M.J., Siegel, I.M., Carr, R.E., Ransohoff, J., Flamm, E., and Shakin, E. 1981. Visual evoked potentials in chiasmal gliomas in four adults. Arch Neurol. 38:362–365.
278. Kupersmith, M.J., Weiss, P.A., and Carr, R.E. 1983. The visual-evoked potential in tobacco-alcohol and nutritional amblyopia. Am. J. Ophthalmol. 95:307–314.
279. Kuroiwa, Y., and Celesia, G.G. 1981. Visual evoked potentials with hemifield pattern stimulation: Their use in the diagnosis of retrochiasmatic lesions. Arch. Neurol. 38:86–90.
280. Kurtzberg, D., and Vaughan, H.G. 1977. Electrophysiological observations on the visuomotor system and visual neurosensorium. In: Visual evoked potentials in man: New developments, ed. J.E. Desmedt, pp. 314–331. Oxford: Clarendon Press.
281. Laget, P., Flores-Guevara, R., d'Allest, A.M., Ostre, C., Raimbault, J., and Mariani, J. 1977. La maturation des potentiels évoqués visuels chez l'enfent normal. Electroencephalogr. Clin. Neurophysiol. 43:732–744.
282. Lee, R.G., and Blair, R.D.G. 1973. Evolution of EEG and visual evoked response changes in Jakob-Creutzfeldt disease. Electroencephalogr. Clin. Neurophysiol. 35:133–142.
283. Lee, S.I., Messenheimer, J.A., Wilkinson, E.C., Brickley, J.J., and Johnson, R.N. 1980. Visual evoked potentials to stimulus trains: Normative data and application to photosensitive seizures. Electroencephalogr. Clin. Neurophysiol. 48:387–394.
284. Lehmann, D., and Julesz, B. 1978. Lateralized cortical potentials evoked in humans by dynamic random-dot stereograms. Vision Res. 18:1265–1271.
285. Lehmann, D., and Skrandies, W. 1979. Multichannel mapping of scalp potential fields evoked by checkerboard reversal to different retinal areas. In: Event-related potentials in man: Applications and problems, ed. D. Lehmann, and E. Callaway, pp. 201–214. New York: Plenum Press.
286. Lehmann, D., and Skrandies, W. 1979. Multichannel evoked potential fields show different properties of human upper and lower hemiretina systems. Exp. Brain Res. 35:151–159.
287. Lesèvre, N., and Joseph, J.P. 1979. Modifications of the pattern-evoked potential (PEP) in relation to the stimulated part of the visual field (clues for the most probable origin of each component). Electroencephalogr. Clin. Neurophysiol. 47:183–203.
288. Lesèvre, N., and Rémond, A. 1972. Potentiels évoqués par l'apparition de patterns: effects de la dimension du pattern et de la densité des contrastes. Electroencephalogr. Clin. Neurophysiol. 32:593–604.
289. Levi, D.M., and Harwerth, R.S. 1978. Contrast evoked potentials in strabismic and anisometropic amblyopia. Invest. Ophthalmol. Visual Sci. 17:571–575.
290. Lewis, E.G., Dustman, R.E., and Beck, E.C. 1970. The effects of alcohol on visual and somato-sensory evoked responses. Electroencephalogr. Clin. Neurophysiol. 28:202–205.
291. Lewis, E.G., Dustman, R.E., and Beck, E.C. 1978. Visual and somatosensory evoked potential characteristics of patients undergoing hemodialysis and kidney transplantation. Electroencephalogr. Clin. Neurophysiol. 44:223–231.
292. Lindsay, K.W., Karlin, J., Kennedy, I., Fry, J., McInnes, A., and Teasdale, G.M. 1981. Evoked potentials in severe

head injury: Analysis and relation to outcome. J. Neurol. Neurosurg. Psychiatry 44:796–802.
293. Livingstone, I.R., Mastaglia, F.L., Edis, R., and Howe, J.W. 1981. Visual involvement in Friedreich's ataxia and hereditary spastic ataxia: A clinical and visual evoked potential study. Arch. Neurol. 38:75–79.
294. Livingstone, I.R., Mastaglia, F.L., Howe, J.W., and Aherne, G.E.S. 1980. Leber's optic neuropathy: Clinical and visual evoked response studies in asymptomatic and symptomatic members of a 4-generation family. Br. J. Ophthalmol. 64:751–757.
295. Lombroso, C.T., Duffy, F.H., and Robb, R.M. 1969. Selective suppression of cerebral evoked potentials to patterned light in amblyopia ex anopsia. Electroencephalogr. Clin. Neurophysiol. 27:238–247.
296. Lowitzsch, K., Göhring, U., Hecking, E., and Köhler, H. 1981. Refractory period, sensory conduction velocity and visual evoked potentials before and after hemodialysis. J. Neurol. Neurosurg. Psychiatry 44:121–128.
297. Lowitzsch, K., and Maurer, K. 1982. Pattern-reversal visual evoked potentials in reclassification of 472 MS patients. Adv. Neurol. 32:487–491.
298. Lowitzsch, K., Rudolph, H.D., Trincker, D., and Müller, E. 1980. Flash and pattern-reversal visual evoked responses in retrobulbar-neuritis and controls: A comparison of conventional and TV stimulation techniques. In: EEG and clinical neurophysiology, Proceedings of the 2nd European congress of EEG and clinical neurophysiology, Salzburg, Austria, September 16–19, 1979, Excerpta Medica International Congress Series No. 526, ed. H. Lechner and A. Aranibar, pp. 451–463. Amsterdam: Elsevier North-Holland.
299. Lücking, C.H., Creutzfeldt, O.D., and Heinemann, U. 1970. Visual evoked potentials of patients with epilepsy and of a control group. Electroencephalogr. Clin. Neurophysiol. 29:557–566.
300. Lüders, H., Kato, M., and Kuroiwa, Y. 1969. Cortical evoked potentials in hepatolenticular degeneration. Electroencephalogr. Clin. Neurophysiol. 27:425–428.
301. Ludlam, W.M., and Meyers, R.R. 1972. The use of visual evoked responses in objective refraction. Trans. N.Y. Acad. Sci. 34:154–170.
302. McKay, D.M. 1977. Adaptation of evoked potentials by patterns of texture-contrast. Exp. Brain Res. 29:149–153.
303. McLeod, J.G., Low, P.A., and Morgan, J.A. 1978. Charcot-Marie-Tooth disease with Leber optic atrophy. Neurology 28:179–184.
304. McSherry, J.W., Walters, C.L., and Horbar, J.D. 1982. Acute visual evoked potential changes in hydrocephalus. Electroencephalogr. Clin. Neurophysiol. 53:331–333.
305. Maffei, L. 1981. Electroretinographic responses to alternating gratings in the cat and in human beings. Doc. Ophthalmol. Proc. Ser. 30:205–209.
306. Maffei, L., and Campbell, F.W. 1970. Neurophysiological localization of the vertical and horizontal visual coordinates in man. Science 167:386–387.
307. Maffei, L., and Fiorentini, A. 1981. Electroretinographic responses to alternating gratings before and after section of the optic nerve. Science 211:953–955.
308. Maitland, C.G., Aminoff, M.J., Kennard, C., and Hoyt, W.F. 1982. Evoked potentials in the evaluation of visual field defects due to chiasmal or retrochiasmal lesions. Neurology 32:986–991.
309. Mamoli, B., Graf, M., and Toifl, K. 1979. EEG, pattern-

evoked potentials and nerve conduction velocity in a family with adrenoleucodystrophy. Electroencephalogr. Clin. Neurophysiol. 47:411–419.
310. Mancuso, R.P., Lawrence, A.F., Hintze, R.W., and White, C.T. 1979. Effect of altered central and peripheral visual field stimulation on correct recognition and visual evoked response. Int. J. Neurosci. 9:113–122.
311. Manny, R.E., and Levi, D.M. 1982. The visually evoked potential in humans with amblyopia: Pseudorandom modulation of uniform field and sine-wave gratings. Exp. Brain Res. 47:15–27.
312. Marg, E., Freeman, D.N., Peltzman, P., and Goldstein, P.J. 1976. Visual acuity development in human infants: Evoked potential measurements. Invest. Ophthalmol. 15:150–153.
313. Markand, O.N., Garg, B.P., and Brandt, I.K. 1982. Nonketotic hyperglycinemia: Electroencephalographic and evoked potential abnormalities. Neurology 32:151–156.
314. Markand, O.N., Garg, B.P., DeMyer, W.E., and Warren, C. 1982. Brain stem auditory, visual and somatosensory evoked potentials in leukodystrophies. Electroencephalogr. Clin. Neurophysiol. 54:39–48.
315. Mastaglia, F.L., Black, J.L., Cala, L.A., and Collins, D.W.K. 1977. Evoked potentials, saccadic velocities, and computerised tomography in diagnosis of multiple sclerosis. Br. Med. J. 1:1315–1317.
316. Mastaglia, F.L., Black, J.L., and Collins, D.W.K. 1976. Visual and spinal evoked potentials in diagnosis of multiple sclerosis. Br. Med. J. 3:732.
317. Matthews, W.B., Read, D.J., and Pountney, E. 1979. Effect of raising body temperature on visual and somatosensory evoked potentials in patients with multiple sclerosis. J. Neurol. Neurosurg. Psychiatry 42:250–255.
318. Matthews, W.B., and Small, D.G. 1979. Serial recording of visual and somatosensory evoked potentials in multiple sclerosis. J. Neurol. Sci. 40:11–21.
319. Matthews, W.B., and Small, M. 1983. Prolonged followup of abnormal visual evoked potentials in multiple sclerosis: Evidence for delayed recovery. J. Neurol. Neurosurg. Psychiatry 46:639–642.
320. Matthews, W.B., Small, D.G., Small, M., and Pountney, E. 1977. Pattern reversal evoked visual potential in the diagnosis of multiple sclerosis. J. Neurol. Neurosurg. Psychiatry 40:1009–1014.
321. Matthews, W.B., Wattam-Bell, J.R.B., and Pountney, E. 1982. Evoked potentials in the diagnosis of multiple sclerosis: A follow up study. J. Neurol. Neurosurg. Psychiatry 45:303–307.
322. May, J.G., Forbes, W.B., and Piantanida, T.P. 1971. The visual evoked response obtained with an alternating barred pattern: Rate, spatial frequency and wave length. Electroencephalogr. Clin. Neurophysiol. 30:222–228.
323. May, J.G., McCullen, J.K., Moskowitz-Cook, A., and Siegfried, J.B. 1979. Effects of meridional variation on steady-state visual evoked potentials. Vision Res. 19:1395–1401.
324. May, J.G., Ralston, J.V., Reed, J.L., and Van Dyk, H.J.L. 1982. Loss in pattern-elicited electroretinograms in optic nerve dysfunction. Am. J. Ophthalmol. 93:418–422.
325. Mayles, W.P.M., and Mulholland, W.V. 1980. The response to pattern reversal in amblyopia. In: Evoked potentials: Proceedings of an international evoked potentials

symposium held in Nottingham, England, ed. C. Barber, pp. 243–249. Lancaster: MTP Press.
326. Meienberg, O., Hemphill, G., Rosenberg, M., and Hoyt, W.F. 1980. Visually evoked response asymmetries in a family with congenital nystagmus: Possible evidence of abnormal visual projections. Arch. Neurol. 37:697–698.
327. Meredith, J.T., and Celesia, G.G. 1982. Pattern-reversal visual evoked potentials and retinal eccentricity. Electroencephalogr. Clin. Neurophysiol. 53:243–253.
328. Miller, J.R., Burke, A.M., and Bever, C.T. 1983. Occurrence of oligoclonal bands in multiple sclerosis and other CNS diseases. Ann. Neurol. 13:53–58.
329. Millidot, M. 1977. The use of visual evoked potentials in optometry. In: Visual evoked potentials in man: New developments, ed. J.E. Desmedt, pp. 401–409. Oxford: Clarendon Press.
330. Millidot, M., and Newton, I. 1981. VEP measurement of the amplitude of accommodation. Br. J. Ophthalmol. 65:294–298.
331. Millidot, M., and Riggs, L.A. 1970. Refraction determined electrophysiologically: Responses to alternation of visual contours. Arch. Ophthalmol. 84:272–278.
332. Milner, B.A., Regan, D., and Heron, J.R. 1974. Differential diagnosis of multiple sclerosis by visual evoked potential recording. Brain 97:755–772.
333. Mintz, M., Tomer, R., and Myslobodsky, M.S. 1981. Visual evoked potentials in hemiparkinsonism. Electroencephalogr. Clin. Neurophysiol. 52:611–616.
334. Mirsky, A.F., and Tecce, J.J. 1968. The analysis of visual evoked potentials during spike and wave EEG activity. Epilepsia 9:211–220.
335. Moskowitz, A., and Sokol, S. 1980. Spatial and temporal interaction of pattern-evoked cortical potentials in human infants. Vision Res. 20:699–707.
336. Moskowitz, A., and Sokol, S. 1983. Developmental changes in the human visual system as reflected by the latency of the pattern reversal VEP. Electroencephalogr. Clin. Neurophysiol. 56:1–15.
337. Mukuno, K., Ishikawa, S., and Okamura, R. 1981. Grating test of contrast sensitivity in patients with Minamata disease. Br. J. Ophthalmol. 65:284–290.
338. Müller, W. 1963. Die corticale Antwort bei ausgelöschtem Elektroretinogramm. Albrecht von Graefes Arch. Ophthalmol. 166:383–386.
339. Müller-Jensen, A., Zschocke, S., and Dannheim, F. 1981. VER analysis of the chiasmal syndrome. J. Neurol. 225:33–40.
340. Nakamura, Z., and Biersdorf, W.R. 1971. Localization of the human visual evoked response. Early components specific to visual stimulation. Am. J. Ophthalmol. 72:988–997.
341. Namerow, N.S., and Enns, N. 1972. Visual evoked responses in patients with multiple sclerosis. J. Neurol. Neurosurg. Psychiatry 35:829–833.
342. Niazy, H.M.A., and Lundervold, A. 1982. Correlation of evoked potentials (SEP and VEP), EEG and CT in the diagnosis of brain tumors and cerebrovascular disease. Clin. Electroencephalogr. 13:71–81.
343. Nikoskelainen, E., and Falck, B. 1982. Do visual evoked potentials give relevant information to the neuro-ophthalmological examination in optic nerve lesions? Acta Neurol. Scand. 66:42–57.

344. Nishitani, H., and Kooi, K.A. 1968. Cerebral evoked responses in hypothyroidism. Electroencephalogr. Clin. Neurophysiol. 24:554–560.
345. Njiokiktjien, C.J., Visser, S.L., and De Rijke, W. 1977. EEG and visual evoked responses in children with learning disorders. Neuropaediatrie 8:134–147.
346. Nuwer, M.R., Perlman, S.L., Packwood, J.W., and Kark, R.A.P. 1983. Evoked potential abnormalities in the various inherited ataxias. Ann. Neurol. 13:20–27.
347. Ochs, A.L., and Aminoff, M.J. 1980. The effect of adaptation to the stimulating pattern on the latency and wave form of visual evoked potentials. Electroencephalogr. Clin. Neurophysiol. 48:502–508.
348. Oepen, G., Brauner, C., Doerr, M., and Thoden, U. 1981. Visual evoked potentials (VEP) elicited by checkerboard versus foveal stimulation in multiple sclerosis: A clinical study in 235 patients. Arch. Psychiatr. Nervenkr. 229:305–313.
349. Oepen, G., Doerr, M., and Thoden, U. 1982. Huntington's disease: Alterations of visual and somatosensory cortical evoked potentials in patients and offspring. Adv. Neurol. 32:141–147.
350. Onofrj, M., Bodis-Wollner, I., and Mylin, L. 1982. Visual evoked potential diagnosis of field defects in patients with chiasmatic and retrochiasmatic lesions. J. Neurol. Neurosurg. Psychiatry 45:294–302.
351. Oosterhuis, H.J.G.H., Ponsen, L., Jonkman, E.J., and Magnus, O. 1969. The average visual response in patients with cerebrovascular disease. Electroencephalogr. Clin. Neurophysiol. 27:23–34.
352. Padmos, P., Haaijman, J.J., and Spekreijse, H. 1973. Visually evoked cortical potentials to patterned stimuli in monkey and man. Electroencephalogr. Clin. Neurophysiol. 35:153–163.
353. Papakostopoulos, D., Hart, C.D., Cooper, R., and Natsikos, V. 1984. Combined electrophysiological assessment of the visual system in central serous retinopathy. Electroencephalogr. Clin. Neurophysiol. 59:77–80.
354. Parker, D.M., Salzen, E.A., and Lishman, J.R. 1982. Visual-evoked responses elicited by the onset and offset of sinusoidal gratings: Latency, waveform, and topographic characteristics. Invest. Ophthalmol. Visual Sci. 22:675–680.
355. Parry-Jones, N.O., and Fenwick, P. 1979. Coloured pattern displacement and VEP amplitude. Electroencephalogr. Clin. Neurophysiol. 46:49–57.
356. Paty, D.W., Blume, W.T., Brown, W.F., Jaatoul, N., Kertesz, A., and McInnis, W. 1979. Chronic progressive myelopathy: Investigation with CSF electrophoresis, evoked potentials, and CT scan. Ann. Neurol. 6:419–424.
357. Paty, J., Bonnaud, E., Latinville, D., Brenot, P., Vital, C., Henry, P., and Faure, J.M.A. 1978. A case of Jakob-Creutzfeldt's disease (form of Heidenhain) electrophysiological, clinical and anatomical correlations. Rev. Neurol. 134:223–231.
358. Paty, J., Brenot, P., Henry, P., and Faure, J.M.A. 1976. Visual evoked potentials and multiple sclerosis. Rev. Neurol. 132:605–621.
359. Paty, J., Narvarte, M.M.. Bensch, C., and Faure, J.M.A. 1974. Diagnostic électrophysiologique des cécités corticales. Bordeaux Med. 8:1143–1153.
360. Peachey, N.S., Sokol, S., and Moskowitz, A. 1983. Recording the contralateral PERG: Effect of different electrodes. Invest. Ophthalmol. Visual Sci. 24:1514–1516.
361. Pedersen, L., and Trojaborg, W., 1981. Visual, auditory

and somatosensory pathway involvement in hereditary cerebellar ataxia, Friedreich's ataxia and familial spastic paraplegia. Electroencephalogr. Clin. Neurophysiol. 52:283–297.

362. Perry, N.W., and Childers, D.G. 1969. The human visual evoked response: Method and theory. Springfield, Ill.: Thomas.

363. Persson, H.E., and Sachs, C. 1981. Visual evoked potentials elicited by pattern reversal during provoked visual impairment in multiple sclerosis. Brain 104:369–382.

364. Petrig, B., Julesz, B., Kropfl, W., Baumgartner, G., and Anliker, M. 1981. Development of stereopsis and cortical binocularity in human infants: Electrophysiological evidence. Science 213:1402–1405.

365. Petry, H.M., Donovan, W.J., Moore, R.K., Dixon, W.B., and Riggs, L.A. 1982. Changes in the human visually evoked cortical potential in response to chromatic modulation of a sinusoidal grating. Vision Res. 22:745–755.

366. Phillips, K.R., Potvin, A.R., Syndulko, K., Cohen, S.N., Tourtellotte, W.W., and Potvin, J.H. 1983. Multimodality evoked potentials and neurophysiological tests in multiple sclerosis: Effects of hyperthermia on test results. Arch. Neurol. 40:159–164.

367. Pirchio, M., Spinelli, D., Fiorentini, A., and Maffei, L. 1978. Infant contrast sensitivity evaluated by evoked potentials. Brain Res. 141:179–184.

368. Plant, G.T. 1983. Transient visual evoked potentials to sinusoidal gratings in optic neuritis. J. Neurol. Neurosurg. Psychiatry 46:1125–1133.

369. Plant, G.T., Simmern, R.L., and Durden, K. 1983. Transient visual evoked potentials to the pattern reversal and onset of sinusoidal gratings. Electroencephalogr. Clin. Neurophysiol. 56:147–158.

370. Posthuma, J., and Visser, S.L. 1982. Visual evoked potentials and alcohol-induced brain damage. Adv. Neurol. 32:149–155.

371. Pratt, H., Bleich, N., and Berliner, E. 1982. Short latency visual evoked potentials in man. Electroencephalogr. Clin. Neurophysiol. 54:55–62.

372. Preston, M.S., Guthrie, J.T., Kirsch, I., Gertman, D., and Childs, B. 1977. VERs in normal and disabled adult readers. Psychophysiology 14:8–14.

373. Purves, S.J., Low, M.D., Galloway, J., and Reeves, B. 1981. A comparison of visual, brainstem auditory, and somatosensory evoked potentials in multiple sclerosis. Can. J. Neurol. Sci. 8:15–19.

374. Puvanendran, K., Devathasan, G., and Wong, P.K. 1983. Visual evoked responses in diabetes. J. Neurol. Neurosurg. Psychiatry 46:643–647.

375. Raitta, C., Karhunen, U., Seppäläinen, A.M., and Naukkarinen, M. 1979. Changes in the electroretinogram and visual evoked potentials during general anaesthesia. Albrecht von Graefes Arch. Klin. Exp. Ophthalmol. 211:139–144.

376. Rappaport, M., Hopkins, H.K., Hall, K., and Beleza, T. 1981. Evoked potentials and head injury. 2. Clinical applications. Clin. Electroencephalogr. 12:167–176.

377. Raudzens, P.A. 1982. Intraoperative monitoring of evoked potentials. Ann. N.Y. Acad. Sci. 388:308–326.

378. Regan, D. 1966. Some characteristics of average steady-state and transient responses evoked by modulated light. Electroencephalogr. Clin. Neurophysiol. 20:238–248.

379. Regan, D. 1972. Evoked potentials in psychology, sensory

380. Regan, D. 1973. An evoked potential correlate of colour: Evoked potential findings and single-cell speculations. Vision Res. 13:1933–1941.
381. Regan, D. 1973. Evoked potentials specific to spatial patterns of luminance and colour. Vision Res. 13:2381–2402.
382. Regan, D. 1977. Steady-state evoked potentials. J. Opt. Soc. Am. 67:1475–1489.
383. Regan, D. 1977. Rapid methods for refracting the eye and for assessing visual acuity in amblyopia, using steady-state visual evoked potentials. In: Visual evoked potentials in man: New developments, ed. J.E. Desmedt, pp. 418–426. Oxford: Clarendon Press.
384. Regan, D. 1978. Assessment of visual acuity by evoked potential recording: Ambiguity caused by temporal dependence of spatial frequency selectivity. Vision Res. 18:439–443.
385. Regan, D., Bartol, S., Murray, T.J., and Beverley, K.I. 1982. Spatial frequency discrimination in normal vision and in patients with multiple sclerosis. Brain 105:735–754.
386. Regan, D., Raymond, J., Ginsburg, A.P., and Murray, T.J. 1981. Contrast sensitivity, visual acuity and the discrimination of Snellen letters in multiple sclerosis. Brain 104:333–350.
387. Regan, D., and Richards, W. 1971. Independence of evoked potentials and apparent size. Vision Res. 11:679–684.
388. Regan, D., and Richards, W. 1973. Brightness contrast and evoked potentials. J. Opt. Soc. Am. 63:606–611.
389. Regan, D., and Spekreijse, H. 1974. Evoked potential indications of colour blindness. Vision Res. 14:89–95.
390. Reilly, E.L., Kondo, C., Brunberg, J.A., and Doty, D.B. 1978. Visual evoked potentials during hypothermia and prolonged circulatory arrest. Electroencephalogr. Clin. Neurophysiol. 45:100–106.
391. Rémond, A., and Lesèvre, N. 1965. Distribution topographique et potentiels évoqués visuels occipitaux chez l'homme normal. Rev. Neurol. 112:317–330.
392. Rhodes, L.E., Obitz, F.W., and Creel, D. 1975. Effect of alcohol and task on hemispheric asymmetry of visually evoked potentials in man. Electroencephalogr. Clin. Neurophysiol. 38:561–568.
393. Richey, E.T., Kooi, K.A., and Tourtellotte, W.W. 1971. Visually evoked responses in multiple sclerosis. J. Neurol. Neurosurg. Psychiatry 34:275–280.
394. Riemslag, F.C.C., Spekreijse, H., and Van Walbeek, H. 1982. Pattern evoked potential diagnosis of multiple sclerosis: A comparison of various contrast stimuli. Adv. Neurol. 32:417–426.
395. Riggs, L.A. 1966. Light as a stimulus for vision. In: Vision and visual perception, ed. C.H. Graham, pp. 1–38. New York: Wiley.
396. Ristanović, D., and Hajduković, R. 1981. Effects of spatially structured stimulus fields on pattern reversal visual evoked potentials. Electroencephalogr. Clin. Neurophysiol. 51:599–610.
397. Rizzo, P.A., Pierelli, F., Pozzessere, G., Floris, R., and Morocutti, C. 1983. Subjective posttraumatic syndrome: A comparison of visual and brain stem auditory evoked responses. Neuropsychobiology 9:78–82.
398. Rizzo, P.A., Pierelli, F., Pozzessere, G., Verardi, S., Casciani, C.U., and Morocutti, C. 1982. Pattern visual evoked potentials and brainstem auditory evoked responses in uremic patients. Acta Neurol. Belg. 82:72–79.

399. Ropper, A.H., Miett, T., and Chiappa, K.H. 1982. Absence of evoked potential abnormalities in acute transverse myelopathy. Neurology 32:80–82.
400. Rosén, I., Bynke, H., and Sandberg, M. 1980. Pattern-reversal visual evoked potentials after unilateral optic neuritis. In: Evoked potentials: Proceedings of an international evoked potentials symposium held in Nottingham, England, ed. C. Barber, pp. 567–574. Lancaster: MTP Press.
401. Rossini, P.M., Caltagirone, C., David, P., and Macchi, G. 1979. Jakob-Creutzfeldt disease: Analysis of EEG and evoked potentials under basal conditions and neuroactive drugs. Eur. Neurol. 18:269–279.
402. Rossini, P.M., Pirchio, M., Sollazzo, D., and Caltagirone, C. 1979. Foveal versus peripheral retinal responses: A new analysis for early diagnosis of multiple sclerosis. Electroencephalogr. Clin. Neurophysiol. 47:515–523.
403. Rossini, P.M., Pirchio, M., Treviso M., Gambi, D., Di Paolo, B., and Albertazzi, A. 1981. Checkerboard reversal pattern and flash VEPs in dialyzed and non-dialyzed subjects. Electroencephalogr. Clin. Neurophysiol. 52:435–444.
404. Rowe, M.J. 1982. The clinical utility of half-field pattern reversal visual evoked potential testing. Electroencephalogr. Clin. Neurophysiol. 53:73–77.
405. Rubinstein, M.P., and Harding, G.F.A. 1981. The visually evoked subcortical potential: Is it related to the electroretinogram? Invest. Ophthalmol. Visual Sci. 21:335–344.
406. Samson-Dollfus, D., Mihout, B., Vachon, B., Vachon, T., Bouhali, M., and Samson, M. 1980. Application of correlation factors to the study of visual evoked potentials in multiple sclerosis. Rev. Neurol. 136:391–400.
407. Schafer, E.W.P., and McKean, C.M. 1975. Evidence that monoamines influence human evoked potentials. Brain Res. 99:49–58.
408. Scott, D.F., Heathfield, K.W.G., Toone, B., and Margerison, J.H. 1972. The EEG in Huntington's chorea: A clinical and neuropathological study. J. Neurol. Neurosurg. Psychiatry 35:97–102.
409. Scott, D.F., Moffett, A., and Bickford, R.G. 1981. Comparison of two types of visual evoked potentials: Pattern reversal and eye movement (lambda). Electroencephalogr. Clin. Neurophysiol. 52:102–104.
410. Selbst, R.G., Selhorst, J.B., Harbison, J.W., and Myer, E.C. 1983. Parainfectious optic neuritis: Report and review following varicella. Arch. Neurol. 40:347–350.
411. Seppäläinen, A.M., Savolainen, K., and Kovola, T. 1981. Changes induced by xylene and alcohol in human evoked potentials. Electroencephalogr. Clin. Neurophysiol. 51:148–155.
412. Seyal, M., Sato, S., White, B.G., and Porter, R.J. 1981. Visual evoked potentials and eye dominance. Electroencephalogr. Clin. Neurophysiol. 52:424–428.
413. Shagass, C. 1975. EEG and evoked potentials in the psychoses. Res. Publ. Assoc. Res. Nerv. Ment. Dis. 54:101–127.
414. Shagass, C., Amadeo, M., and Roemer, R.A. 1976. Spatial distribution of potentials evoked by half-field pattern-reversal and pattern-onset stimuli. Electroencephalogr. Clin. Neurophysiol. 41:609–622.
415. Shagass, C., Straumanis, J.J., Roemer, R.A., and Amadeo, M. 1977. Evoked potentials of schizophrenics in several sensory modalities. Biol. Psychiatry 12:221–235.
416. Shahrokhi, F., Chiappa, K.H., and Young, R.R. 1978. Pattern shift visual evoked responses: Two hundred pa-

tients with optic neuritis and/or multiple sclerosis. Arch. Neurol. 35:65–71.

417. Shaw, N.A., and Cant, B.R. 1980. Age-dependent changes in the latency of the pattern visual evoked potential. Electroencephalogr. Clin. Neurophysiol. 48:237–241.

418. Shaw, N.A., and Cant, B.R. 1981. Age-dependent changes in the amplitude of the pattern visual evoked potential. Electroencephalogr. Clin. Neurophysiol. 51:671–673.

419. Shearer, D.E., and Dustman, R.E. 1980. The pattern reversal evoked potential: The need for laboratory norms. Am. J. EEG Tech. 20:185–200.

420. Sherman, J. 1982. Simultaneous pattern-reversal electroretinograms and visual evoked potentials in diseases of the macula and optic nerve. Ann. N.Y. Acad. Sci. 388:214–226.

421. Shibasaki, H., and Kuroiwa, Y. 1982. Pattern reversal visual evoked potentials in Japanese patients with multiple sclerosis. J. Neurol. Neurosurg. Psychiatry 45:1139–1143.

422. Shipley, T., Jones, R.W., and Fry, A. 1966. Intensity and the evoked occipitogram in man. Vision Res. 6:657–667.

423. Simpson, D., Erwin, C.W., and Linnoila, M. 1981. Ethanol and menstrual cycle interactions in the visual evoked response. Electroencephalogr. Clin. Neurophysiol. 52:28–35.

424. Skalka, H.W. 1980. Comparison of Snellen acuity, VER acuity, and Arden grating scores in macular and optic nerve diseases. Br. J. Ophthalmol. 64:24–29.

425. Skrandies, W., and Lehmann, D. 1982. Spatial principal components of multichannel maps evoked by lateral visual half-field stimuli. Electroencephalogr. Clin. Neurophysiol. 54:662–667.

426. Snyder, E.W., Dustman, R.E., and Shearer, D.E. 1981. Pattern reversal evoked potential amplitudes: Life span changes. Electroencephalogr. Clin. Neurophysiol. 52:429–434.

427. Sobotka, K.R., and May, J.G. 1977. Visual evoked potentials and reaction time in normal and dyslexic children. Psychophysiology 14:18–24.

428. Sokol, S. 1977. Visual evoked potentials to checkerboard pattern stimuli in strabismic amblyopia. In: Visual evoked potentials in man: New developments, ed. J.E. Desmedt pp. 410–417. Oxford: Clarendon Press.

429. Sokol, S. 1980. Pattern visual evoked potentials: Their use in pediatric ophthalmology. In: Electrophysiology and psychophysics: Their use in ophthalmic diagnosis, ed. S. Sokol, pp. 251–268. Boston: Little, Brown.

430. Sokol, S. 1980. Visual evoked potentials. In: Electrodiagnosis in clinical neurology, ed. M.J. Aminoff, pp. 348–369. New York: Churchill Livingstone.

431. Sokol, S. 1983. Abnormal evoked potential latencies in amblyopia. Br. J. Ophthalmol. 67:310–314.

432. Sokol, S., and Dobson, V. 1976. Pattern reversal visually evoked potentials in infants. Invest. Ophthalmol. 15:58–62.

433. Sokol, S., Domar, A., Moskowitz, A., and Schwartz, B. 1981. Pattern evoked potential latency and contrast sensitivity in glaucoma and ocular hypertension. Doc. Ophthalmol. Proc. Ser. 27:79–86.

434. Sokol, S., and Jones, K. 1979. Implicit time of pattern evoked potentials in infants: An index of maturation of spatial vision. Vision Res. 19:747–755.

435. Sokol, S., Jones, K., and Nadler, D. 1983. Comparison of the spatial response properties of the human retina and cortex as measured by simultaneously recorded pattern ERGs and VEPs. Vision Res. 23:723–727.

436. Sokol, S., and Moskowitz, A. 1981. Effect of retinal blur on the peak latency of the pattern evoked potential. Vision Res. 21:1279–1286.
437. Sokol, S., Moskowitz, A., and Paul, A. 1983. Evoked potential estimates of visual accommodation in infants. Vision Res. 23:851–860.
438. Sokol, S., Moskowitz, A., and Towle, V.L. 1981. Age-related changes in the latency of the visual evoked potential: Influence of check size. Electroencephalogr. Clin. Neurophysiol. 51:559–562.
439. Sokol, S., and Nadler, D. 1979. Simultaneous electroretinograms and visually evoked potentials from adult amblyopes in response to a pattern stimulus. Invest. Ophthalmol. Visual Sci. 18:848–855.
440. Spehlmann, R. 1965. The averaged electrical responses to diffuse and to patterned light in the human. Electroencephalogr. Clin. Neurophysiol. 19:560–569.
441. Spehlmann, R. 1977. Visual evoked potentials and post-mortem findings in a case of cortical blindness. Ann. Neurol. 2:531–534.
442. Spekreijse, H. 1978. Maturation of contrast EPs and development of visual resolution. Arch. Ital. Biol. 116:358–369.
443. Spekreijse, H. 1983. Comparison of acuity tests and pattern evoked potential criteria: Two mechanisms underlying acuity maturation in man. Behav. Brain Res. 10:107–117.
444. Spekreijse, H., Van der Tweel, L.H., and Zuidema, T. 1973. Contrast evoked responses in man. Vision Res. 13:1577–1601.
445. Sprague, J.B., and Wilson, W.B. 1981. Electrophysiologic findings in bilateral optic nerve hypoplasia. Arch. Ophthalmol. 99:1028–1029.
446. Spraker, T.E., and Arnett, D.W. 1977. An electronic checkerboard pattern generator for vision research. Electroencephalogr. Clin. Neurophysiol. 42:259–263.
447. Spydell, J.D. 1983. A low-cost light-emitting diode photic stimulator. Electroencephalogr. Clin. Neurophysiol. 55:485–486.
448. Stevens, R.A., and Newman, N.M. 1981. Abnormal visual-evoked potentials from eyes with optic nerve head drusen. Am. J. Ophthalmol. 92:857–862.
449. Stockard, J.J., and Iragui, V.J. 1984. Clinically useful applications of evoked potentials in adult neurology. J. Clin. Neurophysiol. 1:159–202.
450. Stockard, J.J., Hughes, J.F., and Sharbrough, F.W. 1979. Visually evoked potentials to electronic pattern reversal: Latency variations with gender, age, and technical factors. Am. J. EEG Tech. 19:171–204.
451. Straumanis, J.J., Shagass, C., and Schwartz, M. 1965. Visually evoked cerebral response changes associated with chronic brain syndromes and aging. J. Gerontol. 20:498–506.
452. Streletz, L.J., Bae, S.H., Roeshman, R.M., Schatz, N.J., and Savino, P.J. 1981. Visual evoked potentials in occipital lobe lesions. Arch. Neurol. 38:80–85.
453. Streletz, L.J., Chambers, R.A., Bae, S.H., and Israel, H.L. 1981. Visual evoked potentials in sarcoidosis. Neurology 31:1545–1549.
454. Symann-Louett, N., Gascon, G.G., Matsumiya, Y., and Lombroso, C.T. 1977. Wave form difference in visual evoked responses between normal and reading disabled children. Neurology 27:156–159.
455. Tackmann, W., and Radü, E.W. 1980. Pattern shift visual

evoked potentials in Charcot-Marie-Tooth disease, HMSN Type I. J. Neurol. 224:71–74.
456. Takahashi, K., and Fujitani, Y. 1970. Somatosensory and visual evoked potentials in hyperthyroidism. Electroencephalogr. Clin. Neurophysiol. 29:551–556.
457. Takahashi, K., Okada, E., and Fujitani, Y. 1972. Somatosensory and visual evoked potentials in Huntington's chorea. Clin. Neurol. 12:381–385.
458. Thompson, C.R., and Harding, G.F. 1978. The visual evoked potential in patients with cataracts. Doc. Ophthalmol. Proc. Ser. 15:193–201.
459. Towle, V.L., and Harter, M.R. 1977. Objective determination of human visual acuity: Pattern evoked potentials. Invest. Ophthalmol. Visual Sci. 16:1073–1076.
460. Towle, V.L., Moskowitz, A., Sokol, S., and Schwartz, B. 1983. The visual evoked potential in glaucoma and ocular hypertension: Effects of check size, field size, and stimulation rate. Invest. Ophthalmol. Visual Sci. 24:175–183.
461. Trojaborg, W., Böttcher, J., and Saxtrup, O. 1981. Evoked potentials and immunoglobulin abnormalities in multiple sclerosis. Neurology 31:866–871.
462. Trojaborg, W., and Jørgensen, E.O. 1973. Evoked cortical potentials in patients with "isoelectric" EEGs. Electroencephalogr. Clin. Neurophysiol. 35:301–309.
463. Trojaborg, W., and Petersen, E. 1979. Visual and somatosensory evoked cortical potentials in multiple sclerosis. J. Neurol. Neurosurg. Psychiatry 42:323–330.
464. Tronconso, J., Mancall, E.L., and Schatz, N.J. 1979. Visual evoked responses in pernicious anemia. Arch. Neurol. 36:168–169.
465. Tyler, C.W., Apkarian, P., Levi, D.M., and Nakayama, K. 1979. Rapid assessment of visual function: An electronic sweep technique for the pattern visual evoked potential. Invest. Ophthalmol. Visual Sci. 18:703–713.
466. Uhl, R.R., Squires, K.C., Bruce, D.L., and Starr, A. 1980. Effect of halothane anesthesia on the human cortical visual evoked response. Anesthesiology 53:273–276.
467. Umezaki, H., and Morrell, F. 1970. Developmental study of photic evoked responses in premature infants. Electroencephalogr. Clin. Neurophysiol. 28:55–63.
468. Van der Tweel, L.H., Estevez, O., and Cavonius, C.R. 1979. Invariance of the contrast evoked potential with changes in retinal illuminance. Vision Res. 19:1283–1287.
469. Van der Tweel, L.H., and Spekreijse, H. 1969. Signal transport and rectification in the human evoked-response system. Ann. N.Y. Acad. Sci. 156:678–695.
470. Van der Tweel, L.H., and Verduyn Lunel, H.F.E. 1965. Human visual responses to sinusoidally modulated light. Electroencephalogr. Clin. Neurophysiol. 18:587–598.
471. Van Lith, G.H.M., and Hekkert-Wiebenga, W. 1983. Cataract, pattern stimulation and visually evoked potentials. Doc. Ophthalmol. 55:107–112.
472. Van Lith, G.H.M., Van der Torren, K., and Vijfvinkel-Bruinenga, S. 1981. ERG and VECPs in retinal detachments. Doc. Ophthalmol. 50:291–297.
473. Van Lith, G.H.M., Van Marle, G.W., and Van Dok-Mak, G.T.M. 1978. Variations in latency times of visually evoked cortical potentials. Br. J. Ophthalmol. 62:220–222.
474. Van Lith, G.H.M., Vijfvinkel-Bruinenga, S., and Graniewski-Wijnands, H. 1982. Pattern evoked cortical potentials and compressive lesions along the visual pathways. Doc. Ophthalmol. 52:347–353.

475. Van Marle, G.W.R.B., Van Lith, G.H.M., and Vijfvinkel-Bruinenga, S. 1978. Influence of superimposing cartoons on TV pattern-evoked cortical potentials. Ophthal. Res. 10:1–6.
476. Vasconetto, C., Floris, V., and Morocutti, C. 1971. Visual evoked responses in normal and psychiatric subjects. Electroencephalogr. Clin. Neurophysiol. 31:77–83.
477. Vassilev, A., Manahilov, V., and Mitov, D. 1983. Spatial frequency and the pattern onset-offset response. Vision Res. 23:1417–1422.
478. Vaughan, H.G. 1966. The perceptual and physiologic significance of visual evoked responses recorded from the scalp in man. In: Clinical electroretinography, ed. H.M. Burian and J.H. Jacobson, pp. 203–223. Oxford: Pergamon Press.
479. Vaughan, H.G. 1975. Electrophysiologic analysis of regional cortical maturation. Biol. Psychiatry 10:513–526.
480. Vaughan, H.G., Katzman, R., and Taylor, J. 1963. Alterations of visual evoked response in the presence of homonymous visual defects. Electroencephalogr. Clin. Neurophysiol. 15:737–746.
481. Visser, S.L., Njiokiktjien, C.J., and De Rijke, W. 1982. Neurological condition at birth in relation to the electroencephalogram (EEG) and visual evoked potential (VEP) at the age of 5. Electroencephalogr. Clin. Neurophysiol. 54:458–464.
482. Visser, S.L., Stam, F.C., Van Tilburg, W., Op den Velde, W., Blom, J.L., and De Rijke, W. 1976. Visual evoked response in senile and presenile dementia. Electroencephalogr. Clin. Neurophysiol. 40:385–392.
483. Vrijland, H.R., and Van Lith, G.H.M. 1983. The value of preoperative electro-ophthalmological examination before cataract extraction. Doc. Ophthalmol. 55:153–156.
484. Walsh, J.C., Garrick, R., Cameron, J., and McLeod, J.G. 1982. Evoked potential changes in clinically definite multiple sclerosis: A two year follow up study. J. Neurol. Neurosurg. Psychiatry 45:494–500.
485. Walter, W., and Arfel, G. 1972. Réponses aux stimulations visuelles dans les états de coma aigu at de coma chronique. Electroencephalogr. Clin. Neurophysiol. 32:27–41.
486. Wanger, P., and Persson, H.E. 1983. Pattern-reversal electroretinograms in unilateral glaucoma. Invest. Ophthalmol. Visual Sci. 24:749–753.
487. Wantanabe, K., Iwase, K., and Hara, K. 1973. Visual evoked responses during sleep and wakefulness in pre-term infants. Electroencephalogr. Clin. Neurophysiol. 34:571–577.
488. Weatherhead, R.G. 1980. Use of the Arden grating test for screening. Br. J. Ophthalmol. 64:591–596.
489. Wenzel, W., Camacho, L., Claus, D., and Aschoff, J. 1982. Visually evoked potentials in Friedreich's ataxia. Adv. Neurol. 32:131–139.
490. Werre, P.F., and Smith, C.J. 1964. Variability of responses evoked by flashes in man. Electroencephalogr. Clin. Neurophysiol. 17:644–652.
491. Westheimer, G. 1966. The Maxwellian view. Vision Res. 6:669–682.
492. White, C.T., Kataoka, R.W., and Martin, J.I. 1977. Colour-evoked potentials: Development of a methodology for the analysis of the processes involved in colour vision. In: Visual evoked potentials in man: New developments, ed. J.E. Desmedt, pp. 250–272. Oxford: Clarendon Press.

493. Whittaker, S.G., and Siegfried, J.B. 1983. Origin of wavelets in the visual evoked potential. Electroencephalogr. Clin. Neurophysiol. 55:91–101.
494. Wijngaarde, R., and Van Lith, G.H.M. 1979. Pattern EPs in endocrine orbitopathy. Doc. Ophthalmol. 48:327–332.
495. Wildberger, H., Van Lith, G., and Mak, G. 1976. Comparative study of flash and pattern evoked VECPs in optic neuritis. Ophthal. Res. 8:179–185.
496. Wildberger, H.G.H., Van Lith, G.H.M., Wijngaarde, R., and Mak, G.T.M. 1976. Visually evoked cortical potentials in the evaluation of homonymous and bitemporal visual field defects. Br. J. Ophthalmol. 60:273–278.
497. Wilson, W.B. 1978. Visual-evoked response differentiation of ischemic optic neuritis from the optic neuritis of multiple sclerosis. Am. J. Ophthalmol. 86:530–535.
498. Wilson, W.B., and Keyser, R.B. 1980. Comparison of the pattern and diffuse-light visual evoked responses in definite multiple sclerosis. Arch. Neurol. 37:30–34.
499. Wilson, W.B., Kirsch, W.M., Neville, H., Stears, J., Feinsod, M., and Lehman, R.A.W. 1976. Monitoring of visual function during parasellar surgery. Surg. Neurol. 5:323–329.
500. Woodson, P.P., Baettig, K., Etkin, M.W., Kallman, W.M., Harry, G.J., Kallman, M.J., and Rosecrans, J.A. 1982. Effects of nicotine on the visual evoked response. Pharmacol. Biochem. Behav. 17:915–920.
501. Wooten, B.R. 1972. Photopic and scotopic contributions to the human visually evoked cortical potential. Vision Res. 12:1647–1660.
502. Wright, J.E., Arden, G., and Jones, B.R. 1973. Continuous monitoring of the visually evoked response during intra-orbital surgery. Trans. Ophthalmol. Soc. U.K. 93:311–314.
503. Wright, M.J., and Johnston, A. 1982. The effects of contrast and length of gratings on the visual evoked potential. Vision Res. 22:1389–1399.
504. Yagi, A. 1981. Averaged cortical potentials (lambda responses) time-locked to onset and offset of saccades. Physiol. Psychol. 9:318–320.
505. Yiannikas, C., and Walsh, J.C. 1983. The variation of the pattern shift visual evoked response with the size of the stimulus field. Electroencephalogr. Clin. Neurophysiol. 55:427–435.
506. Yiannikas, C., Walsh, J.C., and McLeod, J.G. 1983. Visual evoked potentials in the detection of subclinical optic toxic effects secondary to ethambutol. Arch. Neurol. 40:645–648.
507. Yinon, U., Jakobovitz, L., and Auerbach, E. 1974. The visual evoked response to stationary checkerboard patterns in children with strabismic amblyopia. Invest. Ophthalmol. 13:293–296.
508. York, D.H., Pulliam, M.W., Rosenfeld, J.G., and Watts, C. 1981. Relationship between visual evoked potentials and intracranial pressure. J. Neurosurg. 55:909–916.
509. Yoshida, S., Iwahara, S., and Nagamura, N. 1975. The effect of stimulus orientation on the visual evoked potential in human subjects. Electroencephalogr. Clin. Neurophysiol. 39:53–57.
510. Yoshimura, T., and Uenoyama, K. 1980. Visually evoked potential triggered by saccadic eye movement and produced by an afterimage. In: Evoked potentials: Proceedings of an international evoked potentials symposium held in Nottingham, England, ed. C. Barber, pp. 101–107. Lancaster: MTP Press.

C

Auditory evoked potentials

11. AEP types, principles, and general methods of stimulating and recording
    11.1 AEP types
    11.2 The subject
    11.3 Stimulation methods
        11.3.1 Stimulators
            11.3.1.1 Earphones
            11.3.1.2 Loudspeakers
            11.3.1.3 Bone vibrators
            11.3.1.4 Calibration
        11.3.2 Stimulus types
            11.3.2.1 Broadband clicks
            11.3.2.2 Filtered clicks and tone pips
            11.3.2.3 Tone bursts
        11.3.3 Stimulus rate
        11.3.4 Stimulus intensity
            11.3.4.1 Hearing level
            11.3.4.2 Normal hearing level
            11.3.4.3 Sensory level
            11.3.4.4 Sound pressure level
            11.3.4.5 Peak equivalent sound pressure level
        11.3.5 Stimulus polarity: Condensation and rarefaction clicks
        11.3.6 Masking noise
        11.3.7 Monaural and binaural stimulation
    11.4 Recording electrode placements, montages, and polarity convention

12. The normal BAEP
- 12.1 Description of the normal BAEP
- 12.2 Subject variables
  - 12.2.1 Age
    - 12.2.1.1 Before adulthood
    - 12.2.1.2 Aging
  - 12.2.2 Sex
  - 12.2.3 Body temperature
  - 12.2.4 Hearing
- 12.3 Stimulus characteristics
  - 12.3.1 Effect of stimulus type
  - 12.3.2 Effect of stimulus rate
  - 12.3.3 Effect of stimulus polarity
  - 12.3.4 Effect of stimulus intensity
  - 12.3.5 Effect of monaural and binaural stimulation
- 12.4 Recording parameters
  - 12.4.1 Electrode placements: Ipsilateral and contralateral recordings
  - 12.4.2 Filter settings
  - 12.4.3 Other recording parameters
- 12.5 General strategy
  - 12.5.1 The acoustic nerve and cochlear nuclei
  - 12.5.2 Lower and upper brain stem
  - 12.5.3 Subcortical and cortical cerebral connections
  - 12.5.4 Audiological strategies: Hearing loss due to peripheral defects

13. The abnormal BAEP
- 13.1 Criteria distinguishing abnormal BAEPs
  - 13.1.1 Neurological applications
    - 13.1.1.1 Absence of all BAEP waves
    - 13.1.1.2 Absence of all waves following wave I or wave II
    - 13.1.1.3 Abnormal prolongation of IPLs I–V, I–III, and III–V
    - 13.1.1.4 Absence of wave V or abnormal decrease of the amplitude ratio of waves V and I
    - 13.1.1.5 Interaural latency difference of IPL I–V
    - 13.1.1.6 Abnormal increase of latency to rapidly repeating stimuli
    - 13.1.1.7 Abnormal peak latency
  - 13.1.2 Audiological applications
    - 13.1.2.1 Abnormally high threshold
    - 13.1.2.2 Abnormally long latency of wave V with normal IPL I–V
    - 13.1.2.3 Abnormal latency-intensity curve
- 13.2 General clinical interpretation of abnormal BAEPs
  - 13.2.1 Neurological disorders
    - 13.2.1.1 Absence of all BAEP waves
    - 13.2.1.2 Absence of all waves following wave I or wave II
    - 13.2.1.3 Absent wave V or decreased amplitude ratio of waves V and I

13.2.1.4 Increased latency of all BAEP waves
13.2.1.5 Increased IPL I–V or increased interaural difference of IPLs I–V
13.2.1.6 Abnormal increase of latency of wave V to rapidly repeating stimuli
13.2.2 Audiological disorders
13.3 Neurological disorders that cause abnormal transient BAEPs
13.3.1 Multiple sclerosis
13.3.2 Extramedullary and intramedullary brain stem tumors
13.3.2.1 Cerebellopontine angle tumors
13.3.2.2 Intramedullary brain stem tumors
13.3.2.3 Surgical monitoring
13.3.3 Coma
13.3.3.1 Coma due to structural lesions: Assessment of survival
13.3.3.2 Coma due to metabolic and toxic encephalopathies
13.3.4 Other disorders
13.3.4.1 Degenerative and other diseases of unknown cause involving the brain stem
13.3.4.2 Toxic and metabolic disorders without coma
13.3.4.3 Strokes
13.3.4.4 Vascular malformations of the posterior fossa
13.3.4.5 Leukodystrophies
13.3.4.6 Postconcussion syndrome
13.3.4.7 Mental retardation
13.3.4.8 Sleep apnea and sudden infant death syndrome
13.3.4.9 Neurological disorders in infants
13.4 Audiological disorders that cause abnormal BAEPs
13.4.1 Effects of hearing loss on threshold and latency
13.4.1.1 Effect of hearing loss on threshold
13.4.1.2 Effect of hearing loss on latency
13.4.1.3 Effect of hearing loss on the latency-intensity curve
13.4.2 Conductive hearing loss
13.4.2.1 Effect on amplitude and latency
13.4.2.2 Effect on latency-intensity curve
13.4.3 Sensorineural hearing loss
13.4.3.1 Effect on amplitude and latency
13.4.3.2 Effect on latency-intensity curve

14. Other AEPs
14.1 The slow brain stem AEP
14.2 The frequency following potential
14.3 The middle latency AEP
14.3.1 Methods of stimulating and recording
14.3.2 The normal MLAEP
14.3.3 Neurological disorders that cause abnormal MLAEPs
14.3.4 Audiological disorders that cause abnormal MLAEPs
14.4. The 40 Hz AEP
14.5 The long-latency AEP

14.5.1 Methods of stimulating and recording
14.5.2 The normal LLAEP
14.5.3 Neurological disorders that cause abnormal LLAEPs
14.5.4 Audiological disorders that cause abnormal LLAEPs
14.6 The electrocochleogram
14.6.1 Methods of stimulating and recording
14.6.2 The acoustic nerve action potential
14.6.3 The cochlear microphonic
14.6.4 The summating potential

14.7 Sonomotor AEPs
14.7.1 Methods of stimulating and recording
14.7.2 The normal sonomotor AEP
14.7.3 Neurological disorders that cause abnormal sonomotor AEPs
14.7.4 Audiological disorders that cause abnormal sonomotor AEPs

REFERENCES FOR PART C

11

AEP types, principles, and general methods of stimulating and recording

SUMMARY

11.1. AEP types are classified by their latency into the short-latency brain stem AEP (BAEP), the middle latency AEP (MLAEP), and the long-latency AEP (LLAEP). Other types include the slow brain stem AEP, the frequency following potential (FFP), the 40-Hz AEP, the electrocochleogram (ECochG), and the sonomotor AEP. By far the most important AEP for neurological and audiological studies is the BAEP.

11.2. During the recording, the subject should sit or lie comfortably and quietly; relaxation of neck and scalp muscles is very important for most AEP types. Sleep does not affect BAEP and ECochG recordings and may be desirable to minimize artifacts during recordings of these AEPs.

11.3. Stimuli are delivered through earphones and consist usually of unfiltered, or broadband, clicks containing predominantly high frequencies of 1–4 kHz. Filtered clicks and tone pips are occasionally used for a better definition of the tonal frequencies of the stimulus. Longer tone bursts may be used for LLAEPs. Stimulus rate is about 8–10/sec for the BAEP, slow brain stem AEP, MLAEP, ECochG, and sonomotor AEP and about 1/sec for the LLAEP. Faster rates are used for the 40-Hz AEP and the FFP. Stimulus intensity is measured in decibels above various reference levels such as hearing level (HL), normal hearing level (nHL), sensory level (SL), sound pressure level (SPL), and peak equivalent SPL (peSPL). Stimulus polarity distinguishes condensation clicks, which produce an initial movement of air toward the eardrum, from rarefaction clicks which move air initially away from it. Continuous masking noise is used to shield the ear opposite the stimulated one from receiving sound waves conducted through the head.

11.4. Recording electrodes are placed on the vertex and the earlobes or mastoid processes for most AEPs. BAEPs are usually recorded in two channels using linkages between vertex and ear on the stimulated side and between vertex and ear or mastoid on the other side. Increased positivity at the vertex, or increased negativity at the ear or mastoid, is represented by an upward deflection at the output in many laboratories and by a downward deflection in some others. Other AEPs are usually recorded in a single channel and with different polarity conventions.

11.1 AEP TYPES

The most common AEP types are listed in Table 11.1. AEPs are distinguished mainly by recording electrode placement and latency. Recordings between electrodes on the vertex and the earlobe or mastoid process yield AEPs of three latency and amplitude ranges (Figure 11.1). The short-latency AEPs include peaks of up to 10 msec and of about 0.2 μV; they are generated in the brain stem. The middle latency AEPs have several variable peaks with latencies of 10–50 msec and with amplitudes of about 1 μV; they probably reflect early cortical excitation. The long-latency AEPs, beginning after 50 msec and having peaks of 1–10 μV, represent later cortical excitation. These three kinds of potentials are sometimes referred to as early, middle, and late AEPs. The ECochG and the sonomotor AEPs are recorded with electrodes placed near the cochlea and the neck or scalp muscles, respectively. In contrast to VEPs, AEPs are classified by recording methods, not by stimulus characteristics: Most AEPs are elicited with click stimuli.

TABLE 11.1. AEP types

A. Short-latency (up to 10 msec) AEPs
   1. Transient short-latency AEPs
     a. Brain stem AEP (BAEP)
     b. The slow brain stem AEP
   2. Steady-state short-latency AEP: Frequency following potential (FFP)

B. Middle latency (10–50 msec) AEPs
   1. Transient middle latency AEP (MLAEP)
   2. Steady-state middle latency AEP: 40-Hz AEP

C. Long-latency (over 50 msec) AEPs
   1. Transient long-latency AEP at 100–200 msec (LLAEP)
   2. Transient AEPs of longer latency

D. Electrocochleogram (ECochG)
   1. Acoustic nerve action potential (NAP)
   2. Cochlear microphonic (CM)
   3. Summating potential (SP)

E. Sonomotor AEPs
   1. Postauricular AEP
   2. Neck and scalp sonomotor AEP

11.2 THE SUBJECT

During the recording of the AEP, the subject either reclines in a comfortable chair or lies on a bed in a quiet

Figure 11.1. Schematic diagram of the BAEP, MLAEP, and LLAEP. Stimulation of one ear and recording between the ipsilateral earlobe and the vertex produces sequences of several groups of peaks which can be displayed at three different recording speeds and gain settings. Recording with a short time base at a high gain shows the first five peaks (I–V) that occur in the first 10 msec and represent the BAEP. Recording with a medium time base at a slightly lower gain shows the middle latency AEP (MLAEP) which occurs between 10 and 50 msec after the stimulus. Recording with a long time base at an even lower gain shows the long-latency AEP (LLAEP) beginning more than 50 msec after the stimulus. Positivity at the vertex is plotted upward for the BAEP. Note that the MLAEP and LLAEP are displayed with a polarity opposite to that used for the BAEP. The MLAEP and the LLAEP are preceded by the small peaks of the earlier AEPs. The diagrams are characteristic for stimulation with rarefaction clicks at about 60 dB SL and continuous contralateral masking noise.

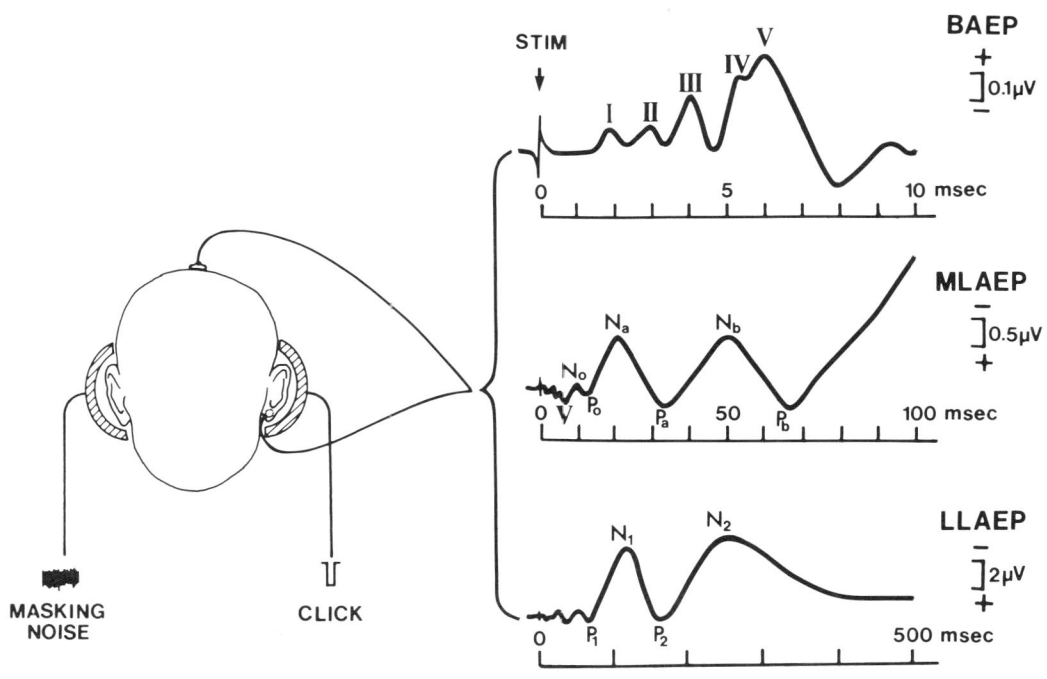

room. In most recordings, special care must be taken to relax neck muscles by placing pillows under the head and adjusting the position of the body. The subject wears earphones or, rarely, listens to a loudspeaker. The recording room should be so quiet that the subject cannot hear any sounds except for the stimulus and the masking noise. Earphones must be applied cautiously in children to avoid collapse of the outer ear canal. Because sleep relaxes scalp muscles and reduces biological artifacts without altering the BAEP and the ECochG,[6,171,387] subjects are encouraged to sleep during these recordings. Sleep is often induced with common sedatives, especially in infants and young children.[387] The MLAEP may be recorded in light sleep, but its threshold increases in deep sleep. LLAEPs vary not only with sleep but even with changes of attention. Sonomotor AEPs must be recorded while the subject is awake and not relaxed.

11.3 STIMULATION METHODS

11.3.1 Stimulators

11.3.1.1 Earphones
The stimulus is most commonly delivered through an earphone. Earphones are usually of the moving-coil, electrodynamic type which has low impedance and, especially at high stimulus intensities, generates electromagnetic fields which induce stimulus artifacts that may require shielding of the earphone with mu metal. Electrostatic and piezoelectric[178,398] earphones, having high impedance, require less current but higher signal voltage, and may not be able to generate sound levels as high as those produced by an electrodynamic earphone. These earphones produce mainly electrostatic stimulus artifacts that are more easily eliminated by shielding.[11] In some laboratories, the stimulus artifact is reduced by separating the source of the sound from the ear by a piece of tubing that introduces a delay between the electric production of the sound and the AEP. Small earphones that fit into the external ear canal may be used for intraoperative monitoring of BAEPs.

11.3.1.2 Loudspeakers
Loudspeakers are used only rarely for acoustic stimulation. This free-field stimulating method does not allow testing of monaural AEPs except in patients with unilateral deafness and in recordings of the ECochG. Furthermore, this stimulating method requires correction of stimulus intensity and response latency for the distance between the stimulus source and the ear.[297]

11.3.1.3 Bone vibrators
Stimulators transmitting sound waves through a pressure foot to the mastoid bone have been used to evaluate bone conduction[35] and to estimate conductive hearing loss. However, the stimulus intensity and frequency reaching the cochlea are difficult to control.[14,152] Bone stimulation is therefore not used routinely but may be helpful in situa-

tions where air stimulation cannot be used, for instance, during operations and in cases of malformation of the external ear.

11.3.1.4 Calibration

The sound stimulus intensity at the tympanic membrane depends on the acoustic coupling between the sound stimulus generator and the ear.[17] Audiometric zero intensities for air and bone conduction are standardized with specific couplers that roughly resemble the human ear and mastoid (artificial ear). The stimulus intensity measured with this method refers to the sound pressure generated by the various earphones and vibrators.[25,203]

11.3.2 Stimulus types

Several types of stimuli may be used to elicit AEPs[11,25,75,203] (Figure 11.2). Most AEPs are produced by clicks or tone pips. Click stimuli are very satisfactory for neurological studies because they produce sudden excitation resulting in a well-defined EP. However, clicks are not very suitable for audiological studies because they contain a wide range of tone frequencies, act mainly by virtue of their high frequency content, and do not test the lower frequency range which is important for speech. Stimulation with tones of lower frequency, although desirable for audiological purposes, creates several problems. Tones act as a stimulus mainly by virtue of their onset and should be short. However, expression of the tonal frequency of a stimulus requires a tone duration of at least a few cycles. Tones of low frequency have a wavelength too long for an effective stimulus, especially in cases of short-latency AEPs. Furthermore, although a stimulus of sudden onset is needed for a clearly defined EP, the sudden onset of a loud tone of any frequency introduces a high-frequency transient and

Figure 11.2. Acoustic waveforms of auditory stimuli. *(a)* Broadband click; *(b)* filtered click at 2 kHz; *(c)* electronically gated sinewave at 2 kHz with a rising phase of 2 cycles, a plateau of 1 cycle, and a falling phase of 2 cycles; *(d)* logon at 2 kHz with an envelope of a probability curve; *(e)* tone burst of 0.5 kHz with rising and falling phases of 2 cycles.

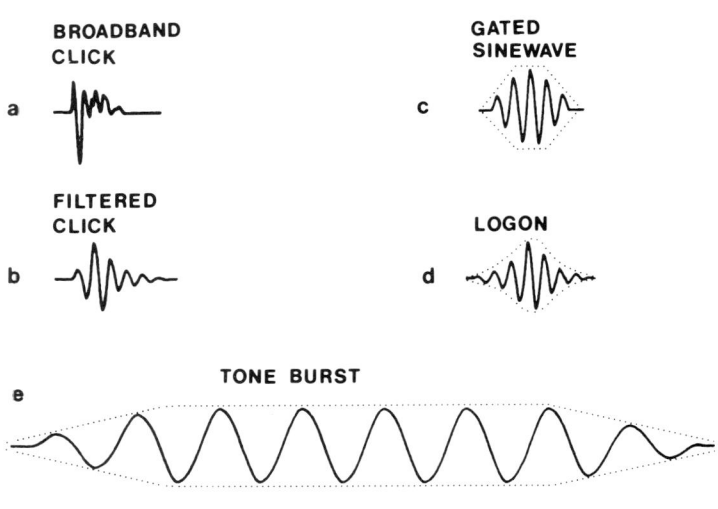

thereby elicits an AEP that is not specific for the tone frequency.[74] A compromise between the requirements of a specific tonal frequency and a well-defined onset is available in the form of the filtered click, the tone pip, and the logon, although low frequencies remain a poor stimulus for elicitation of AEPs.[78] The receptors for low tones, located in the apical part of the cochlea, are not easily explored by AEP methods. Because of the longer travel of sound waves to the apex of the cochlea, responses to low tones are elicited later than responses to high frequencies, which excite the basal part of the cochlea. Also, because low frequency sound waves have a long duration, the responses are dispersed in time and have low amplitude. Response components to the low frequency contents of an auditory stimulus are therefore easily obscured by the earlier and larger response components generated by the more basal parts of the cochlea responding to the high frequency content represented by the sudden onset of the stimulus. Attempts have been made to study BAEPs to low stimulus frequencies by masking high frequency components of the stimulus with continuous noise (12.3.1) or to record slow brain stem AEPs to low frequency tone pips (14.1).

11.3.2.1 Broadband clicks
Clicks may be produced by feeding electric monophasic square pulses into an earphone and thereby deflecting the earphone membrane. Even though the electric pulses are sharp and last only 100 μsec, the earphone membrane reacts to the square pulse only imperfectly and generates pressure changes that are further modified by the material used for coupling the earphone to the head[62] and by the intervening ear structures before they reach the sound receptors in the cochlea.[42,220] These factors convert the original electric square pulse into a sequence of decaying pressure fluctuations with a peak acoustic power at 2–4 kHz (Figure 11.2a). This sound stimulus still contains a fairly wide range of frequencies and is therefore called *broadband,* or *unfiltered,* click. A narrower band of stimulus frequencies can be obtained either by filtering the electric square pulse or by presenting broadband clicks on a background of a continuous masking noise that eliminates the effect of some of the frequencies in the click.

11.3.2.2 Filtered clicks and tone pips
Filtered clicks (Figure 11.2b) are generated by simply passing a rectangular or sinusoidal wave through a filter with a narrow bandpass so as to produce a brief burst of waves of a frequency centered at the filter bandpass.[75,466] Tone pips (Figure 11.2c) have more symmetrical rising and falling phases and are produced either by passing one period of a sine wave through a bandpass filter of the same frequency[77] or by modulating the amplitude of a pure tone electronically to give it the desired rise, fall, and plateau times.[420] Filtered clicks and tone pips are usually given rising and falling phases of two cycles and a plateau of one cycle. A logon (Figure 11.2d) is a tone with an amplitude modulated by the shape of a Gaussian distribution curve, said to give the best compromise between the definitions of stimulus

onset and frequency.[75] Filtered clicks, tone pips, and logons, must always start from the zero level and take off in the same direction to give consistent AEPs.

11.3.2.3 *Tone bursts*
Tone bursts (Figure 11.2e) with rise and fall times of at least 5 msec and durations of at least 30 msec may be used for the LLAEP but are not suited for the MLAEP and BAEP.[75]

11.3.3 Stimulus rate

The stimulus frequency varies for each AEP type. Transient short and medium latency AEPs such as the BAEP, slow brain stem AEP, ECochG, MLAEP, and sonomotor AEP are usually elicited at rates of 8–10/sec. The LLAEP requires stimulation at 1/sec or less. The steady-state 40-Hz AEP is produced by tone bursts of various frequencies repeated at 40/sec; the FFP can be obtained with tones of 100–1,000 Hz. In the choice of the exact stimulus rate, one should avoid synchronization with the power-line frequency which could lead to buildup of interference artifact in the average. While the stimulus frequency is usually not altered during a stimulation, a brief tone of changing frequency[225] and a change of the frequency of a sustained tone[12,13,110,205,356,395] are also capable of eliciting AEPs.

11.3.4 Stimulus intensity

Although the stimulus intensity is usually not changed during the recording of an AEP, a change of the intensity of a continuous tone may also act as a stimulus and elicit an AEP.[13,395]

Stimulus intensity is measured as a ratio between stimulus level and a reference level and is expressed in decibels. The number of decibels equals 20 times the logarithm$_{10}$ of the ratio of amplitude or voltage when expressing amplitude, or 10 times the logarithm$_{10}$ of that ratio when expressing power or the square of voltage. For instance, a stimulus with an amplitude 1,000 times the amplitude of a reference level is 60 dB above that level. The same type of measurement is used to indicate hearing loss: A subject hearing a tone only if it is increased by 40 dB above the normal hearing level is said to have a 40-dB hearing loss for that tone. Several different reference levels are used to describe stimulus intensity in AEP studies.

11.3.4.1 *Hearing level*
The hearing level (HL), routinely used in audiometry, is the average threshold intensity of normally hearing young adults tested with pure tones of at least 0.5 sec.

11.3.4.2 *Normal hearing level*
The normal hearing level (nHL) for pure tones of 0.5 sec or more is the same as the HL. However, the normal hearing level for shorter tones and for clicks and other less

regular sounds used as stimuli for AEPs differs from that for pure tones of the same amplitude. Therefore, the normal hearing level for these stimuli must be defined by measuring the average threshold intensity for the specific stimulus in a group of audiologically normal young adults. If nHL is to be used as a reference for EP studies, the manufacturer or the user of the stimulating equipment must make such measurements to calibrate the equipment and give a valid intensity reference.

11.3.4.3 Sensory level

The sensory level (SL) is the hearing threshold of an individual. In a normally hearing young subject, the SL for sustained pure tones is the same as the HL, and the SL for other stimuli is the same as the nHL. In a subject with reduced hearing, the difference between SL and HL or nHL expresses the magnitude of the hearing loss for the particular stimulus. Hearing threshold is measured by increasing and decreasing stimulus intensity in 5-dB steps and determining the midpoint between the intensities at which the subject begins or ceases to hear the stimulus.

Because SL cannot be measured in infants and young children, one of the other references must be used for stimulus intensity settings. In adults, the use of SL as a reference of stimulus intensity also has some problems (13.2.1.5).

11.3.4.4 Sound pressure level

The sound pressure level (SPL) defines the intensity of the sound stimulus physically but does not relate directly to the physiological effect of the stimulus. An arbitrary zero reference point is conventionally set at 0.0002 dynes/cm^2 (20 microPascal) for a tone of 1,000 Hz. The normal hearing threshold exceeds this level by a number of decibels that varies with the frequency of sustained sound stimuli such as pure tones.

11.3.4.5 Peak equivalent sound pressure level

The pressure of brief sounds is difficult to measure exactly because sound level meters have a minimum response time. One method to overcome this problem is to compare the peak-to-peak amplitude of the stimulus with the amplitude of a sine wave of a pure tone of known SPL. Thus, the SPL of a click equals the SPL of a pure tone having the same peak-to-peak amplitude as the click. Measurements of peSPL do not relate closely to nHL but usually give higher hearing thresholds because shorter stimuli must have higher amplitude to reach hearing threshold.

11.3.5 Stimulus polarity: Condensation and rarefaction clicks

Click stimuli may be produced by electric pulses, which cause an initial deflection of the earphone membrane toward the eardrum, condensing or compressing the air in the ear canal and generating condensation, or compression, clicks. If the polarity of the electric pulse driving the earphone membrane is reversed, the pulse produces an initial deflection of the membrane away from the ear, rarefying the air in the ear canal and causing rarefaction clicks. The manufacturer or user must identify the electric polarity settings that produce rarefaction and compression clicks.[32,46] Stimulus polarity is practically important for short-latency AEPs because condensation and rarefaction clicks produce slightly different BAEPs and ECochGs.

11.3.6 Masking noise

When a sound stimulus is applied through an earphone to one ear, the sound is conducted by the skull and may reach the opposite ear. Although the stimulus is attenuated by about 50–60 dB on its travel across the head,[34,296] it may excite the other ear. Such cross-stimulation is especially likely to occur when the ear to be tested has a higher threshold than the other ear and is exposed to strong stimuli. Cross-stimulation can be avoided by applying a constant masking noise through an earphone on the ear opposite the stimulated one. The masking noise should have an intensity of about 40 dB below the stimulus intensity. Masking intensities of about 60 dB SPL are sufficient for the stimulus intensities used in routine studies. Masking noise should contain either a wide range of frequencies at equal intensities (white noise) or include at least the frequencies of the stimulating sound. Masking is not necessary for the ECochG because the response is generated and recorded only at the stimulated ear.

Masking noise may be used to mask unwanted frequencies of a complex sound stimulus by mixing masking noise and stimulus sound at the same ear (12.3.1).

11.3.7 Monaural and binaural stimulation

Separate stimulation of each ear is needed for studies evaluating hearing, acoustic nerve, and brain stem function on each side. Simultaneous stimulation of both ears produces AEPs that differ from the sum of the AEPs to stimulation of each ear (12.3.5).

11.4 RECORDING ELECTRODE PLACEMENTS, MONTAGES, AND POLARITY CONVENTION

For recordings of most AEPs, electrodes are placed on the vertex and on or near the ears. Ear electrodes may be placed on the lateral or medial surface of the left and right earlobe (A1, A2) or on the scalp over the left and right mastoid bone (M1, M2). Placement on the medial surface of the earlobe can reduce the stimulus artifact of the click. If necessary, the definition of the first wave of the BAEP may be improved by recording the ECochG from an electrode inserted into the ear canal or through the tympanic membrane. Sonomotor AEPs are recorded with electrodes placed over neck or scalp muscles.

The same electrode combinations are used to record BAEPs, MLAEPs, and LLAEPs. For single-channel recordings, the vertex electrode and the electrode at the stimulated ear are connected to the two inputs of the amplifier and the electrode at the opposite ear is connected to the amplifier ground. Dual-channel recordings are preferred for BAEPs, channel 1 recording between vertex and stimulated ear, channel 2 between vertex and nonstimulated ear. A ground electrode is placed anywhere on the head, e.g. at F_2, or elsewhere on the body and is connected to the amplifier ground.

The gain and filter settings, sweep length, and number of responses averaged vary with the AEP type. For every recording, at least two sets of averages must be superimposed to ascertain replication.

The polarity convention (2.3.4.1) for AEPs is not uniform. Many laboratories record the BAEP so that an upward deflection at the output indicates increased positivity at the vertex electrode. MLAEPs and LLAEPs are usually recorded with the opposite polarity convention: Upward deflections indicate increased vertex negativity (Figure 11.1). This habit may change in the future, especially if simultaneous recording of BAEPs and MLEAPs or of all three AEP types will be used more commonly.[293,367]

12

The normal BAEP

SUMMARY

12.1. The normal BAEP consists of a sequence of up to seven waves, named I to VII, which occur during the first 10 msec after the onset of the stimulus and have positive polarity at the vertex. Most reliable are waves I, III, and V. Their latency and amplitude depend on subject variables, stimulus characteristics, and recording parameters but are remarkably constant when these factors are controlled.

12.2. Subject variables that affect the BAEP include age and sex.

12.3. Stimulus characteristics of importance for the BAEP are stimulus type, rate, polarity, and intensity.

12.4. The most important recording parameters are electrode placements and filter settings.

12.5. The general strategy of stimulating and recording is based on the principle that waves I to V are generated by impulses ascending through the lower auditory pathway, mainly ipsilateral to the stimulated ear. Measurements of peak and interpeak latencies are used in the attempt to localize lesions involving the acoustic nerve and lower and upper brain stem. Measurements of the threshold and latency of wave V at different stimulus intensities may be used to identify hearing losses.

12.1 DESCRIPTION OF THE NORMAL BAEP

The most commonly used methods of studying the BAEP are summarized in Table 12.1. Examples of normal BAEPs are shown in Figure 12.1. The BAEP recorded on the side of the stimulated ear contains five wavelets appearing in the first 10 msec after the stimulus and having peaks that are positive at the vertex with reference to the ear. These waves are usually labeled with roman numerals I to V.[185] The BAEP recorded from the side opposite the stimulated ear differs from the ipsilateral BAEP mainly in that it shows no clearcut wave I; the negative peak preceding wave II (I_N) is therefore sometimes used for measurements of interpeak latencies. Other peaks differ slightly but significantly on the two sides (12.4.1).

Not all normal recordings contain all BAEP peaks. Wave V is present most often, waves I and III can usually also be identified. Wave II is often absent and wave IV may merge more or less completely with wave V. Wave V is sometimes followed by waves VI and VII. Peak and interpeak latencies of the BAEP waves are remarkably constant for a given set of subject, stimulus, and recording parameters. They vary little in repeated recordings in the same subject even over many months; they vary only slightly more between recordings from the two sides of the same subject and between different subjects of the same age and sex.[52,91,350,354] Latency is longer and more variable for condensation than for rarefaction clicks (12.3.3). Each laboratory must therefore establish its own normal values for these variables.

Because not all BAEP waves are present in every recording from a normal subject, a few practical maneuvers are often helpful in enhancing and identifying BAEP waves.

Figure 12.1. Normal BAEP to stimulation of the right ear, recorded between vertex (Cz) and right ear (A2) (top tracing) and vertex and left ear (A1) (bottom tracing). The ipsilateral recording shows the vertex positive waves I–V. The contralateral recording shows absence of wave I, presence of a negative wave preceding wave II (I_N), earlier appearance of wave III, and better separation of waves IV and V. Click stimuli of 100 μsec duration and 60 dB SL with 40 dB SL continuous masking noise to the opposite ear. Each tracing is an average of 2,000 responses. Two tracings obtained from successive stimulations are superimposed. Recording bandwidth is 100–3,000 Hz. Relative positivity at the vertex gives an upward deflection. Normal 33-year-old woman.

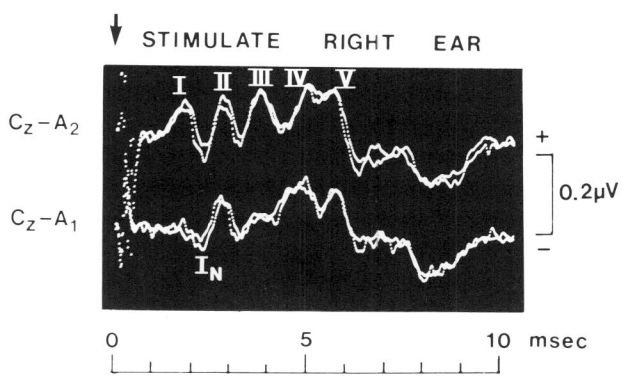

TABLE 12.1. BAEPs to clicks

A. Subject variables
   1. Age: Several separate normal control groups should be used for infants under 18 months and for elderly subjects as needed in each laboratory.
   2. Sex: Separate controls for males and females are used in some laboratories.
   3. Hearing: Hearing loss should be investigated audiologically to help distinguish between peripheral and central causes.
B. Stimulus characteristics
   1. Type: Broadband clicks produced by 100 μsec electric square pulses into an audiometric ear speaker.
   2. Rate: 8–10/sec.
   3. Polarity: Rarefaction clicks; alternating clicks only if stimulus artifact to either polarity is intolerable.
   4. Intensity: 115–120 dB peSPL or 60–70 dB SL; additional intensities for threshold measurements and latency-intensity curves in audiological investigations.
   5. Masking: 60 dB SPL of white noise to the contralateral ear.
   6. Monaural stimulation: Monaural stimuli should be used routinely.
C. Recording parameters
   1. Number of channels: 2.
   2. Electrode placements: Left and right earlobes (A1, A2) or left and right mastoid processes (M1, M2), vertex (Cz); ground electrode at Fz or elsewhere on the head or body.
   3. Montages: Channel 1: Cz-ear or mastoid ipsilateral to stimulus
      Channel 2: Cz-ear or mastoid contralateral to stimulus.
   4. Filter settings: Low frequency filter: 10–30 Hz; 100–200 Hz in case of irreducible EMG or mechanical artifacts; high frequency filter: 2,500–3,000 Hz.
   5. Number of responses averaged: 2,000 (1,000–4,000).
   6. Sweep length: At least 15 msec.
   7. Dwell time and sampling rate: Maximum dwell time is 100 μsec; minimum sampling rate is 10 kHz.
D. Analysis
   1. Normal peaks: I, III, and V in ipsilateral recordings: I_N, III, and V in contralateral recordings.
   2. Criteria for abnormal BAEPs
   Central, or retrochochlear, lesions
      a. Absence of waves I–V, unexplained by extreme hearing loss or technical problems.
      b. Absence of waves following wave I or wave III.
      c. Abnormally prolonged IPL I–V.
      d. Abnormal decrease of the V/I amplitude ratio, especially when associated with other abnormalities.
      e. Abnormally long interaural difference of IPL I–V unexplained by middle or inner ear problems identifiable by audiometric tests.
      f. Questionable criteria: Abnormally long peak latency V on rapid repetitive stimulation; abnormally long peak latency III and V; abnormally increased IPLs I–III, III–V.
   Peripheral, or inner and middle ear, lesions
      a. Increased BAEP threshold.
      b. Increased latency of waves I and V.
      c. Abnormal latency-intensity curves.

1. Wave I may be enhanced by increasing the stimulus intensity and decreasing the stimulus rate. Recording a BAEP to condensation clicks in addition to the BAEP to rarefaction clicks may help to distinguish wave I from mechanical and electric stimulus artifacts and from the cochlear microphonic (14.6): Artifacts and cochlear microphonic reverse polarity with reversal of stimulus polarity whereas wave I does not; however, wave I may change latency with stimulus polarity reversal, especially in subjects with hearing loss. Averaging of responses to alternate rarefaction and condensation stimuli, although not recommended as a routine procedure (12.3.3), may be the only way to resolve wave I in BAEPs to strong stimuli which tend to produce longer stimulus artifacts and to shorten the latency of wave I. Comparisons with recordings between vertex and nonstimulated ear, which do not contain wave I, may help to identify wave I in recordings between vertex and stimulated ear. Recording between the two ears augments wave I relative to other waves. In exceptional cases, simultaneous recording of the ECochG (14.6) may help to identify wave I.
2. Wave II, although often absent in normal subjects, may be of clinical significance if it shows prolonged absolute latency or IPL I–II. Wave II is enhanced in recordings between vertex and contralateral ear that do not show wave I.
3. Wave III may be normally split into two peaks; its latency is then measured to the first peak or to the middle between the two peaks (4.1.3.1). Splitting may disappear if condensation clicks are used instead of rarefaction clicks or vice versa. A split, or bifid, wave III must be distinguished from a partial fusion of waves III and IV or of waves II and III. Fusion of waves II and III occurs especially in recordings between vertex and nonstimulated ear.
4. Wave IV often normally fuses with wave V to form a complex with a peak latency equivalent to that of wave IV, wave V, or an intermediate value. Recording between the vertex and the nonstimulated ear (Figure 12.1), or stimulating with clicks of the opposite phase, may produce better separation of waves IV and V. Wave IV often varies in the same individual with time.
5. Wave V is the most reliable peak. It may be identified by its low threshold, its persistence during repetitive stimulation up to 100/sec, and by the large negativity that commonly follows it. Occasionally, wave V consists of only a small inflection on the downslope of wave IV. Recordings from the nonstimulated ear may give a clearer definition of wave V.[260] An unusually large negativity following wave IV may be a normal variant and may disappear in recordings from the nonstimulated ear or on stimulation with clicks of the opposite polarity.[415] The variation of wave V with click polarity is greater in patients with abnormal BAEPs than in normal subjects.[98]

12.2 SUBJECT VARIABLES

12.2.1 Age

12.2.1.1 Before adulthood
BAEPs change considerably with age. The rate of change, and the variability between subjects of the same age, are largest early in life and decrease with age. BAEPs may be absent in normal premature infants under 30 weeks of conceptional age. BAEP peak and interpeak latencies decrease steadily between conceptional ages of 25 and 44 weeks; at each age, the latency decreases to a different degree with increasing stimulus intensity.[70,82,133,270,297,371,402] The IPL I–V shortens by 2–3 msec over this period, and by about 1 msec during the last 6 weeks before term; the weekly decrease of latency averages 0.45 msec around 32–34 weeks but less than 0.1 msec near term.[408] In full-term newborns, IPL I–V is about 0.8–1 msec longer than in adults, due mainly to a longer latency of wave V.[169,407] Latencies continue to decrease after term. The latency of different waves decreases at different rates, in a manner indicating that the peripheral and central parts of the auditory path mature at a different speed.[167,187,427] The latency of wave I, or peripheral conduction, reaches adult level at about 6 weeks of age, whereas the IPL I–V, or central conduction, reaches that level at about 1.5 years[10,170,360] or somewhat later.[262] The increase of latency with fast repetitive stimulation is also greater at birth than in adults.[169,359,407] Age-specific normative values should therefore be established for intervals of 1 week in the pre-term period, and for the ages of 3 weeks, 6 weeks, 3 months, 6 months, and 1 year thereafter.

12.2.1.2 Aging
Advancing age has often been reported to increase both peak and interpeak latencies.[5,197,253,354] However, various authors report that aging has no effect on peak latencies,[26] increases the latency of wave V,[184] or increases peak latencies without increasing IPLs.[157,291,351] Amplitude is reported to decrease with age.[26,184,197] Because of these conflicting reports, no general rules for age-specific standards have been established. Age-corrected norms are used in laboratories finding significant differences between adult age groups.

12.2.2 Sex

Females, beginning in infancy[362] or as late as about 8 years[283] usually show BAEPs of shorter peak and interpeak latency and of higher amplitude than males, possibly due to sex-related differences in skull and brain anatomy.[5,26,184,194,197,258,351,414] Different sets of normative data are used to evaluate BAEPs of males and females in those laboratories which find that this difference is significant and increases the diagnostic power of BAEPs.

12.2.3 Body temperature

A decrease of body temperature by 1° C may increase the latency of wave V by 160 μsec[315] or by 200 μsec and the

IPL I–V by 160 μsec.[246] Variations of body temperature may explain circadian variations of these latencies.

12.2.4 Hearing

Patients with hearing loss should have an audiological examination to investigate the peripheral or central causes. AEPs may be used to help in this investigation. Peripheral hearing losses tend to increase BAEP peak latencies rather than IPLs and therefore do not necessarily preclude the identification of central lesions (13.2). However, significant hearing losses may make a reliable interpretation of BAEPs impossible.

12.3 STIMULUS CHARACTERISTICS

12.3.1 Effect of stimulus type

Broadband clicks, produced by delivering an electric square pulse of 100 μsec into an audiometric speaker, are the type of stimulus most commonly used in neurological and audiological studies. These clicks act mainly by virtue of their high frequency content (11.3.2). Filtered clicks or tone pips are sometimes used for audiological studies in an attempt to improve the frequency specificity of BAEPs to stimuli of tonal frequencies of 500–2,000 Hz. However, tones in this relatively low frequency range produce BAEPs that are due mainly to the high frequency components produced by the sudden onset of a loud tone. Attempts have been made to study the effects of lower stimulus frequencies by presenting broadband clicks on a background of a continuous masking noise of high frequencies which eliminate the effects of unwanted high frequencies contained in the click stimulus.[221,439] By subtracting the AEP to stimuli presented in the presence of masking noise with a certain high-frequency cutoff from the AEP to the same stimulus presented in the presence of masking noise with a higher frequency, one may derive an AEP representing a narrow range of stimulus frequencies.[68,85,86,95,299] A more direct method than this technique of "derived BAEPs" attempts to obtain frequency-specific BAEPs by mixing the stimulus with masking noise filtered with a notch filter that eliminates a narrow band of frequencies. A broadband transient stimulus given on the background of this continuous masking sound acts by virtue of that narrow part of the auditory spectrum that is not rendered ineffective by the masking sound.[221,314,323]

12.3.2 Effect of stimulus rate

The stimulus rate is usually set at 8–10/sec, avoiding fractions of 60/sec that could lead to synchronization with interference from power lines and build up 60 Hz artifact in the average. An increase of stimulus rate above 30/sec increases the latency and decreases the amplitude of the BAEP. The increase of latency is best seen in wave V since earlier waves may be difficult to distinguish at these high rates.[84,430,468] However, some studies have demonstrated that earlier waves show different degrees of rate-dependent latency shifts, causing an increase of the IPL I–V at high stimulus rates and suggesting that peripheral and central conduction undergo different degrees of adaptation.[157,324,437,459]

Low stimulus rates of about 4/sec may improve the definition of BAEPs in some instances.[197]

12.3.3 Effect of stimulus polarity

BAEPs to rarefaction clicks are used more often because they have shorter latency and clearer definition than BAEPs to condensation clicks.[288,415] The method of averaging responses to clicks of alternating phase, often used to reduce the stimulus artifact and cochlear microphonics, is not generally used as a routine because it leads to the addition of two different kinds of responses and may result in partial cancellation and superimposition of peaks of different latency and shape.[37,63,251,414] In a few patients, abnormal BAEPs appear only with one stimulus polarity but not with the other.[98]

12.3.4 Effect of stimulus intensity

Stimulus intensity affects both amplitude and latency. Wave V has the lowest threshold: It appears at, or slightly above, the hearing threshold; the other waves appear at higher intensities and reach maximum amplitude at different levels, usually at medium intensity. Wave IV may grow with increasing intensity after wave V has reached maximum; if it merges with wave V, it may appear to shift the latency of wave V to an earlier value. The amplitude of wave I continues to grow as stimulus intensity grows to above 60 dB SL. In patients with hearing defects, high stimulus intensities may be needed to elicit a BAEP.

The latency of all BAEP waves decreases with increasing stimulus intensity up to about 60 dB SL. The latency of wave V lies slightly below 6 msec in normal subjects past infancy. The latency decrease is similar for waves II to V so that the IPLs between these waves are generally fairly independent of stimulus intensity. However, the latency of wave I increases rather abruptly, and more than that of other waves, at stimulus intensities below 60 dB SL so that IPLs between wave I and other waves diminish with lower stimulus intensities, especially when stimuli of low tonal frequency are used.[415] Latency and amplitude do not correspond with the subject's estimate of stimulus loudness.[325]

12.3.5 Effect of monaural and binaural stimulation

The BAEP to binaural stimulation has a distribution different from that of BAEPs to monaural stimulation[342] and differs from the sum of BAEPs to monaural stimuli mainly in the later BAEP waves.[2,15,31,83,175,228,458] This binaural interaction may represent activation of brain stem connections not excited by stimulation of either ear. Because binaural stimulation only rarely provides clinically useful information,[320,321,349] and may mask monaurally detected abnormalities,[414] it is not used routinely in most laboratories.

12.4 RECORDING PARAMETERS

12.4.1 Electrode placements: Ipsilateral and contralateral recordings

BAEPs to stimulation of each ear are recorded routinely from both sides of the head with the same electrode placements as used for most other AEPs (11.4, Table 12.1). Recordings between the ear electrodes may be added to better define wave I (12.1).

Recordings from the two sides differ in several regards[18,298,319,404] and seem to be generated mainly by auditory structures ipsilateral to the stimulated ear.[53,284] In addition to the unilateral appearance of wave I at the electrode near the stimulated ear, peak latencies of waves III and IV are shorter and peak latencies of waves II and V are longer in contralateral recordings, causing shortening of IPL II–III and lengthening of IPL IV–V.[241,348,414] This difference is probably not entirely explained by differences in the conducting media on the two sides but is probably due to activity of lateralized structures in the brain stem.

Topographic studies have shown that the BAEP waves following wave I have a wide distribution. Unlike potentials generated in the cortex near one electrode, these far-field potentials cannot be attributed to either the ear electrode or the vertex electrode.[18,185,247,298,428,429,446]

12.4.2 Filter settings

The bandpass should include frequencies between 10–30 Hz and 2,500–3,000 Hz avoiding steep filter roll-offs (2.4.1). The low frequency cutoff may be raised to 100–200 Hz if otherwise irreducible muscle or mechanical artifact obscures the BAEP. Narrowing of filter settings may distort amplitude and latency of BAEP peaks[45,197,222,241,289] (2.4.1). A low frequency filter setting of 100 Hz usually causes no more than minimal distortions of latency,[85] but may be too high for slow components, especially those in BAEPs to low frequency tone pips[419] (14.1). Although filters affect absolute peak latencies more than IPLs because they shift latencies of similar peaks to a similar degree,[43] they cannot help but also change the IPLs because the peaks are not identical in configuration and therefore are not shifted by exactly the same amount.

It seems surprising that low frequency filter settings of only 200 Hz and less should affect the BAEP which consists of waves that have a duration of only 1–2 msec corresponding with wave frequencies of no less than 500 Hz. However, these faster waves are mixed with less obvious slow-wave components which are important for the configuration and the timing of the faster waves. The considerable contribution by slow components to the BAEP is clearly revealed by power spectral analysis of the BAEP.[38,96,223,421]

12.4.3 Other recording parameters

Two channels should be used to record between vertex and stimulated ear and between vertex and nonstimulated ear. About 1,000–4,000 responses must be averaged. The sweep length should be at least 15 msec with a maximum dwell time of 100 μsec per point or a minimum sampling rate of 10 kHz, which can resolve waves up to 5 kHz; even an averager with only 250 points per sweep satisfies these conditions (3.3.4).

12.5 GENERAL STRATEGY

The BAEP is better suited than other AEPs for the detection of lesions in the lower parts of the auditory pathway. It is also the best AEP currently available for electric response audiometry. Although the BAEP is generated in the brain stem, a strict correlation between each wave and a specific brain stem structure in the auditory pathway is not possible. Earlier studies had attempted to relate wave I to the acoustic nerve, wave II to the cochlear nucleus and trapezoid body, wave III to the superior olive, wave IV to the lateral lemniscus, wave V to the inferior colliculus, wave VI to the medial geniculate body of the thalamus, and wave VII to the medial geniculate body or the auditory radiation.[226,403,410] However, later studies have raised much doubt about simple wave-to-point correlations. Even the relationship between wave I and the acoustic nerve is not very clear. There is evidence to suggest that wave II is at least partly generated by the intracranial portion of the acoustic nerve[114,163,217,267,394] and that wave I may be preceded by an earlier wave that may also be due to excitation of the acoustic nerve.[178] Direct recordings from the vicinity of presumed generators in humans have shown that it is difficult to make unequivocal correlations between individual waves and brain stem structures.[160,163,264,265,448] It is now believed that most waves are probably generated by more than one anatomical structure and that each structure may contribute to more than one wave. For clinical purposes, waves I, III, and V have been used to localize lesions to the area between acoustic nerve and upper brain stem and, in some cases, to suggest localization within the upper and lower parts of that area (Figure 12.2). However, these correlations between waves and level of a lesion within the brain stem remain tentative as yet.

The side of a lesion is best determined by recording BAEPs to stimulation of each ear because the BAEP reflects excitation mainly of the brain stem auditory pathways on the stimulated side.

12.5.1 The acoustic nerve and cochlear nuclei

Severe lesions of the acoustic nerve may abolish the BAEP entirely or leave only wave I intact. If necessary, the action potential of the acoustic nerve can be searched for with special procedures (12.1). Less severe lesions may increase the latency of BAEP waves, starting either with wave I or with wave II. If the delay starts with wave I, IPLs are unchanged; if it starts with wave II, IPLs I–III and I–V are increased.

Figure 12.2. Schematic diagram of the strategy attempting to localize lesions with the BAEP. A click stimulus to the ear elicits a sequence of five waves generated by the acoustic nerve and brain stem structures. IPL I–V represents conduction from the acoustic nerve to the upper midbrain; IPLs I–III and III–V represent conduction through the lower and upper brain stem, respectively.

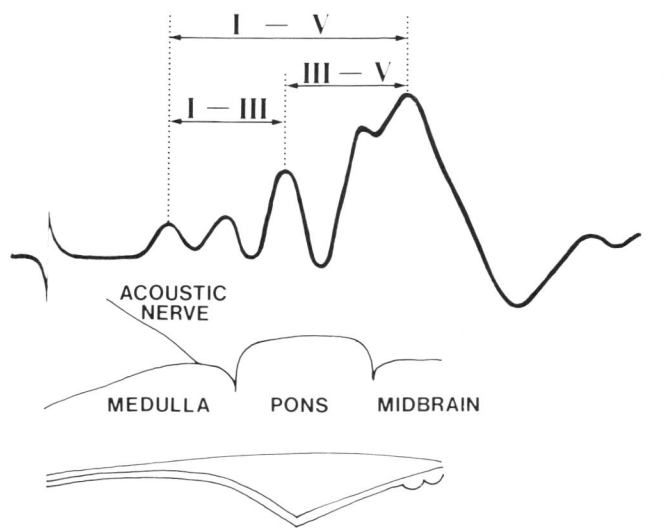

12.5.2 Lower and upper brain stem

Lesions of the lower brain stem reduce and delay BAEP waves starting with wave II or III; they may increase IPLs I–III and I–V. Lesions of the upper brain stem reduce and delay waves IV and V; they may increase IPLs III–V and I–V but spare IPL I–III. However, a clear distinction between these defects and a precise localization of the level of the lesion within the brain stem are often impossible.

The side of unilateral brain stem lesions may be identified by comparing BAEPs to stimulation of each ear. A unilateral lesion may cause abnormalities only in the BAEP elicited by stimulation of the ipsilateral ear and recorded on both sides of the head.

Brain stem lesions may also cause abnormalities of other AEPs, but these abnormalities do not generally help to localize the lesions to the stem.

12.5.3 Subcortical and cortical cerebral connections

Although the BAEP waves VI and VII, like the peaks of the MLAEP and those of the LLAEP, are generated by subcortical and possibly cortical auditory connections, they are quite variable and have not been clearly related to specific structures. Therefore, they have not proved useful in detecting and localizing lesions in the cerebral parts of the auditory pathway.

12.5.4 Audiological strategies: Hearing loss due to peripheral defects

AEPs may be used to evaluate hearing ability. In principle, electric response audiometry (ERA) attempts to detect and to classify hearing losses by determining the threshold stimulus intensity that produces a visible AEP and by measuring the latency of AEPs produced by higher stimulus intensities.[75,124,316] ERA may also be used to select hearing aids.[193]

BAEPs are now used most widely for ERA. A common test procedure first applies a click stimulus of 60–70 dB nHL or of 110–120 dB peSPL (11.3.4) to elicit a BAEP. Wave V is used as the indicator because it has the lowest threshold and highest amplitude of the waves of the conventional BAEP. If wave V can be distinguished at this level, stimuli of 20 or 30 dB nHL, or 75 dB peSPL, are used next (Figure 12.3). If wave V appears at this lower

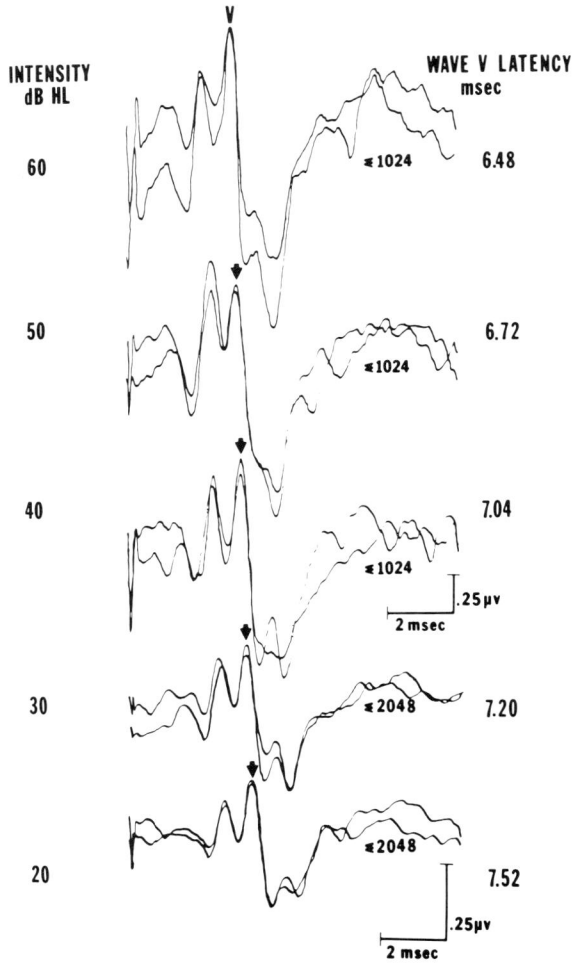

Figure 12.3. Normal BAEPs to click stimuli of decreasing intensity in a 13-month-old infant. Wave V (arrows) increases in latency (numbers to the right of tracings) and decreases in amplitude with decreasing stimulus intensity (note higher sensitivity for the last two tracings). Numbers below the ends of the tracings indicate numbers of responses collected for each BAEP. Two BAEPs are superimposed for each stimulus intensity. Recordings between vertex and mastoid on the side of stimulation. Positivity at the vertex is plotted upward. From Mokotoff et al.,[263] Arch. Otolaryngol. 103:38–43, 1977, Copyright 1977. American Medical Association, with permission by the authors and the American Medical Association.

level, the search for the threshold is discontinued in most laboratories because this finding is consistent with normal hearing and further definition of the threshold is without practical audiological importance.[263,372,408] If no BAEP can be distinguished either at the initial 60- or 70-dB nHL intensity or at the 20- or 30-dB level, averages to stimuli increasing in intensity by 10 dB are obtained until the threshold of wave V is found or a maximum of about 90 or 100 dB nHL is reached. Latencies of wave V at different intensities may be used to plot latency-intensity curves showing the decrease of latency of wave V obtained from increasing the stimulus intensity (Figure 12.4). Separate normal latency-intensity curves must be used for different age groups.

ERA is limited by two problems. (1) AEP thresholds are difficult to determine (4.1.3.1) and exceed the subjec-

Figure 12.4. Normal range of latency-intensity curves derived from BAEPs to click stimuli of different intensity in a group of audiologically normal adults. Horizontal axis: Intensity of the click stimulus in dB nHL. Vertical axis: Latency of wave V in msec.

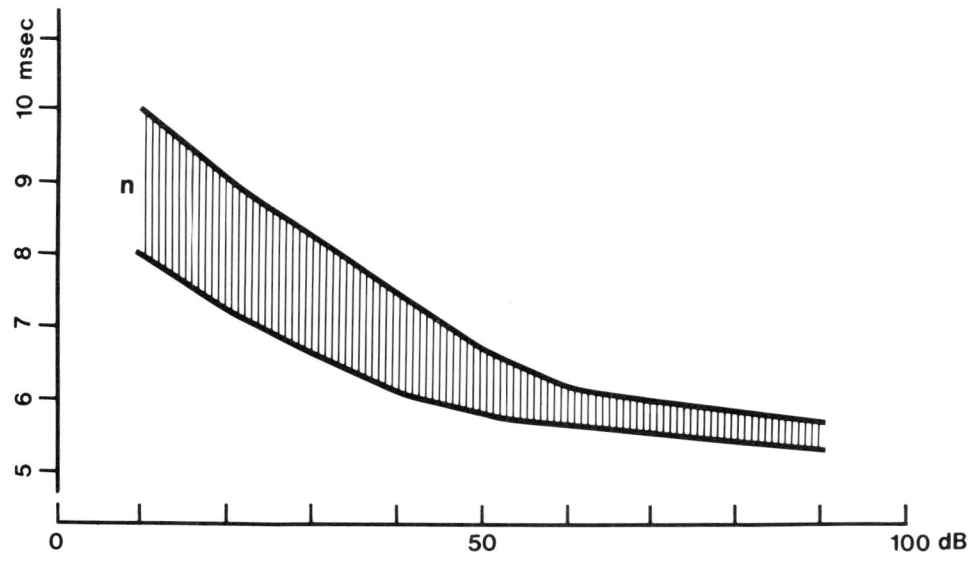

tive hearing level to an extent that varies between subjects. (2) The most effective stimuli, namely, clicks and brief tones, test only the upper part of the audible frequency range. Lower frequencies, especially those of the speech range, cannot be evaluated very effectively with the sudden and brief stimuli required for BAEPs and MLAEPs (11.3.2).

The correlation between the presence or absence of BAEPs and hearing ability is not perfect. Although most subjects showing BAEPs can hear, mild hearing defects may escape detection when only strong stimuli are used. Furthermore, low frequency hearing losses cannot be detected with stimuli of predominantly high frequency content such as clicks. On the other hand, BAEPs may be abnormal, especially in their late peaks, in subjects who can hear normally. This is common in patients with multiple sclerosis and is presumably due either to marked desynchronization of impulses or to abolition of impulses in some parts of the auditory system that may be critical for the production of BAEPs but not for hearing. Very rarely are BAEPs abolished without abolition of hearing (13.4.1.1). BAEPs may be partly mediated by auditory brain stem pathways that are not essential for hearing but subserve such other functions as the spatial localization of sound and the interaural discrimination of pitch and loudness.[164]

13

The abnormal BAEP

SUMMARY

13.1. Major criteria distinguishing abnormal from normal BAEPs are the complete absence of BAEP waves, absence of all waves following wave I or II, and abnormal prolongation of IPLs I–V, I–III, and III–V. Audiological studies use measurements of BAEP threshold and of latency-intensity relationships as indicators of abnormality.

13.2. General clinical interpretation uses abnormal BAEP findings to detect and localize lesions in the acoustic nerve, lower and upper brain stem, and to identify hearing problems.

13.3 Neurological disorders most often studied with BAEPs are multiple sclerosis, cerebellopontine angle tumors, and intramedullary brain stem tumors. In comatose adult patients, abnormal BAEPs indicate structural brain stem lesions, whereas normal BAEPs may be due to reversible metabolic encephalopathies, but do not exclude structural lesions. A variety of other neurological disorders have been found to produce BAEP abnormalities.

13.4. Audiological studies use BAEPs to determine magnitude and type of hearing defects in patients who cannot be examined with standard audiometric tests.

13.1 CRITERIA DISTINGUISHING ABNORMAL BAEPs

13.1.1 Neurological applications

13.1.1.1 Absence of all BAEP waves
The absence of waves I–V is established by a careful search for BAEP waves that includes the use of high stimulus intensities (12.3.4) and of numbers of responses sufficient for a good reduction of residual noise in the recording (3.3.3). Technical problems must be excluded (13.2.1.1).

13.1.1.2 Absence of all waves following wave I or wave II
Wave I must be carefully searched for in tracings showing mainly residual noise (12.1) because preservation of wave I has important clinical implications (13.2.1.2).

13.1.1.3 Abnormal prolongation of IPLs I–V, I–III, and III–V
The IPL I–V is calculated by subtracting the peak latency of wave I from the peak latency of wave V or of the wave IV–V complex. If wave I cannot be clearly distinguished in recordings between vertex and stimulated ear, the negative wave preceding wave II may be more clearly defined and should be used for IPL measurements. This wave may also be used for IPL measurements in recordings between vertex and nonstimulated ear that do not contain wave I. Normative data for these measurements should therefore be collected.

IPLs I–III and III–V are determined by subtracting the latency of wave I from that of wave V and by subtracting the latency of wave III from that of wave V. An increase of one or both of these IPLs, usually associated with an increase of IPL V, may help to further localize the underlying lesion (12.5.2).

13.1.1.4 Absence of wave V or abnormal decrease of the amplitude ratio of waves V and I
Peak-to-peak amplitude measurements (4.1.4.1.1) of wave I and of wave V or of the IV–V wave complex are used to calculate the amplitude ratio of wave V and wave I. This ratio is a more reliable measure than the amplitude of wave V itself and should be determined routinely except in cases of complete absence of wave V. The evaluation of the ratio is complicated by its dependence on click intensity and its variation with hearing loss. Because the ratio has a non-Gaussian distribution in the normal population, the normal range must be defined in terms other than those of mean and standard deviation (5.2.2).

13.1.1.5 Interaural latency difference of IPL I–V
The IPLs I–V to stimulation of each ear may differ to an extent that exceeds the normal range for this interaural latency difference even if neither IPL itself exceeds the normal range for IPL I–V.

*13.1.1.6 Abnormal increase of latency
to rapidly repeating stimuli*
The latency of wave V at stimulus rates of 60–100/sec is usually longer than that at 10–30/sec (12.3.2). An increase of this rate-dependent latency shift beyond the normal range suggests a lesion (13.2.1.6).

13.1.1.7 Abnormal peak latency
Because measurements of IPL are relatively immune to peripheral hearing loss, they are preferred to measurements of peak latencies in neurological studies. However, the latency to the peak of wave I is useful in evaluating peripheral lesions. Furthermore, peak latency of wave V and III must be used instead of IPLs in those instances where wave I and other waves preceding wave V cannot be distinguished.

13.1.2 Audiological applications

13.1.2.1 Abnormally high threshold
An increase of BAEP threshold above 30 dB nHL or 75 dB peSPL is abnormal.

*13.1.2.2 Abnormally long latency of wave V
with normal IPL I–V*
In normal subjects of any age past infancy, the latency of wave V at 60 dB nHL does not exceed 6 msec. The peak latency of wave V is used in some laboratories to estimate hearing loss. This is justified only if the IPL I–V at the same stimulus intensity is not prolonged because such a prolongation may indicate a brain stem lesion.

13.1.2.3 Abnormal latency-intensity curve
The absolute value of latency of wave V and the slope of the latency-intensity curve depend on the subject's age and vary between laboratories. At any age past infancy, the decrease of latency with increasing stimulus intensity should not exceed 40 μsec/dB at middle intensities and about 30 μsec/dB at higher intensities of 60–70 dB HL.[59,112,170,316,371] To determine the slope reliably, measurements should be made over a range of at least 30 dB.

13.2 GENERAL CLINICAL INTERPRETATION OF ABNORMAL BAEPS

13.2.1 Neurological disorders

13.2.1.1 Absence of all BAEP waves
Inability to record waves I–V on stimulation of either ear may be due to various technical problems such as lack of an effective stimulus, faulty synchronization between stimulus and averager, or defects of recording electrodes, electrode connections, amplifiers, or averaging channels (Table 13.1). Absence of waves I–V on stimulation of one or both ears may be due to severe conductive or sensorineural

TABLE 13.1. Clinical interpretation of abnormal AEPs

Abnormal BAEP findings	Interpretation
A. Technical problems	
1. Absence of entire BAEP	Lack of stimulus; failure of stimulus to reach ear; failure of synchronization between stimulus and averager; faulty recording electrodes or equipment
2. Decreased amplitude, increased latency of entire BAEP	Reduced stimulator output; faulty coupling of stimulator to ear
B. Neurological strategy	
1. Absence of all BAEP waves on stimulation of one side	Unilateral distal acoustic nerve lesion; rule out unilateral peripheral hearing loss
2. Absence of entire BAEP on stimulation of either side	Bilateral distal acoustic nerve lesions or brain death; rule out bilateral peripheral hearing loss
3. Absence of all waves after wave I on stimulation of one side	Unilateral proximal acoustic nerve lesion or unilateral pontomedullary lesion
4. Absence of all waves after wave I on stimulation of either side	Bilateral proximal acoustic nerve lesion, bilateral pontomedullary lesion, or brain death
5. Increased latency of wave I and all subsequent waves	Peripheral hearing loss or acoustic nerve lesion
6. Absent wave I, normal IPL III–V	Suspect peripheral hearing loss; normal conduction between lower pons and midbrain
7. Absent wave V or decreased amplitude ratio V/I	Ipsilateral lower or upper brain stem lesion
8. Increased IPL I–V or interaural latency difference of IPL I–V with	
a. increased IPLs I–III and III–V	Ipsilateral upper and lower brain stem lesion
b. increased IPL I–III, normal IPL III–V	Ipsilateral lower brain stem lesion between acoustic nerve and lower pons
c. increased IPL III–V, normal IPL I–III	Ipsilateral upper brain stem lesion between lower pons and midbrain
9. Abnormal increase of latency of wave V on rapidly repeating stimulation	Suspect ipsilateral lower or upper brain stem lesion

TABLE 13.1. *(continued)*

Abnormal BAEP findings	Interpretation
C. Audiological strategy	
1. Increased BAEP threshold or absence of entire BAEP to strongest stimuli	Suspect peripheral hearing loss; rule out distal acoustic nerve lesion
2. Shift of latency-intensity curve of BAEP wave V	
a. Parallel shift	Conductive hearing loss
b. Shift mainly at low intensities	Sensorineural hearing loss

hearing loss which should be further investigated audiologically. A neurological cause for the complete absence of waves I–V is a lesion of the distal acoustic nerve. Only in a few cases of absent BAEPs is hearing not abolished (13.4.1.1).

13.2.1.2 Absence of all waves following wave I or wave II

Preservation of initial waves indicates that the absence of later waves is not due to technical problems or peripheral lesions and therefore provides clear evidence for a retrocochlear lesion involving the proximal acoustic nerve or the pontomedullary region of the brain stem. Bilateral absence of all waves after wave I may be seen in brain death (13.3.3.1).

13.2.1.3 Absent wave V or decreased amplitude ratio of waves V and I

Selective reduction or abolition of wave V often indicates a midbrain lesion.

13.2.1.4 Increased latency of all BAEP waves

A delay of all recorded BAEP waves may be due to a reduction in stimulus intensity caused by technical problems. Hearing loss, especially conductive hearing loss, may have the same effect. However, a lesion of the distal acoustic nerve may also delay the entire BAEP. The IPL I–V should be normal in those instances. An additional central lesion must be suspected if this IPL is also increased. In the absence of a measurable wave I, a delay of waves III and V may be attributable to a peripheral problem if the IPL III–V is normal.

13.2.1.5 Increased IPL I–V or increased interaural difference of IPLs I–V

Abnormal separation between wave I and V usually indicates a central lesion. The lesion may be further localized by measurements of IPLs I–III and III–V (12.5.2). The increased IPL I–V is of such diagnostic importance that it is often searched for by comparing the IPL I–V on both sides; an abnormal difference may indicate a lesion on the side of the longer IPL I–V (13.1.1.5). Even in cases where wave I and other waves preceding wave V cannot be distinguished, the finding of a normal peak latency of wave V is very useful because it practically excludes both central and peripheral lesions.

The rule that an increased IPL I–V means a central lesion has several exceptions. Acoustic nerve lesions may cause abnormalities beginning with wave I and thus resemble peripheral lesions. Both a reduction of stimulus intensity and a change of stimulus polarity (12.3.3) can increase the IPL I–V.[407] Patients with recurrent middle ear disease may develop increased IPLs.[105] Peripheral hearing losses, causing a reduction of stimulus intensity, may lengthen IPL I–V if patients are tested at stimulus intensities below that used as the laboratory standard. Although mild conductive hearing defects can be compensated for by increasing the stimulus intensity to a standard level above the patient's hearing threshold, it may be impossible to overcome the effect of severe hearing losses. If a peripheral hearing loss reduces the effective stimulus intensity relative to the patient's hearing ability, the IPL may be found to be increased in comparison with normative values for standard stimulus intensities. The situation is even more complicated in high-frequency hearing losses which increase the latency of wave I without increasing the latency of wave V to the same degree, thereby reducing the IPL I–V[63] and theoretically decreasing the ability to detect central lesions which increase the IPL I–V.[415] Another problem arises if wave I, having a higher threshold than wave V, is absent from a tracing that shows an abnormally prolonged latency of wave V; this makes it impossible to use IPL I–V as a criterion for the distinction between central and peripheral lesions. The interpretation of BAEPs for neurological applications must take into account these audiological problems to avoid misinterpretations. Audiological evaluation should be obtained in case of any doubt about the central cause of BAEP abnormalities. Such doubt is especially likely to arise in cases showing absence of all BAEP waves or of all waves after wave I.

13.2.1.6 Abnormal increase of latency of wave V to rapidly repeating stimuli

An abnormal shift of wave V latency with increasing stimulus rate may suggest a brain stem lesion.[97,169,322,415,459] Rate-dependent latency shifts of earlier BAEP peaks can usually not be reliably measured because these peaks rapidly lose amplitude at high stimulus rates.

13.2.2 Audiological disorders

In hearing loss due to cochlear and middle ear disorders (Table 13.1), the BAEP may show an increased threshold and an increased latency of wave I and all subsequent waves. Plotting of the latency-intensity curve (12.5.4) can distinguish conductive, sensorineural, and other hearing losses (13.4).

13.3 NEUROLOGICAL DISORDERS THAT CAUSE ABNORMAL TRANSIENT BAEPs

BAEPs are most useful in the diagnosis of (1) multiple sclerosis, (2) extramedullary and intramedullary brain stem tumors, (3) coma due to metabolic and structural lesions, and (4) several other disorders in which BAEPs have been reported to be abnormal. In infants, BAEPs are usually evaluated both for neurological and for audiological purposes to search for lesions in the central auditory pathway and for peripheral hearing defects.

13.3.1 Multiple sclerosis

Abnormal BAEPs are found in many patients with multiple sclerosis; abnormalities occur most often in patients who have a diagnosis of definite multiple sclerosis and in patients who show clinical signs of brain stem involvement. However, abnormal BAEPs are found often enough in patients with probable or possible multiple sclerosis who do not show evidence of brain stem lesions to make the test diagnostically valuable.[19,53,97,104,196,198,248,250,340,413,423] The BAEP may suggest the diagnosis of multiple sclerosis in patients who present with retrobulbar neuritis[422] or transverse myelitis.[120,344]

No particular BAEP abnormality is specific for multiple sclerosis: The type of abnormality varies with the location and the severity of the demyelinating lesions. Figure 13.1 shows an increased IPL I–V and, probably, an increase of IPL I–III, suggesting a lesion in the lower brain stem. Figure 13.2, from another patient, shows nearly complete abolition of wave V with preserved waves I–III, suggesting a lesion in the upper brain stem. Such findings can help to distinguish central lesions from vestibular lesions producing vertigo and nystagmus and thus mimicking brain stem involvement. Sequential recordings of BAEPs in patients with multiple sclerosis show that a BAEP abnormality may increase or decrease with time, and may occasionally disappear completely. Although such fluctuations are often related to clinical changes,[341] they have also been found to occur in clinically stable patients and are therefore not very useful for serial monitoring of changes in clinical condition.[195]

The effectiveness of the BAEP in detecting abnormalities in patients with multiple sclerosis has been compared with that of other EPs. BAEPs have generally been reported to be less effective than VEPs to checkerboard stimulation and SEPs.[53,141,198,208,259,330]

13.3.2 Extramedullary and intramedullary brain stem tumors

13.3.2.1 Cerebellopontine angle tumors

Acoustic neurinomas and meningiomas impinging on the brain stem nearly always produce BAEP abnormalities at the time of diagnosis.[20,29,48,57,129,159,176,252,343,378,431,434] Whereas false negative results are very rare, false positive results are more common because BAEP abnormalities are not specific and may result from damage to the proximal or distal parts of the acoustic nerve and from compression and distortion of the brain stem. The high incidence of abnormal findings has made the BAEP, in combination with the CT scan, the most powerful tool in the diagnosis of acoustic neurinomas.

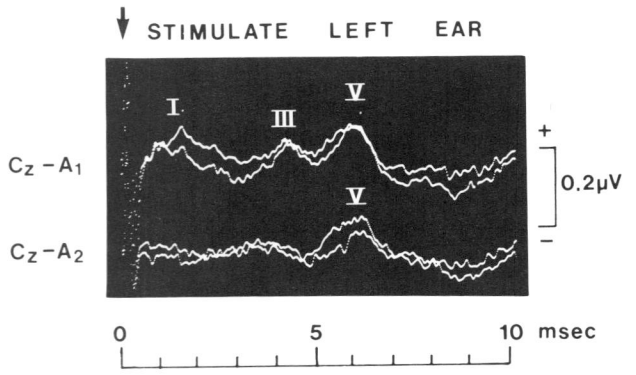

Figure 13.1. Abnormal BAEPs with increased IPL I–V ipsilateral to the stimulated ear and increased peak latency V on the other side. The wave between I and V in the ipsilateral BAEP probably represents wave III and suggests an increased IPL I–III. This would be consistent with a lesion in the lower brain stem. This 49-year-old man has multiple sclerosis manifested by progressive spastic paraparesis of two years and recent onset of bilateral cerebellar signs.

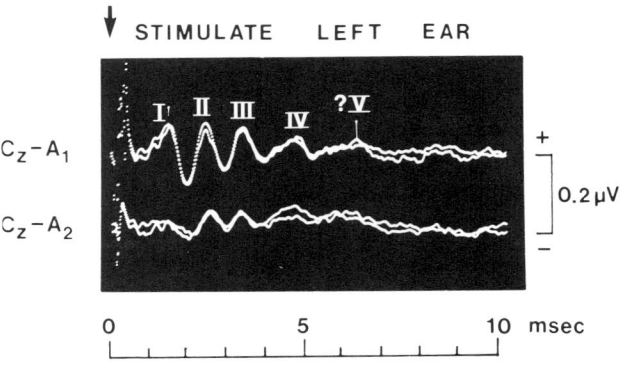

Figure 13.2. Abnormal BAEPs to stimulation of the left ear showing only a questionable wave V (second cursor) which, if used for measuring, indicates abnormal IPLs I–V and III–V, and abnormally low amplitude ratio V/I ipsilateral to the stimulated ear; in the contralateral recording, wave V is completely abolished. These findings suggest a lesion in the upper brain stem. This 27-year-old woman was suspected of having multiple sclerosis because of a right central scotoma, present for five years, and the recent onset of horizontal and rotatory nystagmus.

A frequent BAEP abnormality in cases of acoustic neurinomas is the absence, increased latency, or duration of wave I (Figure 13.3 *A,B*). Subsequent waves are usually severely abnormal or absent, although they may be preserved, but delayed (Figure 13.3*C,D*). Wave V often has an increased peak latency in comparison either with normal ears or with the opposite ear of the same subject. If waves I and III are preserved, the IPL I–III is often prolonged as a result of pontine compression; this prolongation is a very sensitive indicator of cerebellopontine angle tumors (Figure 13.3*E*). An increased IPL III–V may be found on

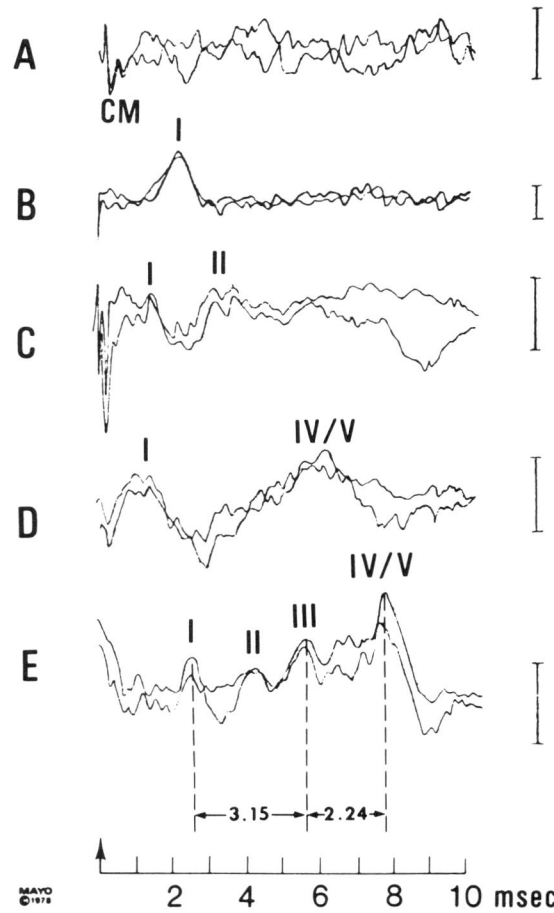

Figure 13.3. Five abnormal BAEP patterns in cerebellopontine angle tumors. *(A)* Absence of all BAEP waves and preservation of the cochlear microphonic (CM) in a large intracanalicular acoustic neurinoma extending into the cerebellopontine angle and compressing the brain stem. *(B)* Absence of waves after a delayed wave I in a small petrous ridge meningioma compressing the extracanalicular portion of the eighth nerve and the pontomedullary junction of the brain stem. *(C)* Absence of waves after wave II with increased IPL I–II in a lesion near the entrance of the acoustic nerve into the brain stem. *(D)* Increased IPL I–V, decreased amplitude ratio V/I, and absence of wave III in a large extracanalicular neurinoma causing severe pontine compression and bilateral cranial nerve defects. *(E)* Increased latency of wave I and increased IPL I–III in a small cholesteatoma involving the eighth nerve at its entrance into the cerebellopontine angle. Click stimulation at 60 dB SL of the ear on the side of the tumor. Recording between vertex and stimulated ear. Positivity at the vertex is plotted upward. Calibration 0.1 μV. From Stockard et al.[415] with permission by the authors and the Mayo Clinic.

recording from the opposite side, probably as a result of distortion of the lower midbrain.[415,453] The BAEP may normalize after surgical removal of the tumor.[238,415]

13.3.2.2 Intramedullary brain stem tumors
Wave I, generated by the acoustic nerve, is usually preserved, whereas later waves are very often abnormal (Figure 13.4). Waves III, IV, and V may be absent; if present, they may show increased latency and decreased amplitude. The IPL I–III is increased if the tumor is located in the pontomedullary region or pons, and the IPL III–V is increased if it involves the pontomesencephalic junction or midbrain. Several IPLs of the BAEPs to stimulation of each ear may be increased by diffuse brain stem tumors; the BAEP is a very sensitive indicator of infiltrating pontine gliomas.[104,278,401,403,413,415]

13.3.2.3 Surgical monitoring
BAEPs have been useful for monitoring the condition of the acoustic nerve and brain stem during surgery, especially since the BAEP is not affected by ordinary anesthetics (13.3.4.2). BAEPs have therefore been used in various types of operations of the posterior fossa.[4,148,153,161,335,336,394]

13.3.3 Coma

BAEPs may help to distinguish coma due to structural lesions from coma due to metabolic, toxic, and other potentially reversible encephalopathies: Structural lesions may cause abnormal BAEPs, whereas metabolic and toxic encephalopathies are generally associated with normal BAEPs unless they cause irreversible damage. However, normal BAEPs cannot definitely exclude the possibility of structural damage in comatose patients, especially damage involving higher cerebral structures.

13.3.3.1 Coma due to structural lesions: Assessment of survival
In comatose patients with head injuries, the BAEP has moderate prognostic value;[8,189,261,376,443] it is less useful than other EPs.[144,229,231,334,443] Survivors have been reported to show normal BAEPs after several months.[370] The BAEP is abnormal in the majority of patients in coma due to meningoencephalitis, but the BAEP is not accurate for prognosis.[183] In spontaneous intracerebral hemorrhage, the BAEP has been found to indicate brain stem damage and to predict outcome.[230]

Widespread postanoxic or posttraumatic cerebral damage is often associated with abnormal BAEPs (Figure 13.5), but may be associated with normal BAEPs, especially in patients who show clinical or laboratory evidence for widespread damage to cortical, subcortical, and diencephalic structures but have preserved brain stem function.[143,189,229,443,445] In contrast, patients in coma due to pontomesencephalic encephalomalacia show abnormalities of early waves corresponding with the level and side of the lesion or show no recognizable waves at all.[104,155,403]

Patients with sufficient brain stem damage to fulfill the criteria of cerebral death show no BAEP waves after wave I or, occasionally, after wave II.[132,202,376,400,403,415] The BAEP can suggest the diagnosis of cerebral death only if wave I is preserved because the complete absence of BAEPs may be due to deafness caused by peripheral auditory problems,

Figure 13.4. BAEPs in a patient with intramedullary brain stem ependymoma. The BAEP to stimulation of the right ear (R) shows only waves I and II. The BAEP to stimulation of the left ear (L) shows waves I, II, and III with normal IPLs. Click stimuli at 80 dB nHL. Recordings between vertex and stimulated ear. Negativity at the vertex is plotted upward; the BAEP waves point down. CT scan shows an enhancing lesion filling the fourth ventricle and situated to the right of the midline. From Nodar et al.[278] with permission by the authors and Laryngoscope Co.

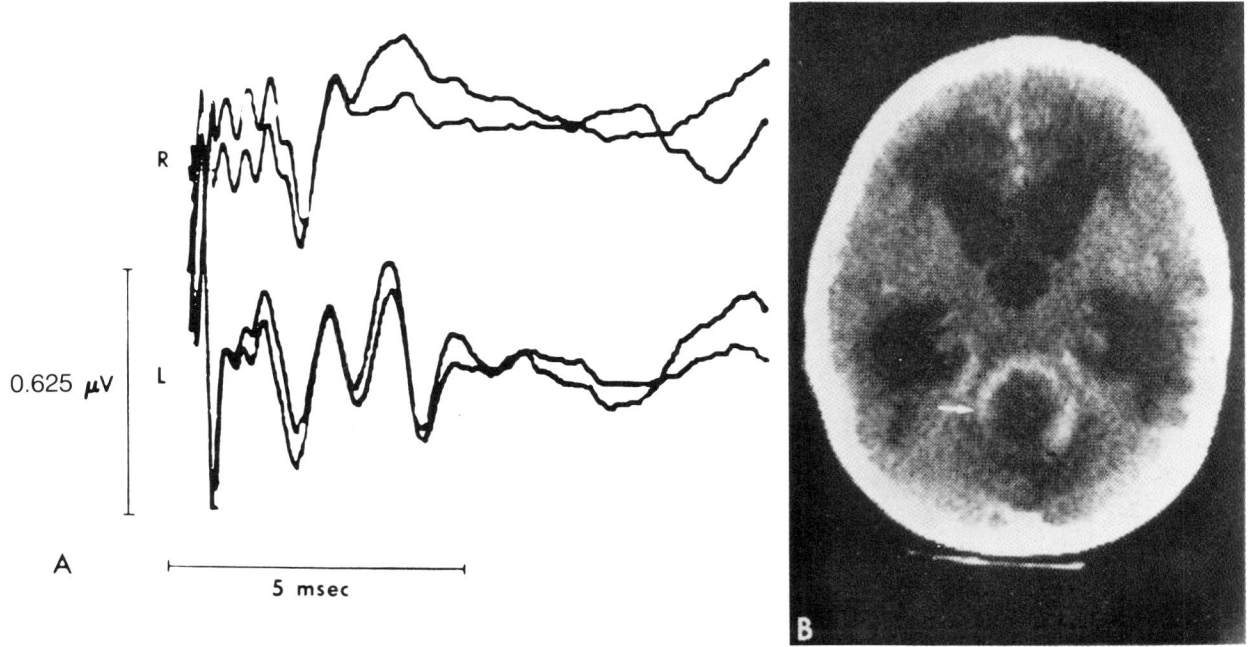

Figure 13.5. BAEPs in a 3-month-old patient with postanoxic encephalopathy. Top tracings show normal adult BAEPs. Middle and bottom tracings show BAEPs to stimulation of the right and left ear of the patient; waves I have normal latency for the patient's age, but later waves, although visible, cannot be identified by comparison with normal waves. Stimulation with clicks of 75 dB nHL. Recordings between vertex and stimulated ear. Positivity at the vertex is plotted upward. Two tracings are superimposed for each condition. Damaged areas are indicated by stippling in the diagram at the left. From Starr and Hamilton[403] with permission by the authors and Elsevier Scientific Publishers Ireland Ltd.

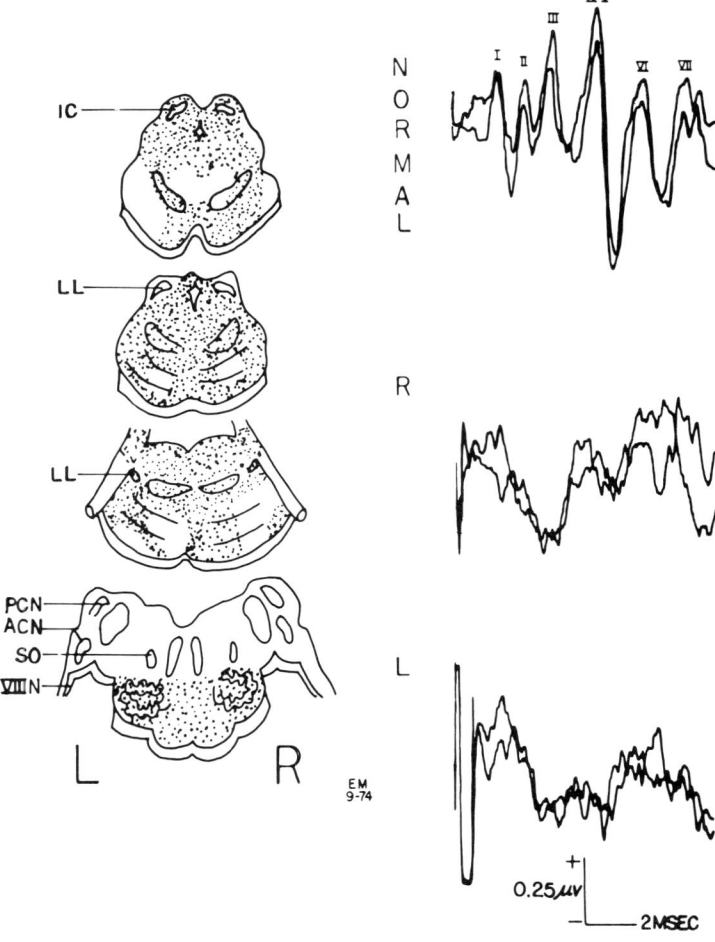

lesions of the acoustic nerve, or malfunction of stimulating and recording equipment.[132] The disappearance of the BAEP does not necessarily indicate brain death but has occasionally been observed to be transient even in cases of structural damage.[424] On the other hand, preservation of the BAEP may help to refute the diagnosis of cerebral death.

13.3.3.2 Coma due to metabolic and toxic encephalopathies

In coma due to metabolic derangements such as uremia, hepatic failure, and diabetic ketoacidosis, or coma due to CNS depressant drugs,[387,415] the BAEP is nearly always normal. Anesthesia with halothane and sodium thiopental does not affect the BAEP.[362] Exceptions are enflurane anesthesia[415] and imipramine overdose,[329] both of which increase IPLs. The BAEP may persist even if the EEG shows electrocerebral silence and the clinical condition indicates abnormal brain stem function.[415] However, the BAEP may be abnormal or absent in patients who have preexisting hearing loss and in patients who suffer structural brain stem damage in the course of metabolic or toxic coma, although increased IPLs have been reported in irreversible hepatic encephalopathy without anatomically demonstrable brain stem damage.[89] A fall of body temperature (12.2.3) to 32° C or less may occur in metabolic or toxic coma and is another possible cause of abnormal BAEPs in these disorders.[412] In infants, metabolic disorders such as phenylketonuria, Leigh's disease, and maple syrup urine disease may lead to coma associated with BAEP abnormalities that are probably due to faulty myelination.[169]

13.3.4 Other disorders

The BAEP has been reported to be abnormal in many disorders involving the brain stem. Although the BAEP may be of diagnostic use in only a few of them, these associations must be considered in the differential diagnosis to avoid misinterpretation of abnormal BAEPs.

13.3.4.1 Degenerative and other diseases of unknown cause involving the brain stem
Abnormal BAEPs have been reported in several disorders:

1. Olivopontocerebellar degeneration[126,151,232,281] (Figure 13.6)
2. Friedreich's ataxia[169,182,304,365,425]
3. Hereditary cerebellar ataxia[304]
4. Alzheimer's disease[158]
5. Parkinson's disease[118]

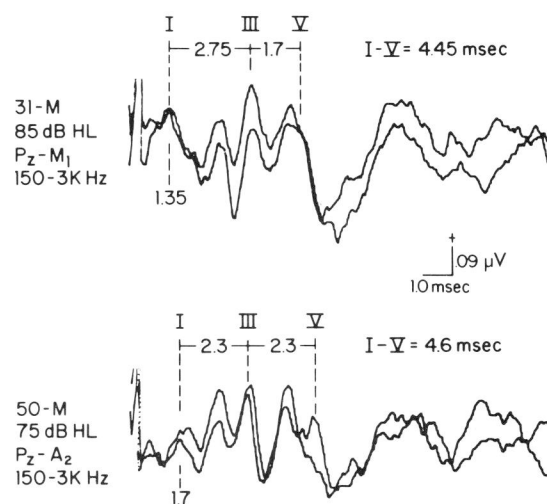

Figure 13.6. Abnormal BAEPs from two patients with olivopontocerebellar degeneration showing abnormally prolonged IPL I–III (top) and IPL III–V (bottom). Two tracings are superimposed for each BAEP. Recordings between vertex and stimulated ear. Positivity at the vertex is plotted upward. From Lynn et al.[232] with permission by the authors and Thieme-Stratton Inc.

6. Hemifacial spasm[268] (13.3.4.4)
7. Myotonic dystrophy[433]
8. Hereditary sensory neuropathy[114,364,415]
9. Dejerine-Sottas disease[16]
10. Albinism[71]

Normal BAEPs were reported in:

1. Supratentorial lesions[162]
2. Spinocerebellar degenerations other than those listed above[109]
3. Progressive supranuclear palsy[415,440]
4. Jakob-Creutzfeldt disease[51]
5. Huntington's chorea[94]
6. Amyotrophic lateral sclerosis[53]
7. Ménière's disease[53]
8. Labyrinthitis and vestibular neuronitis[53]
9. Subacute sclerosing panencephalitis[244]
10. Most cases of palatal myoclonus[452]
11. Tourette's syndrome[212]
12. During seizures induced by electroconvulsive therapy[450]
13. Precocious puberty[432]
14. Narcolepsy[171]
15. Schizophrenia and depression[36]

13.3.4.2 Toxic and metabolic disorders without coma

Drinking of alcohol produces an acute increase of the latency of BAEP waves following wave I,[55,397] which is probably not explained by hypothermia (13.3.3.2). IPLs may also be increased in chronic alcoholics,[54] even after abstinence.[27] A reversible increase of IPL I–V was observed in two alcoholics clinically diagnosed as having central pontine myelinolysis.[411] The BAEP remains normal in Wernicke's encephalopathy.[415]

Most sedatives and anesthetics do not affect BAEPs even in doses causing loss of consciousness (13.3.3.2). A slight increase of IPLs was found in patients on high phenytoin serum levels of over 20 µg/ml[142]; this probably does not significantly interfere with the ability of the BAEP to detect brain stem lesions in patients on therapeutic phenytoin levels. Toluene sniffing was found to abolish BAEP waves following wave II in two patients with cerebellar signs and dementia.[257] Treatment with aminoglycosides may produce BAEP changes.[146]

IPLs may be slightly prolonged in diabetics even at times when the blood sugar is not increased.[87] BAEPs may be abnormal in Leigh's disease,[244] Wilson's disease,[109] and hyperglycinemia.[242] Vitamin B_{12} deficiency caused abnormal BAEPs in one,[213] but not another,[103] study. Only a few patients with chronic renal failure had delays of wave V[180] or increased IPLs.[338] Hypoxia due to breathing air with 9–13 percent oxygen, or hypercapnia due to breathing 7.5–10 percent carbon dioxide, did not affect the BAEP.[388]

13.3.4.3 Strokes

Brain stem strokes may cause abnormal BAEPs[162] (13.3.3.1). The locked-in syndrome may be associated with an increased IPL III–V or other BAEP abnormalities,[127,377,403] but normal BAEPs have also been reported in this syndrome and in Wallenberg's lateral medullary syndrome.[104] Cortical deafness due to bilateral vascular lesions of the temporal lobe was associated with normal BAEPs[138,294] (14.3.3).

13.3.4.4 Vascular malformations of the posterior fossa
Arterial and venous malformations associated with hemifacial spasm (13.3.4.1), trigeminal neuralgia, and facial paresis may cause peak and interpeak latency abnormalities similar to those occurring in cerebellopontine angle tumors,[41] although normal BAEPs have also been reported in trigeminal neuralgia.[53]

13.3.4.5 Leukodystrophies
Abnormal BAEPs were found in patients with Pelizaeus-Merzbacher disease, adrenoleukodystrophy, and metachromatic leukodystrophy,[40,47,115,116,145,243,244] but not in patients with subacute sclerosing panencephalitis and with gray matter degenerations such as Batten's disease and Hallervorden-Spatz disease.[244]

13.3.4.6 Postconcussion syndrome
Prolonged IPLs after head injuries suggest residual organic brain stem damage.[279,337,355]

13.3.4.7 Mental retardation
Adult patients with severe mental retardation due to Down's syndrome or to unknown causes often show BAEP abnormalities indicating central damage or profound hearing loss.[396] The latency of wave V is shorter and the latency-intensity curve is steeper in infants with Down's syndrome than in normal subjects.[106]

13.3.4.8 Sleep apnea and sudden infant death syndrome
Sleep apnea in adults is associated with normal BAEPs[188,272,505] unless it is a symptom of a massive brain stem lesion.[409] A few BAEP investigations were unable to reliably discriminate infants at risk for the sudden infant death syndrome,[147,409] although another study found an increased IPL I–V in a group of premature babies with apnea.[172]

13.3.4.9 Neurological disorders in infants
Even though BAEPs are used in infants mainly to evaluate hearing, brain stem problems may be detected by prolonged IPLs and other abnormalities.[169,393,406] Infants with perinatal problems, such as complications during pregnancy, birth injury, respiratory distress,[360] intraventricular hemorrhage, and apneic syndrome[102] show an abnormal BAEP development. Asphyxic damage leading to neurological handicaps is characterized by abnormal BAEPs.[168] Some children with psychomotor retardation or "minimal brain damage" have BAEP findings suggesting central ab-

normalities.[393] Although some patients with infantile autism show BAEP abnormalities suggesting profound hearing loss, others show evidence for central problems.[280,346,385,393,426] BAEPs were severely abnormal in some infants with Gaucher's disease.[217] In children recovering from bacterial meningitis, BAEPs can detect sensorineural hearing loss.[209,295]

13.4 AUDIOLOGICAL DISORDERS THAT CAUSE ABNORMAL BAEPs

13.4.1 Effects of hearing loss on threshold and latency

13.4.1.1 Effect of hearing loss on threshold
Hearing loss increases the threshold of the BAEP, specifically that of wave V if the loss involves the frequencies of 1–4 kHz through which click stimuli exert their effect. Although the threshold of wave V may be only 5–10 dB above the hearing threshold in many adults and within 10–20 dB above the adult normal hearing level in many normal infants, increases of threshold of less than 30 dB above the normal hearing level cannot be taken as evidence of abnormal hearing.[69,112,263,408] Even then, the method is not absolutely reliable: A few subjects with behavioral evidence for hearing and without indication for brain stem lesions have no recordable BAEP, suggesting that the pathway tested by BAEPs is not entirely the same as that needed for hearing.[77,210,457] On the other hand, a normal threshold to click stimulation does not guarantee normal hearing, especially since this stimulus tests hearing at frequencies of mainly 2–4 kHz. For the same reason, it is not possible to determine the degree of hearing loss from the degree of threshold evaluation. But even with these limitations, BAEPs to clicks are probably the best currently available electrophysiological test of hearing ability.[74,112,166,316] They can be used to detect hearing losses in infants,[82,111,227,263,295,372,373,406,408,460] autistic[393] and difficult-to-test[339] children, and retarded children and adults.[396] In contrast, nonorganic hearing deficits are characterized by much lower BAEP thresholds than predicted by behavioral hearing tests.[386]

13.4.1.2 Effect of hearing loss on latency
Hearing loss increases the latency of wave V above the normal level. This increase has occasionally been used to estimate hearing loss. However, the relationship between these two variables is not simple. Central lesions must be excluded by demonstrating a normal IPL I–V at the stimulus intensity used. Even then, the degree of latency change depends on the type of hearing loss and is not necessarily proportional to the stimulus intensity.

13.4.1.3 Effect of hearing loss on the latency-intensity curve
Hearing loss can be further characterized by plotting the latency of wave V against the stimulus intensity and com-

paring the latency-intensity curve with a plot of the normal range (12.5.4). This method may help to distinguish conductive from sensorineural hearing loss.

13.4.2 Conductive hearing loss

13.4.2.1 Effect on amplitude and latency
Conductive hearing loss interferes with the conduction of sound waves from the ear canal to the cochlea and therefore acts like a reduction of stimulus intensity. The reduction of the click stimulus intensity produces a BAEP of lower amplitude and longer latency of all waves. At low stimulus intensities, the latency of wave I is more increased than that of other waves in conductive hearing loss so that IPLs I–V and I–III are shortened. These effects, if caused by moderate conductive hearing losses, can be compensated for by increasing the stimulus intensity. However, in many cases it may be impossible to completely overcome the hearing defect by increasing the stimulus intensity. In these cases, some laboratories attempt to derive neurologically useful information by correcting latencies for the degree of hearing loss or by using midline loudness matching to determine the stimulus intensity to be used for the ear with decreased hearing. The general rules relating conductive hearing loss to BAEPs do not apply to conductive hearing loss caused by ossicular chain disorders.[237]

Because bone conduction is not affected in conductive hearing losses, some investigators have demonstrated the discrepancy between air and bone conduction by comparing BAEPs to earphone click stimulation with BAEPs produced by a bone vibrator stimulus[249] (11.3.1.3) or by recording BAEPs to clicks masked by bone-conducted noise.[174] Impedance audiometry is probably at least as effective in diagnosing conductive defects as any BAEP test.

13.4.2.2 Effect on latency-intensity curve
Conductive hearing defects increase the latency of wave V equally over the entire range of intensities and therefore shift the latency-intensity curve by an amount equal to the hearing loss; a parallel shift of this sort is therefore characteristic of conductive defects (Figure 13.7). The possibility of a central defect must be excluded by ascertaining that the IPL I–V is normal.

13.4.3 Sensorineural hearing loss

13.4.3.1 Effect on amplitude and latency
In sensorineural hearing loss, the BAEP amplitude is decreased, at least at moderate stimulus intensities, and the amplitude ratio V/I is usually increased.[415] The latency of wave V is increased, in keeping with the degree of hearing loss at 4 kHz.[352] However, wave I, if visible, is at least equally increased in latency, causing a normal or even abnormally short IPL I–V.[63,347,352]

BAEP recordings that include the cochlear microphonic (14.6.3) may be used to subdivide sensorineural hearing losses into neural hearing loss in which wave I is absent while the cochlear microphonic is preserved and sensory hearing loss in which both wave I and cochlear microphonic are absent.[390]

13.4.3.2 Effect on latency-intensity curve
Many causes of high frequency sensorineural hearing loss show the greatest deviation from normal latency and amplitude at low stimulus intensities: The more the stimulus strength exceeds the threshold, the less the disparity between the abnormal and the normal curve; at high intensities, the latency may be normal. This causes a characteristic steepening of the latency-intensity curve[59] which becomes L-shaped, having a very steep slope of up to 240 μsec/dB near threshold and a less steep slope of under 30 μsec/dB

Figure 13.7. Latency-intensity curves in cases of conductive hearing loss (c) and sensorineural hearing loss (s-n) as contrasted with the normal range (n, see Figure 12.4). Horizontal axis: Intensity of the click stimulus in dB nHL. Vertical axis: Latency of wave V in milliseconds.

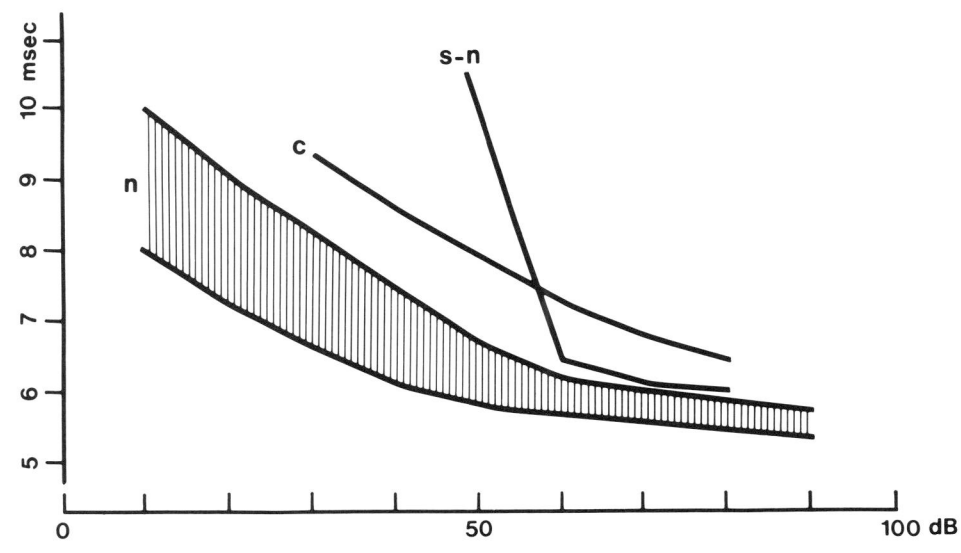

at higher intensities[112,316] (Figure 13.8). The increased threshold and rapid decrease of abnormally prolonged latency of wave V is often found in patients who have a hearing defect characterized by recruitment, i.e., a hearing deficit which is maximal near threshold and decreases with increasing loudness; this kind of deficit is common in many cases of cochlear lesions, but rare in retrocochlear lesions.[112] Electric and behavioral recruitment are characteristic of many disorders causing high-frequency sensorineural hearing loss, for instance, Ménière's disease and presbyacousis.

14
Other AEPs

SUMMARY
Several other AEPs have limited value for audiological purposes and little practical value for neurological purposes.

14.1. The slow brain stem AEP is a relatively slow wave which peaks about 10 msec after brief tone pips and may indicate hearing threshold at frequencies as low as 1,000 Hz and perhaps even 500 Hz.

14.2. The frequency following potential (FFP) is a steady-state AEP that appears during stimulation with tone bursts of 100–2,000 Hz. The FFP consists of a train of rhythmical waves which have the frequency of the stimulus and a latency resembling that of BAEP components. However, the threshold of the FFP is not close to the hearing threshold and not very specifically related to the stimulus frequency.

14.3. The middle latency AEP (MLAEP) has up to five peaks which occur between 10 and 50 msec after the stimulus and probably include the earliest response of the primary auditory cortex. However, the MLAEP is quite variable in normals and is often contaminated by muscle artifact.

14.4. The 40-Hz AEP is a steady-state AEP which is elicited by brief tone bursts that have various frequencies and are repeated at 40/sec. This AEP consists of rhythmical waves at 40/sec that are probably derived from the MLAEP. The role of the 40-Hz AEP as a hearing test has not yet been fully explored.

14.5. The long-latency AEP (LLAEP) occurs 50–300 msec after the stimulus and consists of one or two vertex-negative peaks, each preceded by positive peaks. It reflects activity of both primary auditory and higher cortical areas. The LLAEP may be elicited not only by clicks and tone pips but also by longer tones. It is not very valuable as a hearing test because its threshold varies considerably and changes with the state of alertness.

14.6. The electrocochleogram (ECochG) is recorded from the vicinity of the inner ear with electrodes piercing the eardrum or inserted into the external ear canal. It consists of three components: (1) The nerve action potential of the acoustic nerve (NAP), (2) the cochlear microphonic (CM) representing the response of cochlear sound receptors, and (3) the summating potential (SP) due to vibration of the basilar membrane. The NAP gives a rather accurate measurement of hearing threshold and, in conjunction with CM and SP, may be helpful in some otological studies. Because the first two components of the ECochG can be recorded with surface electrodes on the mastoid process, the BAEP now usually replaces the ECochG as a hearing test.

14.7. Sonomotor AEPs are due to reflex contractions of scalp and neck muscles in response to sound stimuli. They begin 12–15 msec after the stimulus and often contaminate recordings of MLAEPs and LLAEPs, especially when neck and scalp muscles are under tension. Sonomotor responses of the postauricular muscles have been used in audiological testing to demonstrate that a subject can hear. However, sonomotor responses cannot be used to determine hearing threshold, and their absence does not exclude hearing.

14.1 THE SLOW BRAIN STEM AEP

Stimulation with tone pips of gradually increasing amplitude produces a single wave that has longer duration and peak latency than the BAEP produced by clicks; peaks resembling those of the BAEP may be superimposed on the slow wave at high stimulus intensities (Figure 14.1). This slow wave has positive polarity at the vertex[199,200,254,399,418] and is most likely equivalent to a slow negative wave at about 10 msec (SN_{10}) recorded by other authors using higher settings of the low frequency filter.[77,165,190,381] Like the BAEP, this slow potential is probably generated by the brain stem.

The slow brain stem potential is of interest to audiologists because its threshold is close to the subjective hearing threshold and because it seems to be fairly specific for even relatively low tonal stimulus frequencies of 500 Hz and above.

14.2 THE FREQUENCY FOLLOWING POTENTIAL

Stimulation with tone bursts of a frequency of 100–2,000 Hz and a duration of at least a few cycles, repeated at 10–15/sec, and recording between vertex and ipsilateral or contralateral ear electrodes, produce the frequency following potential (FFP) which consists of a sinusoidal rhythm having the frequency of the stimulus and riding on a sustained elevation of the baseline[119,128,245,274] (Figure 14.2) A tran-

Figure 14.1. The normal slow brain stem AEP to tone pips of 0.5 and 1 kHz as contrasted with the BAEP to clicks. Stimulation with tone pips of decreasing intensity produces slowly rising EAPs which decrease in amplitude and increase in latency with decreasing stimulus intensity and can be distinguished to a stimulus level of about 20–30 dB HL. Stimulation with clicks of decreasing intensity induces BAEPs with decreasing numbers of peaks and decreasing amplitude; wave V can barely be distinguished at 30 dB HL. Broadband clicks are produced by electric pulses of 0.1 msec. Tone pips of 0.5 kHz have a duration of 6 msec and rise and fall times of 3 msec. Tone pips of 1 kHz have a 3 msec duration and 1.5 msec rise and fall times. Recording between forehead and mastoid on the stimulated side. Filter bandpass 0.2–2 kHz for BAEPs and 0.02–5 kHz for slow brain stem AEPs. Positivity at the forehead is plotted upward. From Maurizi et al.[254] with permission by the authors and S. Karger AG.

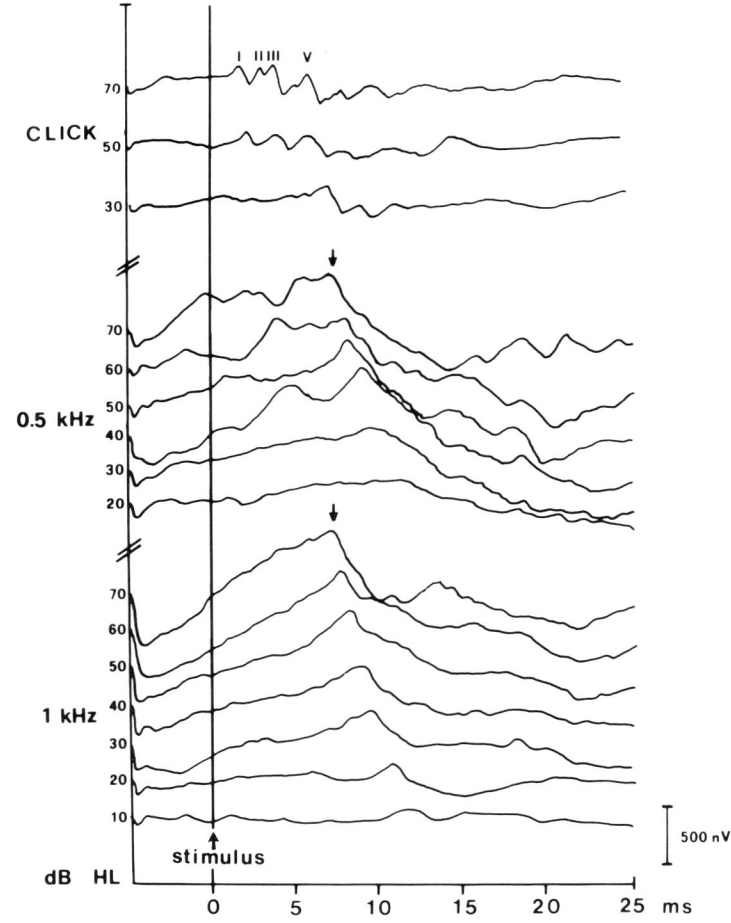

sient BAEP, especially a wave V, may be distinguished during the first few cycles of the FFP.[316,462] The FFP has irregular shape in some subjects; others show a response only at the onset of the stimulus.[76] The FFP varies depending on electrode placement[405] and may be partly obscured at ear electrodes by the cochlear microphonic[99,391,461] (14.6.3) and by postauricular responses[392] (14.7). At a frequency of 500 Hz and an intensity of about 60–70 dB SL, the latency between the onset of the tone burst to the FFP is about 6 msec; it increases with decreasing tone frequency and intensity.[119,128,245,274] The response to clicks given at the same frequency and a similar intensity as the waves of the

Figure 14.2. The normal FFP to stimulation with tones of various frequencies. *(A)* Absence of a FFP to stimulation with tones of 0.5 kHz completely masked by simultaneous application of continuous white noise. *(B-F)* FFPs to stimulation with tones of 2, 1.5, 1, 0.5, and 0.25 kHz. *(G)* Tone stimulus of 0.5 kHz. Recording between vertex and stimulated ear. Positivity at the vertex is plotted upward. Vertical calibration bar indicates 0.5 μV for A-F. From Moushegian et al.[274] with permission by the authors and Elsevier Scientific Publishers Ireland Ltd.

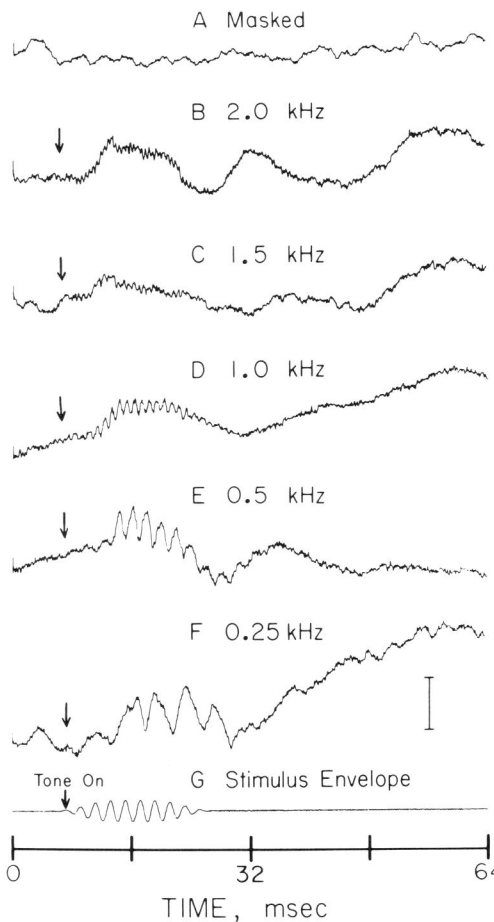

stimulus tone differs from the FFP and shows superimposition of different AEP peaks.[128]

The FFP is generally believed to be generated by brain stem structures which also produce the BAEP,[72,119,274,405] perhaps mainly by structures near the inferior colliculus,[392] although a peripheral origin[99] or contribution[461] has also been proposed.

The audiometric use of the FFP is limited for two reasons: (1) The threshold of the FFP is much higher than the hearing threshold[128,273,274]; (2) even though the FFP has the same frequency as the tone stimulus, it is not entirely specific for the stimulus frequency since FFPs to tone pips of relatively low frequency may be reduced by the simultaneous application of a constant masking sound of higher frequency[76,179] and by high frequency hearing defects. Low frequency hearing losses are especially likely to escape detection.[234] This has prevented the FFP from becoming widely used in electric response audiometry in spite of some encouraging initial results.[72,392]

14.3 THE MIDDLE LATENCY AEP

14.3.1 Methods of stimulating and recording

The subject's neck muscles must be completely relaxed because the slightest muscle tension facilitates sonomotor responses (14.7), which tend to obscure the MLAEP. Changes of attention, light sleep, and mild sedation do not affect the shape or threshold of the MLAEP.[215,256,293,311,384]

Although clicks and high frequency tone bursts are the most effective stimuli, the major audiological interest in the MLAEP lies in its potential to give frequency-specific information. Therefore, tone pips and filtered clicks are often used. However, the response seems to depend at least as much on the sudden onset of the stimulus as on its frequency.[215] Stimuli with a rise time of 5 msec are more effective than stimuli with more slowly increasing amplitude,[28,218] and those with rise times of over 25 msec produce unstable responses.[382] Low frequency tone bursts are less effective in eliciting MLAEPs than are filtered clicks.[466] Stimuli may be repeated at rates of up to 15/sec without a change of the response.[135,236] At higher rates, amplitude begins to decrease.[313]

Recordings are usually made between electrodes on vertex and ear or mastoid. Only one channel is used since MLAEPs to monaural stimuli recorded from the two sides of the head do not differ from each other.[308] Amplification requires a gain almost as high as that of BAEPs because MLAEPs are only slightly larger than BAEPs. To avoid distortion of slow components of the MLAEP, the low frequency filter is usually set at about 10 Hz or less[368] although a low filter of 25 Hz may speed up collection of MLAEPs.[236] High frequency filter settings of less than 150 Hz may distort amplitude and latency of the MLAEP.[219] The sweep length is usually about 100 msec. This results in a dwell time of about 100 μsec per point, or a sampling rate of about 10 kHz, for the usual averager having 1,024 points per channel. Longer dwell times of up to about 1 msec, corresponding with slower sampling rates of down to 1 kHz, are acceptable (3.3.4). Usually 1,000–2,000 responses are collected, although 512 responses[135] or even less[236] may be sufficient. With proper methods, the MLAEP may be recorded simultaneously with the BAEP.[293,367]

14.3.2 The normal MLAEP

Before the discovery of BAEPs, the MLAEP had been called early or early cortical AEP. It consists of several peaks of up to over 1 μV which occur 10–50 msec after the stimulus (Figure 14.3). They are often obscured by,[30] but not entirely due to,[156,357] scalp and neck muscle responses (14.7). MLAEPs vary much more between laboratories and subjects than do BAEPs. Nevertheless, there is some agreement that the MLAEP may show up to five peaks of negative (N) and positive (P) polarity at the vertex: N_o at 8–10 msec, P_o at 10–13 msec, N_a at 16–30 msec, P_a, usually the largest, at 30–45 msec, and N_b at 40–60 msec.[134,218,240,293,313] The first or second peak may be identical with the slow brain stem AEP (14.1). A sixth peak,

P_b, at 50–90 msec, is often found in recordings including the first 100 msec after the stimulus and represents the early peak (P_1) of the late cortical AEP (14.5). A decrease of stimulus intensity reduces the amplitude and increases the latency of the MLAEP.[240,313]

The MLAEP has a wide distribution with a maximum over the frontocentral areas.[131,313,416] An origin of the MLAEP near the auditory cortex has been postulated.[66,447,456] and is supported by direct recordings from the surgically exposed temporal cortex,[49,224,357] although depth recordings have suggested that the earlier peaks N_o, P_o, and N_a may be generated by subcortical structures.[160]

A few maturational studies suggest that the latency of MLAEP peaks decreases only slightly between infancy and adulthood.[17,239,255]

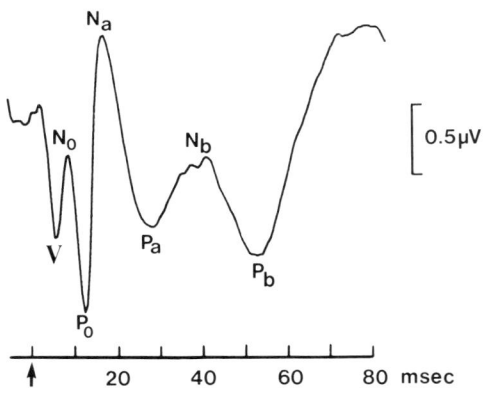

Figure 14.3. The normal MLAEP. This tracing is the grand average of MLAEPs from 35 normal subjects. Wave V of the BAEP precedes the MLAEP waves N_o, P_o, N_a, P_a, N_b, and P_b. Binaural click stimulation at about 75 dB SL (arrow). Recording between vertex and mastoid. Negativity at the vertex is plotted upward; BAEP wave V points down. Slightly modified from Robinson and Rudge[340] with permission by the authors and Oxford University Press.

14.3.3 Neurological disorders that cause abnormal MLAEPs

MLAEPs have been used infrequently in the diagnosis of neurological diseases. When recorded in combination with BAEPs, they have been reported to increase the incidence of abnormal findings in multiple sclerosis[340,341] and acoustic neurinomas.[343,431] Unilateral temporal lesions usually leave the N_a and P_a peaks of the MLAEP intact.[210] Bilateral temporal lobe lesions have been found to abolish the MLAEP without abolishing the BAEP (13.3.4.3) in cases showing cortical deafness[138,294] but to leave the MLAEP intact in a case of auditory agnosia.[303]

14.3.4 Audiological disorders that cause MLAEPs

Audiological interest in the MLAEP is raised by reports suggesting that the MLAEP threshold lies within about 20 dB of the hearing threshold during wakefulness and sleep[215,256] and that MLAEPs elicited by tone pips of different frequencies are specific for the tonal frequency of the stimulus even at relatively low frequencies.[190,215,438] However, the usefulness of the MLAEP in the evaluation of hearing loss has not yet been convincingly proved.[235,369,449] The threshold of the MLAEP may be as high as 60 dB HL in normal sleeping infants[255] and MLAEPs may be absent in some normally hearing subjects,[211,256] especially when tone pips rather than clicks are used.[218] Because of this variable relationship between MLAEP and hearing threshold and because of the frequent contamination by muscle responses, the test is often useless for estimating hearing level, especially in young children where it is most needed.[74] However, recording of the MLAEP in addition to the BAEP has been said to improve threshold evaluation.[108] As a minimum, the MLAEP may be of value in a qualitative manner: Its presence, like that of a postauricular response, may be taken as evidence that conduction from the cochlea to the central nervous system, and therefore hearing, is not entirely absent.

14.4 THE 40-HZ AEP

Stimulation with clicks or tone pips that repeat at 40/sec and recording between electrodes on the forehead and the stimulated ear using a bandwidth of 10–100 Hz produce a steady-state AEP that is probably composed of MLAEP components [113] (Figure 14.4). This 40-Hz AEP has a threshold of up to about 35 dB above hearing threshold during wakefulness. It can be studied with pips of a wide range of tonal frequencies including fairly low frequencies.[199,381] The 40-Hz AEP may therefore become a useful hearing test for low tone frequencies.

14.5 THE LONG-LATENCY AEP

14.5.1 Methods of stimulating and recording

The subject should be alert and relaxed during the recording. Sleep,[192,256,287,290,332] sleep deprivation,[117,327] and changes of attention[137,311,312,358,374,375] change the LLAEP. Infants and children often cannot be examined while awake and must be sedated.[73,361] Relaxation of neck and scalp muscles is not critical because myogenic components are unlikely to obscure LLAEPs. Stimuli may be delivered from loudspeakers unless monaural testing is needed.

Stimuli usually consist of tone bursts of 250–2,000 Hz. To avoid generating a response to the onset of the tone burst, the burst is given a gradual rise over a time of 25–50 msec and a plateau of 30–50 msec.[286,318,383] Stimuli of over 100 msec may produce separate responses to the onset and the end of the tone.[201] Stimuli are usually repeated regularly at 0.5–2/sec,[80] even though larger responses may be obtained at slower rates[277] and with irregular stimulation.[366,447] Stimulation at rates faster than 2/sec may change the wave shape[80] and latency.[417] A LLAEP can also be elicited by changes in tone intensity,[395] frequency,[205,356,395] and apparent location in space.[150]

Recordings are usually made in a single channel be-

Figure 14.4. The normal 40-Hz AEP. *(A)* Stimulation with weak tone bursts repeating at 3.3 and 10/sec produces MLAEP waves with peaks at 25 and 50 msec (top two tracings). Stimulation at 20 and 40/sec produces rhythmical steady-state AEPs with peaks at intervals of 25 msec, i.e., at a frequency of 40/sec (bottom two tracings). *(B)* Intermittent stimulation with stronger tone bursts produces a complete transient MLAEP consisting of waves with a period of about 25 msec (top tracing). Repetition of the stimulus at intervals of 25 msec, or at a frequency of 40/sec, leads to superimposition of MLAEP waves, the presumed mechanism of the 40-Hz AEP (bottom tracings). Each stimulus consists of a 500-Hz tone burst of 6 msec at 25 dB SL in *A* and at 40 dB SL in *B*. Recordings between forehead and stimulated ear. Negativity at the vertex is plotted upward. From Galambos et al.[113] with permission by the authors and the New York Academy of Sciences.

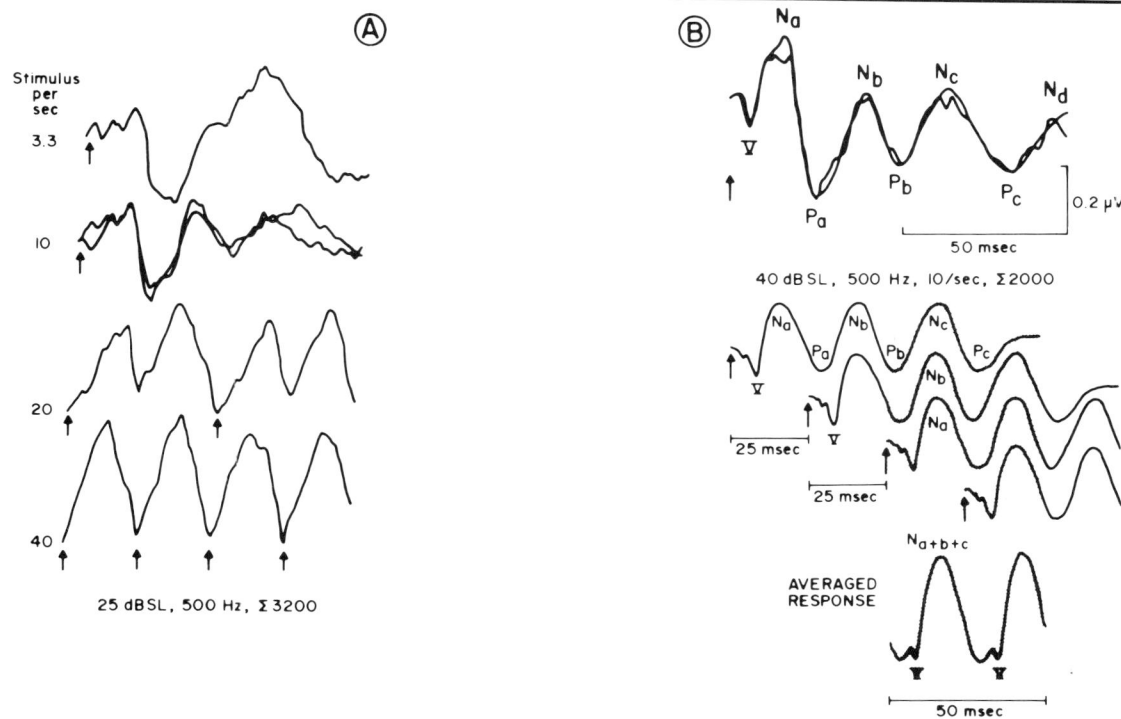

tween vertex and one ear or mastoid. Displacement of the vertex electrode by up to 6 cm does not change the response.[65] Amplification is not as high as for BAEPs or MLAEPs because the LLAEP amplitude is higher, ranging from 1 to 10 µV. Filter settings that include a bandwidth of 0.2–100 Hz are ample; narrower settings of only 2–15 Hz may be used.[121,186,292] The sweep length is usually about 500 msec. Only about 30–100 responses need to be averaged for one LLAEP.

14.5.2 The normal LLAEP

The LLAEP has an inconstant vertex-positive peak P_1 at 50–70 msec, a fairly large negative peak N_1 at 100–150 msec, and a positive peak P_2 at 170–200 msec (Figure 14.5). The prominent N_1–P_2 complex is usually followed by a negative peak N_2.[80,333] A third positive peak P_3 at about 300 msec, or the $\overline{P300}$, and peaks of even longer latency[233] depend on cognitive processes rather than the stimulus (21.1). Long tone stimuli cause a sustained negative potential shift with a delayed onset.[191,317]

The amplitude of the LLAEP decreases and its latency increases with increasing stimulus rates, with increasing stimulus rise time, and with decreasing stimulus duration. Similar changes occur with decreasing stimulus intensity, especially near threshold, and more so for click than for tone stimuli.[24,80,333,444] Tones of low frequency elicit larger LLAEPs than high tones of equal sensation level, and the amplitude of LLAEPs to low tones increases more with increasing stimulus intensity than does the amplitude of LLAEPs to high tones.[9,121,353,444] With increasing stimulus intensity, the LLAEP reaches a maximum after which a further increase of intensity may cause a decrease of amplitude.[310] Latency does not depend on the tonal frequency of the stimulus.[80]

The LLAEP has a wide distribution with a maximum at the vertex.[131,416] The distribution varies with stimulus frequency.[154] Responses recorded from the side opposite

Figure 14.5. The normal LLAEP. This tracing is the grand average of LLAEPs from 35 normal subjects. The LLAEP waves P_1, N_1, P_2, and N_2 are preceded by BAEP and MLAEP waves. The LLAEP wave P_1 is identical with the MLAEP wave P_b. Binaural click stimulation at about 75 dB SL (arrow). Recording between vertex and mastoid. Negativity at the vertex is plotted upward. Slightly modified from Robinson and Rudge[340] with permission by the authors and Oxford University Press.

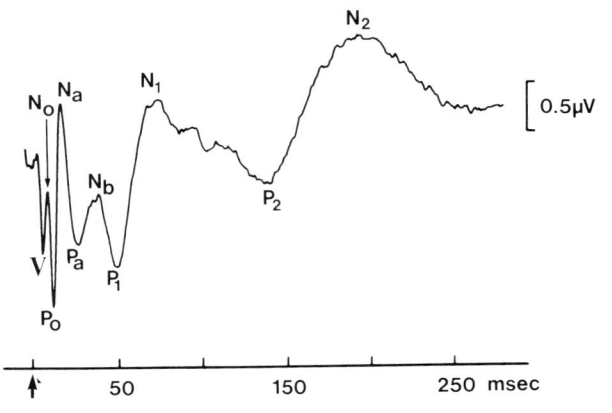

the stimulated ear are slightly larger.[306,454] A local origin of the LLAEP in the sylvian area near the auditory cortex has been suggested,[447] but not confirmed,[206] by scalp recordings. Recordings made directly from the surface of the auditory cortex showed AEPs different from the LLAEPs obtained on the scalp.[49,130] A positive peak at 105 msec followed by negativity at 150–160 msec has been isolated by scalp recordings from the temporal area and was thought to indicate excitation of secondary auditory cortex.[454] The reduction of N_1 by temporoparietal lesions suggested that this peak is generated by posterior-superior temporal and adjacent parietal areas.[204] In general, it seems likely that the scalp-recorded LLAEP is produced by multiple generators, including auditory cortex.[307,456]

The effect of age on the LLAEP has been studied extensively. In premature infants, LLAEP peaks develop in keeping with EEG patterns.[3,140,451] Even in newborns, the LLAEP is quite variable and depends on the sleep stage.[177] The variability decreases with age.[44] In general, the latency decreases and the amplitude increases, mainly during the first year of life.[17,269,285,287,451] In adult and elderly subjects, LLAEP latency increases and amplitude decreases.[136,309]

14.5.3 Neurological disorders that cause abnormal LLAEPs

Multiple sclerosis was found to produce abnormal LLAEPs in a few patients, but LLAEPs were diagnostically less effective than BAEPs and MLAEPs.[340]

Head injuries produced LLAEP changes that reflected severity and predicted outcome.[143,144,189,229] Strokes and tumors in the temporoparietal regions, but not in the frontal regions, reduced the N_1.[204] LLAEPs may be absent in postanoxic encephalopathy and in brain-damaged children.[331] The latency of the LLAEP was found to be increased in Friedreich's ataxia.[425]

LLAEPs of hyperkinetic children differ from those of normal children.[328,363] Amplitude reductions have been described in drowsy narcoleptics.[39] Schizophrenia and affective psychoses may produce nonspecific changes of LLAEPs[67,380] that tend to disappear with antipsychotic drug treatment.[379]

Drugs such as alcohol,[455,467] diazepam, amitriptyline, and dextroamphetamine[173] may alter the LLAEP.

14.5.4 Audiological disorders that cause abnormal LLAEPs

Early studies raised the hope that the LLAEP may become a useful test of hearing at a wide range of frequencies, including low frequencies.[23,79] Although useful measurements may be obtained in healthy adults, the correlation between LLAEP and hearing deteriorates in patients with hearing loss. Large discrepancies between LLAEP threshold and audiometric threshold have been reported.[345] The correlation between LLAEPs and hearing is worse in sleep for adults[65,256,290] and even more so for children.[10,332] The LLAEP is only rarely useful in young children.[74,122] LLAEP studies of hearing loss can therefore give only rather inexact results[301,441] but have been used to detect gross hearing losses in children with cerebral damage and mental retardation[56,64,331] and to distinguish nonorganic,[21,64,386] including hypnotically induced,[149] hearing loss. Grossly, the presence of a LLAEP suggests that a subject can hear, but the absence of LLAEPs in clinical investigations cannot be accepted as definite evidence of hearing loss. The BAEP,[326] ECochG,[181] and MLAEP[74] may provide better audiometric results than the LLAEP, especially in cases where stimulus tone frequency is not essential.

14.6 THE ELECTROCOCHLEOGRAM

In the past, the ECochG has usually been recorded alone mainly for audiological purposes. However, because the ECochG contains the equivalent of wave I of the BAEP, it has been suggested that the ECochG be recorded together with the BAEP in those cases in which a recording of wave I is essential but cannot be obtained by BAEP recording.[7] Many reviews of the ECochG have been published.[22,50,92,121,123,276,463]

14.6.1 Methods of stimulating and recording

Stimuli consist of tone pips or clicks delivered through earphones. Bone vibration may also be used when faulty air conduction is suspected.[216] Masking of the opposite ear is not necessary unless the BAEP is also recorded because the ECochG is restricted to the stimulated ear.

A needle electrode penetrating the eardrum is more effective than electrodes in the lumen or the wall of the auditory canal (2.2.1.5) for recordings of all three components of the ECochG. However, placement of such an electrode requires general anesthesia in children and local anesthesia in adults. The needle electrode is inserted by an otologist through the tympanic membrane so that its tip lies against the promontory close to the round window niche (transtympanic method). Atraumatic electrodes placed into the external auditory canal, or needle electrodes inserted into the wall of the canal, may also be used (extratympanic method). The reference electrode may be placed on the ipsilateral earlobe or mastoid. To obtain recordings that include the BAEP after the ECochG, the reference electrode is placed on the vertex. Single-channel recordings are made between ECochG electrode and mastoid or ear reference. Two-channel recordings may add the combination of ECochG and vertex electrodes for a combined ECochG and BAEP recording. Earphones must have provisions to accommodate the recording electrode. The lead wire of the electrode must be shielded to reduce stimulus artifact. Amplifier gain is adjusted for the signal size, which may vary from less than 1 μV to about 10 μV. Bandpass should range from 10 to 3,000 Hz; narrower limits may be used to isolate the ECochG components. About 1,000 responses are averaged in one ECochG for analysis periods of about 5–10 msec.

The ECochG has three components: (1) the auditory

nerve action potential, (2) the cochlear microphonic, and (3) the summating potential. To separate these three components, condensation and rarefaction stimuli are given alternatingly and the responses are averaged separately. The NAP, which always has the same polarity regardless of stimulus phase, can be separated from the CM, which changes phase with the stimulus, by adding the responses to both stimulus phases. This enhances the NAP in the average and reduces the CM like other noise (Figure 14.6). The NAP may be slightly distorted by this method because NAPs to condensation and rarefaction stimuli may have slightly different latency and shape, especially at low stimulus frequencies. The high frequency filter should be set to at least 3 kHz to avoid distortion of the NAP. High frequency filtering can therefore not be used effectively to reduce the CM. The CM is isolated by subtracting the responses to stimulation with opposite phase from each other; this tends to cancel the NAP and to increase the CM (Figure 14.6). The CM may also be enhanced by filtering the responses with a narrow bandpass filter centered at the stimulus frequency. The CM may be distinguished from electric stimulus artifacts by introducing a piece of tubing between stimulus source and ear. This delays the sound stimulus at the ear against the electric artifact and attenuates the amplitude of the artifact; clamping the tube eliminates the sound stimulus and identifies the artifact. The SP is isolated either by using a high frequency filter of less than 100 Hz to eliminate the NAP and CM or by stimulating at rates of 125–250/sec, which nearly abolishes the NAP without affecting the SP.[92,121,465]

14.6.2 The acoustic nerve action potential

The normal nerve action potential consists of a single peak, negative at the ear electrode, which has a latency of 1–2 msec in response to a tone pip or click of 60–90 dB above threshold (Figure 14.6). The NAP is the compound action potential generated by acoustic nerve fibers. Its latency increases with decreasing stimulus intensity and is 4–6 msec near threshold. The latency to stimuli containing high frequencies is shorter than that to stimuli of lower frequencies.[275] The amplitude of the NAP decreases with decreasing tonal stimulus frequency; responses to 500 Hz and less are unreliable.[275] Increasing the stimulus repetition rate also decreases NAP amplitude.

The NAP threshold is an excellent measure of hearing threshold for tone frequencies of over 1,000 Hz[75,181] and is especially useful in children.[74,300] It is fairly frequency specific[93] but not very useful at low tone frequencies since it has a threshold of up to 60 dB above the hearing threshold at 500 Hz. Specific combinations of threshold values and of relations between stimulus intensity and response latency and amplitude characterize different types of hearing loss.[121,282] Conductive hearing loss has also been evaluated with bone vibration stimuli.[463]

The NAP, in combination with the CM and the SP, has also been used to evaluate sudden hearing loss,[139] the effect of ototoxic drugs, and Ménière's disease.[22,121,463] Even acoustic neurinomas[271] and multiple sclerosis[302] have occasionally been reported to produce NAP abnormalities. In general, however, the NAP is limited to the study of the

Figure 14.6. The normal electrocochleogram (ECochG) to click stimuli. *Top*: Averages of responses to rarefaction (solid line) and compression (dashed line) clicks, showing mainly the N_1 peak of the nerve action potential. Bottom left: Subtraction of responses to compression clicks from responses to rarefaction clicks enhances the cochlear microphonic (CM), which reverses phase with reversal of the stimulus polarity and cancels the action potential and the summating potential, which do not reverse phase with the stimulus polarity. Bottom right: Addition of responses to rarefaction and compression enhances both the N_1 peak of the action potential (AP) and the summating potential (SP) and cancels the CM. Stimulation with clicks of 115 dB peSPL. Recording between an electrode at the wall of the external ear canal near the tympanic membrane and an electrode on the bridge of the nose. Negativity at the ear electrode is plotted upward. From Coats,[60] Arch Otolaryngol. 107:199–208, 1981, Copyright 1981, American Medical Association, with permission by the author and the American Medical Association.

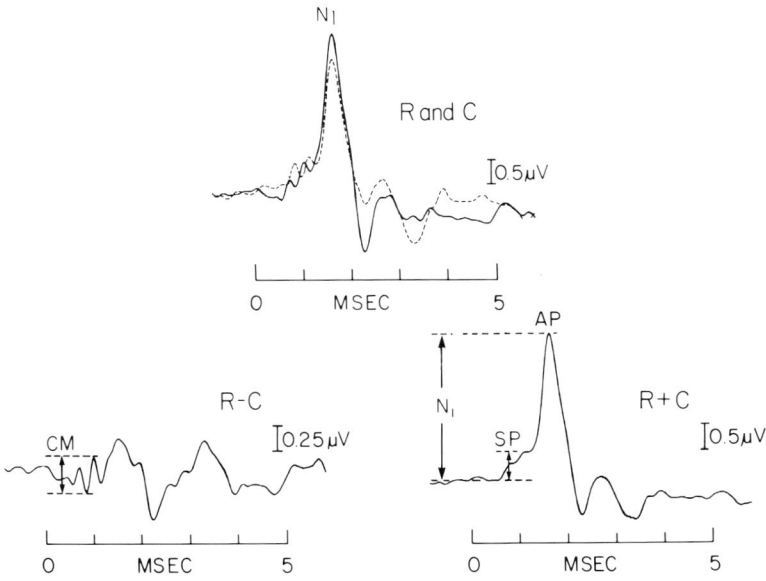

most distal part of the auditory pathway. For investigations of hearing loss, recording of the BAEP is often preferred to recording of the NAP of the ECochG because it gives very similar and nearly equally reliable information with simpler methods.[326] However, BAEP recordings that do not show wave I may be greatly improved by simultaneous ECochG recordings, which have a much better chance to pick up auditory nerve potentials.

14.6.3 The cochlear microphonic

The cochlear microphonic consists of a series of rhythmical low amplitude deflections that coincide with the peaks of the sound wave stimulus and reverse polarity with reversal of the stimulus phase (Figure 14.6). The exact shape of the CM depends on the location of the stimulus electrode.[207] The CM, representing the phasic component of the reaction of the cochlear hair cells to sound waves, provides a crude measure of intactness of the cochlear sound receptors and may help in the qualitative evaluation of hearing loss.[100] The CM may be used to divide sensorineural hearing losses into (1) sensory hearing loss characterized by absence of both CM and NAP and (2) neural hearing loss having a preserved CM and an absent NAP.[390] The CM may be preserved at low stimulus frequencies of about 500 Hz in cases of high frequency hearing loss.[389] The threshold of the CM does not correspond with hearing threshold, and the CM is therefore not useful in estimating the degree of hearing losses.

14.6.4 The summating potential

The summating potential consists of a small deflection of the baseline that has the same polarity as the NAP, which may be superimposed on the SP at low stimulus frequencies (Figure 14.6). The SP may represent a steady component of the receptor potential of cochlear hair cells. The SP cannot indicate hearing but is abnormally increased in Ménière's disease[60,125,214] and normalized by treatment with glycerol.[61]

14.7 SONOMOTOR AEPS

14.7.1 Methods of stimulating and recording

The subject must not be allowed to relax because relaxation reduces scalp and neck muscle tone and may abolish the sonomotor AEP. Tension of neck muscles, intentionally produced by neck extension against resistance, enhances the sonomotor AEP. Effective stimuli have a fast rise time and a fairly high intensity of 50–80 dB HL; tones of 100 Hz and less are ineffective. Stimuli may be repeated at up to 10/sec but the responses may fatigue after two minutes or more.

Sonomotor AEPs may be picked up with electrodes placed near scalp or neck muscles. Postauricular AEPs may be recorded with an electrode up to 2 cm behind the ear with reference to an electrode on the earlobe. Other sonomotor AEPs are often effectively recorded with an electrode on the inion with reference to a distant electrode. The recording methods are otherwise similar to those used for MLAEPs (14.3.1)

14.7.2 The normal sonomotor AEP

Sonomotor AEPs are surface recordings of muscle contractions produced by sudden sound stimuli. As recorded with scalp or neck electrodes, normal sonomotor AEPs vary enormously depending on the recording site and the tone of the underlying muscle. They may begin 6–50 msec after the stimulus and last up to over 100 msec.[30,131,313,416] In contrast to neurogenic AEPs, sonomotor AEPs have a maximum at the periphery of the scalp.[130] Unilateral stimulation produces bilateral sonomotor AEPs. The postauricular AEP, also called crossed acoustic response because of its bilateral distribution, has a peak latency of 12–15 msec[88] (Figure 14.7) but is also quite variable[81] and depends on head position[90] and filter settings.[436] Unlike other sonomotor AEPs, the postauricular AEP may persist in sleep.[416] Sonomotor responses may be recorded even in premature infants.[451]

The postauricular response is probably mediated by excitation of the cochlea[464] and therefore better suited for hearing tests than the response of neck muscles at the inion which is probably mediated by the labyrinth[442] and may be preserved in deaf persons.[30]

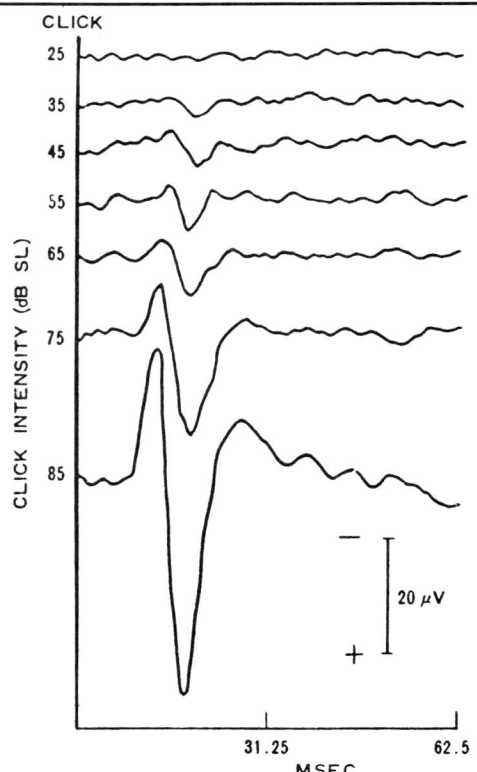

Figure 14.7. Normal postauricular AEPs to stimulation with clicks of increasing intensity. Recording made with the subject sitting upright in a chair without headrest. Electrodes placed behind the stimulated ear and on the earlobe. Negativity behind the ear plotted upward. From Yoshie and Okudaira[464] with permission by Acta Otolaryngologica.

14.7.3 Neurological disorders that cause abnormal sonomotor AEPs

The postauricular AEP has been used to study facial nerve lesions[33] and to detect subclinical lesions in multiple sclerosis.[58]

14.7.4 Audiological disorders that cause abnormal sonomotor AEPs

Although attempts have been made to use the postauricular AEP for auditory threshold determinations,[435] this AEP is generally considered to be too variable for quantitative audiometry.[74] Because it may be absent in subjects with normal hearing,[107] the postauricular AEP can be used only as a screening test, its presence suggesting that hearing is present.[88,107]

References for part C

1. Acoustic Society of America. 1970. American national standard specifications for audiometers, ANSI S3.6–1969 (R1973). New York: American National Standards Institute.
2. Ainslee, P.J., and Boston, J.R. 1980. Comparison of brain stem auditory evoked potentials for monaural and binaural stimuli. Electroencephalogr. Clin. Neurophysiol. 49:291–302.
3. Akiyama, Y., Schulte, F.J., Schultz, M.A., and Parmalee, A.H. 1969. Acoustically evoked responses in premature and full term newborn infants. Electroencephalogr. Clin. Neurophysiol. 26:371–380.
4. Allen, A., Starr, A., and Nudleman, K. 1981. Assessment of sensory function in the operating room utilizing cerebral evoked potentials: A study of fifty-six surgically anesthetized patients. Clin. Neurosurg. 28:457–481.
5. Allison, T., Wood, C.C., and Goff, W.R. 1983. Brain stem auditory, pattern-reversal visual, and short-latency somatosensory evoked potentials: Latencies in relation to age, sex and brain and body size. Electroencephalogr. Clin. Neurophysiol. 55:619–636.
6. Amadeo, M., and Shagass, C. 1973. Brief latency click-evoked potentials during waking and sleep in man. Psychophysiology 10:244–250.
7. American Electroencephalographic Society. 1984. Guidelines for clinical evoked potential studies. J. Clin. Neurophysiol. 1:3–53.
8. Anderson, D.C., Bundlie, S., and Rockswold, G.L. 1984. Multimodality evoked potentials in closed head trauma. Arch. Neurol. 41:369–374.
9. Antinoro, F., Skinner, P.H., and Jones, J.J. 1969. Relation between sound intensity and amplitude of the AER at different stimulus frequencies. J. Acoust. Soc. Am. 46:1433–1436.
10. Appleby, S. 1973. The slow vertex maximal sound evoked response in infants. Acta Oto-Laryngol. Suppl. 206:146–152.
11. Arlinger, S. 1981. Technical aspects of stimulation, recording and signal processing. Scand. Audiol. Suppl. 13:41–53.
12. Arlinger, S.D. 1983. Auditory evoked cortical responses to frequency glides in subjects with retrocochlear hearing impairment. J. Neurol. Neurosurg. Psychiatry 46:917–923.
13. Arlinger, S., and Jerlvall, L. 1981. Early auditory electric responses to fast amplitude and frequency tone glides. Electroencephalogr. Clin. Neurophysiol. 51:624–631.
14. Arlinger, S.D., Kylén, P., and Hellqvist, H. 1978. Skull distortion of bone conducted signals. Acta Oto-Laryngol. 85:318–323.

15. Arslan, E., Prosser, S., and Michelini, S. 1981. The auditory brainstem response to binaural delayed stimuli in man. Scand. Audiol. 10:151–155.
16. Baiocco, F., Testa, D., d'Angelo, A., and Cocchini, F. 1984. Abnormal auditory evoked potentials in Dejerine-Sottas disease: Report of two cases with central acoustic and vestibular impairment. J. Neurol. 231:46–49.
17. Barnet, A.B., Ohlrich, E.S., Weiss, I.P., and Shanks, B. 1975. Auditory evoked potentials during sleep in normal children from ten days to three years of age. Electroencephalogr. Clin. Neurophysiol. 39:29–41.
18. Barratt, H. 1980. Investigation of the mastoid electrode contribution to the brain stem auditory evoked response. Scand. Audiol. 9:203–211.
19. Bartel, D.R., Markand, O.N., and Kolar, O.J. 1983. The diagnosis and classification of multiple sclerosis: Evoked responses and spinal fluid electrophoresis. Neurology 33:611–617.
20. Bauch, C.C., Olsen, W.O., and Harner, S.G. 1983. Auditory brain-stem response and acoustic reflex test. Results for patients with and without tumor matched for hearing loss. Arch. Oto-Laryngol. 109:522–525.
21. Beagley, H.A. 1973. The role of electro-physiological tests in the diagnosis of non-organic hearing loss. Audiology 12:470–480.
22. Beagley, H.A., and Gibson, W.P.R. 1978. Electrocochleography in adults. In: Evoked electrical activity in the auditory nervous system, ed. R.F. Naunton, and C. Fernandez, pp. 259–273. New York: Academic Press.
23. Beagley, H.A., and Kellogg, S.E. 1969. A comparison of evoked response and subjective auditory thresholds. Int. Audiol. 8:345–353.
24. Beagley, H.A., and Knight, J.J. 1967. Changes in auditory evoked response with intensity. J. Laryngol. Otol. 81:861–873.
25. Beagley, H.E., and Knight, J.J. 1979. Sound stimuli for auditory investigations. In: Auditory investigation: The scientific and technological basis, ed. H.E. Beagley, pp. 131–143. London: Clarendon Press.
26. Beagley, H.A., and Sheldrake, J.B. 1978. Differences in brainstem response latency with age and sex. Br. J. Audiol. 12:69–77.
27. Begleiter, H., Porjesz, B., and Chou, C.L. 1981. Auditory brainstem potentials in chronic alcoholics. Science 211:1064–1066.
28. Beiter, R.C., and Hogan, D.D. 1973. Effects of variations in stimulus rise-decay time upon the early components of the auditory evoked response. Electroencephalogr. Clin. Neurophysiol. 34:203–206.
29. Bergenius, J., Borg, E., and Hirsch, A. 1983. Stapedius reflex test, brainstem audiometry and opto-vestibular tests in diagnosis of acoustic neurinomas: A comparison of test sensitivity in patients with moderate hearing loss. Scand. Audiol. 12:3–9.
30. Bickford, R.G., Jacobson, J.L., and Cody, D.T.R. 1964. Nature of average evoked potentials to sound and other stimuli in man. Ann. N.Y. Acad. Sci. 112:204–223.
31. Blegvad, B. 1975. Binaural summation of surface-recorded electrocochleographic responses. Scand. Audiol. 4:233–238.
32. Blunn, R.W., Salt, A.N., and Thornton, A.R.D. 1983. Construction of a simple auditory stimulus polarity indicator. Electroencephalogr. Clin. Neurophysiol. 56:689–690.
33. Bochenek, W., and Bochenek, Z. 1976. Postauricular (12 msec latency) responses to acoustic stimuli in patients with

peripheral, facial nerve palsy. Acta Oto-Laryngol. 81:264–269.
34. Boezeman, E.H.J.F., Kapteyn, T.S., Visser, S.L., and Snel, A.M. 1983. Effect of contralateral and ipsilateral masking of acoustic stimulation on the latencies of auditory evoked potentials from cochlea and brain stem. Electroencephalogr. Clin. Neurophysiol. 55:710–713.
35. Boezeman, E.H.J.F., Kapteyn, T.S., Visser, S.L., and Snel, A.M. 1983. Comparison of the latencies between bone and air conduction in the auditory brain stem evoked potential. Electroencephalogr. Clin. Neurophysiol. 56:244–247.
36. Bolz, J., and Giedke, H. 1982. Brain stem auditory evoked responses in psychiatric patients and healthy controls. J. Neural Transm. 54:285–291.
37. Borg, E., and Lofqvist, L. 1982. Auditory brainstem response (ABR) to rarefaction and condensation clicks in normal and abnormal ears. Scand. Audiol. 11:227–235.
38. Boston, J.R. 1981. Spectra of auditory brainstem responses and spontaneous EEG. IEEE Trans. Biomed. Eng. 28:334–341.
39. Broughton, R., Low, R., Valley, V., Da Costa, B., and Liddiard, S. 1982. Auditory evoked potentials compared to performance measures and EEG in assessing excessive daytime sleepiness in narcolepsy-cataplexy. Electroencephalogr. Clin. Neurophysiol. 54:579–582.
40. Brown, F.R., Shimizu, H., McDonald, J.M., Moser, A.B., Marquis, P., Chen, W.W., and Moser, H.W. 1981. Auditory evoked brainstem response and high-performance liquid chromatography sulfatide assay as early indices of metachromatic leukodystrophy. Neurology 31:980–985.
41. Buettner, U.W., Stöhr, M., and Koletzki, E. 1983. Brainstem auditory evoked potential abnormalities in vascular malformations of the posterior fossa. J. Neurol. 229:247–254.
42. Burkard, R. 1984. Sound pressure level measurement and spectral analysis of brief acoustic transients. Electroencephalogr. Clin. Neurophysiol. 57:83–91.
43. Cacace, A.T., Shy, M., and Satya-Murti, S. 1980. Brainstem auditory evoked potentials: A comparison of two high-frequency filter settings. Neurology 30:765–767.
44. Callaway, E., and Halliday, R.A. 1973. Evoked potential variability: Effects of age, amplitude and methods of measurement. Electroencephalogr. Clin. Neurophysiol. 34:125–133.
45. Campbell, J.A., and Leandri, M. 1984. The effect of high pass filters on computer-reconstructed evoked potentials. Electroencephalogr. Clin. Neurophysiol. 57:99–101.
46. Cann, J., and Knott, J. 1979. Polarity of acoustic click stimuli for eliciting brainstem auditory evoked responses: A proposed standard. Am. J. EEG Technol. 19:125–132.
47. Carlin, L., Roach, E.S., Riela, A., Spudis, E., and McLean, W.T. 1983. Juvenile metachromatic leukodystrophy: Evoked potentials and computed tomography. Ann. Neurol. 13:105–106.
48. Cashman, M.Z., and Rossman, R.N. 1983. Diagnostic features of the auditory brainstem response in identifying cerebellopontine angle tumours. Scand. Audiol. 12:35–41.
49. Celesia, G.G. 1976. Organization of auditory cortical areas in man. Brain 99:403–414.
50. Chatrian, G.E., Wirch, A.L., Lettich, E., Turella, G., and Snyder, J.M. 1982. Click-evoked human electrocochleogram: Noninvasive recording method, origin and physiologic significance. Am. J. EEG Tech. 22:151–174.

51. Chiappa, K.H. 1983. Evoked potentials in clinical medicine. New York: Raven Press.
52. Chiappa, K.H., Gladstone, K.J., and Young, R.R. 1979. Brain stem auditory evoked responses: Studies of waveform variations in 50 normal human subjects. Arch. Neurol. 36:81–87.
53. Chiappa, K.H., Harrison, J.L., Brooks, E.B., and Young, R.R. 1980. Brainstem auditory evoked responses in 200 patients with multiple sclerosis. Ann. Neurol. 7:135–143.
54. Chu, N.S., Squires, K.C., and Starr, A. 1982. Auditory brain stem responses in chronic alcoholic patients. Electroencephalogr. Clin. Neurophysiol. 54:418–425.
55. Church, M.W., and Williams, H.L. 1982. Dose- and time-dependent effects of ethanol on brain stem auditory evoked responses in young adult males. Electroencephalogr. Clin. Neurophysiol. 54:161–174.
56. Claus, H., Ott, A., Handrock, M., and v. Arentsschild, O. 1975. Comparison between conventional audiometry and ERA of deaf children in a serial test. Rev. Laryngol. Otol. Rhinol. 96:133–137.
57. Clemis, J.D., and Mc Gee, T. 1979. Brain stem electric response audiometry in the differential diagnosis of acoustic tumors. Laryngoscope 89:31–42.
58. Clifford-Jones, R.E., Clarke, G.P., and Mayles, P. 1979. Crossed acoustic response combined with visual and somatosensory evoked responses in the diagnosis of multiple sclerosis. J. Neurol. Neurosurg. Psychiatry 42:749–752.
59. Coats, A.C. 1978. Human auditory nerve action potentials and brain stem evoked responses: Latency-intensity functions in detection of cochlear and retrocochlear abnormality. Arch. Otolaryngol. 104:709–717.
60. Coats, A.C. 1981. The summating potential and Ménière's disease. I. Summating potential amplitude in Ménière and non-Ménière ears. Arch. Otolaryngol. 107:199–208.
61. Coats, A.C., and Alford, B.R. 1981. Ménière's disease and the summating potential. III. Effect of glycerol administration. Arch. Otolaryngol. 107:469–473.
62. Coats, A.C., and Kidder, H.R. 1980. Earspeaker coupling effects on auditory action potential and brainstem responses. Arch. Otolaryngol. 106:339–344.
63. Coats, A.C., and Martin, J.L. 1977. Human auditory nerve action potentials and brain stem evoked responses: Effects of audiogram shape and lesion location. Arch. Otolaryngol. 103:605–622.
64. Cody, D.T.R., Griffing, T., and Taylor, W.F. 1968. Assessment of the newer tests of auditory function. Ann. Otol. Rhinol. Laryngol. 77:686–705.
65. Cody, D.T.R., and Klass, D.W. 1968. Cortical audiometry. Potential pitfalls in testing. Arch. Otolaryngol. 88:396–406.
66. Cohen, M.M. 1982. Coronal topography of the middle latency auditory evoked potentials (MLAEPs) in man. Electroencephalogr. Clin. Neurophysiol. 53:231–236.
67. Cohen, R., Sommer, W., and Hermanutz, M. 1981. Auditory event-related potentials in chronic schizophrenics: Effects of electrodermal response type and demands on selective attention. Adv. Biol. Psychiatry 6:180–185.
68. Conraux, C., Dauman, R., and Feblot, P. 1981. Potentiels évoqués auditifs rapides "dérivés." Audiology 20:382–393.
69. Cornacchia, L., Vigliani, E., and Arpini, A. 1982. Comparison between brainstem-evoked response audiometry and behavioral audiometry in 270 infants and children. Audiology 21:359–363.
70. Cox, C., Hack, M., and Metz, D. 1981. Brainstem-evoked

response audiometry: Normative data from the preterm infant. Audiology 20:53–64.
71. Creel, D., Boxer, L.A., and Fauci, A.S. 1983. Visual and auditory anomalies in Chediak-Higashi syndrome. Electroencephalogr. Clin. Neurophysiol. 55:252–257.
72. Daly, D.M., Roeser, R.J., and Moushegian, G. 1976. The frequency-following response in subjects with profound unilateral hearing loss. Electroencephalogr. Clin. Neurophysiol. 40:132–142.
73. Davis, H. 1973. Sedation of young children for electric response audiometry (ERA). Audiology 12:55–57.
74. Davis, H. 1976. Brain stem and other responses in electric response audiometry. Ann. Otol. Rhinol. Laryngol 85:3–14.
75. Davis, H. 1976. Principles of electric response audiometry. Ann. Otol. Rhinol. Laryngol. 85 (Suppl. 28):1–96.
76. Davis, H., and Hirsh, S.K. 1976. The audiometric utility of brain stem responses to low-frequency sounds. Audiology 15:181–195.
77. Davis, H., and Hirsh, S.K. 1979. A slow brain stem response for low-frequency audiometry. Audiology 18:445–461.
78. Davis, H., Hirsch, S.K., Popelka, G.R., and Formby, C. 1984. Frequency selectivity and thresholds of brief stimuli suitable for electric response audiometry. Audiology 23:59–74.
79. Davis, H., Hirsh, S.K., Shelnutt, J., and Bowers, C. 1967. Further validation of evoked response audiometry (ERA). J. Speech Hear. Res. 10:717–732.
80. Davis, H., Mast, T., Yoshie, N., and Zerlin, S. 1966. The slow response of the human cortex to auditory stimuli: Recovery process. Electroencephalogr. Clin. Neurophysiol. 21:105–113.
81. De Grandis, D., and Santoni, P. 1980. The post-auricular response: A single motor unit study. Electroencephalogr. Clin. Neurophysiol. 50:437–440.
82. Despland, P.A., and Galambos, R. 1980. The auditory brainstem response (ABR) is a useful diagnostic tool in the intensive care nursery. Pediatr. Res. 14:154–158.
83. Dobie, R.A., and Norton, S.J. 1980. Binaural interaction in human auditory evoked potentials. Electroencephalogr. Clin. Neurophysiol. 49:303–313.
84. Don, M., Allen, A.R., and Starr, A. 1977. Effect of click rate on the latency of auditory brain stem responses in humans. Ann. Otol. Rhinol. Laryngol. 86:186–195.
85. Don, M., and Eggermont, J.J. 1978. Analysis of the click-evoked brainstem potentials in man using high-pass noise masking. J. Acoust. Soc. Am. 63:1084–1092.
86. Don, M., Eggermont, J.J., and Brackmann, D.E. 1979. Reconstruction of the audiogram using brain stem responses and high-pass noise masking. Ann. Otol. Rhinol. Laryngol. Suppl. 57:1–20.
87. Donald, M.W., Bird, C.E., Lawson, J.S., Letemendia, F.J.J., Monga, T.N., Surridge, D.H.C., Varette-Cerre, P., Williams, D.L., Williams D.M.L., and Wilson, D.L. 1981. Delayed auditory brainstem responses in diabetes mellitus. J. Neurol. Neurosurg. Psychiatry 44:641–644.
88. Douek, E., Gibson, W., and Humphries, K. 1973. The crossed acoustic response. J. Laryngol. Otol. 87:711–726.
89. Drake, M.E., Erwin, C.W., and Massey, E.W. 1982. Ocular bobbing in metabolic encephalopathy: Clinical, patho-

logic, and electrophysiologic study. Neurology 32:1029–1031.

90. Dus, V., and Wilson, S.J. 1975. The click-evoked post-auricular myogenic response in normal subjects. Electroencephalogr. Clin. Neurophysiol. 39:523–525.

91. Edwards, R.M., Buchwald, J.S., Tanguay, P.E., and Schwafel, J.A. 1982. Sources of variability in auditory brain stem evoked potential measures over time. Electroencephalogr. Clin. Neurophysiol. 53:125–132.

92. Eggermont, J.J. 1979. Human electrocochleography. In: Auditory investigation: The scientific and technological basis, ed. H.E. Beagley, pp. 283–301. London: Clarendon Press.

93. Eggermont, J.J., Spoor, A., and Odenthal, D.W. 1967. Frequency specificity of tone-burst electrocochleography. In: Electrocochleography, ed. R.J. Ruben, C. Elberling, and G. Salomon, pp. 215–246. Baltimore: University Park Press.

94. Ehle, A.L., Steward, R.M., Lellelid, N.A., and Leventhal, N.A. 1984. Evoked potentials in Huntington's disease: A comparative and longitudinal study. Arch. Neurol. 41:379–382.

95. Elberling, C. 1974. Action potentials along the cochlear partition recorded from the ear canal in man. Scand. Audiol. 3:13–19.

96. Elberling, C. 1979. Auditory electrophysiology: Spectral analysis of cochlear and brain stem evoked potentials. Scand. Audiol. 8:57–64.

97. Elidan, J., Sohmer, H., Gafni, M., and Kahana, E. 1982. Contribution of changes in click rate and intensity on diagnosis of multiple sclerosis by brainstem auditory evoked potentials. Acta Neurol. Scand. 65:570–585.

98. Emerson, R.G., Brooks, E.B., Parker, S.W., and Chiappa, K.H. 1982. Effects of click polarity on brainstem auditory evoked potentials in normal subjects and patients: Unexpected sensitivity of wave V. Ann. N.Y. Acad. Sci. 388:710–720.

99. Euler, M., and Kiessling, J. 1981. Frequency-following potentials in man by lock-in technique. Electroencephalogr. Clin. Neurophysiol. 52:400–404.

100. Euler, M., and Kiessling, J. 1983. Far-field cochlear microphonics in man and their relation to cochlear integrity. Electroencephalogr. Clin. Neurophysiol. 56:86–89.

101. Fabiani, M., Sohmer, H., Tait, C., Gafni, M., and Kinarti, R. 1979. A functional measure of brain activity: Brain stem transmission time. Electroencephalogr. Clin. Neurophysiol. 47:483–491.

102. Fawer, C.L., Dubowitz, L.M.S., Levene, M.I., and Dubowitz, V. 1983. Auditory brainstem responses in neurologically abnormal infants. Neuropaediatrie 14:88–92.

103. Fine, E.J., and Hallett, M. 1980. Neurophysiological study of subacute combined degeneration. J. Neurol. Sci. 45:331–336.

104. Fischer, C., Blanc, A., Mauguière, F., and Courjon, J. 1981. Diagnostic value of brainstem auditory evoked potentials. Rev. Neurol. 137:229–240.

105. Folsom, R.C., Weber, B.A., and Thompson, G. 1983. Auditory brainstem responses in children with early recurrent middle ear disease. Ann. Otol. Rhinol. Laryngol. 92:249–253.

106. Folsom, R.C., Widen, J.E., and Wilson, W.R. 1983. Au-

ditory brain-stem responses in infants with Down's syndrome. Arch. Otolaryngol. 109:607–610.
107. Fraser, J.G., Conway, M.J., Keene, M.H., and Hazell, J.W.P. 1978. The post-auricular myogenic response: A new instrument which simplifies its detection by machine scoring. J. Laryngol. Otol. 92:293–303.
108. Frye-Osier, H.A., Hirsch, J.E., Goldstein, R., and Weber, K. 1982. Early- and middle-AER components to clicks as response indices for neonatal hearing screening. Ann. Otol. Rhinol. Laryngol. 91:272–276.
109. Fujita, M., Hosoki, M., and Miyazaki, M. 1981. Brainstem auditory evoked responses in spinocerebellar degeneration and Wilson disease. Ann. Neurol. 9:42–47.
110. Funasaka, S., and Yamamoto, E. 1983. Human frequency-specific whole-nerve response by use of frequency-modulated tone. Audiology 22:1–8.
111. Galambos, R., and Despland, P.A. 1980. The auditory brainstem response (ABR) evaluates risk factors for hearing loss in the newborn. Pediatr. Res. 14:159–163.
112. Galambos, R., and Hecox, K.E. 1978. Clinical applications of the auditory brain stem response. Otolaryngol. Clin. North Am. 11:709–722.
113. Galambos, R., Makeig, S., and Tamachoff, P.J. 1981. A 40-Hz auditory potential recorded from the human scalp. Proc. Natl. Acad. Sci. USA 78:2643–2647.
114. Garg, B.P., Markand, O.N., and Bustion, P.F. 1982. Brainstem auditory evoked responses in hereditary motor-sensory neuropathy: Site of origin of wave II. Neurology 32:1017–1019.
115. Garg, B.P., Markand, O.N., and DeMyer, W.E. 1983. Usefulness of BAER studies in the early diagnosis of Pelizaeus-Merzbacher disease. Neurology 33:955–956.
116. Garg, B.P., Markand, O.N., DeMyer, W.E., and Warren, C. 1983. Evoked response studies in patients with adrenoleukodystrophy and heterozygous relatives. Arch. Neurol. 40:356–359.
117. Gauthier, P., and Gottesmann, C. 1983. Influence of total sleep deprivation on event-related potentials in man. Psychophysiology 20:351–355.
118. Gawel, M.J., Das, P., Vincent, S., and Rose, F.C. 1981. Visual and auditory evoked responses in patients with Parkinson's disease. J. Neurol. Neurosurg. Psychiatry 44:227–232.
119. Gerken, G.M., Moushegian, G., Stillman, R.D., and Rupert, A.L. 1975. Human frequency-following responses to monaural and binaural stimuli. Electroencephalogr. Clin. Neurophysiol. 38:379–386.
120. Gibson, J.M., and Kennedy, P. 1984. Does testing of brainstem auditory response aid diagnosis in progressive spastic paraparesis? Acta Neurol. Scand. 69:107–111.
121. Gibson, W.P.R. 1978. Essentials of clinical electric response audiometry. New York: Churchill Livingstone.
122. Gibson, W.P.R. 1980. The auditory evoked potentials. In: Evoked potentials: Proceedings of an international evoked potentials symposium held in Nottingham, England, ed. C. Barber, pp. 43–54. Lancaster: MTP Press.
123. Gibson, W.P.R. 1982. Electrocochleography. In: Evoked potentials in clinical testing, ed. A.M. Halliday, pp. 283–311. Edinburgh: Churchill Livingstone.
124. Gibson, W.P.R. 1982. The use of auditory evoked potentials for the estimation of hearing. In: Evoked potentials in clinical testing, ed. A.M. Halliday, pp. 313–343. Edinburgh: Churchill Livingstone.
125. Gibson, W.P.R., Moffat, D.A., and Ramsden, R.T. 1977.

Clinical electrocochleography in the diagnosis and management of Ménière's disorder. Audiology 16:389–401.

126. Gilroy, J., and Lynn, G.E. 1978. Computerized tomography and auditory-evoked potentials: Use in the diagnosis of olivopontocerebellar degeneration. Arch. Neurol. 35:143–147.

127. Gilroy, J., Lynn, G.E., Ristow, G.E., and Pellerin, R.J. 1977. Auditory evoked brain stem potentials in a case of "locked-in" syndrome. Arch. Neurol. 34:492–495.

128. Glaser, E.M., Suter, C.M., Dasheiff, R., and Goldberg, A. 1976. The human frequency-following response: Its behavior during continuous tone and tone burst stimulation. Electroencephalogr. Clin. Neurophysiol. 40:25–32.

129. Glasscock, M.E., Jackson, C.G., Josey, A.F., Dickins, J.R.E., and Wiet, R.J. 1979. Brain stem evoked response audiometry in a clinical practice. Laryngoscope 89:1021–1035.

130. Goff, W.R. 1978. The scalp distribution of auditory evoked potentials. In: Evoked electrical activity in the auditory nervous system, ed. R.F. Naunton, and C. Fernandez, pp. 505–524. New York: Academic Press.

131. Goff, G.D., Matsumiya, Y., Allison, T., and Goff, W.R. 1977. The scalp topography of human somatosensory and auditory evoked potentials. Electroencephalogr. Clin. Neurophysiol. 42:57–76.

132. Goldie, W.D., Chiappa, K.H., Young, R.R., and Brooks, E.B. 1981. Brainstem auditory and short-latency somatosensory evoked responses in brain death. Neurology 31:248–256.

133. Goldstein, P.J., Krumholz, A., Felix, J.K., Shannon, D., and Carr, R.F. 1979. Brain stem-evoked response in neonates. Am. J. Obstet. Gynecol. 135:622–628.

134. Goldstein, R., and Rodman, L.B. 1967. Early components of averaged evoked responses to rapidly repeated auditory stimuli. J. Speech Hear. Res. 10:697–705.

135. Goldstein, R., Rodman, L.B., and Karlovich, R.S. 1972. Effects of stimulus rate and number on the early components of the averaged electroencephalic response. J. Speech Hear. Res. 15:559–566.

136. Goodin, D.S., Squires, K.C., Henderson, B.H., and Starr, A. 1978. Age-related variations in evoked potentials to auditory stimuli in normal human subjects. Electroencephalogr. Clin. Neurophysiol. 44:447–458.

137. Goodin, D.S., Squires, K.C., and Starr, A. 1983. Variations in early and late event-related components of the auditory evoked potential with task difficulty. Electroencephalogr. Clin. Neurophysiol. 55:680–686.

138. Graham, J., Greenwood, R., and Lecky, B. 1980. Cortical deafness. A case report and review of the literature. J. Neurol. Sci. 48:35–49.

139. Graham, J.M., Ramsden, R.T., Moffatt, D.A., and Gibson, W.P.R. 1978. Sudden sensorineural hearing loss: Electrocochleographic findings in 70 patients. J. Laryngol. Otol. 92:581–589.

140. Graziani, L.J., Katz, L., Cracco, R.Q., Cracco, J.B., and Weitzman, E.D. 1974. The maturation and interrelationship of EEG patterns and auditory evoked responses in premature infants. Electroencephalogr. Clin. Neurophysiol. 36:367–375.

141. Green, J.B., and Walcoff, M.R. 1982. Evoked potentials in multiple sclerosis. Arch. Neurol. 39:696–697.

142. Green, J.B., Walcoff, M.R., and Lucke, J.F. 1982. Comparison of phenytoin and phenobarbital effects on far-field

auditory and somatosensory evoked potential interpeak latencies. Epilepsia 23:417–421.
143. Greenberg, R.P., Becker, D.P., Miller, J.D., and Mayer, D.J. 1977. Evaluation of brain function in severe human head trauma with multimodality evoked potentials. Part 2. Localization of brain dysfunction and correlation with posttraumatic neurological conditions. J. Neurosurg. 47:163–177.
144. Greenberg, R.P., Newlon, P.G., Hyatt, M.S., Narayan, R.K., and Becker, D.P. 1981. Prognostic implications of early multimodality evoked potentials in severely head-injured patients: A prospective study. J. Neurosurg. 55:227–236.
145. Grimes, A.M., Elks, M.L., Grunberger, G., and Pikus, A.M. 1983. Auditory brain-stem responses in adrenomyeloneuropathy. Arch. Neurol. 40:574–576.
146. Guerit, J.M., Mahieu, P., Houben-Giurgea, S., and Herbay, S. 1981. The influence of ototoxic drugs on brainstem auditory evoked potentials in man. Arch. Oto-Rhino-Laryngol. 233:189–199.
147. Gupta, P.R., Guilleminault, C., and Dorfman, L.J. 1981. Brainstem auditory evoked potentials in near-miss sudden infant death syndrome. J. Pediatr. 98:791–794.
148. Hacke, W., Berg-Dammer, E., and Zeumer, H. 1982. Evoked potential monitoring during acute occlusion of the basilar artery and selective local thrombolytic therapy. Arch. Psychiatr. Nervenkr. 232:541–548.
149. Halliday, A.M., and Mason, A.A. 1964. The effect of hypnotic anaesthesia on cortical responses. J. Neurol. Neurosurg. Psychiatry 27:300–312.
150. Halliday, R., and Callaway, E. 1978. Time shift evoked potentials (TSEPs): Methods and basic results. Electroencephalogr. Clin. Neurophysiol. 45:118–121.
151. Hammond, E.J., and Wilder, B.J. 1983. Evoked potentials in olivopontocerebellar atrophy. Arch. Neurol. 40:366–369.
152. Harder, H., Arlinger, S., and Kylén, P. 1983. Electrocochleography with bone-conducted stimulation. A comparative study of different methods of stimulation. Acta Oto-Laryngol. 95:35–45.
153. Hardy, R.W., Kinney, S.E., Lueders, H., and Lesser, R.P. 1982. Preservation of cochlear nerve function with the aid of brain stem auditory evoked potentials. Neurosurgery 11:16–19.
154. Hari, R., Kaila, K., Katila, T., Tuomisto, T., and Varpula, T. 1982. Interstimulus interval dependence of the auditory vertex response and its magnetic counterpart: Implications for their neural generation. Electroencephalogr. Clin. Neurophysiol. 54:561–569.
155. Hari, R., Sulkava, R., and Haltia, M. 1982. Brainstem auditory evoked responses and alpha-pattern coma. Ann. Neurol. 11:187–189.
156. Harker, L.A., Hosick, E., Voots, R.J., and Mendel, M.I. 1977. Influence of succinylcholine on middle component auditory evoked potentials. Arch. Otolaryngol. 103:133–137.
157. Harkins, S.W. 1981. Effects of age and interstimulus interval on the brainstem auditory evoked potential. Int. J. Neurosci. 15:107–118.
158. Harkins, S.W. 1981. Effects of presenile dementia of the Alzheimer's type on brainstem transmission time. Int. J. Neurosci. 15:165–170.
159. Hart, R.G., Gardner, D.P., and Howieson, J. 1983.

Acoustic tumors: Atypical features and recent diagnostic tests. Neurology 33:211–221.

160. Hashimoto, I. 1982. Auditory evoked potentials from the human midbrain: Slow brain stem responses. Electroencephalogr. Clin. Neurophysiol. 53:652–657.

161. Hashimoto, I., Ishiyama, Y., Totsuka, G., and Mizutani, H. 1980. Monitoring brainstem function during posterior fossa surgery with brainstem auditory evoked potentials. In: Evoked potentials: Proceedings of an international evoked potentials symposium held in Nottingham, England, ed. C. Barber, pp. 377–390. Lancaster: MTP Press.

162. Hashimoto, I., Ishiyama, Y., and Tozuka, G. 1979. Bilaterally recorded brain stem auditory evoked responses: Their asymmetric abnormalities and lesions of the brain stem. Arch. Neurol. 36:161–167.

163. Hashimoto, I., Ishiyama, Y., Yoshimoto, T., and Nemoto, S. 1981. Brain-stem auditory-evoked potentials recorded directly from human brain-stem and thalamus. Brain 104:841–859.

164. Hausler, R., and Levine, R.A. 1980. Brain stem auditory evoked potentials are related to interaural time discrimination in patients with multiple sclerosis. Brain Res. 191:589–594.

165. Hawes, M.D., and Greenberg, H.J. 1981. Slow brain stem responses (SN10) to tone pips in normally hearing newborns and adults. Audiology 20:113–122.

166. Hayes, D., and Jerger, J. 1982. Auditory brainstem response (ABR) to tone-pips: Results in normal and hearing-impaired subjects. Scand. Audiol. 11:133–142.

167. Hecox, K. 1975. Electrophysiological correlates of human auditory development. In: Infant perception: From sensation to cognition, vol. 2., ed. L.B. Cohen and P. Salapatek, pp. 151–191. New York: Academic Press.

168. Hecox, K.E., and Cone, B. 1981. Prognostic importance of brainstem auditory evoked responses after asphyxia. Neurology 31:1429–1433.

169. Hecox, K.E., Cone, B., and Blaw, M.E. 1981. Brainstem auditory evoked response in the diagnosis of pediatric neurologic diseases. Neurology 31:832–840.

170. Hecox, K., and Galambos, R. 1974. Brain stem auditory evoked responses in human infants and adults. Arch. Otolaryngol. 99:30–33.

171. Hellekson, C., Allen, A., Greeley, H., Emery, S., and Reeves, A. 1979. Comparison of interwave latencies of brain stem auditory evoked responses in narcoleptics, primary insomniacs and normal controls. Electroencephalogr. Clin. Neurophysiol. 47:742–744.

172. Henderson-Smart, D.J., Pettigrew, A.G., and Campbell, D.J. 1983. Clinical apnea and brain-stem neural function in preterm infants. N. Engl. J. Med. 308:353–357.

173. Herrmann, W.M., Hofmann, W., and Kubicki, S. 1981. Psychotropic drug induced changes in auditory averaged evoked potentials: Results of a double-blind trial using an objective fully automated AEP analysis method. Int. J. Clin. Pharmacol. Ther. Toxicol. 19:56–62.

174. Hicks, G.E. 1980. Auditory brainstem response. Sensory assessment by bone conduction masking. Arch. Otolaryngol. 106:392–395.

175. Hosford-Dunn, H., Mendelson, T., and Salamy, A. 1981. Binaural interactions in the short-latency evoked potentials of neonates. Audiology 20:394–408.

176. House, J.W., and Brackmann, D.E. 1979. Brainstem au-

diometry in neurotologic diagnosis. Arch. Otolaryngol. 105:305–309.
177. Hrbek, A., Hrbková, M., and Lenard, H.G. 1969. Somato-sensory, auditory and visual evoked responses in newborn infants during sleep and wakefulness. Electroencephalogr. Clin. Neurophysiol. 26:597–603.
178. Hughes, J.R., and Fino, J. 1980. Usefulness of piezoelectric earphones in recording the brain stem auditory evoked potentials: A new early deflection. Electroencephalogr. Clin. Neurophysiol. 48:357–360.
179. Huis in't Veld, F., Osterhammel, P., and Terkildsen, K. 1977. The frequency selectivity of the 500 Hz frequency following response. Scand. Audiol. 6:35–42.
180. Hutchinson, J.C., and Klodd, D.A. 1982. Electrophysiologic analysis of auditory, vestibular and brain stem function in chronic renal failure. Laryngoscope 92:833–843.
181. Ino, T. 1967. Comparison of the response threshold between ERA and electrocochleography. In: Electrocochleography, ed. R.J. Ruben, C. Elberling, and G. Salomon, pp. 247–256. Lancaster: MTP Press.
182. Jabbari, B., Schwartz, D.M., MacNeil, D.M., and Coker, S.B. 1983. Early abnormalities of brainstem auditory evoked potentials in Friedreich's ataxia: Evidence of primary brainstem dysfunction. Neurology 33:1071–1074.
183. Jain, S., and Maheshwari, M.C. 1984. Brainstem auditory evoked responses in coma due to meningoencephalitis. Acta Neurol. Scand. 69:163–167.
184. Jerger, J., and Hall, J. 1980. Effects of age and sex on auditory brainstem response. Arch. Otolaryngol. 106:387–391.
185. Jewett, D.L., and Williston, J.S. 1971. Auditory-evoked far fields averaged from the scalp of humans. Brain 94:681–696.
186. Johannsen, H.S. 1971. Elimination of movement artifacts in evoked response audiometry in infants. J. Aud. Res. 11:351–356.
187. Kaga, K., and Tanaka, Y. 1980. Auditory brainstem response and behavioral audiometry: Developmental correlates. Arch. Otolaryngol. 106:564–566.
188. Karnaze, D., Gott, P., Mitchell, F., and Loftin, J. 1984. Brainstem auditory evoked potentials are normal in idiopathic sleep apnea. Ann. Neurol. 15:406.
189. Karnaze, D.S., Marshall, L.F., McCarthy, C.S., Klauber, M.R., and Bickford, R.G. 1982. Localizing and prognostic value of auditory evoked responses in coma after closed head injury. Neurology 32:299–302.
190. Kavanagh, K.T., Harker, L.A., and Tyler, R.S. 1984. Auditory brainstem and middle latency responses. I. Effects of response filtering and waveform identification. II. Threshold responses to a 500-Hz tone pip. Ann. Otol. Rhinol. Laryngol. 93, Suppl. 108:1–12.
191. Keidel, W.D. 1971. D.C.-Potentials in the auditory evoked response in man. Acta Oto-Laryngol. 71:242-248.
192. Kevanishvili, Z.S., and Von Specht, H. 1979. Human slow auditory evoked potentials during natural and drug-induced sleep. Electroencephalogr. Clin. Neurophysiol. 47:280–288.
193. Kiessling, J. 1982. Hearing aid selection by brainstem audiometry. Scand. Audiol. 11:269–275.
194. Kjaer, M. 1979. Differences of latencies and amplitudes of brain stem evoked potentials in subgroups of a normal material. Acta Neurol. Scand. 59:72–79.
195. Kjaer, M. 1980. Variations of brain stem auditory evoked

potentials correlated to duration and severity of multiple sclerosis. Acta Neurol. Scand. 61:157–166.

196. Kjaer, M. 1980. Brain stem auditory and visual evoked potentials in multiple sclerosis. Acta Neurol. Scand. 62:14–19.
197. Kjaer, M. 1980. Recognizability of brain stem auditory evoked potential components. Acta Neurol. Scand. 62:20–33.
198. Kjaer, M. 1980. The value of brain stem auditory, visual and somatosensory evoked potentials and blink reflexes in the diagnosis of multiple sclerosis. Acta Neurol. Scand. 62:220–236.
199. Klein, A.J. 1983. Properties of the brain-stem response slow-wave component. I. Latency, amplitude, and threshold sensitivity. Arch. Otolaryngol. 109:6–12.
200. Klein, A.J. 1983. Properties of the brain-stem response slow-wave component. II. Frequency specificity. Arch. Otolaryngol. 109:74–78.
201. Klingenberg Schweitzer, P. 1977. Auditory evoked brain responses: Comparison of ON and OFF responses at long and short durations. Percept. Psychophys. 22:87–94.
202. Klug, N. 1982. Brainstem auditory evoked potentials in syndromes of decerebration, the bulbar syndrome and in central death. J. Neurol. 227:219–228.
203. Knight, J.J. 1979. Acoustic measurements. In: Auditory investigation: The scientific and technological basis, ed. H.E. Beagley, pp. 144–169. London: Clarendon Press.
204. Knight, R.T., Hillyard, S.A., Woods, D.L., and Neville, H.J. 1980. The effects of frontal and temporal-parietal lesions on the auditory evoked potential in man. Electroencephalogr. Clin. Neurophysiol. 50:112–124.
205. Kohn, M., Lifshitz, K., and Litchfield, D. 1978. Averaged evoked potentials and frequency modulation. Electroencephalogr. Clin. Neurophysiol. 45:236–243.
206. Kooi, K.A., Tipton, A.C., and Marshall, R.E. 1971. Polarities and field configurations of the vertex components of the human auditory evoked response: A reinterpretation. Electroencephalogr. Clin. Neurophysiol. 31:166–169.
207. Köpcke, J., Hoke, M., and Lütkenhöner, B. 1980. Influence of the intratympanic recording site on the frequency response of cochlear microphonics. Scand. Audiol. Suppl. 11:65–71.
208. Koshbin, S., and Hallett, M. 1981. Multimodality evoked potentials and blink reflex in multiple sclerosis. Neurology 31:138–144.
209. Kotagal, S., Rosenberg, C., Rudd, D., Dunkle, L.M., and Horenstein, S. 1981. Auditory evoked potentials in bacterial meningitis. Arch. Neurol. 38:693–695.
210. Kraus, N., Özdamar, Ö., Hier, D., and Stein, L. 1982. Auditory middle latency responses (MLRs) in patients with cortical lesions. Electroencephalogr. Clin. Neurophysiol. 54:275–287.
211. Kraus, N., Özdamar, Ö., Stein, L., and Reed, N. 1984. Absent auditory brain stem response: Peripheral hearing loss or brain stem dysfunction. Laryngoscope 94:400–406.
212. Krumholz, A., Singer, H.S., Niedermeyer, E., Burnite, R., and Harris, K. 1983. Electrophysiological studies in Tourette's syndrome. Ann. Neurol. 14:638–641.
213. Krumholz, A., Weiss, H.D., Goldstein, P.J., and Harris, K.C. 1981. Evoked responses in vitamin B_{12} deficiency. Ann. Neurol. 9:407–409.
214. Kumagami, H., Nishida, H., and Baba, M. 1982. Electrocochleographic study of Ménière's disease. Arch. Otolaryngol. 108:284–288.

215. Kupperman, G.L., and Mendel, M.I. 1974. Threshold of the early components of the averaged electroencephalic response determined with tone pips and clicks during drug-induced sleep. Audiology 13:379–390.
216. Kylén, P., Harder, H., Jerlvall, L., and Arlinger, S. 1982. Reliability of bone-conducted electrocochleography: A clinical study. Scand. Audiol. 11:223–226.
217. Lacey, D.J., and Terplan, K. 1984. Correlating auditory evoked and brainstem histologic abnormalities in infantile Gaucher's disease. Neurology 34:539–541.
218. Lane, R.H., Kupperman, G.L., and Goldstein, R. 1971. Early components of the averaged electroencephalic response in relation to rise-decay time and duration of pure tones. J. Speech Hear. Res. 14:408–415.
219. Lane, R.H., Mendel, M.I., Kupperman, G.L., Vivion, M.C., Buchanan, L.H., and Goldstein, R. 1974. Phase distortion of averaged electroencephalic response. Arch. Otolaryngol. 99:428–432.
220. Laukli, E. 1983. Stimulus waveforms used in brainstem response audiometry. Scand. Audiol. 12:83–89.
221. Laukli, E. 1983. High-pass and notch masking in suprathreshold brainstem response audiometry. Scand. Audiol. 12:109–115.
222. Laukli, E., and Mair, I.W.S. 1981. Early auditory-evoked responses: Filter effects. Audiology 20:300–312.
223. Laukli, E., and Mair, I.W.S. 1981. Early auditory-evoked responses: Spectral content. Audiology 20:453–464.
224. Lee, Y.S., Lueders, H., Dinner, D.S., Lesser, R.P., Hahn, J., and Klemm, G. 1984. Recording of auditory evoked potentials in man using chronic subdural electrodes. Brain 107:115–131.
225. Lenhardt, M.L. 1982. Wave V latency and chirp (linear frequency ramp): Repetition rate. Audiology 21:425–432.
226. Lev, A., and Sohmer, H. 1972. Sources of averaged neural responses recorded in animal and human subjects during cochlear audiometry (electro-cochleogram). Arch. Klin. Exp. Ohr. Nas. Kehlkopfheilkd. 201:79–90.
227. Levi, H., Tell, L., Feinmesser, M., Gafni, M., and Sohmer, H. 1983. Early detection of hearing loss in infants by auditory nerve and brain stem responses. Audiology 22:181–188.
228. Levine, R.A. 1981. Binaural interaction in brainstem potentials of human subjects. Ann. Neurol. 9:384–393.
229. Lindsay, K.W., Karlin, J., Kennedy, I., Fry, J., McInnes, A., and Teasdale, G.M. 1981. Evoked potentials in severe head injury: Analysis and relation to outcome. J. Neurol. Neurosurg. Psychiatry 44:796–802.
230. Lumenta, C.B. 1984. Measurements of brain-stem auditory evoked potentials in patients with spontaneous intracerebral hemorrhage. J. Neurosurg. 60:548–552.
231. Lütschg, J., Pfenninger, J., Ludin, H.P., and Vassella, F. 1983. Brain-stem auditory evoked potentials and early somatosensory evoked potentials in neurointensively treated comatose children. Am. J. Dis. Child. 137:421–426.
232. Lynn, G.E., Cullis, P.A., and Gilroy, J. 1983. Olivopontocerebellar degeneration: Effects on auditory brainstem responses. Semin. Hear. 4:375–383.
233. McCallum, W.C., and Curry, S.H. 1981. Late slow wave components of auditory evoked potentials: Their cognitive significance and interaction. Electroencephalogr. Clin. Neurophysiol. 51:123–137.
234. McDermott, J.C. 1983. Cochlear initiation sites of the frequency following potential. Scand. Audiol. 12:97–102.

235. McFarland, W.H., Vivion, M.C., and Goldstein, R. 1977. Middle components of the AER to tone-pips in normal-hearing and hearing-impaired subjects. J. Speech Hear. Res. 20:781–798.
236. McFarland, W.H., Vivion, M.C., Wolf, K.E., and Goldstein, R. 1975. Reexamination of effects of stimulus rate and number on the middle components of the averaged electroencephalic response. Audiology 14:456–465.
237. McGee, T.J., and Clemis, J.D. 1982. Effects of conductive hearing loss on auditory brainstem response. Ann. Otol. Rhinol. Laryngol. 91:304–309.
238. Mackay, I.S., and King, I.J. 1977. Pre- and post-operative brainstem responses in a case of acoustic neuroma, sparing the VIII nerve. Clin. Otolaryngol. 2:233–238.
239. McRandle, C.C., Smith, M.A., and Goldstein, R. 1974. Early averaged electroencephalic responses to clicks in neonates. Ann. Otol. Rhinol. Laryngol. 83:695–702.
240. Madell, J.R., and Goldstein, R. 1972. Relation between loudness and the amplitude of the early components of the averaged electroencephalic response. J. Speech Hear. Res. 15:134–141.
241. Mair, I.W.S., and Laukli, E. 1980. Identification of early auditory-evoked responses. Audiology 19:384–394.
242. Markand, O.N., Garg, B.P., and Brandt, I.K. 1982. Nonketotic hyperglycinemia: Electroencephalographic and evoked potential abnormalities. Neurology 32:151–156.
243. Markand, O.N., Garg, B.P., DeMyer, W.E., and Warren, C. 1982. Brain stem auditory, visual and somatosensory evoked potentials in leukodystrophies. Electroencephalogr. Clin. Neurophysiol. 54:39–48.
244. Markand, O.N., Ochs, R., Worth, R.M., and DeMeyer, W.E. 1980. Brainstem auditory evoked potentials in chronic degenerative central nervous system disorders. In: Evoked potentials: Proceedings of an international evoked potentials symposium held in Nottingham, England, ed. C. Barber, pp. 367–375. Lancaster: MTP Press.
245. Marsh, J.T., Brown, W.S., and Smith, J.C. 1975. Far-field recorded frequency-following responses: Correlates of low pitch auditory perception in humans. Electroencephalogr. Clin. Neurophysiol. 38:113–119.
246. Marshall, N.K., and Donchin, E. 1981. Circadian variation in the latency of brainstem responses and its relation to body temperature. Science 212:356–358.
247. Martin, M.E., and Moore, E.J. 1977. Scalp distribution of early (0 to 10 msec) auditory evoked responses. Arch. Otolaryngol. 103:326–328.
248. Matthews, W.B., Wattam-Bell, J.R.B., and Pountney, E. 1982. Evoked potentials in the diagnosis of multiple sclerosis: A follow up study. J. Neurol. Neurosurg. Psychiatry 45:303–307.
249. Mauldin, L., and Jerger, J. 1979. Auditory brain stem evoked responses to bone-conducted signals. Arch. Otolaryngol. 105:656–661.
250. Maurer, K., and Lowitzsch, K. 1982. Brainstem auditory evoked potentials in reclassification of 143 MS patients. Adv. Neurol. 32:481–486.
251. Maurer, K., Schafer, E., and Leitner, H. 1980. The effect of varying stimulus polarity (rarefaction vs. condensation) on early auditory evoked potentials (EAEPs). Electroencephalogr. Clin. Neurophysiol. 50:332–334.
252. Maurer, K., Strümpel, D., and Wende, S. 1982. Acoustic tumour detection with early auditory evoked potentials and neuroradiological methods. J. Neurol. 227:177–185.
253. Maurizi, M., Altissimi, G., Ottaviani, F., Paludetti, G.,

and Bambini, M. 1982. Auditory brainstem responses (ABR) in the aged. Scand. Audiol. 11:213–221.
254. Maurizi, M., Paludetti, G., Ottaviani, F., and Rosignoli, M. 1984. Auditory brainstem responses to middle- and low-frequency tone pips. Audiology 23:75–84.
255. Mendel, M.I., Adkinson, C.D., and Harker, L.A. 1977. Middle components of the auditory evoked potentials in infants. Ann. Otol. Rhinol. Laryngol. 86:293–299.
256. Mendel, M.I., Hosick, E.C., Windman, T., Davis, H., Hirsh, S.K., and Dinges, D.F. 1975. Audiometric comparison of the middle and late components of the adult auditory evoked potentials awake and asleep. Electroencephalogr. Clin. Neurophysiol. 38:27–33.
257. Metrick, S.A., and Brenner, R.P. 1982. Abnormal brainstem auditory evoked potentials in chronic paint sniffers. Ann. Neurol. 12:553–556.
258. Michalewski, H.J., Thompson, L.W., Patterson, J.V., Bowman, T.E., and Litzelman, D. 1980. Sex differences in the amplitudes and latencies of the human auditory brain stem potential. Electroencephalogr. Clin. Neurophysiol. 48:351–356.
259. Miller, J.R., Burke, A.M., and Bever, C.T. 1983. Occurrence of oligoclonal bands in multiple sclerosis and other CNS diseases. Ann. Neurol. 13:53–58.
260. Mizrahi, E.M., Maulsby, R.L., and Frost, J.D. 1983. Improved wave V resolution by dual-channel brain stem auditory evoked potential recording. Electroencephalogr. Clin. Neurophysiol. 55:105–107.
261. Mjøen, S., Nordby, H.K., and Torvik, A. 1983. Auditory evoked brainstem responses (ABR) in coma due to severe head trauma. Acta Oto-Laryngol. 95:131–138.
262. Mochizuki, Y., Go, T., Ohkubo, H., and Motomura, T. 1983. Development of human brainstem auditory evoked potentials and gender differences from infants to young adults. Prog. Neurobiol. 20:273–285.
263. Mokotoff, B., Schulman-Galambos, C., and Galambos, R. 1977. Brain stem auditory evoked responses in children. Arch. Otolaryngol. 103:38–43.
264. Møller, A.R., and Jannetta, P.J. 1982. Evoked potentials from the inferior colliculus in man. Electroencephalogr. Clin. Neurophysiol. 53:612–620.
265. Møller, A.R., and Jannetta, P.J. 1983. Interpretation of brainstem auditory evoked potentials: Results from intracranial recordings in humans. Scand. Audiol. 12:125–133.
266. Møller, A.R., and Jannetta, P.J. 1983. Auditory evoked potentials recorded from the cochlear nucleus and its vicinity in man. J. Neurosurg. 59:1013–1018.
267. Møller, A.R., Jannetta, P., and Møller, M.B. 1982. Intracranially recorded auditory nerve response in man. Arch. Otolaryngol. 108:77–82.
268. Møller, M.B., Møller, A.R., and Jannetta, P.J. 1982. Brain stem auditory evoked potentials in patients with hemifacial spasm. Laryngoscope 92:484–852.
269. Monod, N., and Garma, L. 1971. Auditory responsivity in the human premature. Biol. Neonate 17:292–316.
270. Morgon, A., and Salle, B. 1980. A study of brain stem evoked response in prematures. Acta Oto-Laryngol. 89:370–375.
271. Morrison, A.W., Gibson, W.P., and Beagley, H.A. 1976. Trans-tympanic electrocochleography in the diagnosis of retro-cochlear tumours. Clin. Otolaryngol. 1:153–167.
272. Mosko, S.S., Pierce, S., Holowach, J., and Sassin, J.F. 1981. Normal brain stem auditory evoked potentials recorded in sleep apneics during waking and as a function of

arterial oxygen saturation during sleep. Electroencephalogr. Clin. Neurophysiol. 51:477–482.

273. Moushegian, G. 1977. The frequency following potentials in man. Prog. Clin. Neurophysiol. 2:20–29.

274. Moushegian, G., Rupert, A.L., and Stillman, R.D. 1973. Scalp-recorded early responses in man to frequencies in the speech range. Electroencephalogr. Clin. Neurophysiol. 35:665–667.

275. Naunton, R.F., and Zerlin, S. 1976. Human whole-nerve response to clicks of various frequency. Audiology 15:1–9.

276. Naunton, R.F., and Zerlin, S.S. 1978. Electrocochleography: Behavioral threshold comparisons. In: Evoked electrical activity in the auditory nervous system, ed. R.F. Naunton and C. Fernandez, pp. 221–236. New York: Academic Press.

277. Nelson, D.A., and Lassman, F.M. 1968. Effects of intersignal interval on the human auditory evoked response. J. Acoust. Soc. Am. 44:1529–1532.

278. Nodar, R.H., Hahn, J., and Levine, H.L. 1980. Brain stem auditory evoked potentials in determining site of lesion of brain stem gliomas in children. Laryngoscope 90:258–265.

279. Noseworthy, J.H., Miller, J., Murray, T.J., and Regan, D. 1981. Auditory brainstem responses in postconcussion syndrome. Arch. Neurol. 38:275–278.

280. Novick, B., Vaughan, H.G., Kurtzberg, D., and Simson, R. 1980. An electrophysiologic indication of auditory processing defects in autism. Psychiatry Res. 3:107–114.

281. Nuwer, M.R., Perlman, S.L., Packwood, J.W., and Kark, R.A.P. 1983. Evoked potential abnormalities in the various inherited ataxias. Ann. Neurol. 13:20–27.

282. Odenthal, D.W., and Eggermont, J.J. 1974. Clinical electrocochleography. Acta Oto-Laryngol. Suppl. 316:62–74.

283. O'Donovan, C.A., Beagley, H.A., and Shaw, M. 1980. Latency of brainstem response in children. Br. J. Audiol. 14:23–29.

284. Oh, S.J., Kuba, T., Soyer, A., Choi, I.S., Bonikowski, F.P., and Vitek, J. 1981. Lateralization of brainstem lesions by brainstem auditory evoked potentials. Neurology 31:14–18.

285. Ohlrich, E.S., Barnet, A.B., Weiss, I.P., and Shanks, B.L. 1978. Auditory evoked potential development in early childhood: A longitudinal study. Electroencephalogr. Clin. Neurophysiol. 44:411–423.

286. Onishi, S., and Davis, H. 1968. Effects of duration and rise time of tone bursts on evoked V potentials. J. Acoust. Soc. Am. 44:582–591.

287. Ornitz, E.M., Ritvo, E.R., Carr, E.M., Panman, L.M., and Walter, R.D. 1967. The variability of the auditory averaged evoked response during sleep and dreaming in children and adults. Electroencephalogr. Clin. Neurophysiol. 22:514–524.

288. Ornitz, E.M., and Walter, D.O. 1975. The effect of sound pressure waveform on human brain stem auditory evoked responses. Brain Res. 92:490–498.

289. Osterhammel, P. 1981. The unsolved problems in analog filtering on the auditory brain stem responses. Scand. Audiol. Suppl. 13:69–74.

290. Osterhammel, P.H., Davis, H., Wier, C.C., and Hirsh, S.K. 1973. Adult auditory evoked vertex potentials in sleep. Audiology 12:116–128.

291. Otto, W.C., and McCandless, G.A. 1982. Aging and the auditory brain stem response. Audiology 21:466–473.

292. Owen, J.H., and Matsusaka, E.K. 1982. Influence of test parameter combinations on the auditory evoked late response. Electroencephalogr. Clin. Neurophysiol. 53:329–339.
293. Özdamar, Ö., and Kraus, N. 1983. Auditory middle-latency responses in humans. Audiology 22:34–49.
294. Özdamar, Ö., Kraus, N., and Curry, F. 1982. Auditory brain stem and middle latency responses in a patient with cortical deafness. Electroencephalogr. Clin. Neurophysiol. 53:224–230.
295. Özdamar, Ö., Kraus, N., and Stein, L. 1983. Auditory brainstem responses in infants recovering from bacterial meningitis. Audiologic evaluation. Arch. Otolaryngol. 109:13–18.
296. Özdamar, Ö., and Stein, L. 1981. Auditory brain stem response (ABR) in unilateral hearing loss. Laryngoscope 41:565–574.
297. Paludetti, G., Maurizi, M., Ottaviani, F., and Rosignoli, M. 1981. Reference values and characteristics of brain stem audiometry in neonates and children. Scand. Audiol. 10:177–186.
298. Parker, D.J. 1981. Dependence of the auditory brainstem response on electrode location. Arch. Otolaryngol. 107:367–371.
299. Parker, D.J., and Thornton, A.R.D. 1978. Frequency specific components of the cochlear nerve and brainstem evoked responses of the human auditory system. Scand. Audiol. 7:53–60.
300. Parving, A., Elberling, C., and Salomon, G. 1981. ECochG and psychoacoustic tests compared in identification of hearing loss in young children. Audiology 20:365–381.
301. Parving, A., Elberling, C., and Salomon, G. 1981. Slow cortical responses and the diagnosis of central hearing loss in infants and young children. Audiology 20:465–479.
302. Parving, A., Elberling, C., and Smith, T. 1981. Auditory electrophysiology: Findings in multiple sclerosis. Audiology 20:123–142.
303. Parving, A., Salomon, G., Elberling, C., Larsen, B., and Lassen, N.A. 1980. Middle components of the auditory evoked response in bilateral temporal lobe lesions. Scand. Audiol. 9:161–167.
304. Pedersen, L., and Trojaborg, W., 1981. Visual, auditory and somatosensory pathway involvement in hereditary cerebellar ataxia, Friedreich's ataxia and familial spastic paraplegia. Electroencephalogr. Clin. Neurophysiol. 52:283–297.
305. Peled, R., Pratt, H., Scharf, B., and Lavie, P. 1983. Auditory brainstem evoked potentials during sleep apnea. Neurology 33:419–423.
306. Peronnet, F., and Giard, M.H. 1980. Inter-hemispheric and inter-aural differences in the human auditory evoked potential. In: Evoked potentials: Proceedings of an international evoked potentials symposium held in Nottingham, England, ed. C. Barber, pp. 317–324. Lancaster: MTP Press.
307. Peronnet, F., Giard, M.H., Bertrand, O., and Pernier, J. 1984. The temporal component of the auditory evoked potential: A reinterpretation. Electroencephalogr. Clin. Neurophysiol. 59:67–71.
308. Peters, J.F., and Mendel, M.I. 1974. Early components of the averaged electroencephalic response to monaural and binaural stimulation. Audiology 13:195–204.
309. Pfefferbaum, A., Ford, J.M., Roth, W.T., Hopkins, W.F., and Kopell, B.S. 1979. Event-related potential changes in

healthy aged females. Electroencephalogr. Clin. Neurophysiol. 46:81–86.
310. Picton, T.W., Goodman, W.S., and Bryce, D.P. 1970. Amplitude of evoked responses to tones of high intensity. Acta Oto-Laryngol. 70:77–82.
311. Picton, T.W., and Hillyard, S.A. 1974. Human auditory evoked potentials. II. Effects of attention. Electroencephalogr. Clin. Neurophysiol. 36:191–199.
312. Picton, T.W., Hillyard, S.A., and Galambos, R. 1976. Habituation and attention in the auditory system. In: Handbook of sensory physiology, Vol. V/3 Auditory system, Part 3 (Clinical and special topics), ed. W.D. Keidel and W.D. Neff, pp. 343–389. Berlin: Springer Verlag.
313. Picton, T.W., Hillyard, S.A., Krausz, H.I., and Galambos, R. 1974. Human auditory evoked potentials. I. Evaluation of components. Electroencephalogr. Clin. Neurophysiol. 36:179–190.
314. Picton, T.W., Ouellette, J., Hamel, G., and Durieux Smith, A. 1979. Brain stem evoked potentials to tonepips in notched noise. J. Otolaryngol. 8:289–313.
315. Picton, T.W., Stapells, D.R., and Campbell, K.B. 1982. Auditory evoked potentials from the human cochlea and brainstem. J. Otolaryngol. Supp 9:1–41.
316. Picton, T.W., Woods, D.L., Baribeau-Braun, J., and Healey, T.M.G. 1977. Evoked potential audiometry. J. Otolaryngol. 6:90–119.
317. Picton, T.W., Woods, D.L., and Proulx, G.B. 1978. Human auditory sustained potentials. I. The nature of the response. Electroencephalogr. Clin. Neurophysiol. 45:186–197.
318. Prasher, D.K. 1980. The influence of stimulus spectral content on rise time effects in cortical-evoked responses. Audiology 19:355–362.
319. Prasher, D.K., and Gibson, W.P.R. 1980. Brain stem auditory evoked potentials: Significant latency differences between ipsilateral and contralateral stimulation. Electroencephalogr. Clin. Neurophysiol. 50:240–246.
320. Prasher, D.K., and Gibson, W.P.R. 1980. Brain stem auditory evoked potentials: A comparative study of monaural versus binaural stimulation in the detection of multiple sclerosis. Electroencephalogr. Clin. Neurophysiol. 50:247–253.
321. Prasher, D.K., Sainz, M., and Gibson, W.P.R. 1982. Binaural voltage summation of brainstem auditory evoked potentials: An adjunct to the diagnostic criteria for multiple sclerosis. Ann. Neurol. 11:86–91.
322. Pratt, H., Ben-David, Y., Peled, R., Podoshin, L., and Scharf, B. 1981. Auditory brain stem evoked potentials: Clinical promise of increasing stimulus rate. Electroencephalogr. Clin. Neurophysiol. 51:80–90.
323. Pratt, H., and Bleich, N. 1982. Auditory brain stem potentials evoked by clicks in notch-filtered masking noise. Electroencephalogr. Clin. Neurophysiol. 53:417–426.
324. Pratt, H., and Sohmer, H. 1976. Intensity and rate functions of cochlear and brainstem evoked responses to click stimuli in man. Arch. Oto-Rhino-Laryngol. 212:85–92.
325. Pratt, H., and Sohmer, H. 1977. Correlations between psychophysical magnitude estimates and simultaneously obtained auditory nerve, brain stem and cortical responses to click stimuli in man. Electroencephalogr. Clin. Neurophysiol. 43:802–812.
326. Pratt, H., and Sohmer, H. 1978. Comparison of hearing threshold determined by auditory pathway electric re-

326. sponses and by behavioural responses. Audiology 17:285–292.
327. Pressman, M.R., Spielman, A.J., Pollak, C.P., and Weitzman, E.D. 1982. Long-latency auditory evoked responses during sleep deprivation and in narcolepsy. Sleep 5 (Suppl. 2):147–156.
328. Prichep, L.S., Sutton, S., and Hakerem, G. 1976. Evoked potentials in hyperkinetic and normal children under certainty and uncertainty: A placebo and methylphenidate study. Psychophysiology 13:419–428.
329. Pulst, S.M., and Lombroso, C.T. 1983. External ophthalmoplegia, alpha and spindle coma in imipramine overdose: Case report and review of the literature. Ann. Neurol. 14:587–590.
330. Purves, S.J., Low, M.D., Galloway, J., and Reeves, B. 1981. A comparison of visual, brainstem auditory, and somatosensory evoked potentials in multiple sclerosis. Can. J. Neurol. Sci. 8:15–19.
331. Rapin, I., and Graziani, L.J. 1967. Auditory-evoked responses in normal, brain-damaged, and deaf infants. Neurology 17:881–894.
332. Rapin, I., Schimmel, H., and Cohen, M.M. 1972. Reliability in detecting the auditory evoked response (AER) for audiometry in sleeping subjects. Electroencephalogr. Clin. Neurophysiol. 32:521–528.
333. Rapin, I., Schimmel, H., Tourk, L.M., Krasnegor, N.A., and Pollak, C. 1966. Evoked responses to clicks and tones of varying intensity in waking adults. Electroencephalogr. Clin. Neurophysiol. 21:335–344.
334. Rappaport, M., Hopkins, H.K., Hall, K., and Beleza, T. 1981. Evoked potentials and head injury. 2. Clinical applications. Clin. Electroencephalogr. 12:167–176.
335. Raudzens, P.A. 1982. Intraoperative monitoring of evoked potentials. Ann. N.Y. Acad. Sci. 388:308–326.
336. Raudzens, P.A., and Shetter, A.G. 1982. Intraoperative monitoring of brain-stem auditory evoked potentials. J. Neurosurg. 57:341–348.
337. Rizzo, P.A., Pierelli, F., Pozzessere, G., Floris, R., and Morocutti, C. 1983. Subjective posttraumatic syndrome: A comparison of visual and brain stem auditory evoked responses. Neuropsychobiology 9:78–82.
338. Rizzo, P.A., Pierelli, F., Pozzessere, G., Verardi, S., Casciani, C.U., and Morocutti, C. 1982. Pattern visual evoked potentials and brainstem auditory evoked responses in uremic patients. Acta Neurol. Belg. 82:72–79.
339. Robier, A., Lemaire, M.C., Garreau, B., Ployet, M.J., Martineau, J., Delvert, J.C., and Reynaud, J. 1983. Auditory brain stem responses and cortical auditory-evoked potentials in difficult-to-test children. Audiology 22:219–228.
340. Robinson, K., and Rudge, P. 1977. Abnormalities of the auditory evoked potentials in patients with multiple sclerosis. Brain 100:19–40.
341. Robinson, K., and Rudge, P. 1978. The stability of the auditory evoked potentials in normal man and patients with multiple sclerosis. J. Neurol. Sci. 36:147–156.
342. Robinson, K., and Rudge, P. 1981. Wave form analysis of the brain stem auditory evoked potential. Electroencephalogr. Clin. Neurophysiol. 52:583–594.
343. Robinson, K., and Rudge, P. 1983. The differential diagnosis of cerebello-pontine angle lesions: A multidisciplinary approach with special emphasis on the brainstem auditory evoked potential. J. Neurol. Sci. 60:1–21.
344. Ropper, A.H., Miett, T., and Chiappa, K.H. 1982. Ab-

sence of evoked potential abnormalities in acute transverse myelopathy. Neurology 32:80–92.
345. Rose, D.E., Keating, L.W., Hedgecock, L.D., Miller, K.E., and Schreurs, K.K. 1972. A comparison of evoked response audiometry and routine clinical audiometry. Audiology 11:238–243.
346. Rosenblum, S.M., Arick, J.R., Krug, D.A., Stubbs, E.G., Young, N.B., and Pelson, R.O. 1980. Auditory brainstem evoked responses in autistic children. J. Autism Dev. Disord. 10:215–225.
347. Rosenhamer, H. 1981. The auditory evoked brainstem electric response (ABR) in cochlear hearing loss. Scand. Audiol. Suppl. 13:83–93.
348. Rosenhamer, H., and Holmkvist, C. 1982. Bilaterally recorded auditory brainstem responses to monaural stimulation. Scand. Audiol. 11:197–202.
349. Rosenhamer, H.J., and Holmkvist, C. 1983. Latencies of ABR (waves III and V) to binaural clicks: Effects of interaural time and intensity differences. Scand. Audiol. 12:201–207.
350. Rosenhamer, H.J., Lindstrom, B., and Lundborg, T. 1978. On the use of click-evoked electric brainstem responses in audiological diagnosis. I. The variability of the normal response. Scand. Audiol. 7:193–205.
351. Rosenhamer, H.J., Lindstrom, B., and Lundborg, T. 1980. On the use of click-evoked electric brainstem responses in audiological diagnosis. II. The influence of sex and age upon the normal response. Scand. Audiol. 9:93–100.
352. Rosenhamer, H.J., Lindstrom, B., and Lundborg, T. 1981. On the use of click-evoked electric brainstem responses in audiological diagnosis. III. Latencies in cochlear hearing loss. Scand. Audiol. 10:3–11.
353. Rothman, H.H. 1970. Effects of high frequencies and intersubject variability on the auditory-evoked cortical response. J. Acoust. Soc. Am. 47:569–573.
354. Rowe, M.J. 1978. Normal variability of the brain-stem auditory evoked response in young and old adult subjects. Electroencephalogr. Clin. Neurophysiol. 44:459–470.
355. Rowe, M.J., and Carlson, C. 1980. Brainstem auditory evoked potentials in postconcussion dizziness. Arch. Neurol. 37:679–683.
356. Ruhm, H.B. 1970. Rate of frequency change and the acoustically evoked response. J. Aud. Res. 10:29–34.
357. Ruhm, H., Walker, E., and Flanigin, H. 1967. Acoustically-evoked potentials in man: Mediation of early components. Laryngoscope 77:806–822.
358. Salamy, A., and McKean, C.M. 1977. Habituation and dishabituation of cortical and brainstem evoked potentials. Int. J. Neurosci. 7:175–182.
359. Salamy, A., McKean, C.M., Pettett, G., and Mendelson, T. 1978. Auditory brainstem recovery processes from birth to adulthood. Psychophysiology 15:214–220.
360. Salamy, A., Mendelson, T., Tooley, W.H., and Chaplin, E.R. 1980. Differential development of brainstem potentials in healthy and high-risk infants. Science 210:553–555.
361. Salomon, G., Beck, O., and Elberling, C. 1973. The role of sedation in ERA from the vertex. Audiology 12:150–166.
362. Sanders, R.A., Duncan, P.G., and McCullough, D.W. 1979. Clinical experience with brain stem audiometry performed under general anesthesia. J. Otolaryngol. 8:24–31.
363. Satterfield, J.H., and Braley, B.W. 1977. Evoked potentials and brain maturation in hyperactive and normal children. Electroencephalogr. Clin. Neurophysiol. 43:43–51.

364. Satya-Murty, S., Cacace, A.T., and Hanson, P.A. 1979. Abnormal auditory evoked potentials in hereditary motor-sensory neuropathy. Ann. Neurol. 5:445–448.
365. Satya-Murti, S., Cacace, A., and Hanson, P. 1980. Auditory dysfunction in Friedreich ataxia: Result of spiral ganglion degeneration. Neurology 30:1047–1053.
366. Schaefer, E.W.P., Amochaev, A., and Russell, M.J. 1981. Knowledge of stimulus timing attenuates human evoked cortical potentials. Electroencephalogr. Clin. Neurophysiol. 52:9–17.
367. Scherg, M. 1982. Simultaneous recording and separation of early and middle latency auditory evoked potentials. Electroencephalogr. Clin. Neurophysiol. 54:339–341.
368. Scherg, M. 1982. Distortion of the middle latency auditory response produced by analog filtering. Scand. Audiol. 11:57–60.
369. Scherg, M., and Volk, S.A. 1983. Frequency specificity of simultaneously recorded early and middle latency auditory evoked potentials. Electroencephalogr. Clin. Neurophysiol. 56:443–452.
370. Scherg, M., von Cramon, D., and Elton, M. 1984. Brainstem auditory-evoked potentials in post-comatose patients after severe closed head trauma. J. Neurol. 231:1–5.
371. Schulman-Galambos, C., and Galambos, R. 1975. Brain stem auditory-evoked responses in premature infants. J. Speech Hear. Res. 18:456–465.
372. Schulman-Galambos, C., and Galambos, R. 1979. Brain stem evoked response audiometry in newborn hearing screening. Arch. Otolaryngol. 105:86–90.
373. Schulman-Galambos, C., and Galambos, R. 1979. Assessment of hearing. In: Infants born at risk: Behavior and development, ed. T.M. Field, M. Sostek, S. Goldberg, and H. Shuman, pp. 91–119. New York: Medical Science Books.
374. Schwent, V.L., and Hillyard, S.A. 1975. Evoked potential correlates of selective attention with multi-channel auditory inputs. Electroencephalogr. Clin. Neurophysiol. 38:131–138.
375. Schwent, V.L., Hillyard, S.A., and Galambos, R. 1976. Selective attention and the auditory vertex potential. I. Effects of stimulus delivery rate. Electroencephalogr. Clin. Neurophysiol. 40:604–614.
376. Seales, D.M., Rossiter, V.S., and Weinstein, M.E. 1979. Brainstem auditory evoked responses in patients comatose as a result of blunt head trauma. J. Trauma 19:347–352.
377. Seales, D.M., Torkelson, R.D., Shuman, R.M., Rossiter, V.S., and Spencer, J.D. 1981. Abnormal brainstem auditory evoked potentials and neuropathology in "locked-in" syndrome. Neurology 31:893–896.
378. Selters, W.A., and Brackmann, D.E. 1977. Acoustic tumor detection with brain stem electric response audiometry. Arch. Otolaryngol. 103:181–187.
379. Shagass, C., and Straumanis, J.J. 1978. Drugs and human sensory evoked potentials. In: Psychopharmacology: A generation of progress, ed. M.A. Lipton, A. DiMascio, and K.F. Killam, pp. 699–709. New York: Raven Press.
380. Shagass, C., Straumanis, J.J., Roemer, R.A., and Amadeo, M. 1977. Evoked potentials of schizophrenics in several sensory modalities. Biol. Psychiatry 12:221–235.
381. Shallop, J.K., and Osterhammel, P.A. 1983. A comparative study of measurements of SN-10 and the 40/sec middle latency responses in newborns. Scand. Audiol. 12:91–95.
382. Skinner, P.H., and Antinoro, F. 1971. The effects of signal rise time and duration on the early components of the au-

ditory evoked cortical response. J. Speech Hear. Res. 14:552–558.
383. Skinner, P.H., and Jones, H.C. 1968. Effects of signal duration and rise time on the auditory evoked potential. J. Speech Hear. Res. 11:301–306.
384. Skinner, P., and Shimota, J. 1975. A comparison of the effects of sedatives on the auditory evoked cortical response. J. Am. Audiol. Soc. 1:71–78.
385. Skoff, B.F., Mirsky, A.F., and Turner, D. 1980. Prolonged brainstem transmission time in autism. Psychiatry Res. 2:157–166.
386. Sohmer, H., Feinmesser, M., Bauberger-Tell, L., and Edelstein, E. 1977. Cochlear, brain stem, and cortical evoked responses in nonorganic hearing loss. Ann. Otol. Rhinol. Laryngol. 86:227–234.
387. Sohmer, H., Gafni, M., and Chisin, R. 1978. Auditory nerve and brain stem responses. Comparison in awake and unconscious subjects. Arch. Neurol. 35:228–230.
388. Sohmer, H., Gafni, M., and Chisin, R. 1982. Auditory nerve-brain stem potentials in man and cat under hypoxic and hypercapnic conditions. Electroencephalogr. Clin. Neurophysiol. 53:506–512.
389. Sohmer, H., Kinarti, R., and Gafni, M. 1980. The source along the basilar membrane of the cochlear microphonic potential recorded by surface electrodes in man. Electroencephalogr. Clin. Neurophysiol. 49:506–514.
390. Sohmer, H., and Pratt, H. 1976. Recording of the cochlear microphonic potential with surface electrodes. Electroencephalogr. Clin. Neurophysiol. 40:253–260.
391. Sohmer, H., and Pratt, H. 1977. Identification and separation of acoustic frequency following responses (FFRs) in man. Electroencephalogr. Clin. Neurophysiol. 42:493–500.
392. Sohmer, H., Pratt, H., and Kinarti, R. 1977. Sources of frequency following responses (FFR) in man. Electroencephalogr. Clin. Neurophysiol. 42:656–664.
393. Sohmer, H., and Student, M. 1978. Auditory nerve and brain-stem evoked responses in normal, autistic, minimal brain dysfunction and psychomotor retarded children. Electroencephalogr. Clin. Neurophysiol. 44:380–388.
394. Spire, J.P., Dohrmann, G.J., and Prieto, P.S. 1982. Correlation of brainstem evoked response with direct acoustic nerve potential. Adv. Neurol. 32:159–167.
395. Spoor, A., Timmer, F., and Odenthal, D.W. 1969. The evoked auditory response (EAR) to intensity modulated and frequency modulated tones and tone bursts. Int. Audiol. 8:410–415.
396. Squires, N., Aine, C., Buchwald, J., Norman, R., and Galbraith, G. 1980. Auditory brain stem response abnormalities in severely and profoundly retarded adults. Electroencephalogr. Clin. Neurophysiol. 50:172–185.
397. Squires, K.C., Chu, N.S., and Starr, A. 1978. Acute effects of alcohol on auditory brainstem potentials in humans. Science 201:174–176.
398. Stakenburg, M., and Wit, H.P. 1983. Piezoelectric earphones for artefact-free recording of auditory brainstem responses (ABR). Scand. Audiol. 12:79–80.
399. Stapells, D.R., and Picton, T.W. 1981. Technical aspects of brainstem evoked potential audiometry using tones. Ear Hear. 2:20–29.
400. Starr, A. 1976. Auditory brain-stem responses in brain death. Brain 99:543–554.
401. Starr, A., and Achor, L.J. 1975. Auditory brain stem responses in neurological disease. Arch. Neurol. 32:761–768.
402. Starr, A., Amlie, R.N., Martin, W.H., and Sanders, S.

1977. Development of auditory function in newborn infants revealed by auditory brainstem potentials. Pediatrics 60:831–839.
403. Starr, A., and Hamilton, A.E. 1976. Correlation between confirmed sites of neurological lesions and abnormalities of far-field auditory brainstem responses. Electroencephalogr. Clin. Neurophysiol. 41:595–608.
404. Starr, A., and Squires, K. 1982. Distribution of auditory brainstem potentials over the scalp and nasopharynx in humans. Ann. N.Y. Acad. Sci. 388:427–442.
405. Stillman, R.D., Crow, G., and Moushegian, G. 1978. Components of the frequency-following potential in man. Electroencephalogr. Clin. Neurophysiol. 44:438–446.
406. Stockard, J.E., Stockard, J.J., Kleinberg, F., and Westmoreland, B.F. 1983. Prognostic value of brainstem auditory evoked potentials in neonates. Arch. Neurol. 40:360–365.
407. Stockard, J.E., Stockard, J.J., Westmoreland, B.F., and Corfits, J.L. 1979. Brainstem auditory-evoked responses. Normal variation as a function of stimulus and subject characteristics. Arch. Neurol. 36:823–831.
408. Stockard, J.E., and Westmoreland, B.F. 1981. Technical considerations in the recording and interpretation of the brainstem auditory evoked potential for neonatal neurologic diagnosis. Am. J. EEG Tech. 21:31–54.
409. Stockard, J.J. 1982. Brainstem auditory evoked potentials in adult and infant sleep apnea syndromes, including sudden infant death syndrome and near-miss for sudden infant death. Ann. N.Y. Acad. Sci. 388:443–465.
410. Stockard, J.J., and Rossiter, V.S. 1977. Clinical and pathologic correlates of brain stem auditory response abnormalities. Neurology 27:316–325.
411. Stockard, J.J., Rossiter, V.S., Wiederholt, W.C., and Kobayashi, R.M. 1976. Brain stem auditory-evoked responses in suspected central pontine myelinolysis. Arch. Neurol. 33:726–728.
412. Stockard, J.J., Sharbrough, F.W., and Tinker, J.A. 1978. Effects of hypothermia on the human brainstem auditory response. Ann. Neurol. 3:368–370.
413. Stockard, J.J., Stockard, J.E., and Sharbrough, F.W. 1977. Detection and localization of occult lesions with brainstem auditory responses. Mayo Clin. Proc. 52:761–769.
414. Stockard, J.J., Stockard, J.E., and Sharbrough, F.W. 1978. Nonpathological factors influencing brainstem auditory evoked potentials. Am. J. EEG Tech. 18:177–209.
415. Stockard, J.J., Stockard, J.E., and Sharbrough, F.W. 1980. Brainstem auditory evoked potentials in neurology: Methodology, interpretation, clinical application. In: Electrodiagnosis in clinical neurology, ed. M.J. Aminoff, pp. 370–413. New York: Churchill Livingstone.
416. Streletz, L.J., Katz, L., Hohenberger, M., and Cracco, R.Q. 1977. Scalp recorded auditory evoked potentials and sonomotor responses: An evaluation of components and recording techniques. Electroencephalogr. Clin. Neurophysiol. 43:192–206.
417. Surwillo, W.W. 1977. Cortical evoked response recovery functions: Physiological manifestations of the psychological refractory period? Psychophysiology 14:32–39.
418. Suzuki, T., Hirai, Y., and Horiuchi, K. 1977. Auditory brain stem responses to pure tone stimuli. Scand. Audiol. 6:51–56.
419. Suzuki, T., and Horiuchi, K. 1977. Effect of high-pass filter on auditory brain stem responses to tone pips. Scand. Audiol. 6:123–126.

420. Suzuki, T., and Horiuchi, K. 1981. Rise time of pure-tone stimuli in brain stem response audiometry. Audiology 20:101–112.
421. Suzuki, T., Sakabe, N., and Miyashita, Y. 1982. Power spectral analysis of auditory brain stem responses to pure tone stimuli. Scand. Audiol. 11:25–30.
422. Tackmann, W., Ettlin, T., and Strenge, H. 1982. Multimodality evoked potentials and electrically elicited blink reflex in optic neuritis. J. Neurol. 227:157–163.
423. Tackmann, W., Strenge, H., Barth, R., and Sojka-Raytscheff, A. 1980. Auditory brain stem evoked potentials in patients with multiple sclerosis: Investigations in patients with different degrees of diagnostic probability. Eur. Neurol. 19:396–401.
424. Taylor, M.J., Houston, B.D., and Lowry, N.J. 1983. Recovery of auditory brain-stem responses after a severe hypoxic ischemic insult. N. Eng. J. Med. 309:1169–1170.
425. Taylor, M.J., McMenamin, J.B., Andermann, E., and Watters, G.V. 1982. Electrophysiological investigation of the auditory system in Friedreich's ataxia. Can. J. Neurol. Sci. 9:131–135.
426. Taylor, M.J., Rosenblatt, B., and Linschoten, L. 1982. Auditory brainstem response abnormalities in autistic children. Can. J. Neurol. Sci. 9:429–433.
427. Teas, D.C., Klein, A.J., and Kramer, S.J. 1982. An analysis of auditory brainstem responses in infants. Hear. Res. 7:19–54.
428. Terkildsen, K., and Osterhammel, P. 1981. The influence of reference electrode position on recordings of the auditory brainstem responses. Ear Hear. 2:9–14.
429. Terkildsen, K., Osterhammel, P., and Huis in't Veld, F. 1974. Far-field electrocochleography, electrode positions. Scand. Audiol. 3:123–129.
430. Terkildsen, K., Osterhammel, P., and Huis in't Veld, F. 1975. Far-field electrocochleography: Adaptation. Scand. Audiol. 4:215–220.
431. Terkildsen, K., Osterhammel, P., and Thomsen, J. 1981. The ABR and the MLR in patients with acoustic neuromas. Scand. Audiol. Suppl. 13:103–107.
432. Theodore, W.H., Comite, F., Sato, S., Loriaux, L., and Cutler, G. 1983. EEG and evoked potentials in precocious puberty. Electroencephalogr. Clin. Neurophysiol. 55:69–72.
433. Thompson, D.S., Woodward, J.B., Ringel, S.P., and Nelson, L.M. 1983. Evoked potential abnormalities in myotonic dystrophy. Electroencephalogr. Clin. Neurophysiol. 56:453–456.
434. Thomsen, J., Terkildsen, K., and Osterhammel, P. 1982. Auditory brain stem responses in patients with acoustic neuromas. Acta Oto-Laryngol. Suppl. 386:20–22.
435. Thornton, A.R.D. 1975. The use of post-auricular muscle responses. J. Laryngol. Otol. 89:997–1010.
436. Thornton, A.R.D. 1975. Distortion of averaged post-auricular muscle responses due to system bandwidth limits. Electroencephalogr. Clin. Neurophysiol. 39:195–197.
437. Thornton, A.R.D., and Coleman, M.J. 1975. The adaptation of cochlear and brainstem auditory evoked potentials in humans. Electroencephalogr. Clin. Neurophysiol. 39:399–406.
438. Thornton, H.A.R., Mendel, M.I., and Anderson, C. 1977. Effect of stimulus frequency and intensity on the middle components of the averaged auditory electroencephalic response. J. Speech Hear. Res. 20:81–94.

439. Thummler, I., Tietze, G., and Matkei, P. 1981. Brain-stem responses when masking with side-band and high-pass filtered noise. Scand. Audiol. 10:255–259.
440. Tolosa, E.S., and Zeese, J.A. 1979. Brainstem auditory evoked responses in progressive supranuclear palsy. Ann. Neurol. 6:369.
441. Townsend, G.L., and Cody, D.T.R. 1970. Vertex response: Influence of lesions in the auditory system. Laryngoscope 80:979–999.
442. Townsend, G.L., and Cody, D.T.R. 1971. The averaged inion response evoked by acoustic stimulation: Its relation to the saccule. Ann. Otol. Rhinol. Laryngol. 80:121–132.
443. Tsubokawa, T., Nishimoto, H., Yamamoto, T., Kitamura, M., Katayama, Y., and Moriyasu, N. 1980. Assessment of brainstem damage by the auditory brainstem response in acute severe head injury. J. Neurol. Neurosurg. Psychiatry 43:1005–1011.
444. Tyberghein, J., and Forrez, G. 1969. Cortical audiometry in normal hearing subjects. Acta Oto-Laryngol. 67:24–32.
445. Uziel, A., and Benezech, J. 1978. Auditory brain-stem responses in comatose patients: Relationship with brain-stem reflexes and levels of coma. Electroencephalogr. Clin. Neurophysiol. 45:515–524.
446. Van Olphen, A.R., Rodenburg, M., and Verwey, C. 1978. Distribution of brain stem responses to acoustic stimuli over the human scalp. Audiology 17:511–518.
447. Vaughan, H.G., and Ritter, W. 1970. The sources of auditory evoked responses recorded from the human scalp. Electroencephalogr. Clin. Neurophysiol. 28:360–367.
448. Velasco, M., Velasco, R., Almanza, X., and Coats, A.C. 1982. Subcortical correlates of the auditory brain stem potentials in man: Bipolar EEG and multiple unit activity and electrical stimulation. Electroencephalogr. Clin. Neurophysiol. 53:133–142.
449. Vivion, M.C., Wolf, K.E., Goldstein, R., Hirsch, J.C., and McFarland, W.H. 1979. Toward objective analysis for electroencephalic audiometry. J. Speech Hear. Res. 22:88–102.
450. Weiner, R.D., Erwin, C.W., and Weber, B.A. 1981. Acute effects of electroconvulsive therapy on brain stem auditory-evoked potentials. Electroencephalogr. Clin. Neurophysiol. 52:202–204.
451. Weitzman, E.D., and Graziani, L.J. 1968. Maturation and topography of the auditory evoked response of the prematurely born infant. Dev. Psychobiol. 1:79–89.
452. Westmoreland, B.F., Sharbrough, F.W., Stockard, J.J., and Dale, A.J.D. 1983. Brainstem auditory evoked potentials in 20 patients with palatal myoclonus. Arch. Neurol. 40:155–158.
453. Wielaard, R., and Kemp, B. 1979. Auditory brainstem evoked responses in brainstem compression due to posterior fossa tumors. Clin. Neurol. Neurosurg. 81:185–193.
454. Wolpaw, J.R., and Penry, J.K. 1977. Hemispheric differences in the auditory evoked response. Electroencephalogr. Clin. Neurophysiol. 43:99–102.
455. Wolpaw, J.R., and Penry, J.K. 1978. Effects of ethanol, caffeine, and placebo on the auditory evoked response. Electroencephalogr. Clin. Neurophysiol. 44:568–574.
456. Wood, C.C., and Wolpaw, J.R. 1982. Scalp distribution of human evoked potentials. II. Evidence for overlapping sources and involvement of auditory cortex. Electroencephalogr. Clin. Neurophysiol. 54:25–38.

457. Worthington, D.W., and Peters, J.F. 1980. Quantifiable hearing and no ABR: Paradox or error? Ear Hear. 1:281–285.
458. Wrege, K.S., and Starr, A. 1981. Binaural interaction in human auditory brainstem evoked potentials. Arch. Neurol. 38:572–580.
459. Yagi, T., and Kaga, K. 1979. The effect of the click repetition rate on the latency of the auditory evoked brain stem response and its clinical use for a neurological diagnosis. Arch. Oto-Rhino-Laryngol. 222:91–97.
460. Yamada, O., Ashikawa, H., Kodera, K., and Yamane, H. 1983. Frequency-selective auditory brain-stem response in newborns and infants. Arch. Otolaryngol. 109:79–82.
461. Yamada, O., Marsh, R.R., and Handler, S.D. 1982. Contributing generator of frequency-following response in man. Scand. Audiol. 11:53–56.
462. Yamada, O., Yamane, H., and Kodera, K. 1977. Simultaneous recordings of the brain stem response and the frequency-following response to low-frequency tone. Electroencephalogr. Clin. Neurophysiol. 43:362–370.
463. Yoshie, N. 1973. Diagnostic significance of the electrocochleogram in clinical audiometry. Audiology 12:504–539.
464. Yoshie, N., and Okudaira, T. 1969. Myogenic evoked potential responses to clicks in man. Acta Oto-Laryngol. Suppl. 252:89–103.
465. Zerlin, S., and Naunton, R.F. 1978. Recording and analysis techniques. In: Evoked electrical activity in the auditory nervous system, ed. R.F. Naunton and C. Fernandez, pp. 195–208. New York: Academic Press.
466. Zerlin, S., Naunton, R.F., and Mowry, H.J. 1973. The early evoked cortical response to third-octave clicks and tones. Audiology 12:242–249.
467. Zilm, D.H., Kaplan, H.L., and Capell, H. 1981. Electroencephalographic tolerance and abstinence phenomena during repeated alcohol ingestion by nonalcoholics. Science 212:1175–1177.
468. Zöllner, C., Karnahl, T., and Stange, G. 1976. Input-output function and adaptation behaviour of the five early potentials registered with the earlobe-vertex pick-up. Arch. Oto-Rhino-Laryngol. 212:23–33.

D

Somatosensory evoked potentials

15. SEP types, principles, and general methods of stimulating and recording
    15.1 SEP types
    15.2 The subject
    15.3 Stimulation methods
       15.3.1 Stimulus types
          15.3.1.1 Electric shocks
             15.3.1.1.1 Types and application of stimulus electrodes
             15.3.1.1.2 Stimulus electrode placements
             15.3.1.1.3 Stimulus intensity
             15.3.1.1.4 Stimulus duration
          15.3.1.2 Other sensory stimuli
       15.3.2 Stimulus rate
       15.3.3 Unilateral and bilateral stimulation
    15.4 General recording parameters
       15.4.1 Recording electrode placements
       15.4.2 Filter settings
       15.4.3 Other recording parameters

16. Normal SEPs to arm stimulation
    16.1 Normal SEPs at different recording sites
       16.1.1 The clavicular SEP (Erb's point potential)
       16.1.2 The cervical SEP
       16.1.3 The scalp SEP
          16.1.3.1 The scalp SEP recorded between parietal and midfrontal electrodes
          16.1.3.2 The scalp SEP recorded between parietal and ear or other distant cephalic electrodes

16.1.3.3 The scalp SEP recorded between scalp and noncephalic reference electrodes
16.2 Subject variables
  16.2.1 Age
    16.2.1.1 From premature to adult age
    16.2.1.2 From adult to old age
  16.2.2 Sex
  16.2.3 Limb length
  16.2.4 Limb temperature
  16.2.5 Sensation
16.3 Stimulus characteristics
  16.3.1 Stimulus electrode placement
  16.3.2 Stimulus intensity, rate, and duration
16.4 Recording parameters
16.5 General strategy of stimulating and recording

17. ABNORMAL SEPs TO ARM STIMULATION
17.1 Criteria distinguishing abnormal SEPs
  17.1.1 Absence of clavicular, cervical, and scalp SEPs
  17.1.2 Slow peripheral conduction velocity
  17.1.3 Prolonged central conduction times
  17.1.4 Abnormal latency differences to stimulation of either side of the body
  17.1.5 Decreased amplitude
17.2 General clinical interpretation of abnormal SEPs to arm stimulation
  17.2.1 Technical problems
  17.2.2 Peripheral and central lesions of the somatosensory pathway
17.3 Peripheral lesions that cause abnormal SEPs to arm stimulation
  17.3.1 Polyneuropathies
  17.3.2 Chronic renal failure
  17.3.3 Guillain-Barré syndrome
  17.3.4 Charcot-Marie-Tooth disease, Friedreich's ataxia, Adie's syndrome, and tabes dorsalis
  17.3.5 Thoracic outlet syndrome
  17.3.6 Brachial plexus lesions
  17.3.7 Carpal tunnel syndrome
  17.3.8 Congenital insensitivity to pain
  17.3.9 Hereditary pressure-sensitive neuropathy
17.4 Lesions of cervical roots and of the cervical cord that cause abnormal SEPs to arm stimulation
  17.4.1 Cervical root lesions
  17.4.2 Cervical cord lesions
17.5 Lesions of the brain stem and cerebral hemispheres that cause abnormal SEPs to arm stimulation
  17.5.1 Multiple sclerosis
  17.5.2 Brain stem strokes and tumors
  17.5.3 Thalamic lesions
  17.5.4 Parietal infarcts and tumors
  17.5.5 Lesions in other cerebral areas
  17.5.6 Hemispherectomy
  17.5.7 Diffuse cerebral disorders with reduced consciousness
    17.5.7.1 Cerebral death

17.5.7.2 Head injury
17.5.7.3 Chronic vegetative state
17.5.7.4 Reye syndrome
17.5.7.5 Subarachnoid hemorrhage
17.5.7.6 Perinatal asphyxia
17.5.7.7 Surgical monitoring
17.5.8 Diffuse cerebral disorders without reduced consciousness
    17.5.8.1 Myoclonus epilepsy, Jakob-Creutzfeldt disease, Ramsay Hunt syndrome, and other conditions with myoclonus
    17.5.8.2 Friedreich's and other hereditary cerebellar ataxias, familial spastic paraplegia, and amyotrophic lateral sclerosis
    17.5.8.3 Huntington's chorea
    17.5.8.4 Wilson's disease
    17.5.8.5 Myotonic dystrophy
    17.5.8.6 Leukodystrophies
    17.5.8.7 Down's syndrome
    17.5.8.8 Hepatic encephalopathy
    17.5.8.9 Hyperthyroidism
    17.5.8.10 Minamata disease
    17.5.8.11 Tourette's syndrome
    17.5.8.12 Drug effects
    17.5.8.13 Psychiatric disorders

18. NORMAL SEPs TO LEG STIMULATION
  18.1 Normal SEPs at different recording sites
    18.1.1 Popliteal fossa potential
    18.1.2 Lumbar and low thoracic SEPs
    18.1.3 The scalp SEP
  18.2 Subject variables
    18.2.1 Age
    18.2.2 Leg length, body height, temperature, and sensation
  18.3 Effect of stimulus parameters
    18.3.1 Stimulus electrode placements
    18.3.2 Stimulus intensity, rate, and duration
  18.4 Recording parameters
  18.5 General strategy of stimulating and recording

19. ABNORMAL SEPs TO LEG STIMULATION
  19.1 Criteria distinguishing abnormal SEPs
    19.1.1 Absence of SEPs
    19.1.2 Slow peripheral conduction velocity
    19.1.3 Slow central conduction velocity
  19.2 General clinical interpretation of abnormal SEPs to leg stimulation
  19.3 Peripheral nerve and root lesions that cause abnormal SEPs to leg stimulation
    19.3.1 Peripheral nerve lesions
    19.3.2 Radiculopathy
  19.4 Lesions of spinal cord and brain stem that cause abnormal SEPs to leg stimulation
    19.4.1 Multiple sclerosis
    19.4.2 Spinal cord injury
    19.4.3 Spinal cord compression
    19.4.4 Charcot-Marie-Tooth disease, Friedreich's ataxia, olivopontocerebellar degeneration, Adie's syndrome, and tabes dorsalis

- 19.4.5 Myotonic dystrophy
- 19.4.6 Subacute combined degeneration of the spinal cord
- 19.4.7 Diabetes mellitus
- 19.4.8 Subacute myelo-opticoneuropathy
- 19.4.9 Degenerative CNS diseases in children
- 19.4.10 Surgical monitoring of spinal cord condition
- 19.5 Cerebral lesions that cause abnormal SEPs to leg stimulation

20. Other SEPs
- 20.1 Sensory nerve action potentials
- 20.2 SEPs to trigeminal nerve stimulation
- 20.3 SEPs to pudendal and bladder stimulation
- 20.4 SEPs to adequate stimuli
  - 20.4.1 Touch
  - 20.4.2 Vibrotactile stimuli
  - 20.4.3 Joint movement
  - 20.4.4 Muscle stretch
  - 20.4.5 Pain stimuli
  - 20.4.6 Warm and cold stimuli
- 20.5 Somatomotor EPs

References for part D

15

SEP types, principles, and general methods of stimulating and recording

SUMMARY

15.1. SEP types consist mainly of SEPs to electric stimulation of arm or leg nerves. SEPs to arm stimulation are usually recorded simultaneously from clavicular, cervical, and scalp electrodes; SEPs to leg stimulation are recorded from lumbar, low thoracic, and scalp electrodes. SEPs can also be elicited by stimulation of other parts of the body and by stimuli other than electric shocks.

15.2. The subject should lie or recline comfortably during the recording. Sleep does not alter the clinically important short-latency peaks. Limb temperature should be controlled, and distance between stimulus and recording electrodes must be measured for calculation of peripheral nerve conduction velocities.

15.3. Stimuli consist mainly of electric shocks administered, with surface or needle electrodes, to sensory nerves at a rate of 4–7/sec.

15.4. Recordings are made in four channels using high and low filter settings of about 20 Hz and 3,000 Hz, respectively. Electrode placements, sweep length, and number of responses averaged vary for each SEP type.

15.1 SEP TYPES

Table 15.1 gives a breakdown of SEPs. They are distinguished mainly by the location of the stimulus and recording electrodes. In this regard, classification of SEPs differs from that of VEPs and AEPs, although it resembles AEP classification in that it depends more on response latency than on stimulus type. Most SEPs are produced by electric shocks applied to nerves of the arm or leg. SEPs to arm nerve stimulation are usually elicited by stimulating the median nerve at the wrist. They are recorded simultaneously with electrodes placed at Erb's point above the clavicle *(clavicular SEP, or Erb's point potential)*, the neck *(cervical SEP)*, and the parietal scalp *(scalp SEP)*, reflecting activity generated mainly in the brachial plexus, the upper cervical cord, and the somatosensory cortex, respectively. In a different approach, the effect of arm nerve stimulation may be recorded by far-field techniques using two widely spaced electrodes to record a single SEP that consists of a sequence of peaks representing excitation of the brachial plexus, spinal cord, and somatosensory cortex. SEPs to stimulation of arm nerves may be further characterized by the nerve stimulated: Stimulation of different peripheral nerves produces SEPs of slightly different latency and distribution.

SEPs to leg stimulation are usually elicited by stimulating the posterior tibial nerve at the ankle or the common peroneal nerve at the knee. They are recorded over the lumbar *(lumbar SEP)* and lower thoracic *(thoracic SEP)* spine and the scalp *(scalp SEP)*, reflecting activity of the cauda equina, lower spinal cord, and somatosensory cortex, respectively. Far-field recordings may show peaks of subcortical origin preceding the cortical SEP. Recordings from the upper thoracic and the cervical spine usually do not show reliable SEPs to leg stimulation. In the case of posterior tibial nerve stimulation, recording over the popliteal fossa may be added to better evaluate peripheral sensory conduction. Sensory nerve action potentials (20.1) may

TABLE 15.1. SEP types

A. SEPs to electric stimulation of arm nerves
   1. Clavicular SEP (Erb's point potential)
   2. Cervical SEP
   3. Scalp SEP
   4. Far-field SEP
B. SEPs to electric stimulation of leg nerves
   1. Lumbar SEP
   2. Low thoracic SEP
   3. Scalp SEP
   4. Far-field SEP
C. Sensory nerve action potential (SNAP)
D. SEP to electric stimulation of the trigeminal nerve
E. SEP to electric stimulation of pudendal and bladder mucosa
F. Steady-state SEP to repetitive electric stimulation
G. SEPs to touch, joint position change, and other stimuli
H. Somatomotor SEP

be studied by stimulating a purely sensory nerve and recording from the same nerve proximal to the stimulus point. SEPs not routinely used in clinical practice include SEPs to stimulation of the trigeminal nerve, skin of dermatomes, pudendal and bladder mucosa, SEPs to stimuli other than electric shocks, and steady-state SEPs of any kind.

Most, if not all, types of SEPs are mediated by fibers in the dorsal columns of the spinal cord, the medial lemniscal system of the brain stem, and the nucleus ventralis posterolateralis and posteromedialis of the thalamus. This is suggested by the frequent clinical observation that SEPs are rendered abnormal by lesions that cause loss of vibration and position sense rather than by lesions that reduce pain and temperature sensation. Muscle afferents traveling with cutaneous afferents in the dorsal columns probably contribute to the shortest latency peaks elicited by stimulation of mixed motor and sensory nerves.[36,119,304] Fibers which travel in the anterolateral part of the spinal cord and in extralemniscal systems of the brain stem and mediate pain and temperature sensation may contribute to some cortical peaks of long latency.

15.2 THE SUBJECT

The subject should lie comfortably on a bed or in a reclining chair. Muscles at the recording sites, especially scalp and neck muscles, should be relaxed to avoid contamination of SEPs by muscle potentials. If muscle artifact cannot be eliminated by comfortable positioning of the subject and by other attempts at relaxation, a sedative or mild hypnotic may be used. The subcortical and short-latency cortical SEPs recorded in routine clinical studies are not significantly altered by changes of attention, sleep, or CNS depressant drugs. These factors generally affect only peaks of longer latency. Changes of attention may alter cortical SEP peaks occurring no sooner than 30 msec after arm stimulation,[79,193] even though operant conditioning has been reported to be capable of altering even peaks of shorter latency.[116] Sleep may change the amplitude, but not the latency of early cortical peaks; both the amplitude and latency of late cortical peaks are affected by sleep,[75,127,131,334,340] especially in newborns.[69,75,153] Alcohol[93,279] and CNS depressant drugs[258,285] have been reported to alter scalp SEPs,[199,285] although central conduction time was found to be independent of phenobarbital at serum levels of up to 146 μg/ml[159] and of thiopental at levels suppressing the spontaneous EEG.[124] Phenytoin at toxic serum levels has been reported to increase central conduction time.[138] Active or passive movement of a stimulated finger may decrease the amplitude of cortical peaks appearing about 50 msec after the stimulus.[250,271]

The limbs should not be allowed to become cool because decreases of temperature prolong peak latencies. Deep central hypothermia[305] and hyperthermia of 42° C[91] may alter SEPs. The distances between stimulating and recording electrodes should be measured to evaluate peripheral and central conduction velocities (17.1.2, 19.1.2, 19.1.3).

15.3 STIMULATION METHODS

15.3.1 Stimulus types

15.3.1.1 Electric shocks
Electric shocks to a peripheral nerve are the clinically most useful stimulus type. The shocks may be delivered through surface or needle electrodes.

15.3.1.1.1 Types and application of stimulus electrodes. Surface stimulus electrodes consist of EEG disc electrodes, bipolar EMG stimulating electrodes embedded in a plastic strip, or ring electrodes for finger stimulation. The skin at contact points over a nerve is slightly abraded. Disc electrodes are glued into place with collodion and filled with conductive jelly. Attachment with sticky conductive paste is less secure (2.2.1.1). EMG stimulating electrodes may be held in place with adhesive tape or a Velcro strip. Ring electrodes are moistened with conductive jelly and placed on individual fingers for stimulation of the digital nerves. Electrode impedance should be less than 10 kΩ to reduce discomfort and stimulus artifact.

Needle electrodes consist of sharp wires of stainless steel or other inert metal, less than 1 mm in diameter and about 1 cm long. They are inserted perpendicularly through the cleansed and disinfected skin so that the electrode tips lie close to the nerve. The lead wires should be taped to the skin to keep the needles from slipping out. Although insertion causes brief pain, needle electrodes are generally well tolerated and have the advantage of generating less stimulus artifact in the recording because they require much less stimulus current than do surface electrodes.[34,67,346] Because of the danger of infection, needle electrodes cannot be used during long procedures such as surgical monitoring.

For nerve stimulation, the two stimulus electrodes of a pair are placed along the course of a nerve. Surface electrodes are separated from each other by 2–3 cm, needle electrodes by only about 1 cm. For dermatomal stimulation, both electrodes are placed a few centimeters apart along the middle of a dermatome. For both nerve and skin stimulation, the proximal stimulus electrode is connected to the negative pole of the stimulator output and becomes the cathode; the distal electrode is connected to the positive pole and becomes the anode. The two output terminals of the stimulator are isolated from ground through a radiofrequency or optoelectric isolation unit to avoid current flow from either stimulus electrode to the ground electrode, thereby restricting the stimulus current flow to the area between the two stimulus electrodes. Spread of stimulus current could increase the stimulus artifact and be a hazard to the subject. The subject's safety also requires that high stimulus intensity, long stimulus pulses, and repetitive stimulation at high rates be avoided, especially in patients who cannot feel the stimulus.

15.3.1.1.2 Stimulus electrode placements. Arm stimulation commonly uses the median nerve at the wrist. Leg stimulation is usually applied either to the posterior tibial nerve at the ankle or to the common peroneal nerve at the knee. Because stimulation of these nerves yields relatively large and fairly constant SEPs, it is the best method for most clinical studies. In principle, however, stimulation of mixed motor and sensory nerves is less than ideal because it excites (1) heterogeneous afferents from skin, joints, muscle, and deep tissues producing orthodromically conducted SEP components in various pathways up to the level of the scalp, and (2) motor fibers generating antidromically conducted SEP components at the clavicular and lumbar levels. Some of these SEP components may persist with normal latency in cases of sensory neuropathies in which the SEP components generated by somatosensory input may be delayed or abolished; this may result in combinations of SEPs that are difficult to interpret. Therefore, in cases where testing of purely somatosensory input is essential, stimuli must be applied to sensory nerves or nerve branches. The precise electrode placement for stimulation of mixed and sensory nerves is described in chapter 16 for arm (16.3.1) and leg (18.3.1) nerves.

Stimulation of sensory nerves or nerve branches may also be used to detect the location of peripheral lesions, or the level of spinal cord lesions. This requires investigation of several inputs, which may tax the patience of the patient. Furthermore, SEPs to stimulation of small nerves differ from each other in latency and distribution on the scalp[92] and require establishment of normal SEP values for each stimulus location. Also, they often have very low amplitude.

15.3.1.1.3 Stimulus intensity. For stimulation of mixed nerves, the intensity is usually set slightly above the motor threshold, i.e., at an intensity sufficient to produce a muscle contraction causing a visible twitch. If no twitch can be elicited, as in severe peripheral neuropathy, the stimulus should have an intensity known to produce a visible twitch in normal subjects. This usually requires stimulus currents of 5–15 mA for stimulation with surface electrodes and weaker currents for stimulation with needle electrodes.

For sensory nerve stimulation, the stimulus intensity is set at 2.5 or 3 times the sensory threshold. When stimulating an anesthetic area, stimulus intensity for purely sensory nerves may be set by the effect on simultaneously recorded sensory nerve action potentials (20.1). Especially in cases of superficial sensory loss and of unconscious patients, stimulus intensity must be kept moderate to avoid local damage.

The definition of stimulus intensity by motor and sensory thresholds is more reliable than the use of stimulus voltage or current. Stimulus voltage is not a precise measure of stimulus intensity because the effect of electric stimulation depends on current flow; the current generated by a stimulus voltage depends on the impedance of the stimulus electrodes and of the tissue between them. Even stimulus current is not an entirely satisfactory measure because the fraction of current crossing the nerve fiber membranes and acting as a stimulus varies with the distance between, and the orientation of, stimulating electrodes and nerve. However, in situations where constant stimulus electrode impedance is difficult to maintain, such as during surgical monitoring, a stimulator capable of delivering constant current output can compensate for changes of stimulus electrode impedance and thereby help to maintain constant stimulus conditions.

15.3.1.1.4 Stimulus duration. The duration of the electric pulse is usually 200 μsec, especially for stimulation of mixed nerves. Shorter stimuli tend to excite a larger proportion of motor than of sensory fibers.[339]

15.3.1.2 Other sensory stimuli
Touch, vibration, joint movement, muscle stretch, pain, and temperature changes are rarely used for clinical SEP studies (20.4).

15.3.2 Stimulus rate

Transient SEPs are usually elicited with stimulus rates of 4–7/sec. Rates up to 10/sec do not alter the clinically important short-latency peaks, including early scalp SEP peaks, but may be uncomfortable for the patient.[182,255] Slower rates must be used for studies of the clinically less important late peaks of the scalp SEP. Steady-state cortical SEPs may be elicited at rates of 12–200/sec but have been used only rarely.[239] The stimulus rate should not be an integral of 60 Hz to avoid buildup of power-line artifact.

15.3.3 Unilateral and bilateral stimulation

For arm stimulation, only one side of the body is usually stimulated at a time to aid proper lateralization of abnormalities. Simultaneous bilateral stimulation may give better defined SEPs[182] and, in conjunction with bilateral scalp recording, has been said to detect hemispheric lesions not detected by unilateral stimulation of either side alone.[355] However, bilateral stimulation can obscure unilateral abnormalities and does not permit lateralization of abnormalities in subcortical structures. Routine studies therefore use unilateral stimulation. The same is generally true for leg stimulation except when lateralization of a lesion is not critical, for instance, during surgical monitoring of spinal cord function. Here, nerves of both legs may be stimulated simultaneously to obtain larger scalp SEPs. However, simultaneous bilateral stimulation is acceptable only if peripheral conduction distances and velocities are equal because a delay of one peripheral afferent can distort the scalp SEPs to combined stimulation and suggest central problems.

15.4 GENERAL RECORDING PARAMETERS

15.4.1 Recording electrode placements

The volley of impulses ascending in the stimulated somatosensory pathway is usually recorded with electrode pairs that are placed at different levels of that pathway, one electrode in each pair being located as closely as possible to the structures generating the SEP peaks at that level (1.3.4.1). On the other hand, widely spaced electrodes may be used to record a far-field SEP containing several peaks generated by the various relays of the somatosensory pathway (1.3.4.2). Specific recording electrode placements and combinations are described in the sections on arm (16.1) and leg (18.1) SEPs. The subject is grounded through a large metal plate or band that is covered with electrode jelly or saline-soaked gauze and placed between the stimulating and recording electrodes to minimize stimulus artifact. The ground electrode is connected to the ground input of the recording amplifier or the groundpost of the stimulator so that the subject, the recording equipment, and the stimulator all lie at the same ground potential level. The ground electrode should have an impedance of less than 10 kΩ. As in all EP recordings, a ground electrode is needed to protect the subject and to reduce the chance for pickup of interference artifacts (2.3.4.2). In SEP recordings in particular, proper grounding is needed to keep the stimulus artifact to a minimum. If the stimulus artifact is excessive, some relief may be obtained by reducing the impedance of the ground electrode or by changing its connection from amplifier input to stimulator groundpost or vice versa. Other measures apt to reduce the stimulus artifact include reducing the impedance of the stimulus electrodes, changing the orientation of stimulus electrodes to recording electrodes, and changing the filter settings, especially of the low frequency filter.[211]

15.4.2 Filter settings

A wide bandpass of from 5–30 Hz to 2,500–4,000 Hz is usually used for all channels. Reduction of the high frequency filter can substantially increase the latency (2.4.1.4). If filters are set differently for each channel, the high frequencies could be cut more for cortical SEPs and the low frequencies could be cut more for the other SEPs. Filter roll-off slopes should not exceed 12 dB/octave for low frequencies and 24 dB/octave for high frequencies (2.4.1).

15.4.3 Other recording parameters

Averagers with four channels are commonly used to record SEPs from different locations simultaneously. Different sweep lengths and numbers of sweeps are used for arm (16.4) and leg (18.4) SEPs. At least two sets of averages must be superimposed to ascertain replication.

16

Normal SEPs to arm stimulation

SUMMARY

16.1. Electric stimulation of the median nerve at the wrist produces a clavicular SEP with a negative peak at about 9 msec ($\overline{N9}$), a cervical SEP with a major negative peak at about 13 msec ($\overline{N13}$), and a scalp SEP with a fairly large negative wave at 20 msec ($\overline{N20}$). Recordings between scalp and ear electrodes pick up a small positive peak at 13–14 msec ($\overline{P13/14}$) preceding the $\overline{N20}$, whereas recordings between scalp and noncephalic reference electrodes show even earlier SEP peaks. Absolute SEP latencies vary with subject variables, stimulus characteristics, and recording parameters.

16.2. Subject variables of importance for the SEPs to arm stimulation include age, limb length, and temperature.

16.3. Stimulus characteristics affecting SEPs to arm stimulation are the placement of the stimulus electrodes and the intensity, rate, and duration of the stimulus.

16.4. The most important recording parameter is the location of the recording electrodes.

16.5. General strategy localizes lesions in different parts of the somatosensory pathway by looking for the presence of SEPs at each level of recording and by evaluating the conduction times between them.

16.1 NORMAL SEPs AT DIFFERENT RECORDING SITES

A great variety of stimulating and recording methods are used in different laboratories. Older studies often recorded from only one or two sites. Because averagers with four channels have now become widely available, current practice favors simultaneous recordings from three or four points along the somatosensory pathway. A commonly used technique uses stimulation of the median nerve at the wrist and recording from clavicular, cervical, and scalp electrodes;[12] (Figure 16.1). Other methods use different stimulus and recording sites. SEPs to arm stimulation vary depending on subject variables, stimulus characteristics, and recording parameters. Commonly used conditions are summarized in Table 16.1.

16.1.1 The clavicular SEP (Erb's point potential)

Clavicular SEPs may be recorded from Erb's point 2–3 cm above the clavicle at the posterior border of the clavicular head of the sternocleidomastoid muscle, which can be easily seen when the subject bends his head against resistance. Electrode placements more lateral over the clavicle may give recordings of higher amplitude and only slightly different latency.[165] The clavicular electrode on the stimulated side may be referred to the opposite clavicular electrode. If the clavicular electrode on the stimulated side is referred to a midfrontal electrode, i.e., Fz in the International 10-20 System of EEG electrode placements, the recording may contain later peaks that reflect activity of the central parts of the somatosensory pathway and follow the early peak picked up by the clavicular electrode.

Clavicular SEPs have a major negative peak, often preceded and followed by a smaller positive wave; smaller negative waves may follow this first complex (Figure 16.2). Latency is measured to the peak, or, less commonly, to the onset of the negative wave. In normal adults, peak latency amounts to about 9 msec for stimulation of the median nerve at the wrist ($N\overline{9}$) and about 11 or 12 msec for digital nerve stimulation, depending on age and arm length. The negative peak is the benchmark from which central conduction times to cervical and scalp SEP peaks are measured (16.5).

The clavicular SEP is generated by the peripheral nerve fibers contained in the brachial plexus. In case of sensory nerve stimulation, the clavicular SEP is generated by somatosensory fibers only. When a mixed nerve is stimulated, orthodromic responses of muscle afferents and an antidromic motor volley are added to the somatosensory component.

16.1.2 The cervical SEP

Cervical SEPs may be recorded from the neck at the level of the C5 or C2 vertebra, i.e., two or five spinous processes above C7 which has the most prominent spinous process at the base of the neck; C2 lies at the point of the deepest

Figure 16.1. Schematic diagram of normal SEPs to arm stimulation. Tracings, from bottom to top, show clavicular SEP (Erb's potential point, $N\overline{9}$), cervical SEP ($N\overline{13}$), far-field SEP recorded between scalp and noncephalic reference electrodes, and scalp SEP recorded between scalp and cephalic reference electrodes ($N\overline{20}$). Negativity at the electrodes connected with a solid line in the diagram at the left is plotted upward.

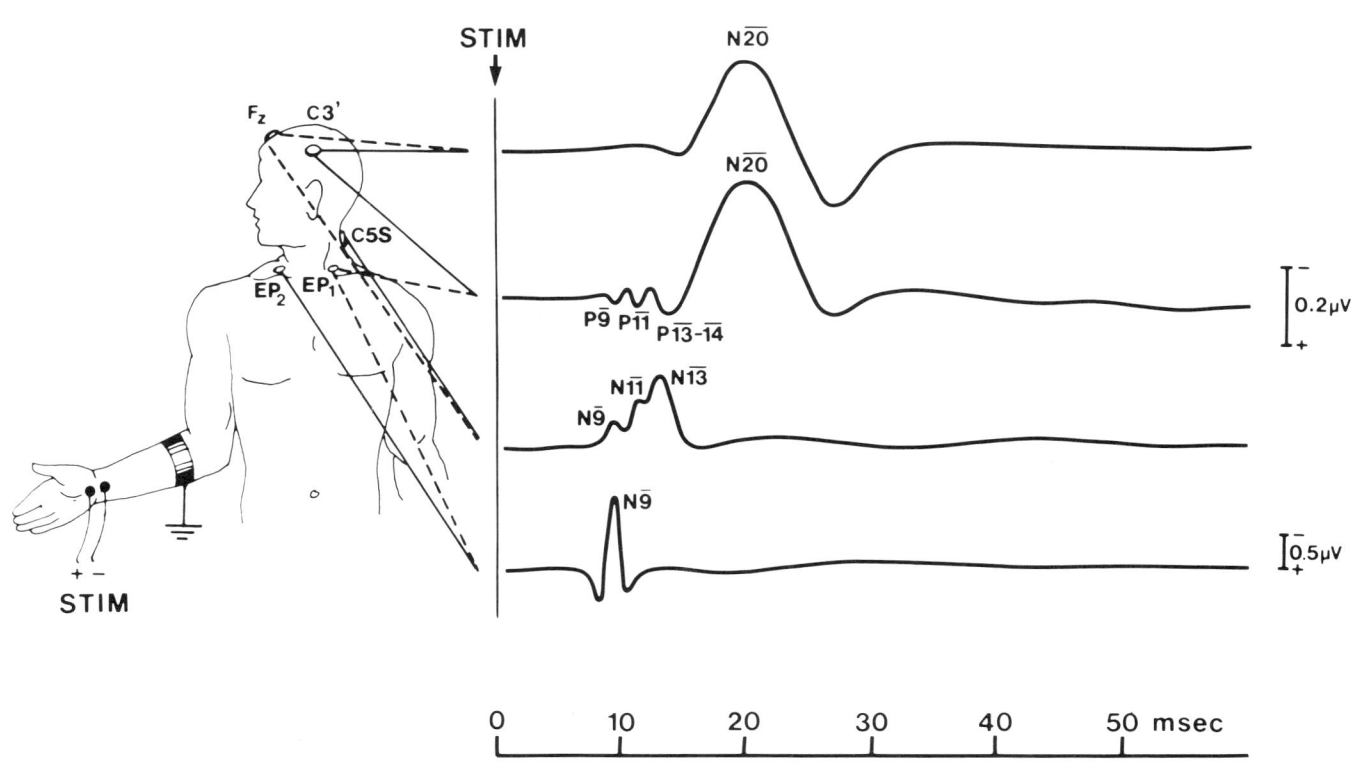

TABLE 16.1. SEPs to stimulation of the median nerve at the wrist

A. Subject variables
   1. Age: Several separate normal control groups are needed for children up to about 8 years; requirements for old age have not yet been defined.
   2. Sex: Separate control groups for adult males and females may be used as needed in each laboratory.
   3. Limb length: Distance from stimulus to recording electrode must be measured, especially for determination of peripheral conduction velocity.
   4. Temperature: To avoid cooling, the room temperature should be kept at 22–24° C.
   5. Sensory disturbances: Loss of sensation must be defined in terms of distribution, modality, and degree. Stimulus intensity must be adjusted cautiously in anesthetic areas.
B. Stimulus
   1. Type: Electric shocks from electrically isolated constant voltage or constant current stimulator.
   2. Electrodes: EEG disc electrodes of less than 10 kΩ impedance or needle electrodes.
   3. Electrode placement: Cathode over median nerve at the wrist; anode 2–3 cm distal or on dorsum of wrist.
   4. Rate: 4–7/sec.
   5. Duration: 200–300 µsec.
   6. Intensity: Above threshold for thumb abduction twitch.
   8. Unilateral stimulation: Each side should be stimulated separately in routine studies.
C. Recording
   1. Number of channels: 4.
   2. Electrode placements: Left and right Erb's point (EP1, EP2), over C2 or C5 spinous processes on the neck (C5S, C2S), on the scalp 2 cm posterior to the C3 and C4 positions of the International 10–20 System of EEG electrode placements (C3', C4'), on the midfrontal scalp (Fz); ground electrode on the forearm.
   3. Montage for stimulation of the left median nerve (opposite recording side of lateral electrodes for stimulation of the right median nerve):
      Channel 1: C4'–Fz
      Channel 2: C4'–EP2
      Channel 3: C5S or C2S–Fz
      Channel 4: EP1–EP2
   4. Filter settings: Low-frequency filter: 5–30 Hz; high-frequency filter: 2,500–4,000 Hz.
   5. Number of responses averaged: 500–2,000.
   6. Sweep length: 40 msec; 60–100 msec for delayed SEPs.
D. Analysis
   1. Normal peaks: Erb's point potential ($\overline{N9}$) in channel 4; $\overline{N13}$ in channel 3; $\overline{P9}$, $\overline{P13/14}$, and $\overline{N20}$ in channel 2; $\overline{N20}$ with or without preceding $\overline{P13/14}$ in channel 1.
   2. Criteria for abnormal
      a. Absence of normal peaks.
      b. Slow peripheral conduction velocity from stimulus cathode to Erb's point.
      c. Increased central conduction times between Erb's point potential and cervical SEP ($\overline{N9}$–$\overline{N13}$), between cervical and scalp SEP ($\overline{N13}$–$\overline{N20}$), between brain stem far-field and scalp SEP ($\overline{P13/14}$–$\overline{N20}$) and between Erb's point potential and scalp SEP ($\overline{N9}$–$\overline{N20}$).
      d. Questionable criteria: Abnormal latency differences to stimulation of each side of the body; abnormal amplitude differences to stimulation of each side; abnormal amplitude ratios between central and peripheral SEPs.

indentation a few centimeters below the inion. In some laboratories, cervical SEPs are recorded from an electrode on C7. Recordings from neck electrodes commonly use a midfrontal reference electrode.

SEPs recorded from the neck have, as their largest and most consistent part, a negative peak with a latency at about 13 msec ($\overline{N13}$) after stimulation of the median nerve at the wrist (Figures 16.1 and 16.2). This peak may be preceded and followed by smaller negative peaks at 9 msec, 11 msec, and 14 msec ($\overline{N9}$, $\overline{N11}$, and $\overline{N14}$). The latencies are longer after stimulation of the digital nerves. Latencies and conduction times vary between laboratories, mostly depending on the location of the stimulating and recording electrodes.[16,49,67,70,105,121,157,165,179,194,221,222,299,351]

In general, the generators of these peaks have not yet been clearly identified. However, it is now commonly accepted that the $\overline{N9}$ peak is due to excitation of peripheral nerve fibers between axilla and spinal cord. $\overline{N11}$ is probably generated at the level of the cervical cord segments by the dorsal root entry zone, gray matter of the dorsal horn, and nearby dorsal columns, whereas $\overline{N13}$ and $\overline{N14}$ probably have several sources in the spinal cord, including the dorsal columns and nucleus cuneatus of the dorsal columns. The brain stem makes little or no contribution to these peaks.[10,49,108,113,123,125,165,229,351,352] Direct recordings from electrodes near the cord[111,218,292,293] indicate that at least part of the $\overline{N13}$ and $\overline{N14}$ waves are generated locally by a stationary source, probably the dorsal horn of the spinal cord or the dorsal column nuclei.[72,157,194,197] Recordings from surface electrodes, showing a rostral increase of latency of some components, suggest that another part of the $\overline{N13}$ and $\overline{N14}$ peaks may be due to impulses ascending in the medial lemniscus,[16] although lemniscal and higher parts of the somatosensory pathway are more clearly reflected in the far-field peak of $\overline{P13/14}$ recorded with widely spaced electrodes (16.1.3.2, 16.1.3.3).

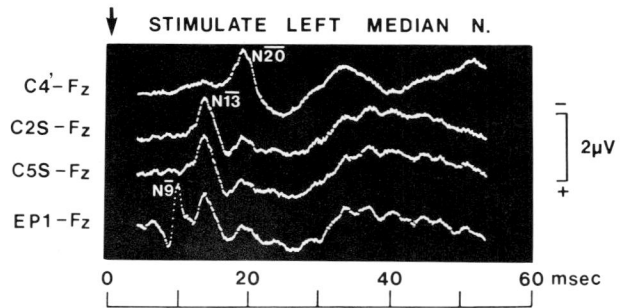

Figure 16.2. Normal SEP to median nerve stimulation. Stimulation of the left median nerve at the wrist produces clavicular (bottom tracing), cervical (two middle tracings), and scalp (top tracing) SEPs. Recording electrodes are on Erb's point (bottom tracing), over the spinous processes of C5 and C2 (third and second tracings, respectively) and on the scalp over the hand area on the right side. Negativity at these electrodes is plotted upward. Midfrontal reference electrode. The sweep is delayed by 4 msec to eliminate the stimulus artifact.

16.1.3 The scalp SEP

Scalp SEPs to arm stimulation are recorded with electrodes over the contralateral parietal scalp, often placed 2 cm behind the C3 or C4 electrode positions of the International 10–20 System of EEG electrode placement (C3', C4'). These electrodes record a relatively large SEP to stimulation of the opposite upper limb. The early peaks vary depending on the location of the reference electrode.

16.1.3.1 The scalp SEP recorded between parietal and midfrontal electrodes

The near-field scalp SEP to stimulation of the median nerve at the contralateral wrist consists of a series of waves. The clinically most important negative peak has a latency of about 20 msec (Figures 16.1 and 16.2). The $\overline{N20}$ may be preceded by earlier peaks which have very low amplitude in recordings between parietal and midfrontal electrodes and can be better analyzed with widely spaced electrodes. The $\overline{N20}$ is followed by a positive peak at about 30 msec. This $\overline{N20}$-$\overline{P30}$ sequence is the first part of a W-shaped complex that is recordable in only about one-half of all young normal adults[71]; the second part of the "W" consists of a negative peak at 30–40 msec and a positive peak at about 40–50 msec. This complex may be followed by a slower negative wave with a peak at about 50–70 msec and a positive peak at 80–90 msec. Later waves include a negative-positive sequence at 140–170 msec and 190–270 msec and a positive peak at about 300 msec. The latencies of these peaks increase from the front of the head to the back[350] and are 2–3 msec longer on stimulation of the digital nerves than on stimulation of the median nerve at the wrist.[67,127,128]

While later peaks are most likely generated by cortical relays, the generator of the $\overline{N20}$ is still debated. An origin at the parietal cortex is supported by the limited distribution of the $\overline{N20}$ in the parietal area contralateral to the stimulated arm,[9,70,127,132,133] by the effect of cortical lesions,[60,226,308] and by recordings made directly from the cortex.[9,31,83,127,177,206] However, it has also been claimed that the $\overline{N20}$ is generated in the thalamocortical radiation[40,232] or the nucleus ventralis lateralis of the thalamus,[48,49,238] although this does not seem to be supported by direct recordings from the thalamus[40,112,118,136,188] or by the effect of thalamic lesions.[228] Multiple cortical and subcortical generators have also been proposed.[209]

SEPs recorded over the frontal scalp or cortex may show much independence of amplitude and latency from SEPs recorded simultaneously in the parietal areas,[40,56,311] suggesting that frontal and parietal SEPs are generated separately,[71] perhaps via independent cortical input from the thalamus.[73,230,251]

Small SEP peaks of short latency can be recorded from the scalp ipsilateral to the stimulated arm and are probably conducted electrotonically from the contralateral hemisphere.[190] Later peaks with latencies of over 40–60 msec may be due to activation of ipsilateral cortical areas through sensory input from bilateral pathways[127,335,353] or through callosal connections between the hemispheres.[326,348] The late $\overline{N140}$-$\overline{P190}$ complex and the $\overline{P300}$ have a maximum at the vertex or posterior and lateral to it. Cerebral SEPs may be distorted by myogenic potentials that are triggered by the stimulus, begin at latencies under 20 msec, and last over 100 msec (20.5).

16.1.3.2 The scalp SEP recorded between parietal and ear or other distant cephalic electrodes

Recordings between the parietal electrode and electrodes on the ear or other scalp areas may pick up small peaks before the $N\overline{20}$ that are generated by subcortical structures.[332] An early positive wave occurs 13–15 msec after stimulation of the median nerve at the wrist ($P\overline{13/14}$) and has a maximum contralateral to the stimulated side. This wave probably originates mainly from the lemniscal input to the thalamus[72] with possible contributions by the nucleus ventralis posterolateralis thalami and its thalamocortical outflow.[9,31,56] This origin is supported by recordings from patients with lesions at different levels of the sensory pathway[229,235] and by direct recordings from these structures in humans.[40,112,118,275] Negative waves at 16 and 17 msec have also been thought to be generated by the thalamus or its radiation.[4] Even earlier peaks, reflecting the activity of spinal cord structures usually recorded with cervical electrodes, may be picked up between parietal and various scalp reference electrodes.[5] Recordings between parietal and ear electrodes may distort later peaks because the ear is not electrically indifferent since it picks up some SEP components.

16.1.3.3 The scalp SEP recorded between scalp and noncephalic reference electrodes

Recordings between a scalp electrode and an extracranial reference electrode may show up to six peaks: $P\overline{9}$, $P\overline{11}$, $P\overline{13}$, $P\overline{14}$, $N\overline{20}$, and $P\overline{27}$ (Figure 16.3). The three earlier ones are far-field SEPs that correspond with the negative peaks recorded at the neck. The $P\overline{14}$ corresponds with the $P\overline{13/14}$ recorded with widely spaced scalp electrodes. The $N\overline{20}$ and $P\overline{27}$ are probably near-field recordings of cortical potentials picked up mainly by the scalp electrode.[10,15,58,70,107,141,182,231,238,347,351] A widely distributed $N\overline{18}$ may be recorded before the $N\overline{20}$ and is probably generated below the cortex.[73] Most of the earlier peaks can be more unequivocally analyzed with clavicular and cervical recordings.[67] The latency of the $P\overline{9}$ in far-field recordings varies with shoulder position.[80]

16.2 SUBJECT VARIABLES

16.2.1 Age

16.2.1.1 From premature to adult age

Scalp SEPs of premature babies of over 24 weeks conceptional age show a slow, large negative wave of over 200 msec latency. With increasing age, the amplitude of this wave decreases while earlier peaks emerge.[156] The scalp SEP appears first over primary sensory cortex and later over frontal association areas.[337] Full-term newborns have SEPs that vary more with sleep stages than do those of

Figure 16.3. Normal far-field SEPs to arm nerve stimulation in four young adults (*A-D*). Stimulation of the left median nerve at the wrist and recording between the contralateral parietal (*A-C*) or midfrontal (*D*) electrode and an electrode on the right hand produces a sequence of positive peaks (P$\overline{9}$, P$\overline{11}$, P$\overline{13}$, and P$\overline{14}$) followed by a larger negative peak (N$\overline{20}$). Two tracings are superimposed in *C*. Negativity at the scalp electrodes is plotted upward. From Desmedt and Cheron[73] with permission by the authors and Elsevier Scientific Publishers Ireland Ltd.

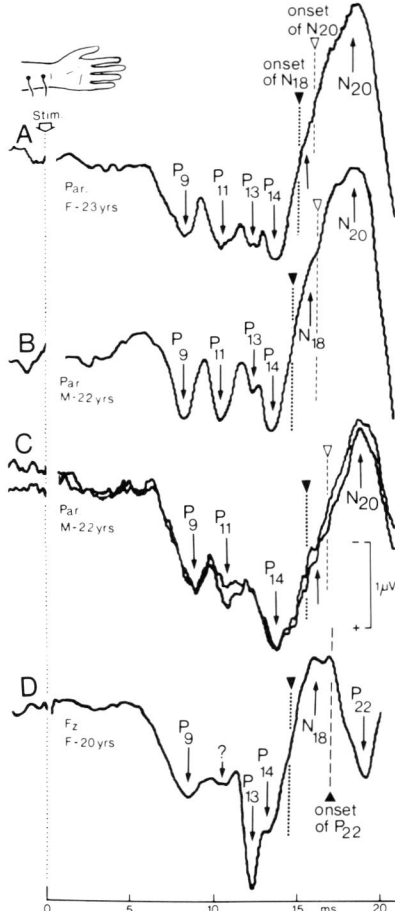

adults[75,154] but often show an N$\overline{30}$ on stimulation of arm nerves and a P$\overline{50}$ after stimulation of leg nerves; each SEP is narrowly restricted to the cortical representation of the stimulated extremity.[69] The SEP gradually reaches adult form and latency at an age between 3 years[185] and 10 years.[151] However, the latencies of SEPs of children cannot be equated with those of adults: Because children have smaller bodies, latencies of the same length as in adults reflect much slower conduction velocities. Central conduction velocity increases from about 10 m/sec at birth and reaches adult values of about 50 m/sec at the age of about 8 years.[68,69] In contrast, peripheral conduction velocity matures faster, namely, from 20–35 m/sec at birth to the adult value of 60–80 m/sec at the age of 12–18 months.[78] Changes of latency between ipsilateral and contralateral SEP peaks have suggested maturation of callosal connections.[272]

16.2.1.2 From adult to old age

Aging seems to affect SEPs less than other EPs. In octogenarians, all cortical peaks have slightly longer latency due to slowing of peripheral conduction velocity which was found to decrease from 71 m/sec at the age of 22 years to 61 m/sec at the age of 82 in one study.[71] Another study showed peripheral conduction velocity to decrease throughout adult life at an annual rate of 0.16 m/sec.[86] Central conduction between cervical and scalp electrodes slows only slightly,[11,160,314] if at all.[71,180] The amplitude of the cervical SEP decreases progressively after the age of 40.[160] The amplitude of the parietal SEP[160,202,284] and the clear definition of the W shape[73] increase with age. Age-dependent changes in the precentrally recorded SEP differ from those in the parietal area.[70] Differential aging of separate parts of the somatosensory cortex has been suggested by the finding of different latency increases of the $\overline{N20}$ and the subsequent positive peak.[297]

16.2.2 Sex

Several studies report that SEPs have longer latencies in men than in women[11,161,180,284]; however, in one study the latency differences disappeared when differences in height were taken into account.[1] Most laboratories do not use separate normal controls for the two sexes.

16.2.3 Limb length

Absolute latencies depend on the distance between stimulus and recording electrodes which therefore must be measured for calculation of conduction velocities (17.1.2).

16.2.4 Limb temperature

The stimulated limb must be kept warm by maintaining the room temperature at 20–22° C, by covering the limb, or by applying a heating pad as needed.

16.2.5 Sensation

Numbness is often present in patients with abnormal SEPs and requires careful adjustment of stimulus intensity (15.3.1.1.3).

16.3 STIMULUS CHARACTERISTICS

16.3.1 Stimulus electrode placement

Stimulus electrodes are most often placed on the median nerve at the wrist. One electrode is positioned 2 cm proximal to the wrist crease between the tendons of the palmaris longus and the flexor carpi radialis muscles which can be easily seen when the subject flexes his wrist against resistance. This electrode is connected to the negative pole of the stimulator and thereby becomes the cathode. Another electrode is placed 2–3 cm distal to the cathode or on the dorsal surface of the wrist and is connected to the positive pole of the stimulator, thereby becoming the anode. The ground electrode consists of a plate on the palmar surface of the forearm or a band electrode around the forearm (Figure 16.1). Stimulation of the ulnar and radial nerves at the wrist has been used to detect lesions of other parts of the sensory input from the upper extremity.[141] For stimulation of purely sensory nerves, stimuli may be applied (1) to median nerve fibers through ring electrodes placed over the digital nerves distal to the interphalangeal joint of the second or third finger or of both these fingers, (2) to ulnar nerve fibers through ring electrodes over digital nerves of the fifth finger, (3) to the superficial branch of the radial nerve at the radial side of the dorsal part of the wrist, and (4) to the cutaneous branch of the musculocutaneous nerve two finger breadths below the lateral part of the cubital crease. The cervical segments C5-C8 can be tested by stimulating various nerves or nerve branches at different points: C5 by stimulating the musculocutaneous nerve on the radial side of the forearm, C6 by stimulating the skin of the thumb, C7 by stimulating the adjoining surfaces of the second and third fingers, C8 by stimulating the skin of the fifth finger.[99] C5 and C6 have also been studied by stimulating the musculocutaneous nerve near the wrist.[320] Other segments have been studied by dermatomal stimulation.[28,327]

16.3.2 Stimulus intensity, rate, and duration

In the case of mixed nerve stimulation, the intensity is set slightly above the twitch threshold for a visible thumb twitch for the median nerve. In most laboratories, this requires an intensity of at least several milliamperes for surface electrodes, or of a few milliamperes for needle electrodes. The threshold for sensory nerves is determined as described earlier (15.3.1.1.3).

A change of stimulus intensity produces little[196] or no[49,127,157,182,299] significant change of the clinically important short-latency SEPs except that the amplitude decreases at low intensities. In general, peaks of longer latency have slightly higher thresholds than peaks of shorter latency.

Stimulus rate is 4–7/sec (15.3.2). Stimulus duration is 200–300 μsec (15.3.1.1.4).

16.4 RECORDING PARAMETERS

Most recording parameters (15.4), including electrode placements (16.1), have been described and summarized (Table 16.1). About 500–2,000 responses must be collected for an SEP to stimulation of the median nerve at the wrist. Larger numbers are needed for the smaller SEPs produced by stimulation of smaller nerves. The sweep length is about 40 msec for most SEPs to arm stimulation. Longer sweeps may be required for SEPs of abnormally long latencies.

The importance of the sweep length for the wave frequencies in the SEP is discussed in section 18.4.

16.5 GENERAL STRATEGY OF STIMULATING AND RECORDING

The strategy of testing the various parts of the somatosensory pathway from the arm to the cortex is straightforward for either of the two recording methods used. The more common method of recording SEPs simultaneously from clavicular, cervical, and scalp electrodes yields the peaks $\overline{N9}$, $\overline{N13}$, and $\overline{N20}$ which indicate excitation of the brachial plexus, the upper spinal cord, and the cerebral cortex or its thalamic afferents, respectively. The less common method of recording between scalp and noncephalic reference electrodes reflects activity of the same structures in the successive peaks of a single SEP. With both methods, a lesion between these structures either abolishes or delays and diminishes the peaks representing structures at the level of the lesion and proximal to it. This rule permits distinction of lesions that involve the three segments of the somatosensory pathway between stimulation point and brachial plexus, between brachial plexus and upper cervical cord, and between upper cervical cord and somatosensory cortex or thalamocortical afferents. (1) Peripheral nerve and plexus lesions delay or abolish all SEP peaks. The delay of the peak representing Erb's point is best expressed as a decrease in the peripheral nerve conduction velocity. (2) Lesions of the cervical roots and cervical cord leave intact the clavicular SEP but delay or abolish the cervical SEP and, in most instances, also the scalp SEP. The delay causes an increase of the clavicular-cervical ($\overline{N9}$-$\overline{N13}$) and the clavicular-scalp ($\overline{N13}$-$\overline{N20}$) conduction times in separate recordings from these points. Recordings between scalp and noncephalic electrodes show increased separation between $\overline{P9}$ and subsequent peaks. (3) Lesions of the brain stem and cerebrum delay or abolish the scalp SEP without interfering with the clavicular and cervical SEP. The delay causes an increase of the cervical-scalp ($\overline{N13}$-$\overline{N20}$) and clavicular-scalp ($\overline{N9}$-$\overline{N20}$) conduction times in separate recordings from these points, and an increased separation between $\overline{N20}$ and the preceding peaks in recordings between scalp and noncephalic electrodes. The role of the $\overline{P13/14}$ in SEP strategy is not certain yet, but if this peak is generated by the thalamus or nearby structures, it may become useful for a more precise localization of lesions between medulla and cerebral cortex. Lesions in the lower part of this segment may render abnormal both the $\overline{P13/14}$ and the $\overline{N20}$, and cause an increase of the $\overline{N13}$–$\overline{P13/14}$ conduction time, whereas higher lesions may affect only the $\overline{N20}$ and prolong the $\overline{P13/14}$–$\overline{N20}$ conduction time.

Calculation of peripheral nerve conduction velocity has the advantage of giving a quantitative measure of abnormality that relates directly to pathology. Measurements of central conduction time largely eliminate the effects of peripheral lesions and of changes in temperature,[159] reduce the variation due to different arm length, and help to localize lesions to the peripheral and central segments of the somatosensory pathway. These measurements are also more sensitive to pathology than the absolute latencies.[122,315]

17

Abnormal SEPs to arm stimulation

SUMMARY

17.1. Criteria distinguishing abnormal from normal SEPs to arm stimulation are absent peaks, slowed peripheral conduction velocity, and prolonged central conduction times between peaks generated by the brachial plexus, upper cervical cord, and sensory cortex or its thalamic input. Absolute peak latencies, amplitudes, and differences of latency or amplitude of SEPs to stimulation of each arm are less reliable indicators of abnormality.

17.2. General clinical interpretation of abnormal SEPs uses decrease of peripheral conduction velocity, increase of central conduction time, and abolition of SEPs recorded from different levels to identify lesions of peripheral nerves, plexus, nerve root, spinal cord, brain stem, and cerebral parts of the somatosensory pathway.

17.3. Peripheral lesions such as peripheral neuropathies and lesions of the brachial plexus abolish or delay all SEPs. The delay is due to slowing of peripheral conduction velocity; central conduction times are not prolonged.

17.4. Lesions of cervical roots and the cervical cord such as spondylotic myelopathy, trauma, tumor, and infarct may alter the $\overline{N13}$ and the $\overline{N20}$.

17.5. Lesions of the brain stem and cerebral hemispheres such as multiple sclerosis, strokes, and tumors affect the $\overline{N20}$ and may also affect the $\overline{P13/14}$.

17.1 CRITERIA DISTINGUISHING ABNORMAL SEPs

17.1.1 Absence of clavicular, cervical, and scalp SEPs

The absence of the $\overline{N9}$ peak in recordings from Erb's point, the absence of the $\overline{N13}$ peak in recordings from the neck, and the absence of the $\overline{N20}$ peak in scalp recordings is abnormal if technical problems are excluded. The absence of the corresponding peaks in recordings between scalp and noncephalic electrodes is likely to be abnormal. The absence of the $\overline{N13}$ with preservation of an $\overline{N20}$ of normal latency is of doubtful diagnostic significance. Absence of a $\overline{P13/14}$ is of importance only in laboratories using methods that consistently produce this peak. An absent $\overline{N9}$ with normal $\overline{N13}$ and $\overline{N20}$ may occur in normal subjects.

17.1.2 Slow peripheral conduction velocity

Peripheral conduction velocity is calculated from recordings made with a clavicular electrode by measuring the conduction distance as a straight line between stimulating cathode, or the midpoint between cathode and anode, and the recording electrode at Erb's point and by dividing this distance by the latency from the onset of the stimulus pulse to the peak of the clavicular SEP. Measurement of the peak is preferred over measurement to the onset because peaks are usually better defined than onsets. Peripheral conduction velocity derived with this method does not reflect the maximum velocity but comes closer to the velocity of the majority of the fast-fiber group causing the deflection.

The finding of a decreased peripheral conduction velocity is abnormal unless explained by low arm temperature or inaccurate measurement of arm length. The abnormality can be further investigated by recordings of sensory nerve action potentials (20.1) and of motor nerve conduction velocity. Although these tests generally do not reflect the function of the proximal part of the peripheral system, they can help to identify lesions of the distal part and to distinguish motor from sensory defects.

17.1.3 Prolonged central conduction times

Central conduction times are derived by calculating latency differences between peaks recorded at Erb's point, neck, and scalp. Most important are the $\overline{N9}$-$\overline{N13}$ and the $\overline{N13}$-$\overline{N20}$ conduction times. The $\overline{N9}$-$\overline{N20}$ conduction time is usually increased when one of the other two central conduction times is increased. In SEPs recorded between scalp and noncephalic reference electrodes, central conduction times are measured as the separation between successive peaks.

17.1.4 Abnormal latency differences to stimulation of either side of the body

The difference between latencies or central conduction times of SEP peaks to stimulation of either side of the body may be abnormal in some patients even when the latencies and central conduction times themselves are not abnormally increased. Abnormally large latency differences suggest a lesion on the side of the pathway responsible for the longer latency. Evaluation of latency asymmetries between SEPs to stimulation of the two sides may increase the sensitivity of the test.[3,101]

17.1.5 Decreased amplitude

Because SEP amplitude varies considerably and has a non-Gaussian distribution in normal subjects, its use as an indicator of abnormality is limited.[3] Amplitude differences between two SEPs recorded at the same level in response to stimulation of each side may suggest a lesion in the pathway yielding the SEP of lower amplitude.[18] Since lesions in the peripheral and central parts of the sensory pathway selectively reduce peripheral or central SEPs, measurements of amplitude ratios between central and peripheral SEPs may have diagnostic value.[103]

17.2 GENERAL CLINICAL INTERPRETATION OF ABNORMAL SEPs TO ARM STIMULATION

17.2.1 Technical problems

SEPs to arm stimulation may be absent at all recording levels due to lack of an effective stimulus, to failure of proper synchronization between stimulator and averager, or to faulty recording electrodes, amplifiers, or averaging channels (Table 17.1). The latency of SEPs may be increased at all recording levels due to low body temperature, especially of the stimulated extremity, or due to inaccurate measurement of the length of the extremity.

17.2.2 Peripheral and central lesions of the somatosensory pathway

Absence of the clavicular, cervical, and scalp SEPs indicates a lesion in the peripheral nerve or brachial plexus unless explained by technical problems. Slowing of peripheral conduction velocity, as indicated by an abnormal delay of Erb's point potential, has the same implication.

Absence of the cervical SEP, almost always associated with absence of the scalp SEP,[299] and preservation of the clavicular SEP suggest a lesion involving the spinal cord or roots if technical problems with the cervical and scalp recordings are excluded. An increase of clavicular-cervical ($N\overline{9}$-$N\overline{13}$) and clavicular-scalp ($N\overline{9}$-$N\overline{20}$) conduction times combined with normal peripheral nerve conduction velocity is a rather reliable indicator of a lesion of spinal roots or spinal cord below the lower medulla. An absent scalp SEP with preserved clavicular and cervical SEPs suggests a lesion above the lower medulla if technical problems with the scalp recording are excluded. An increase of clavicular-scalp ($N\overline{9}$-$N\overline{20}$) and cervical-scalp ($N\overline{13}$-$N\overline{20}$) conduction time with normal clavicular-cervical ($N\overline{9}$-$N\overline{13}$) conduction

TABLE 17.1. Clinical interpretation of abnormal SEPs to arm stimulation

Abnormal SEP finding	Interpretation
A. Technical problems	
1. Absent SEPs to arm stimulation at all recording levels	Lack of stimulus; lack of synchronization between stimulator and averager; faulty recording electrodes or equipment
2. Increased latency of SEPs at all recording levels	Hypothermia; inaccurate measurement of the distance between stimulating and recording electrodes
B. Lesions of the nervous system	
1. Absent $\overline{N9}$ with normal $\overline{N13}$ and $\overline{N20}$	Normal
2. Absent $\overline{N9}$ with absent or delayed $\overline{N13}$ and $\overline{N20}$	Peripheral nerve or plexus lesion; rule out technical problems
3. Increased latency of $\overline{N9}$ with equally increased latency of $\overline{N13}$ and $\overline{N20}$: Decreased peripheral conduction velocity with normal central conduction times	Peripheral nerve or plexus lesion; rule out technical problems
4. Increased $\overline{N9}$–$\overline{N13}$ conduction time with normal $\overline{N13}$ amplitude and shape, normal peripheral conduction velocity, normal $\overline{N13}$–$\overline{N20}$ conduction time	Defect above the brachial plexus and below the lower medulla
5. Absent $\overline{N13}$ and absent or delayed $\overline{N20}$	Defect above the brachial plexus and below or at the lower medulla
6. Increased $\overline{N13}$–$\overline{N20}$ central conduction time with normal $\overline{N9}$–$\overline{N13}$ conduction time and normal peripheral conduction velocity	Defect above the lower medulla and at or below the somatosensory cortex
7. Absent $\overline{N20}$ and normal $\overline{N9}$–$\overline{N13}$ conduction time and normal peripheral conduction velocity	Defect above the lower medulla and at or below the somatosensory cortex
8. Decreased peripheral conduction velocity and increased central conduction times	Combination of peripheral nerve or plexus lesion and central defect

time and normal peripheral conduction velocity indicates a lesion above the lower medulla and below the somatosensory cortex. A combination of decreased peripheral conduction velocity and of increased central conduction times suggests lesions in both the peripheral and the central parts of the somatosensory pathway.

17.3 PERIPHERAL LESIONS THAT CAUSE ABNORMAL SEPs TO ARM STIMULATION

In general, peripheral sensory nerve lesions cause abnormalities of SEPs recorded at all levels of the afferent pathway. While complete lesions of peripheral nerves abolish both peripheral and central SEPs, partial peripheral lesions may make peripheral recordings such as sensory nerve action potentials more difficult to obtain than central SEPs.[77,127] In such instances, scalp SEPs have been used to evaluate nerve lesions.[19,76,77,97,127]

17.3.1 Polyneuropathies

Evaluation of mixed and sensory neuropathies with averaging techniques shows that these neuropathies slow, reduce, or abolish SEPs depending on the degree of neuronal damage rather than on its cause; cervical-scalp ($\overline{N13}$-$\overline{N20}$) conduction time remains intact.[7,24,77,98,127,146,289] Clavicular-cervical ($\overline{N9}$-$\overline{N13}$) conduction time may be increased if the dorsal roots are also involved.[289]

17.3.2 Chronic renal failure

Scalp SEPs, like VEPs, show an increase of latency and amplitude of long-latency peaks that has no consistent relationship to blood chemistries but tends to return to normal after successful kidney transplantation.[200] Increased latencies and absence of peaks at different recording points can be explained by peripheral neuropathy in most cases, but involvement of the central somatosensory pathway cannot be excluded.[280,338]

17.3.3 Guillain-Barré syndrome

Clavicular and cervical SEPs are delayed.[344] Sensory conduction is slowed proximally more than distally[24,32,98]; cervical radiculopathy may be indicated by an abnormal delay between clavicular and cervical SEP.[307]

17.3.4 Charcot-Marie-Tooth disease, Friedreich's ataxia, Adie's syndrome, and tabes dorsalis

Peripheral nerve lesions account for most[169] or all[25,243] SEP abnormalities seen in Charcot-Marie-Tooth disease and for the SEP abnormalities in Adie's syndrome and tabes dorsalis.[25] Friedreich's ataxia may reduce or abolish clavicular SEPs and sensory nerve action potentials (Figure 17.1); cervical and scalp SEPs may be preserved, but show some evidence for a central conduction defect in addition to the peripheral one (17.5.8.2).

17.3.5 Thoracic outlet syndrome

Cervical SEPs to ulnar nerve stimulation have been found to be abnormal in patients with cervical ribs and neurolog-

Figure 17.1. SEPs in a patient with Friedreich's ataxia. Top two tracings show delays of the $\overline{N20}$ in the parietal scalp SEPs to stimulation of the contralateral median nerves in the right upper extremity (RUE) and left upper extremity (LUE). Bottom tracings show a cervical SEP (third tracing) with a marginal delay of the $\overline{N13}$ and a clavicular SEP (fourth tracing) with an $\overline{N9}$ of low amplitude, obtained after stimulation of the right arm. Stimulation of the median nerve at the wrist. Recordings from electrodes on the left and right parietal scalp (top two tracings), over the seventh cervical vertebra (third tracing) and Erb's point (bottom tracing). Negativity at these electrodes is plotted upward. Frontal reference electrode. From Nuwer et al.[245] with permission by the authors and Little, Brown and Company.

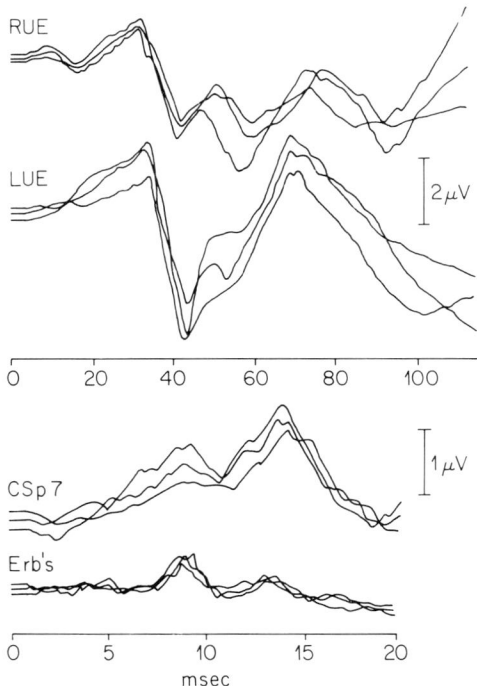

ical findings. Clavicular SEPs were also abnormal in some of these patients.[130,295,356]

17.3.6 Brachial plexus lesions

Damage to the brachial plexus abolishes or delays SEPs at clavicular and more proximal recording points.[15,77] Partial plexus lesions, such as lesions of the lower plexus, leave SEPs to stimulation of the unaffected portions intact. SEPs may therefore be useful in the evaluation and management of brachial plexus lesions[171,187,316,321,356] (Figure 17.2). Complete investigation requires stimulation of several of the nerves contributing to the plexus (16.3.1). Brachial plexus lesions are often associated with cervical root lesions (17.4.1).

17.3.7 Carpal tunnel syndrome

Scalp SEPs to median nerve stimulation may indicate slowing of conduction even in cases where stimulation of sensory nerve fibers in the digital nerves and recordings from the median nerve showed no sensory nerve action potentials.[76]

17.3.8 Congenital insensitivity to pain

Cortical SEPs to the usual sensory nerve stimuli remain intact.[7,146] SEPs to ordinarily painful electric tooth pulp stimulation (20.4.5) have been found to be altered in one patient,[44] but not in another.[214]

17.3.9 Hereditary pressure-sensitive neuropathy

Peripheral conduction defects increase the absolute latencies but not the central conduction times.[94]

17.4 LESIONS OF CERVICAL ROOTS AND OF THE CERVICAL CORD THAT CAUSE ABNORMAL SEPs TO ARM STIMULATION

17.4.1 Cervical root lesions

Lesions of cervical roots are distinguished by preservation of the clavicular SEP and of sensory nerve action potentials, unless there is additional damage to the brachial plexus (Figure 17.3). Postganglionic, but not preganglionic, root damage is followed by retrograde degeneration of sensory nerve fibers and eventual disappearance of the clavicular SEP and of sensory nerve action potentials.[171] Cervical and scalp SEPs are absent in complete avulsions and may be delayed or reduced in incomplete lesions (Figure 17.3) such as spondylotic radiculopathy.[49,106,120,296,316,320,324] Herpes zoster radiculitis delays the cervical SEP and may reduce the amplitude of the clavicular SEP.[307] As with plexus lesions, clear definition of the level of the lesion requires multiple nerve stimulation[99] (16.3.1).

Figure 17.2. SEPs in a patient with a right brachial plexus traction lesion. Top three tracings on left: Stimulation of the left median nerve and simultaneous recording from right parietal, midline cervical, and left clavicular electrodes with reference to a midfrontal electrode produces normal SEPs. Bottom tracing on left: Stimulation of the left radial nerve produces a normal sensory nerve action potential. Top three tracings on right: Stimulation of the right median nerve produces no peaks in the left parietal, midcervical, and right clavicular recordings. Bottom tracing on right: Stimulation of the right median nerve produces no sensory nerve action potential, probably because of retrograde degeneration. Diagram at bottom: Surgical exploration showed postganglionic lesions of the C5-C7 roots on the right. Positivity at the reference electrode is plotted upward. From Jones et al.[171] with permission by the authors and John Wright and Sons Ltd.

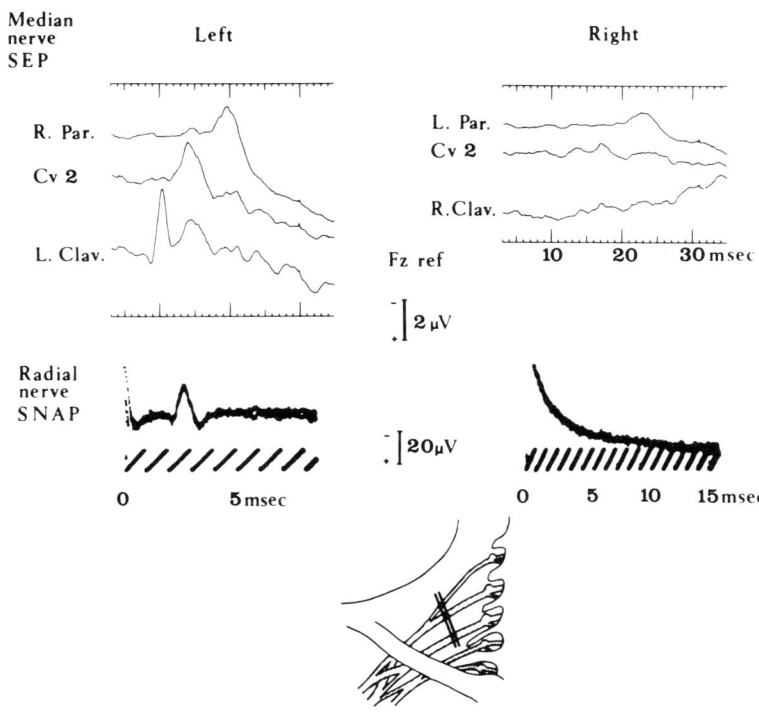

Figure 17.3. SEPs in a patient with C5-C6 root lesions. The nerve action potential (ARM) and clavicular SEP (ERB POINT) have normal latency by the standards of the particular laboratory. Lower (CVII) and upper (CII) cervical SEPs are delayed. The clavicular and cervical SEPs are also attenuated. Stimulation of the median nerve at the wrist on the involved side. Recording from electrodes on the arm, at Erb's point, and over C7 and C2. Negativity at these electrodes is plotted upward. Midfrontal reference electrode. From Synek and Cowan[321] with permission by the authors and Modern Medicine Publications, Inc.

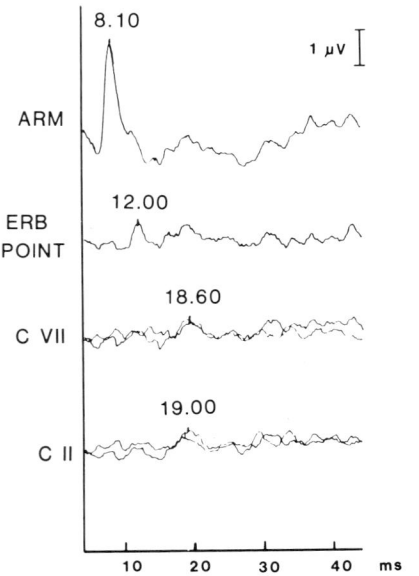

17.4.2 Cervical cord lesions

Although lower spinal cord lesions must be studied with SEPs to leg stimulation, lesions of the cervical cord may be detected with SEPs to arm stimulation. These lesions may cause abnormal cervical and scalp SEPs with preserved clavicular SEPs. Such abnormalities have been found in many cases of cervical spondylotic myelopathy with or without radiculopathy[106,120,221,296,306] (Figure 17.4) and in some cases of subacute combined degeneration of the spinal cord due to vitamin B_{12} deficiency,[115] cervical cord injury, tumors, syringomyelia, and hydromyelia.[15,221,342,306] In contrast, infarcts in the distribution of the anterior spinal artery, which do not involve the posterior columns, leave the SEP intact.[15,221]

17.5 LESIONS OF THE BRAIN STEM AND CEREBRAL HEMISPHERES THAT CAUSE ABNORMAL SEPs TO ARM STIMULATION

17.5.1 Multiple sclerosis

Because multiple sclerosis may interrupt fibers at any level of the central somatosensory pathway, it may produce abnormalities of the cervical and scalp SEPs (Figure 17.5) or of the scalp SEP only (Figure 17.6); sometimes only the cervical SEP is abnormal. Delays of these SEPs lead to prolongation of clavicular-cervical ($\overline{N9}$-$\overline{N13}$), clavicular-

Figure 17.4. Abnormal SEPs in a patient with cervical myelopathy at C4/C5. Stimulation of the right (r) and left (l) median nerve produced normal SEPs at C7 (top two tracings) but no definite SEPs at C2 (bottom two tracings). Negativity at the neck electrodes is plotted upward. Midfrontal reference electrode. From Stöhr et al.[306] with permission by the authors and Elsevier Scientific Publishers Ireland Ltd.

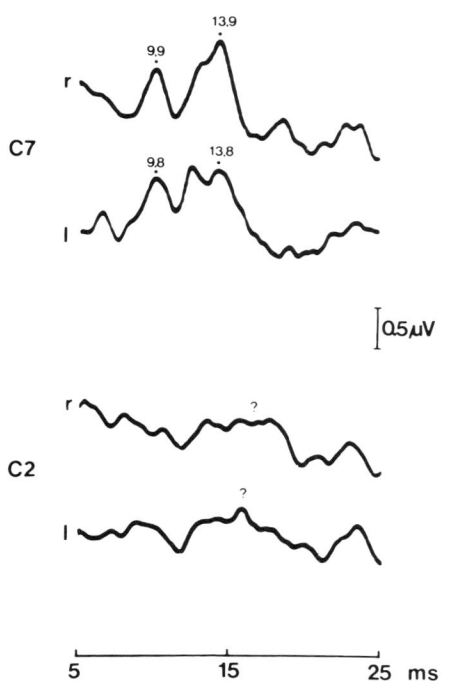

scalp ($\overline{N9}$-$\overline{N20}$), and cervical-scalp ($\overline{N13}$-$\overline{N20}$) conduction times. The chance of finding SEP abnormalities in multiple sclerosis is increased by using SEPs to leg stimulation which test the entire length of the spinal cord and are more often affected by the diffuse and scattered lesions of multiple sclerosis.

SEPs to arm stimulation are abnormal in the majority of patients with multiple sclerosis, more often in definite than in probable or possible cases.[18,29,101,104,122,225,301,315,355]

Figure 17.5. Abnormal SEPs to stimulation of the median nerve in a patient with multiple sclerosis. Stimulation of the left median nerve produces a clavicular SEP of normal latency (bottom tracing). Recordings at C5 and C2 (middle two tracings) show no definite peaks in the normal latency range. The scalp recording shows an abnormally delayed $\overline{N20}$ indicating a defect above the level of the brachial plexus. This 31-year-old man has multiple sclerosis without clinically detectable sensory involvement.

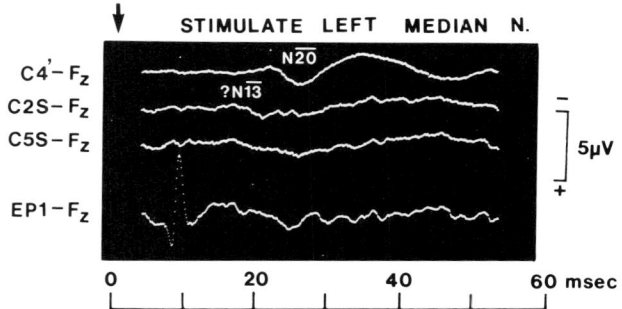

While the incidence of abnormal SEPs is very high in patients who show defects of vibration, position, and touch sensation in the stimulated limb, a substantial portion of multiple sclerosis patients with abnormal SEPs have normal sensation[18,51,104,243] and may have clinically silent plaques.[223] Abnormal SEPs can be found in patients presenting with only optic neuritis.[301,323] The incidence of SEP abnormalities in patients with multiple sclerosis increases with progression of the disease,[343] but in patients without relapses, the SEP was found to remain stable.[224] Steady-

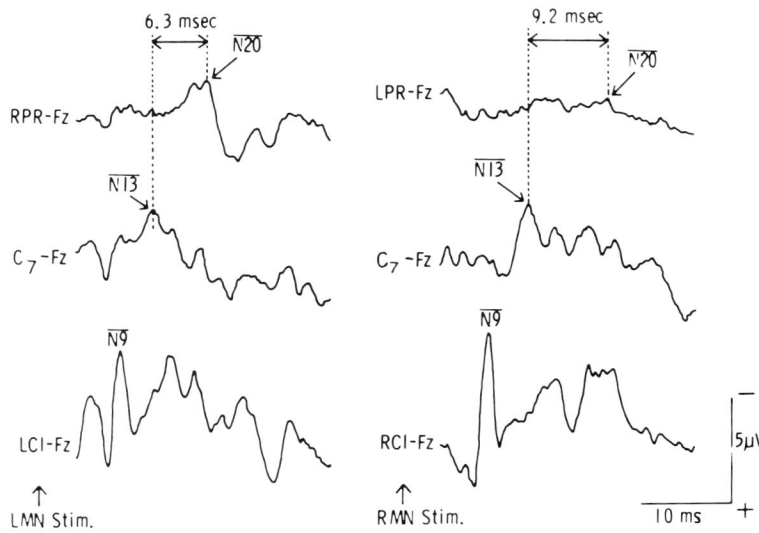

Figure 17.6. SEPs to median nerve stimulation in a patient with multiple sclerosis. Stimulation of the left median nerve (left half of figure) produces normal clavicular (LC1-Fz), cervical (C_7-Fz), and right parietal scalp (RPR-Fz) SEPs (bottom to top). Stimulation of the right median nerve (right half of figure) produces normal clavicular and cervical SEPs, but the scalp SEP is attenuated and delayed indicating a conduction defect above the level of the spinal cord. This 21-year-old woman suffers from multiple sclerosis producing complete proprioceptive sensory loss of the right hand. From Shibasaki et al.[288] with permission by the authors and Elsevier Biomedical Press B.V.

state scalp SEPs may be abnormal in multiple sclerosis.[278] A normal SEP to median nerve stimulation distinguishes acute transverse myelitis of other causes, involving the spinal cord below the cervical level, from the myelopathic form of multiple sclerosis, which may be associated with abnormal SEPs to arm stimulation due to clinically silent lesions at or above the cervical cord.[265]

The usefulness of the SEP in the diagnosis of multiple sclerosis has been evaluated by comparing the proportions of abnormal SEPs with those of abnormal VEPs and AEPs in patients with possible, probable, and definite multiple sclerosis. In general, SEPs and VEPs are more effective than BAEPs. SEPs are abnormal more often than VEPs if both arm and leg stimulation are used (19.4.1). Studies using SEPs to arm stimulation only have reported the SEP to be abnormal more frequently than the VEP in patients with possible or probable multiple sclerosis,[137,219] or with definite and possible[180] multiple sclerosis. However, some studies of the SEP to arm stimulation found the VEP to be more effective in most diagnostic groups.[51,259]

17.5.2 Brain stem strokes and tumors

Strokes and tumors of the brain stem cause scalp SEP abnormalities in many cases.[15,49,235] In the "locked-in syndrome" the SEP may be abnormal if the pontine lesion extends into the medial lemnisci of the tegmentum.[242] In contrast, infarcts causing the lateral medullary syndrome of Wallenberg or the peduncular syndrome of Weber, neither of which usually involves lemniscal fibers, leave the SEP intact.[15,220,242]

17.5.3 Thalamic lesions

Lesions of the thalamus may cause delays or reductions of the $\overline{N20}$-$\overline{P30}$ peaks of the scalp SEP; the $\overline{N13}$, and often also the $\overline{P13/14}$, remains intact[15,49,60,125,228,235] (Figure 17.7). Abnormal scalp SEPs were found to result from lesions which involve the nucleus ventralis posterolateralis and encroach on the nucleus ventralis posteromedialis, but not from lesions of the nucleus ventralis lateralis.[84] Scalp SEPs are likely to be abnormal in cases of thalamic lesions that affect position, vibration, and touch sensation,[146,326,335] including cases of the thalamic syndrome of Dejerine and Roussy that show impairment in these modalities.[146,326,342]

17.5.4 Parietal infarcts and tumors

Mass lesions involving the primary sensory receiving areas produce abnormalities of the scalp SEP which are usually associated with contralateral sensory defects; only a few patients with sensory defects have normal SEPs.[60,127,189,233,241,247,348,353] Most often, parietal lesions reduce or abolish the $\overline{N20}$-$\overline{P30}$ and later peaks while leaving earlier peaks intact (Figure 17.8). An increase of latency is less characteristic of cerebral lesions than of lesions of afferent pathways.[15,127,230,235,243,308,326,348] Central conduction time has been shown to increase transiently with pathological reductions in cerebral blood flow.[318] Amplitude reductions of the $\overline{N20}$ peak in patients paralyzed by strokes have been said to suggest an unfavorable prognosis.[186]

17.5.5 Lesions in other cerebral areas

Cerebral lesions outside the primary sensory area may reduce parietal SEP peaks of latencies longer than those of

Figure 17.7. SEPs in a patient with right thalamic hemorrhage. Stimulation of the right median nerve (left half of figure) produces normal scalp (top three tracings) and cervical (bottom tracing) SEPs. Stimulation of the left median nerve (right half of figure) produces only an early positive scalp peak corresponding with the negative neck SEP. The peaks $\overline{N16}$, $\overline{P16}$, and $\overline{N21}$ in this figure correspond with the normal $\overline{N13}$, $\overline{P13/14}$, and $\overline{N20}$. Recordings from electrodes 2 cm behind Cz (top tracing), 5 cm (second tracing), and 7 cm (third tracing) lateral to that electrode, and from an electrode on the C2 spinous process (bottom tracing). Negativity at these electrodes is plotted upward. Reference electrodes on interconnected ears for top three tracings and at Fz for bottom tracing. This 70-year-old woman has a complete left hemiplegia and hemianesthesia. The CT scan shows a hemorrhage in the right internal capsule and thalamus. From Mauguière and Courjon[228] with permission by the authors and Little, Brown and Company.

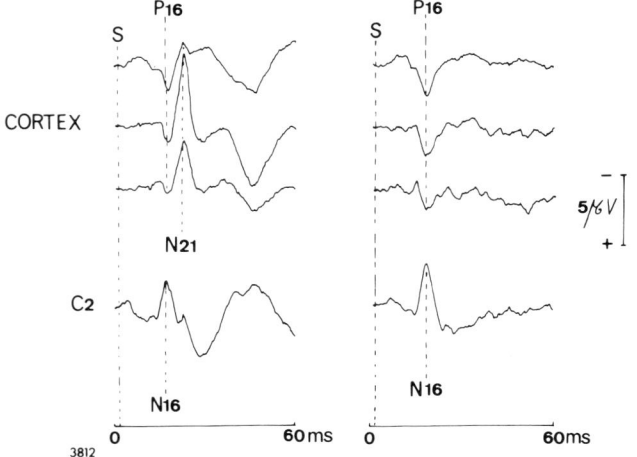
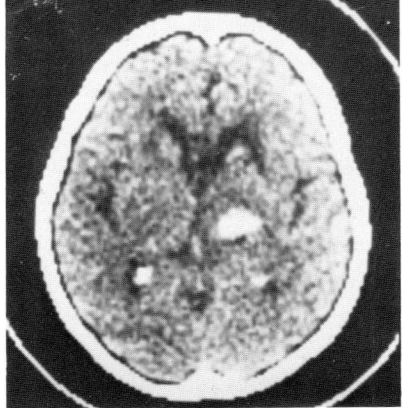

N20 and P30.[291,335,353] Lesions in the frontal area may selectively reduce SEP peaks recorded from that area[230,243] (Figure 17.9). Lesions outside the primary sensory area may increase amplitude and duration of the parietal SEP,[247] whereas chronic parietal lesions may increase the SEP in the frontal area.[230] Unilateral cerebral lesions alter SEPs recorded with different latency on both sides of the head after stimulation of the contralateral arm, but these lesions do not affect the SEPs produced on both sides by stimulation of the ipsilateral arm.[236,335,348]

17.5.6 Hemispherectomy

Removal of one hemisphere eliminates scalp SEP peaks of medium latency recorded over both sides of the head in

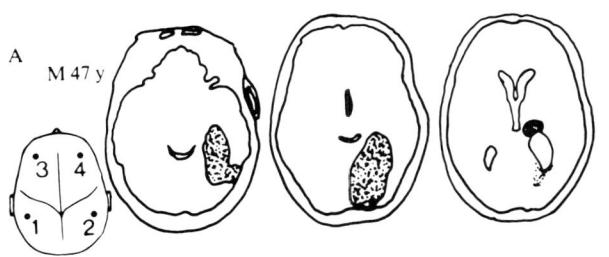

Figure 17.8. Scalp SEPs in a patient with a right parietal lesion. Stimulation of the right median nerve (R MN in *B*) produces normal prerolandic (3) and parietal (1) SEPs, including an early far-field peak (FF). Stimulation of the left median nerve (L MN in *C*) produces a fairly normal prerolandic SEP (4) but no clear parietal peaks after the far-field peak (2). Recording with scalp electrodes as indicated in the diagram (*A*). Negativity at the scalp electrodes is plotted upward. Ipsilateral ear reference electrodes. This 47-year-old man had suffered a sudden cerebrovascular lesion five years earlier and has residual left astereognosis, loss of graphesthesia, position sense and two-point discrimination, and left upper quadrantanopsia. The CT scan showed an area of reduced density in the right parieto-occipital and posterior thalamic regions (*A*). From Mauguière et al.[230] with permission by the authors and Oxford University Press.

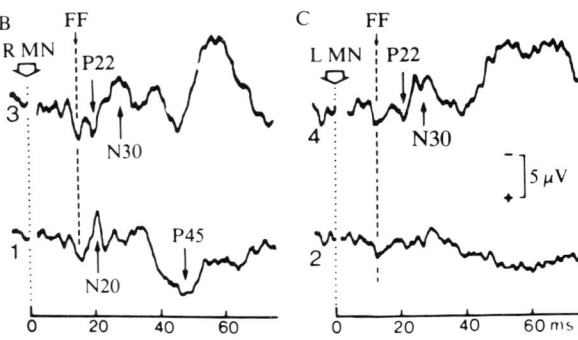

Figure 17.9. Scalp SEPs in a patient with a left prerolandic lesion. Stimulation of the right median nerve (R MN in *B*) produces a normal parietal SEP (1), including an early far-field potential (FF); the prerolandic recording shows no peaks after the far-field peak (3). Stimulation of the left median nerve (LM N in *C*) produces normal parietal (2) and prerolandic (4) SEPs. The tracings of *B* and *C* are superimposed in *D*, the heavier lines representing the involved side. Recordings from scalp electrodes as indicated in diagram (*A*). Negativity at the scalp electrodes is plotted upward. Ipsilateral ear reference electrodes. This 10-year-old girl suffered an intracerebral hemorrhage from an arteriovenous malformation one year earlier and has residual right spastic hemiplegia and moderate dysphasia and dysarthria but no sensory abnormalities. The CT scan shows a left rolandic lesion (*A*). From Mauguière et al.[230] with permission by the authors and Oxford University Press.

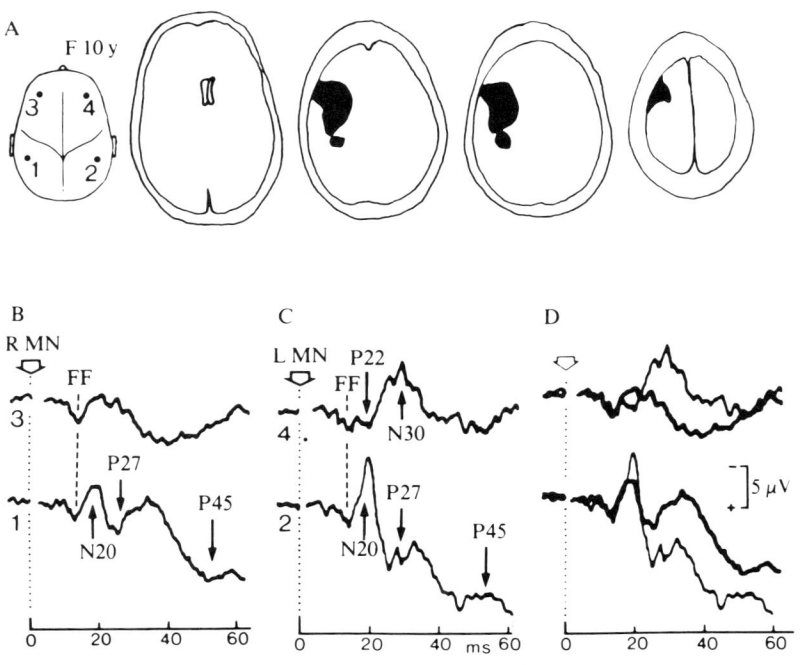

response to stimulation of the arm contralateral to the hemispherectomy. Early peaks, generated by preserved subcortical structures,[243] and late peaks, presumably generated in the ipsilateral hemisphere via extralemniscal afferents,[152,274] may be preserved.

17.5.7 Diffuse cerebral disorders with reduced consciousness

17.5.7.1 Cerebral death
The precise role of SEPs, like that of other EPs, in evaluating cerebral death has not yet been clarified. In general, patients who fulfill the criteria for cerebral death and have no electrocerebral activity in the EEG show no $\overline{N20}$ and $\overline{P30}$ in the scalp SEP,[14,135] even though subcortically originating peaks may persist in scalp[14,332] and neck[135,332] SEPs (Figure 17.10). The persistence of cervical SEPs in cases without cortical SEPs proves that the sensory input has reached the central nervous system so that the absence of the higher SEPs can safely be attributed to absent cerebral functioning.[135] Patients showing the alpha frequency coma pattern in the EEG may retain SEPs until the EEG further deteriorates or disappears.[246] However, the absence of central SEPs does not correspond precisely with brain death: An $\overline{N20}$ has been reported to persist in the SEP of a case where cerebral death was diagnosed with angiography but without EEG recording.[14]

17.5.7.2 Head injury
SEPs alone[63,64,158,159,270] or in combination with other EPs[13,139,140,201,207,260] have been used to evaluate the severity of cerebral injuries and to make an early prognosis. Serial recordings showing a reduction in central conduction time or an increase in the number of peaks of the scalp SEP generally indicated an imminent return of function.

17.5.7.3 Chronic vegetative state
Various combinations of scalp and cervical SEP abnormalities have been described. In general, the cervical SEP is preserved; complete absence of cortical SEPs suggests a poor outcome.[15,135]

17.5.7.4 Reye syndrome
Scalp SEPs, nearly or completely abolished initially, may be of prognostic value: Early recovery of short-latency peaks indicates survival; progressive recovery of peaks with latencies over 100 msec precedes satisfactory clinical recovery.[134]

17.5.7.5 Subarachnoid hemorrhage
Central conduction time has been found to be increased in patients with aneurysmal subarachnoid hemorrhage in poor condition and with poor prognosis.[345]

Figure 17.10. SEPs from three brain-dead patients (*A-C*) and a comatose but not brain-dead patient (*D*). The three tracings for each patient represent scalp (top), neck (middle), and clavicular (bottom) recordings. The three brain-dead patients have no scalp SEPs. Neck and clavicular SEPs are preserved although the neck SEP in patient *A* is abnormal. The comatose patient in *D* has cervical and scalp peaks preserved although distorted and delayed, possibly due to hypothermia of less than 92° F. Stimulation of the median nerve at the wrist. EP = Erb's point potential; A, B = Cervical peaks; N2 = $\overline{N20}$. Negativity at the scalp, neck, and clavicular electrodes is plotted upward. Midfrontal reference electrode. Calibration bars are 10 msec and 0.25 μV. From Goldie et al.[135] with permission by the authors and Modern Medicine Publications, Inc.

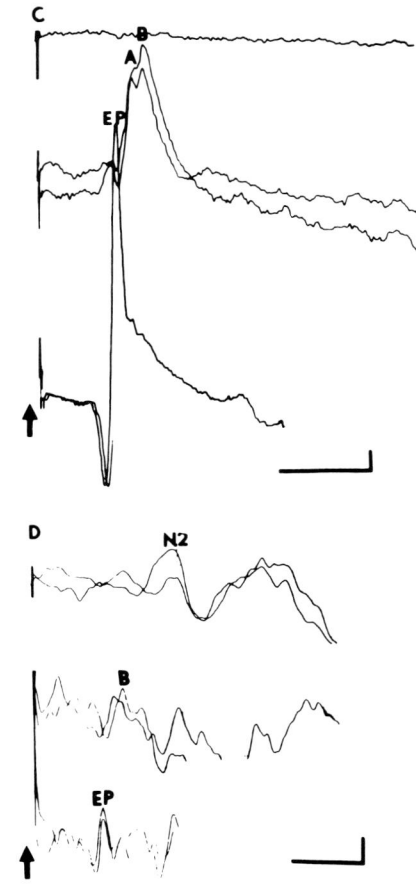

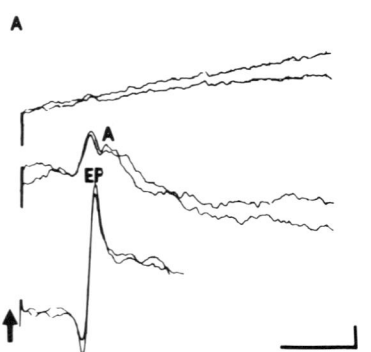

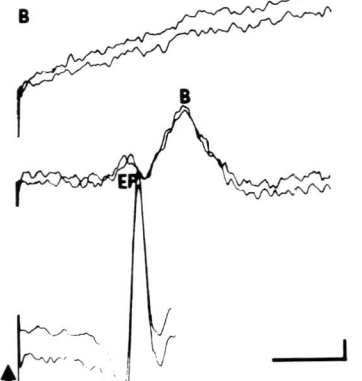

17.5.7.6 Perinatal asphyxia
Scalp SEPs have been reported to be abnormal in the majority of asphyxiated newborn infants, showing absence, reduced number, or low amplitude of peaks and increased latency. The degree of these abnormalities corresponded with the severity of the asphyxia.[155]

17.5.7.7 Surgical monitoring
Scalp SEPs have been used to monitor the condition of the brain during carotid endarterectomy[215] and aneurysmal surgery.[319]

17.5.8 Diffuse cerebral disorders without reduced consciousness

17.5.8.1 Myoclonus epilepsy, Jakob-Creutzfeldt disease, Ramsay Hunt syndrome, and other conditions with myoclonus
The scalp SEP is greatly increased in amplitude in many patients with myoclonus epilepsy, especially if myoclonus is present at the time of the recording,[88,145,178] and is similar to the EEG spike that precedes spontaneous myoclonic jerks in this disorder.[268,290] The neck SEP in three such patients was normal.[221]

In Jakob-Creutzfeldt disease, the amplitude of the scalp SEP has been found to be increased[178] in some studies and not increased[290] or decreased[266] in others. No peaks after $\overline{N20}$ were found in one study.[48] The cervical SEP in another case report was normal.[221]

The SEP amplitude has been reported to be increased in the dyssynergia cerebellaris myoclonica of Ramsay Hunt,[41,227,290] in familial startle disease,[217] in a few cases of postanoxic myoclonus,[41] and in one case of unilateral myoclonus after a stroke.[317]

17.5.8.2 Friedreich's and other hereditary cerebellar ataxias, familial spastic paraplegia, and amyotrophic lateral sclerosis
Friedreich's ataxia often causes a decrease of amplitude and temporal dispersion of the $\overline{N20}$-$\overline{P30}$ of the scalp SEP, suggesting cerebral conduction defects in addition to the peripheral conduction defects[49,168,220,243,245,252] (17.3.2). Abnormal SEPs were found in olivopontocerebellar degeneration[149] and in some cases of hereditary cerebellar ataxia[17,252] and familial spastic paraplegia,[252,328] but not in isolated cases of cerebellar ataxia.[221] A normal SEP generally distinguishes patients with amyotrophic lateral sclerosis,[221] although an abnormal SEP has been reported in one case.[17]

17.5.8.3 Huntington's chorea
The scalp SEP has low amplitude and may have slightly increased latency in patients and many subjects at risk.[17,96,174,244,248]

17.5.8.4 Wilson's disease
The SEP may show reduction of scalp SEP peaks following the $\overline{N20}$[205] or increased central conduction times.[17]

17.5.8.5 Myotonic dystrophy
A few patients show increased central conduction times.[329]

17.5.8.6 Leukodystrophies
Pelizaeus-Merzbacher disease, adrenoleukodystrophy, and metachromatic leukodystrophy may reduce or abolish the cervical SEP and delay or abolish the scalp SEP.[38,126,216]

17.5.8.7 Down's syndrome
The amplitude and latency of scalp SEPs may be increased.[22,27,313]

17.5.8.8 Hepatic encephalopathy
Central conduction times may be increased.[17]

17.5.8.9 Hyperthyroidism
The scalp SEP, like the VEP, has been reported to have increased amplitude but essentially normal latency.[325]

17.5.8.10 Minamata disease
The $\overline{N20}$ has been reported to be absent; clavicular and cervical SEPs were preserved.[330]

17.5.8.11 Tourette's syndrome
SEPs have been found to be normal.[183]

17.5.8.12 Drug effects
Like other EPs, SEPs show diverse changes of late peaks of the scalp SEP in patients taking drugs that influence behavior, whereas the early peaks commonly used in clinical diagnosis remain unaffected. Depressant drugs generally increase latency and decrease amplitude.[285]

17.5.8.13 Psychiatric disorders
SEPs, like other EPs, have been reported to differ from normal in patients with schizophrenia and affective psychoses, but no specific diagnostic SEP abnormalities have been isolated.[282] Schizophrenia tends to reduce the amplitude of scalp SEP peaks of over 100 msec latency; in chronic paranoid patients, a negative peak at 60 msec was found to be increased.[286] SEP measures may distinguish chronic schizophrenia from psychiatric depression.[283]

Hysterical absence of pain sensation has been reported not to cause scalp SEP abnormalities.[7,348] Hypnotically induced anesthesia produced no significant changes in the scalp SEP in one study[147] but reduced SEP amplitude in another.[198]

18

Normal SEPs to leg stimulation

SUMMARY

18.1 SEPs to stimulation of the posterior tibial nerve at the ankle may be recorded at the knee, lumbar or lower thoracic spine, and scalp; the scalp SEP shows P37 and N45 peaks. Stimulation of the common peroneal nerve at the knee produces lumbar and thoracic SEPs and an early scalp SEP consisting of P27 and N35 peaks. Stimulation of the sural and saphenous nerves elicits SEPs of lower amplitude and different latency. Recordings with widely spaced electrodes may show cortical SEPs preceded by subcortical SEPs. SEPs to leg stimulation vary with stimulus characteristics, subject variables, and recording parameters.

18.2. Subject variables that have practical importance for SEPs to leg stimulation are age, leg length, body height, and temperature.

18.3. Stimulus parameters affecting SEPs to leg stimulation include stimulus electrode placement and stimulus intensity, rate, and duration.

18.4. The most important recording parameter is the location of the recording electrodes.

18.5 General strategy localizes lesions in different segments of the somatosensory pathway by looking for the presence of potentials at the popliteal, low spinal, and scalp level and by calculating peripheral and central conduction velocities.

18.1 NORMAL SEPs AT DIFFERENT RECORDING SITES

SEPs to leg stimulation may be produced by stimulating the posterior tibial nerve at the ankle (Figure 18.1) or the common peroneal nerve at the knee (Figure 18.2) and by recording from the lumbothoracic spine and the scalp. Recordings from the neck do not reliably show SEPs in response to leg stimulation. Nerve action potentials can be recorded at the knee when the posterior tibial nerve is stimulated. Other nerves are used occasionally for stimulation. Like SEPs to arm stimulation, SEPs to leg stimulation vary depending on subject variables, stimulus characteristics, and recording parameters (Table 18.1).

18.1.1 Popliteal fossa potential

To record the action potential of tibial nerve fibers at the knee after stimulation of the posterior tibial nerve at the ankle, a popliteal fossa electrode is placed 4–6 cm above the popliteal crease, midway between the combined tendons of the semimembranosus and semitendinosus muscles medially and the tendon of the biceps femoris laterally. These tendons can be brought out by having the subject bend his knee against resistance. The popliteal electrode may be referred to an electrode on the medial surface of the knee.

Like the clavicular SEP, the popliteal fossa potential consists of a major negative peak which may be preceded and followed by smaller positive peaks (Figure 18.3). The latency, measured to the negative peak, depends on the distance between stimulus and recording electrodes and amounts to about 9–10 msec in normal adults.

18.1.2 Lumbar and low thoracic SEPs

Recording electrodes may be placed over the spinous processes of L3, T12, and T6 (electrodes L3S, T12S, and T6S). The process of L3 lies above a line connecting both iliac crests. Reference electrodes may be placed 4 cm above each of these electrodes. Alternatively, an electrode over the iliac crest or at Fz may be used as a common reference for the spinal electrodes. All three spinal electrode pairs may be used, in addition to a scalp electrode pair, in four-channel recordings of the SEPs to common peroneal nerve stimulation. Only the lower two pairs are used in four-channel recordings of SEPs to posterior tibial nerve stimulation, the other two channels being allocated to scalp and popliteal fossa recordings (Table 18.1).

Lumbar SEPs have a negative peak that may be preceded by a small positive peak.[55] These peaks may be followed by a second negative peak that is best recorded at a slightly higher level.[59,65,81] The latency of the lumbar peaks depends on the stimulation and recording sites, the length and temperature of the leg, and the peripheral conduction speed. The first negative peak recorded at the lumbosacral area normally has a latency of about 17–21 msec after stimulation of the posterior tibial nerve at the ankle (Figure

Figure 18.1. Schematic diagram of normal SEPs to stimulation of the posterior tibial nerve at the ankle. Tracings, from bottom to top, show popliteal fossa potential, lumbar and low thoracic spinal potentials, and scalp SEP. Negativity at the electrodes connected with a solid line in the diagram at the left is plotted upward.

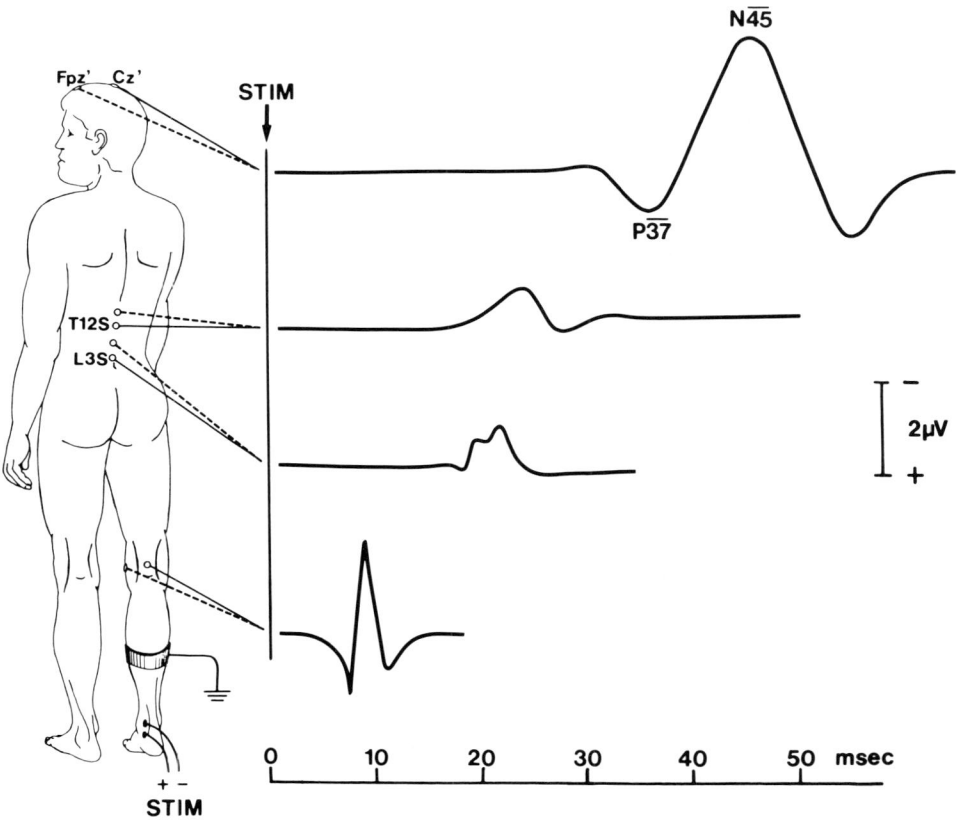

Figure 18.2. Schematic diagram of the normal SEPs to stimulation of the common peroneal nerve at the knee. Tracings, from bottom to top, show lumbar, low thoracic and middle thoracic spinal potentials, and scalp SEP. Negativity at the electrodes connected with a solid line in the diagram at the left is plotted upward.

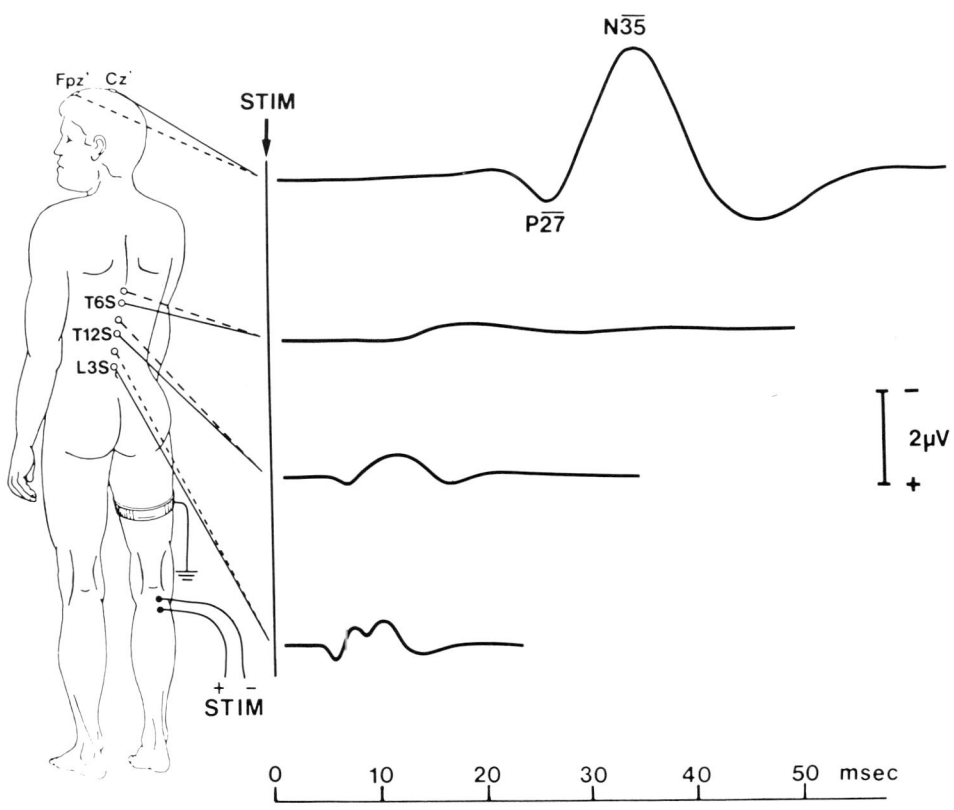

TABLE 18.1. SEPs to stimulation of the common peroneal nerve (CPN) at the knee or the posterior tibial nerve (PTN) at the ankle

A. Subject variables as in Table 16.1 except for:
 1. Age: Conduction velocities vary in children; age ranges for control groups have not yet been standardized.
B. Stimulus as in Table 16.1 except for:
 1. Electrode placement
   a. For PTN stimulation: Cathode behind medial malleolus; anode 3 cm distal.
   b. For CPN stimulation: Cathode in lateral popliteal fossa; anode 3 cm distal.
 2. Intensity
   a. For PTN stimulation: Above threshold for twitch causing plantar toe flexion.
   b. For CPN stimulation: Above threshold for muscle twitch causing plantar flexion and eversion of the foot.
 3. Unilateral stimulation: Each side should be stimulated separately in routine recordings. Simultaneous bilateral stimulation is used during surgical monitoring to enhance SEP amplitude.
C. Recording. As in Table 16.1 except for:
 1. Electrode placement
   a. For PTN stimulation: Over the tibial nerve in the upper middle popliteal fossa (PF) with a reference electrode on the medial surface of the knee; over spinous processes of L3 and T12 vertebrae and on the scalp as for CPN stimulation except for the top spinal pair; ground electrode on the calf.
   b. For CPN stimulation: Over L3, T12, and T6 spinous processes (L3S, T12S, T6S) with reference electrodes 4 cm rostral to each of these three electrodes; on the scalp 2 cm posterior to Cz (Cz′) with a reference electrode midway between Fpz and Fz (Fpz′); ground electrode at midthigh level.
 2. Montages
   a. For PTN stimulation:
     Channel 1: Cz′–Fpz′
     Channel 2: T12S–electrode 4 cm rostral
     Channel 3: L3S–electrode 4 cm rostral
     Channel 4: PF–medial surface of knee
   b. For CPN stimulation:
     Channel 1: Cz′–Fpz′
     Channel 2: T6S–electrode 4 cm rostral
     Channel 3: T12s–electrode 4 cm rostral
     Channel 4: L3S–electrode 4 cm rostral
 3. Number of responses averaged: 1,000–4,000.
 4. Sweep length:
   a. For PTN stimulation: 60–80 msec; 100–200 msec for delayed SEPs.
   b. For CPN stimulation: 40–60 msec; 100–200 msec for delayed SEPs.
D. Analysis
 1. Normal peaks
   a. For PTN stimulation: PF potential, L3 and T12 spine potentials, P37 and N45 peaks of scalp SEPs.
   b. For CPN stimulation: L3, T12, and T6 spine potential, P27 and N35 peaks of scalp SEPs.
 2. Criteria of abnormal
   a. Absence of all spine and scalp SEPs in recordings including 100–200 msec sweeps.

TABLE 18.1. *(continued)*

b. Extremely slow peripheral conduction velocity from stimulus cathode to PF peak and to L3 peak for PTN stimulation and from stimulus cathode to L3 peak for CPN stimulation.

c. Abnormally slow central conduction velocity from L3 and T12 to scalp $P\overline{37}$ peak for PTN stimulation and from L3, T12, and T6 spine potential to scalp $P\overline{27}$ peak for CPN stimulation.

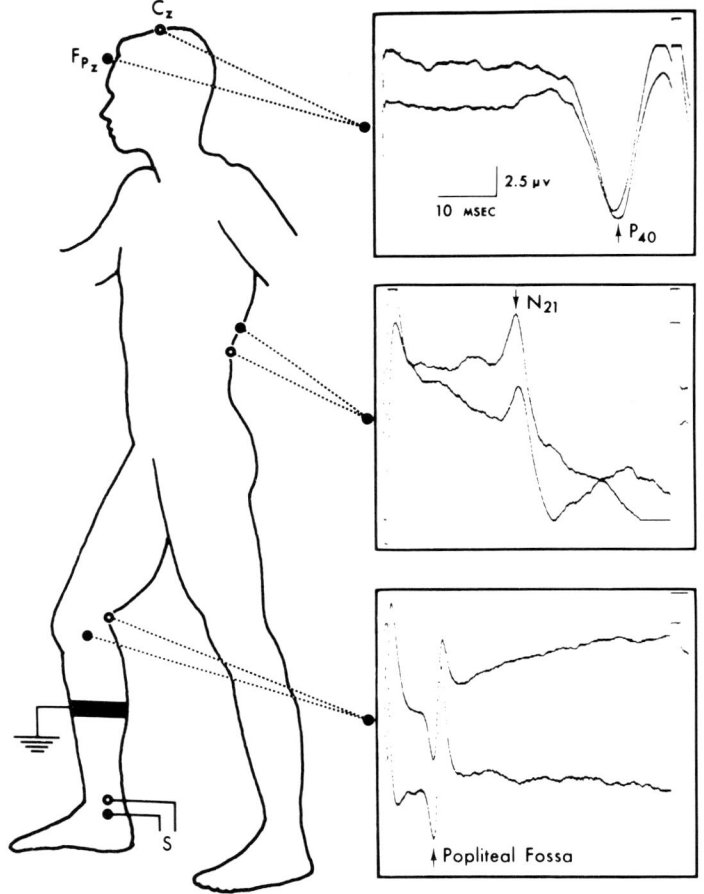

18.3) and of about 9–12 msec after stimulation of the common peroneal nerve at the knee (Figure 18.4). This peak is used to calculate the lumbar-scalp conduction time.

The small positive peak,[172] and perhaps also the first negative wave,[59,65] of the lumbar SEP probably originate from the dorsal roots of the cauda equina. The second

Figure 18.3. Normal SEPs to stimulation of the posterior tibial nerve at the ankle. Recordings from the popliteal fossa (bottom tracings) show a negative peak at about 10 msec preceded by a positive peak. Recordings from the low thoracic spine (middle tracings) contain a negative peak at about 21 msec (N_{21}). Scalp recordings (top tracings) show a P_{40} peak. Two tracings are superimposed for each recording site. Negativity at the recording electrodes marked with open circles is plotted upward. From Eisen and Odusote[101] with permission by the authors and Elsevier Scientific Publishers Ireland Ltd.

negative peak probably represents postsynaptic cord elements[65,81,82,254]; ventral root discharges can be recorded with special methods.[261] The generators of these SEPs have been further investigated with epidural and intrathecal recordings.[59,82,111,170,293]

The low thoracic SEP consists of one or more peaks which probably reflect activity of intramedullary continuations of dorsal root fibers followed by synaptic and postsynaptic spinal activity.

Figure 18.4. Normal SEPs to stimulation of the common peroneal nerve at the knee. Recordings from lumbar and thoracolumbar electrodes show small negative peaks at about 11 and, possibly, 13 msec (bottom two tracings). Scalp recordings show $\overline{P27}$ and $\overline{N34}$ peaks (top two tracings). The beginning of the tracing is delayed by a few milliseconds against the stimulus to eliminate the stimulus artifact. Negativity at the first electrode in the pairs indicated at the left margin is plotted upward.

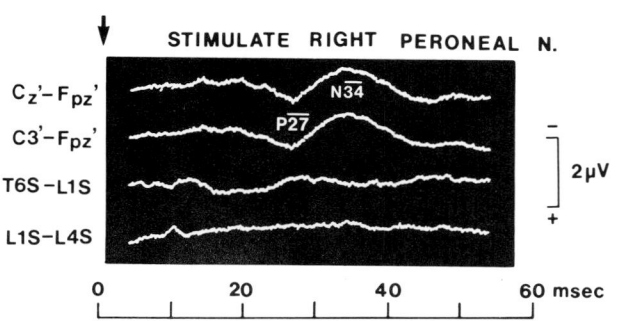

18.1.3 The scalp SEP

Scalp SEPs to leg stimulation are recorded with an electrode in an area up to 3 cm behind the vertex, in the midline, or up to a few centimeters from the midline, usually opposite to the stimulated leg. The guidelines of the American EEG Society[12] recommend midline electrode placements, one 2 cm behind the Cz position of the International 10–20 System (Cz'), the other one midway between Fpz and Fz (Fpz').

Stimulation of the posterior tibial nerve at the ankle produces a parietal $\overline{P37}$-$\overline{N45}$ complex (Figure 18.3); stimulation of the common peroneal nerve at the knee and recording from the scalp produces a positive-negative complex, called $\overline{P27}$-$\overline{N35}$ (Figure 18.4). The actual normal values vary considerably with body height and other factors. In some instances, the first positive peak is preceded by a small negative potential.[26,101,263,341] The early positive-negative complex may be followed by later, more variable peaks.[25,127,334]

The distribution of the positive-negative complex is restricted to the postcentral area, while later peaks have a wider distribution.[334] Although usually recorded at the midline or slightly contralateral to the stimulated leg, the scalp SEP to leg stimulation has been reported to have higher amplitude over the ipsilateral scalp, suggesting that the electric field generated in the contralateral leg area of the parietal cortex is so oriented that it produces higher potential gradients over the ipsilateral hemisphere.[62,334,336]

Recordings between a central scalp electrode and extracranial reference electrodes show small and variable earlier peaks, probably representing far-field SEPs from subcortical cerebral structures.[74,175,203,267,281]

18.2 SUBJECT VARIABLES

18.2.1 Age

Measurements of SEPs at different levels of the spinal cord in infants and children have suggested that conduction velocities of peripheral nerves and spinal cord in newborns are about half those of adults; adult velocities are reached at the age of about 3 years for peripheral conduction and at about 4–5 years for cord conduction.[54,59] Although peripheral conduction velocity decreases steadily throughout adulthood, spinal cord conduction velocity has been reported to show little change up to the age of 60 and then to decrease sharply.[86]

18.2.2 Leg length, body height, temperature, and sensation

These variables have the same effect on leg SEPs as on arm SEPs (16.2), except that body height is more important for leg SEPs than for arm SEPs and therefore must be taken into account by calculating central conduction velocities (18.5).

18.3 EFFECT OF STIMULUS PARAMETERS

18.3.1 Stimulus electrode placements

Stimulus electrode placements for each nerve are fairly well standardized. For stimulation of the posterior tibial nerve, one electrode is placed midway between the medial border of the Achilles tendon and the posterior border of the medial malleolus. This electrode is connected to the negative pole of the stimulator and serves as the cathode. Another electrode is placed 3 cm distal to the cathode and connected to the positive pole of the stimulator, serving as the anode. A band electrode around the calf or a plate electrode on the calf is used as a ground lead (Figure 18.1). For stimulation of the common peroneal nerve at the knee, the cathode is placed over the lateral part of the popliteal fossa, just medial to the tendon of the biceps femoris and below the leg crease. The tendon of the biceps femoris can be easily seen when the subject bends his knee against resistance. The anode is placed 3 cm distal to the cathode. A plate or band electrode on the midthigh is used as a ground lead (Figure 18.2). Purely sensory fibers may be excited by stimulating the sural nerve at the lateral malleolus, the superficial peroneal nerve a handbreadth above the lateral malleolus, or the saphenous nerve above and anterior to the medial malleolus.[98] Segmental input from L2-S1 can be tested by stimulating various nerve branches; L2, L3 can be tested by stimulating the lateral femoral cutaneous nerve at the thigh; L3, by stimulating the saphenous nerve at the medial side of the knee[322]; L4, by stimulating the saphenous nerve below the inner malleolus; L5, by stimulating the superficial peroneal nerve on the lateral aspect of the leg; and S1, by stimulating the sural nerve above the lateral malleolus.[99] Segments not well represented by sensory branches may be studied by direct stimulation of the skin of the corresponding dermatomes.[173,277]

18.3.2 Stimulus intensity, rate, and duration

The motor threshold for posterior tibial nerve stimulation at the ankle is marked by plantar flexion of the toes; the threshold for stimulation of the common peroneal nerve at the knee is indicated by plantar flexion and eversion of the foot. Procedures for selecting stimulus intensity for mixed nerves in case of peripheral nerve lesions, and for sensory nerve stimulation, are described above (15.3.1.1.3).

An increase of stimulus intensity does not change the latency of the lumbar SEP but increases its amplitude to a maximum.[254]

Stimulus rate is 4–7/sec; stimulus duration is usually 200–300 μsec (15.3).

18.4 RECORDING PARAMETERS

Most recording parameters (15.4), including electrode placements (18.1), have been described and summarized (Table 18.1). About 1,000–4,000 responses are averaged for SEPs to stimulation of the posterior tibial and the common peroneal nerve. Larger numbers may be needed for the small SEPs to stimulation of small nerves or dermatomes. The sweep length is 60–80 msec for posterior tibial nerve stimulation and 40–60 msec for common peroneal nerve stimulation. Longer sweeps are needed for abnormally delayed SEPs. The requirements for the temporal resolution of SEP waves during these sweeps pose no problems for larger averagers, but come close to the limits of averagers with a small number of points: In an averager with only 250 points per channel, a sweep length of 100 msec results in a dwell time of 0.4 msec, or a sampling rate of 2,500 Hz, which can resolve signals up to 1,250 Hz (3.3.4); this is barely sufficient for the brief waves contained in SEPs.

18.5 GENERAL STRATEGY OF STIMULATING AND RECORDING

SEPs to posterior tibial nerve stimulation recorded from the popliteal fossa, lumbothoracic spine, and scalp allow distinction of lesions in three segments: (1) Lesions of the distal peripheral nerve affect the popliteal fossa potential and later SEPs; (2) lesions of the proximal peripheral nerve and cauda equina leave the popliteal fossa potential intact but affect the lumbothoracic and scalp SEPs; and (3) lesions of the spinal cord, brain stem, or cerebral hemispheres leave the lumbar SEP intact but delay or abolish the scalp SEP. SEPs to stimulation of the common peroneal nerve, recorded at lumbothoracic and scalp electrodes, can distinguish lesions only in segments (2) and (3). In any segment, lesions produce SEP abnormalities only if they involve specific parts of the somatosensory pathway, namely, the sensory fibers contained in peripheral nerves, plexus, or spinal roots, the dorsal columns of the spinal cord, the medial lemniscus of the brain stem and diencephalon, the nucleus ventralis posterolateralis of the thalamus, the thalamic radiation, and the leg area of the somatosensory cortex.

A lesion in any segment either abolishes or delays all potentials generated proximally. The increase in latency is best referred to the length of the segment and expressed in terms of peripheral and central conduction velocity (19.1). Whereas peripheral conduction velocity approximately equals the speed of conduction of the majority of the fast fibers in a peripheral nerve, the term central conduction velocity is not entirely appropriate in this context because it is used to denote values that represent not only axonal conduction but also synaptic transmission (19.1.3). Nevertheless, it is necessary to use a measure that relates central conduction time to the conduction distance. In contrast to the rather short central conduction distances for SEPs to arm stimulation, the central conduction distances for SEPs to leg stimulation, extending from the lower spine to the vertex, are fairly long and vary considerably with the subject's height; these variations affect both the latencies and central conduction times of leg SEPs.[192]

Recordings from electrodes over the cervical and upper thoracic spine do not usually yield SEPs in adults although they often show well-defined peaks in infants[54] and then permit calculation of spinal conduction times separately from brain stem and cerebral conduction times. In adults, conduction times through the higher segments of the somatosensory pathway have been estimated indirectly from F and M waves[85,100] or H responses[294] recorded in conjunction with the scalp SEP. Conduction times between lumbar and thoracic electrodes are too variable in normals to be of clinical value for estimating conduction velocity between cord segments. Epidural, intrathecal, and esophageal recordings suggest spinal conduction velocities of 35–85 m/sec.[74,110,170,212,218]

The level of a lesion may be evaluated by recording both arm and leg SEPs. Lesions of the upper cervical cord or higher structures may cause abnormalities of both kinds of SEPs, whereas lesions below the upper cervical cord may render abnormal the SEPs to leg stimulation while leaving intact the SEPs to arm stimulation. More detailed information on the level of lesions can be obtained with the painstaking study of segmental input from various nerves and dermatomes (18.3.1).

19
Abnormal SEPs to leg stimulation

SUMMARY

19.1. Criteria distinguishing abnormal from normal SEPs to leg stimulation are the absence of potentials at all recording sites and the delay of peaks at different sites expressed relative to conduction distances as peripheral and central conduction velocities.

19.2. General clinical interpretation of abnormal SEPs uses abolition of SEPs and slowing of conduction velocities to detect lesions involving somatosensory fibers in the peripheral nerve, cauda equina, spinal cord, brain stem, and hemispheres.

19.3. Peripheral neuropathies and radiculopathies are occasionally studied with SEPs to leg stimulation.

19.4. Disorders of the spinal cord detected by abnormalities of these SEPs include multiple sclerosis, spinal cord injury, tumors, and degenerative diseases. SEPs can be used to monitor the condition of the spinal cord during spinal surgery.

19.5. Cerebral lesions have been studied only rarely.

19.1 CRITERIA DISTINGUISHING ABNORMAL SEPs

19.1.1 Absence of SEPs

The absence of SEPs at all recording levels, namely, the popliteal fossa, lumbar and thoracic spinal column, and the scalp, is abnormal if technical problems are excluded and if abnormally delayed peaks have been looked for with sweeps of up to 200 msec. However, the absence of spinal SEPs cannot be considered abnormal if scalp SEPs of normal latency can be recorded. The absence of scalp SEPs with preserved normal spinal SEPs, although consistent with a lesion above the spinal recording level, cannot be considered a definite abnormality because scalp SEPs often have very low amplitude, making them difficult to record.

19.1.2 Slow peripheral conduction velocity

Peripheral conduction velocity of the common peroneal nerve is calculated by measuring the straight-line distance between the stimulating cathode in the popliteal fossa and the L3S recording electrode, and dividing this distance by the latency from the leading edge of the stimulus pulse to the peak of the L3 potential. Three peripheral conduction velocities may be calculated for posterior tibial nerve stimulation. One is obtained by dividing the distance between the stimulus point and the recording electrode in the popliteal fossa by the latency of the popliteal fossa potential. The other one is determined by dividing the distance between the stimulating electrode and the L3S electrode by the latency of the L3 potential. The third is derived by subtracting the distance to the popliteal electrode from that to the lumbar electrode and dividing this difference by the difference in latencies of the popliteal and lumbar potentials. For instance, if posterior tibial nerve stimulation produces a popliteal potential with a peak at 9 msec and an L3 potential with a latency of 21 msec, and if the distance between stimulus electrode and popliteal fossa electrode measures 40 cm and the distance between stimulus electrode and L3S electrode is 90 cm, the conduction velocity between stimulus electrode and popliteal fossa equals 400 mm divided by 9 msec, or 44 m/sec, and the conduction velocity between stimulus electrode and L3S equals 900 mm divided by 21 msec, or 43 m/sec. The conduction velocity between popliteal fossa and lumbar electrodes equals the distance between these electrodes, i.e., 500 mm, divided by the latency difference of 12 msec, or 42 m/sec.

19.1.3 Slow central conduction velocity

Central conduction velocity is determined by dividing the conduction distances between recording electrodes on the spinal cord and scalp by the differences in the peak latencies at these points, i.e., by the central conduction times. The resulting values do not represent true conduction velocities because the measurements on the skin are somewhat longer than those of the spinal cord[74] and because the travel between the recording points includes synaptic transmission in addition to axonal conduction (18.5). Besides, like peripheral conduction velocity, central conduction velocity is measured to the peak rather than the onset of a deflection and therefore does not reflect the maximum speed of conduction. Furthermore, the electric potentials recorded at different points are composed of different fractions of presynaptic and postsynaptic elements and do not necessarily represent equivalent indicators of the passage of nerve impulses (1.4).

Three central conduction velocities can be calculated for common peroneal nerve stimulation. Straight-line distances are measured from the L3S, T12S, and T6S electrodes to the Cz' electrode and are used as central conduction distances. These central conduction distances are divided by the corresponding central conduction times obtained by subtracting from the peak latency of the $\overline{P27}$ the peak latencies of the L3, T12, and T6 potentials. Two central conduction velocities can be calculated for posterior tibial nerve stimulation by measuring the distances from the two spinal recording electrodes used with this stimulus, namely, L3S and T12S, to Cz' and dividing these distances by the latency differences between the L3 and the $\overline{P37}$ peaks and between the T12 and the $\overline{P37}$ peaks, respectively. For instance, if posterior tibial nerve stimulation produces an L3 potential at a latency of 19 msec, a T12 potential at 21 msec, and a scalp SEP with a $\overline{P37}$ at 37 msec, and if the distance from L3S to Cz' measures 70 cm and that from T12S to Cz' is 60 cm, the central conduction velocity between L3S and Cz' equals 700 mm divided by 18 msec, or 39 m/sec, and the central conduction velocity between T12S and Cz' equals 600 mm divided by 16 msec, or 38 m/sec.

19.2 GENERAL CLINICAL INTERPRETATION OF ABNORMAL SEPs TO LEG STIMULATION

Absence of all SEPs, unless explained by technical problems, indicates a lesion at or below the cauda equina (Table 19.1). A decrease of peripheral conduction velocity has the same significance when technical problems are excluded. Stimulation of the posterior tibial nerve affords recordings from two points over the nerve and therefore permits distinction between distal peripheral neuropathy and proximal peripheral nerve or plexus neuropathy. In the case of distal involvement, the popliteal fossa potential and all subsequent SEPs may be absent, or peripheral conduction velocity to the popliteal fossa electrode may be slowed, whereas in the case of more proximal lesions, SEPs at the lumbar electrode may be absent, or peripheral conduction to that electrode may be abnormally slow and popliteal fossa potentials may be normal. With either posterior tibial or common peroneal nerve stimulation, the abolition of spinal potentials above a normal L3 potential raises the suspicion of a high lumbar or low thoracic spinal lesion but cannot prove such a lesion because cord potentials are highly

TABLE 19.1. Clinical interpretation of abnormal SEPs to stimulation of the common peroneal nerve (CPN) at the knee or the posterior tibial nerve at the ankle

Abnormal SEP finding	Interpretation
A. Technical problems	
1. Absent SEPs to leg stimulation at all recording levels	Lack of stimulus; lack of synchronization between stimulus and averager; faulty recording electrodes or equipment
2. Increased latency of SEPs at all recording levels	Hypothermia; inaccurate measurement of the distance between stimulating and recording electrodes
B. Lesions of the nervous system	
1. PTN stimulation	
a. Absent PF potential with absent or normal L3 potential and normal scalp SEP	Normal
b. Absent PF potential with either absent spinal potentials and absent scalp SEPs or normal central conduction velocities	Lesion between ankle and PF
c. Decreased peripheral conduction velocity to PF and	Defect below cauda equina
(1) equally decreased peripheral conduction velocity to L3	Defect of both distal and proximal peripheral nerve
(2) no decrease of peripheral conduction velocity between PF and L3S	Defect between ankle and PF
d. Decreased peripheral conduction velocity to L3S with normal peripheral conduction velocity to PF	Lesion between PF and cauda equina
e. Absent L3 and T12 potentials, absent or delayed scalp SEP with normal peripheral conduction velocity to PF	Probably lesion between PF and cauda equina
2. CPN stimulation	
a. Absent L3 potential with present or absent T12 and T6 potentials and normal scalp SEP	Normal
b. Absent L3, T12, and T6 potentials, delayed or absent scalp SEPs	Defect at or above the cauda equina, or both
c. Decreased peripheral conduction velocity to L3	Peripheral defect between PF and cauda equina

TABLE 19.1. *(continued)*

Abnormal SEP finding	Interpretation
3. Either PTN or CPN stimulation	
a. Decreased central conduction velocity	Defect above the cauda equina and below or at the somatosensory cortex
b. Absent scalp SEP	Suspect defect above the cauda equina and below or at the somatosensory cortex
c. Decreased peripheral conduction velocity and decreased central conduction velocity	Lesions above and below the cauda equina, or a single lesion at the cauda equina or lower spinal cord

variable in normal subjects and more difficult to obtain at higher levels. For the same reason, slowing of conduction between cord segments cannot be taken as evidence for lesions between these segments. The absence of scalp SEPs, or a slowing of central conduction velocity, with preserved spinal potentials and normal peripheral conduction velocity, may be due to a lesion above the lumbar spinal cord, in the brain stem, or the cerebral hemisphere opposite the stimulated leg; however, absent scalp SEPs may be due to technical problems with the recording of these normally often very small potentials.

19.3 PERIPHERAL NERVE AND ROOT LESIONS THAT CAUSE ABNORMAL SEPs TO LEG STIMULATION

19.3.1 Peripheral nerve lesions

Peripheral nerve lesions are most conveniently diagnosed by a decreased peripheral conduction velocity. However, scalp SEPs may be easier to record than the diminished distal SEPs and have been used to evaluate peripheral conduction.[24,98,342]

19.3.2 Radiculopathy

Nerve root compression has been found not to alter SEPs reliably.[342] Stimulation of several cutaneous nerves[98,99] or of dermatomes[277] may be needed for a clear identification of the involved roots, but conventional EMG probably has greater diagnostic power, especially in cases with motor defects.[99]

19.4 LESIONS OF SPINAL CORD AND BRAIN STEM THAT CAUSE ABNORMAL SEPs TO LEG STIMULATION

19.4.1 Multiple sclerosis

Leg SEPs are often abnormal due to lesions at spinal or supraspinal levels which may cause an increased latency of the scalp SEP with a normal lumbar SEP, i.e., an increased lumbar-scalp conduction time and decreased central conduction velocity (Figures 19.1 and 19.2). The scalp SEP may be entirely abolished.[29,87,100,101] SEPs to leg stimulation are more often abnormal in multiple sclerosis than are SEPs to arm stimulation or other EPs, probably because they test a longer pathway which is more likely to be affected by the scattered lesions of multiple sclerosis.[21,101,181,288,331,333]

19.4.2 Spinal cord injury

Complete interruption of ascending spinal pathways abolishes SEPs above the lesion.[55] Incomplete cord lesions, especially those reducing joint position sense, abolish or delay SEPs.[89] SEPs both elicited and recorded either below or above the lesion may be preserved. For instance, the lumbar SEP to leg stimulation and SEPs to arm stimulation may be present in lesions above the cauda equina and below the cervical spinal cord. However, in some cases the SEP is abnormal below the level of a lesion.[195]

SEPs may recover before the clinical condition improves. Presence of an SEP soon after injury, early return, and progressive normalization of the SEP waveform usually indicate a favorable prognosis.[46,253,269,303,357]

19.4.3 Spinal cord compression

Compression of the spinal cord by cervical spondylosis, extramedullary and intramedullary tumors, and Hodgkin's disease may reduce[148,243,294] or delay[243] the cortical SEP if these lesions interfere with position sense, but the SEP remains intact in many cases of extramedullary spinal lesions.[264,342] In the Brown-Sequard syndrome, the SEP is abnormal on stimulation of the side with the decreased vibration and position sense, but not on stimulation of the other side.[127,148] The abnormal SEP may persist after re-

Figure 19.1. Scalp SEPs to stimulation of the posterior tibial nerve in a patient with possible multiple sclerosis. Stimulation of the right tibial nerve produces a normal SEP (top tracings). Stimulation of the left tibial nerve produces an SEP showing an absent or delayed first positive peak (bottom tracing). Two tracings are superimposed for each SEP. Negativity at Cz is plotted upward. This 54-year-old woman has progressive spinal multiple sclerosis. From Eisen and Odusote[101] with permission by the authors and Elsevier Scientific Publishers Ireland Ltd.

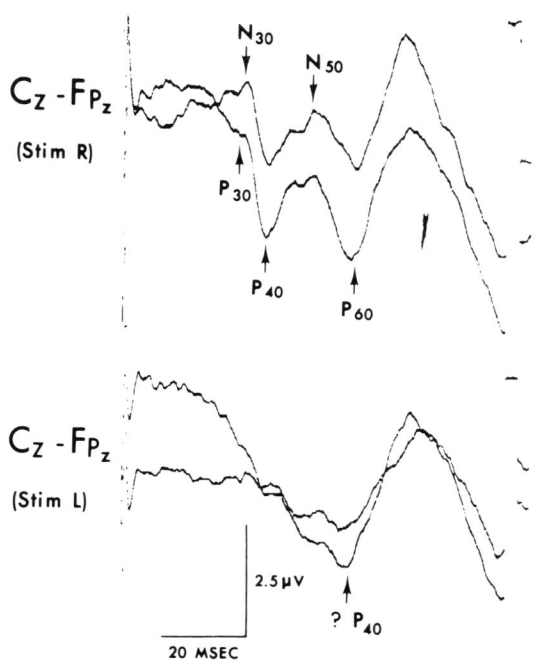

Figure 19.2. SEPs to common peroneal nerve stimulation in a patient with possible multiple sclerosis. Stimulation of the left common peroneal nerve (arrow) elicits a negative peak of over 10 msec in the lumbar recording (bottom tracing) and of about 15 msec in the lumbothoracic recording (third tracing). Scalp recordings show only a late negative wave, with a peak at over 40 msec (top two tracings). Stimulation and recording methods as in Figure 18.4. This 16-year-old boy developed progressive weakness and numbness of both legs and urinary retention. He had hyperreflexia of both legs, slightly decreased touch and pain sensation up to the level of L3, and markedly decreased vibration and position sense. Myelogram and CSF were normal.

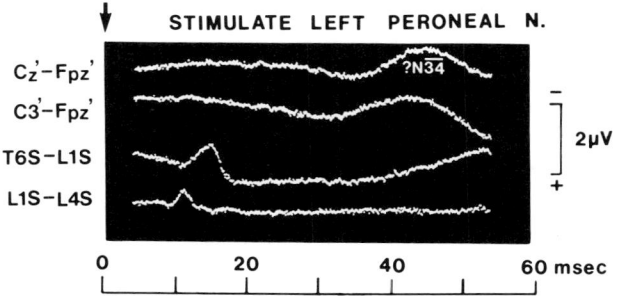

moval of the lesion and after recovery of normal sensory function.[148]

19.4.4 Charcot-Marie-Tooth disease, Friedreich's ataxia, olivopontocerebellar degeneration, Adie's syndrome, and tabes dorsalis

Scalp SEPs are often severely abnormal.[25,149] Central conduction times have been found to be normal in peroneal muscular atrophy but increased in Friedreich's ataxia.[276]

19.4.5 Myotonic dystrophy

Increases of SEP latency suggest motor, sensory, or mixed nerve involvement in many cases.[234]

19.4.6 Subacute combined degeneration of the spinal cord

SEPs are commonly abnormal.[115] They show significant delays[342] or are abolished with marked myelopathy and return after treatment.[184]

19.4.7 Diabetes mellitus

Diabetics with mild or no clinical signs of polyneuropathy may show slowing of conduction not only in peripheral nerves but also in the spinal cord, suggesting subclinical dysfunction of the posterior columns.[53,143]

19.4.8 Subacute myelo-opticoneuropathy

Clioquinol intoxication, reported from Japan, leads to widespread CNS involvement, including marked posterior column damage; this affects central conduction of leg SEPs.[287]

19.4.9 Degenerative CNS diseases in children

Spinal cord involvement has been reported to produce abnormal SEPs in children with various degenerative diseases[52] (Figure 19.3). Scalp SEPs are often delayed in adrenoleukodystrophy[126] and Friedreich's ataxia (19.4.4).

19.4.10 Surgical monitoring of spinal cord condition

Compression, traction, and ischemia of the spinal cord during surgery may impair conduction through sensory pathways and cause SEP changes even before producing lasting damage. The scalp SEP to leg stimulation may therefore be used as a monitor of the condition of the spinal cord during spinal surgery. A deterioration of the SEP can indicate impending cord damage if other causes such as

Figure 19.3. Spinal (bottom 4 tracings) and scalp (top tracing) SEPs in a normal child (left) and a child with degenerative CNS disease (right). In the normal child, SEPs can be seen at all spinal levels and in the scalp recording. In the patient, SEPs disappear at rostral spinal levels, and no scalp SEP is recorded. Bilateral peroneal nerve stimulation at the knee. Negativity at the first electrode named in each pair at the left of the tracings is plotted upward. SEPs in the patient have shorter latency because of a shorter recording distance. From Cracco et al.[52] with permission by the authors and Elsevier Scientific Publishers Ireland Ltd.

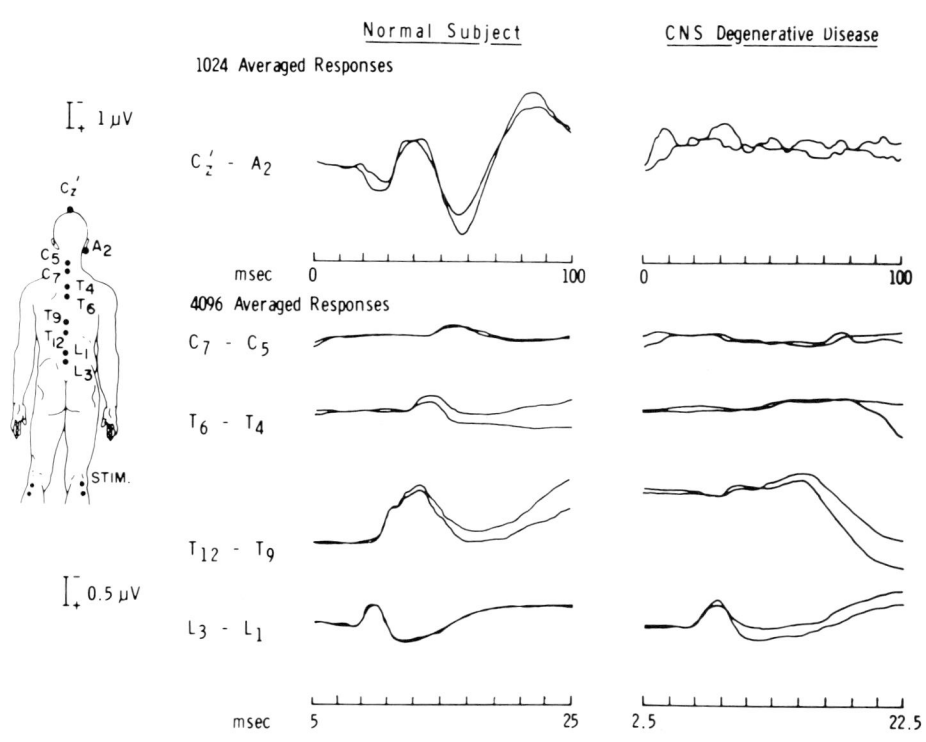

changes in the level of anesthesia, of blood pressure, and of ventilation are excluded.[8,33,109,142,210,240,349] Recordings from electrodes placed into the epidural space,[170] inserted into the interspinal ligament,[204] or implanted into the spinous processes[208] above the operative site have also been used for monitoring and are less susceptible to changes of blood pressure and anesthetic levels than scalp recordings.[167,213,262]

19.5 CEREBRAL LESIONS THAT CAUSE ABNORMAL SEPs TO LEG STIMULATION

Parasagittal tumors have been reported to abolish the $\overline{N70}$ of the scalp SEP to leg stimulation.[95] The amplitude of the scalp SEP to leg stimulation is reduced in patients with Huntington's disease and in many subjects at risk[244] for this disease.

20
Other SEPs

SUMMARY

20.1. Sensory nerve action potentials are elicited by stimulation of sensory nerve fibers and recording at a point proximal to the stimulus. They may be used to evaluate peripheral sensory conduction.

20.2. SEPs to stimulation of the trigeminal nerve reflect excitation of that nerve and its central connections.

20.3. SEPs to pudendal and bladder stimulation may become useful in the analysis of sexual, bowel, and bladder dysfunctions.

20.4. SEPs to adequate stimuli include SEPs to touch and vibrotactile stimuli, joint movement, muscle stretch and vibration, pain, and temperature changes.

20.5. Somatomotor EPs are due to muscle contractions elicited by sensory stimuli and are similar to sonomotor AEPs in that they may contaminate recordings of cerebral and spinal SEPs from scalp and neck.

20.1 SENSORY NERVE ACTION POTENTIALS

Sensory nerve action potentials (SNAPs) are not used routinely with SEP recordings but may be studied in cases requiring investigation of sensory nerve function (Figure 17.2). In these cases, stimulus electrodes are placed over sensory nerves or sensory branches of mixed nerves, and recording electrodes are placed proximally over the same nerve.[34,129] The amplitude of SNAPs is usually low even after averaging large numbers of responses and may be increased by recording with fine needle electrodes inserted subcutaneously close to the nerve.[66]

SNAPs have a major negative peak, usually preceded and followed by minor positive peaks, with a latency depending on conduction distance and fiber speed. This may be followed by later peaks representing fibers of lower conduction speed. The latency of SNAPs may decrease with increasing stimulus intensity, presumably because of wider spread of stimulus current along the stimulated nerve. Sensory conduction velocity is calculated by dividing the distance from the stimulating to the recording electrode by the latency of the negative SNAP peak. Conduction in sensory nerve fibers can also be measured during recording of arm or leg SEPs if sensory nerves or nerve branches are used for stimulation. Recordings from clavicular and lumbosacral electrodes then represent SNAPs.

20.2 SEPs TO TRIGEMINAL NERVE STIMULATION

Electric stimulation of the lips, gums, or mental nerve produces scalp SEPs consisting of a series of peaks which probably represent the sensory pathway from Gasser's ganglion to the cerebral cortex[23,90,114,298,309] (Figures 20.1 and 20.2). Trigeminal SEPs may be abnormal in multiple sclerosis,[35,102] trigeminal neuralgia[35,310] (Figure 20.2), brain stem infarcts and tumors,[35] after Gasserian thermocoagulation leading to hypesthesia,[273] in acoustic neurinoma, and in sarcoidosis involving the trigeminal nerve.[298]

20.3 SEPs TO PUDENDAL AND BLADDER STIMULATION

SEPs to stimulation of mucous membrane in the distribution of the pudendal nerve[144] and inside the bladder[20] have been explored in normal subjects with the expectation that these SEPs may become useful in the evaluation of patients with sexual, bladder, or bowel dysfunctions.

20.4 SEPs TO ADEQUATE STIMULI

In contrast to the routine use of adequate stimuli for VEPs and AEPs, such stimuli have been used only rarely for

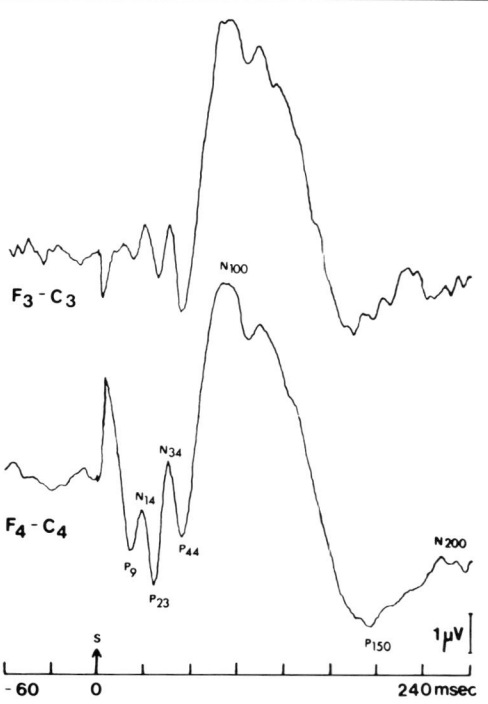

Figure 20.1. Normal trigeminal SEPs. Stimulation of the left trigeminal nerve produces a complex left (top tracing) and right (bottom tracing) scalp SEP with several deflections representing subcortical and cortical peaks. Stimulation of the third trigeminal branch at the mental foramen. Recordings between frontal and central electrodes. Negativity at the frontal electrode is plotted upward. From Drechsler[90] with permission by the author and MTP Press Ltd.

SEPs. Since SEPs are mediated through the lemniscal system, touch is the most effective adequate sensory stimulus. Even though recordings have been made in a few neurological disorders, no routine diagnostic tests have as yet been developed.

20.4.1 Touch

Sudden displacement of the skin or of a fingernail elicits clavicular, cervical, and scalp SEPs similar to those produced by electric stimulation of the corresponding nerve except that the peaks are generally less clearly defined.[162,163,176,191,237,255,256,257,302] Short stimuli of 5 msec are as effective as longer ones, and stimulus rates of 8/sec and more may be used.[255] Latency varies with stimulus site and is very short for the tongue.[162] Stimulation of the trunk and the proximal parts of arms and legs tends to give bilateral responses.[162,191,236,237] SEPs to mechanical stimulation may be abnormal in patients with peripheral nerve and central sensory lesions[236] but are normal in persons with hypnotically induced anesthesia.[147]

20.4.2 Vibrotactile stimuli

Vibratory stimulation of the skin is less suitable for producing discrete short-latency SEPs than is an intermittent touch stimulus,[150,163,164] but the effect of vibratory stimuli can be studied by their interference with SEPs to concurrent electric nerve stimuli.[2,166]

20.4.3 Joint movement

Passive displacement of fingers produces scalp SEPs with latencies comparable to those of electrically induced SEPs and probably mediated by joint capsule afferents.[249] Active and passive joint movements modulate SEPs to electric stimulation.[6]

Figure 20.2. Trigeminal SEPs in a normal subject (left) and a patient with trigeminal neuralgia (right). In the normal subject, SEPs from the right side (RS) and left side (LS) of the scalp to stimulation of the contralateral lips show a positive peak at about 20 msec. In the patient with trigeminal neuralgia, stimulation of the intact left side produces a normal SEP over the right side of the head (RS) whereas stimulation of the affected right side produces an SEP with a peak of lower amplitude and longer latency over the left side (LS). Two tracings are shown for each SEP. Simultaneous stimulation of upper and lower lips opposite the recording site. Recordings between C5 or C6 and an Fz reference electrode of the International 10–20 System. Negativity at the central electrodes is plotted upward. From Stöhr et al.[310] with permission by the authors and Little, Brown and Company.

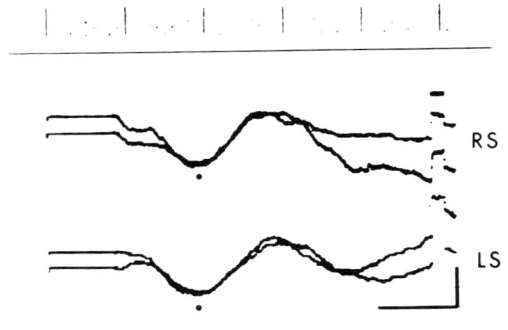
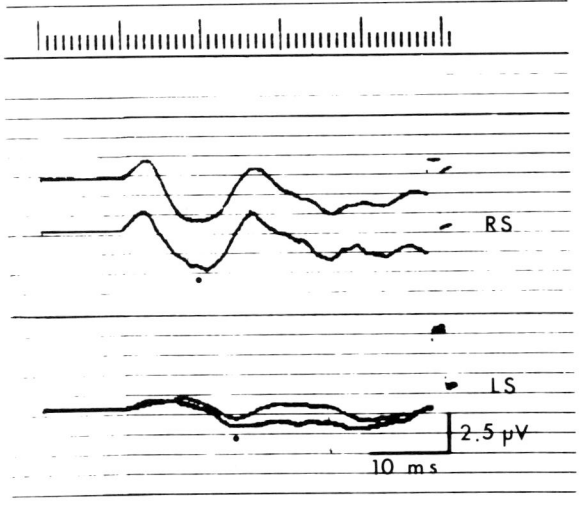

20.4.4 Muscle stretch

Sudden muscle stretch elicits scalp SEPs of short latency, probably mediated by proprioceptive afferents.[153,304]

20.4.5 Pain stimuli

Painful radiant heat to the skin[30,39] or electric stimuli to peripheral nerves[312] produce SEPs different from those to painless stimuli. However, electric stimulation of the skin can probably not excite C-fibers in numbers sufficient for an SEP without producing tissue damage. Electric shocks to the tooth pulp, generating a pure pain stimulus, cause long-latency cortical SEPs[42,43,47] that do not correlate precisely with perceived pain.[50,61]

20.4.6 Warm and cold stimuli

Changes in skin temperature, although difficult to control precisely, produce SEPs.[45,117]

20.5 SOMATOMOTOR EPs

Like click stimuli, electric stimulation of mixed or sensory nerves can induce muscle responses that may begin about 15 msec after the stimulus, especially in contracted muscles, and may contaminate recordings of cerebral and spinal SEPs from scalp and neck.[37,56,57] These potentials can often be distinguished from SEPs by their disappearance with relaxation and by their greater prominence at the periphery of the scalp near larger muscle groups.

References for part D

1. Abbruzzese, G., Abbruzzese, M., Cocito, L., Favale, E., Leandri, M., and Ratto, S. 1980. Conduction time of the lemniscal pathway in males and females. Acta Neurol. Scand. 62:132–136.
2. Abbruzzese, G., Abbruzzese, M., Favale, E., Ivaldi, M., Leandri, M., and Ratto, S. 1980. The effect of hand muscle vibration on the somatosensory evoked potential in man: An interaction between lemniscal and spino-cerebellar inputs? J. Neurol. Neurosurg. Psychiatry 43:433–437.
3. Abbruzzese, G., Cocito, L., Ratto, S., Abbruzzese, M., Leandri, M., and Favale, E. 1981. A reassessment of sensory evoked potential parameters in multiple sclerosis: A discriminant analysis approach. J. Neurol. Neurosurg. Psychiatry 44:133–139.
4. Abbruzzese, M., Favale, E., Leandri, M., and Ratto, S. 1978. Spinal components of the cerebral somatosensory evoked response in normal man: The "S wave." Acta Neurol. Scand. 58:213–220.
5. Abbruzzese, M., Favale, E., Leandri, M., and Ratto, S. 1978. New subcortical components of the cerebral somatosensory evoked potential in man. Acta Neurol. Scand. 58:325–332.
6. Abbruzzese, G., Ratto, S., Favale, E., and Abbruzzese, M. 1981. Proprioceptive modulation of somatosensory evoked potentials during active or passive finger movements in man. J. Neurol. Neurosurg. Psychiatry 44:942–949.
7. Alajouanine, T., Scherrer, J., Barbizet, J., Calvet, J., and Verley, R. 1958. Potentiels évoqués corticaux chez des sujets atteints de trouble somesthésiques. Rev. Neurol. 98:757–762.
8. Allen, A., Starr, A., and Nudleman, K. 1981. Assessment of sensory function in the operating room utilizing cerebral evoked potentials: A study of fifty-six surgically anesthetized patients. Clin. Neurosurg. 28:457–481.
9. Allison, T., Goff, W.R., Williamson, P.D., and VanGilder, J.C. 1980. On the neural origin of early components of the human somatosensory evoked potential. Prog. Clin. Neurophysiol. 7:51–68.
10. Allison, T., and Hume, A.L. 1981. A comparative analysis of short-latency somatosensory evoked potentials in man, monkey, cat, and rat. Exp. Neurol. 72:592–611.
11. Allison, T., Wood, C.C., and Goff, W.R. 1983. Brain stem auditory, pattern-reversal visual, and short-latency somatosensory evoked potentials: Latencies in relation to age, sex and brain and body size. Electroencephalogr. Clin. Neurophysiol. 55:619–636.
12. American Electroencephalographic Society. 1984. Guide-

lines for clinical evoked potential studies. J. Clin. Neurophysiol. 1:3–53.
13. Anderson, D.C., Bundlie, S., and Rockswold, G.L. 1984. Multimodality evoked potentials in closed head trauma. Arch. Neurol. 41:369–374.
14. Anziska, B.J., and Cracco, R.Q. 1980. Short latency somatosensory evoked potentials in brain dead patients. Arch. Neurol. 37:222–225.
15. Anziska, B., and Cracco, R.Q. 1980. Short latency somatosensory evoked potentials: Studies in patients with focal neurological disease. Electroencephalogr. Clin. Neurophysiol. 49:227–239.
16. Anziska, B.J., and Cracco, R.Q. 1981. Short latency SEPs to median nerve stimulation: Comparison of recording methods and origin of components. Electroencephalogr. Clin. Neurophysiol. 52:531–539.
17. Anziska, B.J., and Cracco, R.Q. 1983. Short-latency somatosensory evoked potentials to median nerve stimulation in patients with diffuse neurologic disease. Neurology 33:989–993.
18. Anziska, B., Cracco, R.Q., Cook, A.W., and Feld, E.W. 1978. Somatosensory far field potentials: Studies in normal subjects and patients with multiple sclerosis. Electroencephalogr. Clin. Neurophysiol. 45:602–610.
19. Assmus, H. 1980. Somatosensory evoked cortical potentials in peripheral nerve lesions. In: Evoked potentials: Proceedings of an international evoked potentials symposium held in Nottingham, England, ed. C. Barber, pp. 437–442. Lancaster: MTP Press.
20. Badr, G., Carlsson, C.A., Fall, M., Friberg, S., Lindstrom, L., and Ohlsson, B. 1982. Cortical evoked potentials following stimulation of the urinary bladder in man. Electroencephalogr. Clin. Neurophysiol. 54:494–498.
21. Bartel, D.R., Markand, O.N., and Kolar, O.J. 1983. The diagnosis and classification of multiple sclerosis: Evoked responses and spinal fluid electrophoresis. Neurology 33:611–617.
22. Beck, E.C., Dustman, R.E., and Lewis, E.G. 1975. The use of the averaged evoked potential in the evaluation of central nervous system disorders. Int. J. Neurol. 9:211–232.
23. Bennett, M.H., and Jannetta, P.J. 1980. Trigeminal evoked potentials in humans. Electroencephalogr. Clin. Neurophysiol. 48:517–526.
24. Bergamini, L., Bergamasco, B., Fra, L., Gandiglio, G., Mombelli, A.M., and Mutani, R. 1965. Somato-sensory evoked cortical potentials in subjects with peripheral nervous lesions. Electromyography 5:121–130.
25. Bergamini, L., Bergamasco, B., Fra, L., Gandiglio, G., Mombelli, A.M., and Mutani, R. 1966. Réponses corticales et périphériques évoquées par stimulation du nerf dans la pathologie des cordons postérieurs. Rev. Neurol. 115:99–112.
26. Berić, A., and Prevec, T.S. 1981. The early negative potential evoked by stimulation of the tibial nerve in man. J. Neurol. Sci. 50:299–306.
27. Bigum, H.B., Dustman, R.E., and Beck, E.C. 1970. Visual and somato-sensory evoked responses from mongoloid and normal children. Electroencephalogr. Clin. Neurophysiol. 28:576–585.
28. Blair, A.W. 1971. Sensory examinations using electrically induced somatosensory potentials. Dev. Med. Child Neurol. 13:447–455.

29. Bøttcher, J., and Trojaborg, W. 1982. Follow-up of patients with suspected multiple sclerosis: A clinical and electrophysiological study. J. Neurol. Neurosurg. Psychiatry 45:809–814.
30. Bromm, B., Neitzel, H., Tecklenburg, A., and Treede, R.D. 1983. Evoked cerebral potential correlates of C-fibre activity in man. Neurosci. Lett. 43:109–114.
31. Broughton, R., Rasmussen, T., and Branch, C. 1981. Scalp and direct cortical recordings of somatosensory evoked potentials in man (circa 1967). Can. J. Psychol. 35:136–158.
32. Brown, W.F., and Feasby, T.E. 1984. Sensory evoked potentials in Guillain-Barré polyneuropathy. J. Neurol. Neurosurg. Psychiatry 47:288–291.
33. Brown, R.H., and Nash, C.L. 1979. Current status of spinal cord monitoring. Spine 4:466–470.
34. Buchthal, F., Rosenfalck, A., and Behse, F. 1975. Sensory potentials of normal and diseased nerves. In Peripheral neuropathy, ed. P.J. Dyck, P.K. Thomas, and E.H. Lambert, vol. 1, pp. 442–464. Philadelphia: Saunders.
35. Buettner, U.W., Petruch, F., Scheglmann, K., and Stöhr, M. 1982. Diagnostic significance of cortical somatosensory evoked potentials following trigeminal nerve stimulation. Adv. Neurol. 32:339–345.
36. Burke, D., Skuse, N.F., and Lethlean, A.K. 1981. Cutaneous and muscle afferent components of the cerebral potential evoked by electrical stimulation of human peripheral nerves. Electroencephalogr. Clin. Neurophysiol. 51:579–588.
37. Calmes, R.L., and Cracco, R.Q. 1971. Comparison of somatosensory and somatomotor evoked responses to median nerve and digital nerve stimulation. Electroencephalogr. Clin. Neurophysiol. 31:547–562.
38. Carlin, L., Roach, E.S., Riela, A., Spudis, E., and McLean, W.T. 1983. Juvenile metachromatic leukodystrophy: Evoked potentials and computed tomography. Ann. Neurol. 13:105–106.
39. Carmon, A., Mor, J., and Goldberg, J. 1976. Evoked cerebral responses to noxious thermal stimuli in humans. Exp. Brain Res. 25:103–107.
40. Celesia, G.G. 1979. Somatosensory evoked potentials recorded directly from human thalamus and Sm I cortical area. Arch. Neurol. 36:399–405.
41. Chadwick, D., Hallett, M., Harris, R., Jenner, P., Reynolds, E.H., and Marsden, C.D. 1977. Clinical, biochemical, and physiological features distinguishing myoclonus responsive to 5-hydroxytryptophan, tryptophan with a monoamine oxidase inhibitor, and clonazepam. Brain 100:455–487.
42. Chapman, C.R., Chen, A.C.N., and Harkins, S.W. 1979. Brain evoked potentials as correlates of laboratory pain: A review and perspective. Adv. Pain Res. Ther. 3:791–803.
43. Chatrian, G.E., Canfield, R.C., Knauss, T.A., and Lettich, E. 1975. Cerebral responses to electrical tooth pulp stimulation in man: An objective correlate of acute experimental pain. Neurology 25:745–757.
44. Chatrian, G.E., Farrell, D.F., Canfield, R.C., and Lettich, E. 1975. Congenital insensitivity to noxious stimuli. Arch. Neurol. 32:141–145.
45. Chatt, A.B., and Kenshalo, D.R. 1977. Cerebral evoked responses to skin warming recorded from human scalp. Exp. Brain Res. 28:449–455.
46. Chehrazi, B., Parkinson, J., and Bucholz, R. 1981. Evoked somatosensory potentials to common peroneal nerve stimulation in man. J. Neurosurg. 55:733–741.

47. Chen, A.C.N., Chapman, C.R., and Harkins, S.W. 1979. Brain evoked potentials are functional correlates of induced pain in man. Pain 6:365–374.
48. Chiappa, K.H. 1982. Physiologic localization using evoked responses: Pattern shift visual, brainstem auditory, and short latency somatosensory. In: New perspectives in cerebral localization, ed. R.A. Thompson and J.R. Green, pp. 63–114. New York: Raven Press.
49. Chiappa, K.H., Choi, S.K., and Young, R.R. 1980. Short-latency somatosensory evoked potentials following median nerve stimulation in patients with neurological lesions. Prog. Clin. Neurophysiol. 7:264–281.
50. Chudler, E.H. 1983. The assessment of pain by cerebral evoked potentials. Pain 16:221–244.
51. Clifford-Jones, R.E., Clarke, G.P., and Mayles, P. 1979. Crossed acoustic response combined with visual and somatosensory evoked responses in the diagnosis of multiple sclerosis. J. Neurol. Neurosurg. Psychiatry 42:749–752.
52. Cracco, J.B., Bosch, V.V., and Cracco, R.Q. 1980. Cerebral and spinal somatosensory evoked potentials in children with CNS degenerative disease. Electroencephalogr. Clin. Neurophysiol. 49:437–445.
53. Cracco, J., Castells, S., and Mark, E. 1984. Spinal somatosensory evoked potentials in juvenile diabetes. Ann. Neurol. 15:55–58.
54. Cracco, J.B., Cracco, R.Q., and Stolove, R. 1979. Spinal evoked potential in man: A maturational study. Electroencephalogr. Clin. Neurophysiol. 46:58–64.
55. Cracco, R.Q. 1973. Spinal evoked response: Peripheral nerve stimulation in man. Electroencephalogr. Clin. Neurophysiol. 35:379–386.
56. Cracco, R.Q. 1980. Scalp-recorded potentials evoked by median nerve stimulation: Subcortical potentials, traveling waves and somatomotor potentials. Prog. Clin. Neurophysiol. 7:1–14.
57. Cracco, R.Q., and Bickford, R.G. 1968. Somatomotor and somatosensory evoked responses. Arch. Neurol. 18:52–68.
58. Cracco, R.Q., and Cracco, J.B. 1976. Somatosensory evoked potential in man: Far field potentials. Electroencephalogr. Clin. Neurophysiol. 41:460–466.
59. Cracco, R.Q., Cracco, J.B., Sarnowski, R., and Vogel, H.B. 1980. Spinal evoked potentials. Prog. Clin. Neurophysiol. 7:87–104.
60. Crespi, V., Mandelli, A., and Minoli, G. 1982. Short-latency somatosensory-evoked potentials in patients with acute focal vascular lesions of the supratentorial somesthesic pathways. Acta Neurol. Scand. 65:274–279.
61. Cruccu, G., Fornarelli, M., Inghilleri, M., and Manfredi, M. 1983. The limits of tooth pulp evoked potentials for pain quantitation. Physiol. Behav. 31:339–342.
62. Cruse, R., Klem, G., Lesser, R.P., and Lueders, H. 1982. Paradoxical lateralization of cortical potentials evoked by stimulation of posterior tibial nerve. Arch Neurol. 39:222–225.
63. De la Torre, J.C. 1981. Evaluation of brain death using somatosensory evoked potentials. Biol. Psychiatry 16:931–935.
64. De la Torre, J.C., Trimble, J.L., Beard, R.T., Hanlon, K., and Surgeon, J.W. 1978. Somatosensory evoked potentials for the prognosis of coma in humans. Exp. Neurol. 60:304–317.
65. Delbeke, J., McComas, A.J., and Kopec, S.J. 1978. Analysis of evoked lumbosacral potentials in man. J. Neurol. Neurosurg. Psychiatry 41:293–302.

66. Desmedt, J.E. 1970. Somatosensory cerebral evoked potentials in man. In: Handbook of electroencephalography and clinical neurophysiology, vol. 9, ed. A. Rémond, pp. 55–82. Amsterdam: Elsevier.
67. Desmedt, J.E., and Brunko, E. 1980. Functional organization of far-field and cortical components of somatosensory evoked potentials in normal adults. Prog. Clin. Neurophysiol. 7:27–50.
68. Desmedt, J.E., Brunko, E., and Debecker, J. 1976. Maturation of the somatosensory evoked potentials in normal infants and children, with special reference to the early N_1 component. Electroencephalogr. Clin. Neurophysiol. 40:43–58.
69. Desmedt, J.E., Brunko, E., and Debecker, J. 1980. Maturation and sleep correlates of the somatosensory evoked potential. Prog. Clin. Neurophysiol. 7:146–161.
70. Desmedt, J.E., and Cheron, G. 1980. Central somatosensory conduction in man: Neuronal generators and interpeak latencies of the far-field components recorded from neck and right or left scalp and earlobes. Electroencephalogr. Clin. Neurophysiol. 50:382–403.
71. Desmedt, J.E., and Cheron, G. 1980. Somatosensory evoked potentials to finger stimulation in healthy octogenarians and in young adults: Wave forms, scalp topography and transit times of parietal and frontal components. Electroencephalogr. Clin. Neurophysiol. 50:404–425.
72. Desmedt, J.E., and Cheron, G. 1981. Prevertebral (oesophageal) recording of subcortical somatosensory evoked potentials in man: The spinal P_{13} component and the dual nature of the spinal generators. Electroencephalogr. Clin. Neurophysiol. 52:257–275.
73. Desmedt, J.E., and Cheron, G. 1981. Non-cephalic reference recording of early somatosensory potentials to finger stimulation in adult or aging normal man: Differentiation of widespread N18 and contralateral N20 from the prerolandic P22 and N30 components. Electroencephalogr. Clin. Neurophysiol. 52:533–570.
74. Desmedt, J.E., and Cheron, G. 1983. Spinal and far-field components of human somatosensory evoked potentials to posterior tibial nerve stimulation analysed with oesophageal derivations and non-cephalic reference recording. Electroencephalogr. Clin. Neurophysiol. 56:635–651.
75. Desmedt, J.E., and Manil, J. 1970. Somatosensory evoked potentials of the normal human neonate in REM sleep, in slow wave sleep and in waking. Electroencephalogr. Clin. Neurophysiol. 29:113–126.
76. Desmedt, J.E., Manil, J., Borenstein, S., Debecker, J., Lambert, C., Franken, L., and Danis, A. 1966. Evaluation of sensory nerve conduction from averaged cerebral evoked potentials in neuropathies. Electromyography 6:263–269.
77. Desmedt, J.E., and Noël, P. 1973. Average cerebral evoked potentials in the evaluation of lesions of the sensory nerves and of the central somatosensory pathway. In: New developments in electromyography and clinical neurophysiology, vol. 2, ed. J.E. Desmedt, pp. 352–371. Basel: Karger.
78. Desmedt, J.E., Noël, P., Debecker, J., and Namèche, J. 1973. Maturation of afferent conduction velocity as studied by sensory nerve potentials and by cerebral evoked potentials. In: New developments in electromyography and clinical neurophysiology, vol. 2, ed. J.E. Desmedt, pp. 52–63. Basel: Karger.
79. Desmedt, J.E., and Robertson, D. 1977. Differential en-

hancement of early and late components of the cerebral somatosensory evoked potentials during forced-paced cognitive tasks in man. J. Physiol. (Lond.) 271:761–782.
80. Desmedt, J.E., Tran Huy, N., and Carmeliet, J. 1983. Unexpected latency shifts of the stationary P9 somatosensory evoked potential far field with changes in shoulder position. Electroencephalogr. Clin. Neurophysiol. 56:628–634.
81. Dimitrijevic, M.R., Larsson, L.E., Lehmkuhl, D., and Sherwood, A. 1978. Evoked spinal cord and nerve root potentials in humans using a non-invasive recording technique. Electroencephalogr. Clin. Neurophysiol. 45:331–340.
82. Dimitrijevic, M.R., Lehmkuhl, L.D., Segwick, E.M., Sherwood, A.M., and McKay, W.B. 1980. Characteristics of spinal cord-evoked responses in man. Appl. Neurophysiol. 43:118–127.
83. Domino, E.F., Matsuoka, S., Waltz, J., and Cooper, I. 1964. Simultaneous recordings of scalp and epidural somatosensory-evoked responses in man. Science 145:1199–1200.
84. Domino, E.F., Matsuoka, S., Waltz, J., and Cooper, I.S. 1965. Effects of cryogenic thalamic lesions on the somesthetic evoked response in man. Electroencephalogr. Clin. Neurophysiol. 19:127–138.
85. Dorfman, L.J. 1977. Indirect estimation of spinal cord conduction velocity in man. Electroencephalogr. Clin. Neurophysiol. 42:26–34.
86. Dorfman, L.J., and Bosley, T.M. 1979. Age-related changes in peripheral and central nerve conduction in man. Neurology 29:38–44.
87. Dorfman, L.J., Bosley, T.M., and Cummins, K.L. 1978. Electrophysiological localization of central somatosensory lesions in patients with multiple sclerosis. Electroencephalogr. Clin. Neurophysiol. 44:742–753.
88. Dorfman, L.J., Pedley, T.A., Tharp, B.A., and Scheithauer, B.W. 1978. Juvenile neuraxonal dystrophy: Clinical, electrophysiological, and neuropathological features. Ann. Neurol. 3:419–428.
89. Dorfman, L.J., Perkash, I., Bosley, T.M., and Cummins, K.L. 1980. Use of cerebral evoked potentials to evaluate spinal somatosensory function in patients with traumatic and surgical myelopathies. J. Neurosurg. 52:654–660.
90. Drechsler, F. 1980. Short and long latency cortical potentials following trigeminal nerve stimulation in man. In: Evoked potentials: Proceedings of an international evoked potentials symposium held in Nottingham, England, ed. C. Barber, pp. 415–422. Lancaster: MTP Press.
91. Dubois, M., Coppola, R., Buchsbaum, M.S., and Lees, D.E. 1981. Somatosensory evoked potentials during whole body hyperthermia in humans. Electroencephalogr. Clin. Neurophysiol. 52:157–162.
92. Duff, T.A. 1980. Topography of scalp recorded potentials evoked by stimulation of the digits. Electroencephalogr. Clin. Neurophysiol. 49:452–460.
93. Dustman, R.E., Schenkenberg, T., and Beck, E.C. 1976. The development of the evoked response as a diagnostic and evaluative procedure. In: Developmental psychophysiology of mental retardation, ed. R. Karrer, pp. 247–310. Springfield, Ill.: Thomas.
94. Ebner, A., Dengler, R., and Meier, C. 1981. Peripheral and central conduction times in hereditary pressure-sensitive neuropathy. J. Neurol. 226:85–99.

95. Ebner, A., Einsiedel-Lechtape, H., and Lücking, C.H. 1982. Somatosensory tibial nerve evoked potentials with parasagittal tumours: A contribution to the problem of generators. Electroencephalogr. Clin. Neurophysiol. 54:508–515.
96. Ehle, A.L., Steward, R.M., Lellelid, N.A., and Leventhal, N.A. 1984. Evoked potentials in Huntington's disease: A comparative and longitudinal study. Arch. Neurol. 41:379–382.
97. Eisen, A. 1982. The somatosensory evoked potential. Can. J. Neurol. Sci. 9:65–77.
98. Eisen, A., and Elleker, G. 1980. Sensory nerve stimulation and evoked cerebral potentials. Neurology 30:1097–1105.
99. Eisen, A., Hoirch, M., and Moll, A. 1983. Evaluation of radiculopathies by segmental stimulation and somatosensory evoked potentials. Can. J. Neurol. Sci. 10:178–182.
100. Eisen, A., and Nudleman, K. 1979. Cord to cortex conduction in multiple sclerosis. Neurology 29:189–193.
101. Eisen, A., and Odusote, K. 1980. Central and peripheral conduction times in multiple sclerosis. Electroencephalogr. Clin. Neurophysiol. 48:253–265.
102. Eisen, A., Paty, D., Purves, S., and Hoirch, M. 1981. Occult fifth nerve dysfunction in multiple sclerosis. Can. J. Neurol. Sci. 8:221–225.
103. Eisen, A., Purves, S., and Hoirch, M. 1982. Central nervous system amplification: Its potential in the diagnosis of early multiple sclerosis. Neurology 32:359–364.
104. Eisen, A., Stewart, J., Nudleman, K., and Cosgrove, J.B.R. 1979. Short-latency somatosensory responses in multiple sclerosis. Neurology 29:827–834.
105. El-Negamy, E., and Sedgwick, E.M. 1978. Properties of spinal somatosensory evoked potential recorded in man. J. Neurol. Neurosurg. Psychiatry 41:762–768.
106. El-Negamy, E., and Sedgwick, E.M. 1979. Delayed cervical somatosensory potentials in cervical spondylosis. J. Neurol. Neurosurg. Psychiatry 42:238–241.
107. Emerson, R.G., and Pedley, T.A. 1984. Generator sources of median somatosensory evoked potentials. J. Clin. Neurophysiol. 1:203–218.
108. Emerson, R.G., Seyal, M., and Pedley, T.A. 1984. Somatosensory evoked potentials following median nerve stimulation. I. The cervical components. Brain 107:169–182.
109. Engler, G.L., Spielholz, N.I., Bernhard, W.N., Danziger, F., Merkin, H., and Wolff, T. 1978. Somatosensory evoked potentials during Harrington instrumentation for scoliosis. J. Bone Joint Surg. [Am.] 60A:528–532.
110. Ertekin, C. 1976. Studies on the human evoked electrospinogram. II. The conduction velocity along the dorsal funiculus. Acta Neurol. Scand. 53:21–38.
111. Ertekin, C. 1978. Comparison of the human evoked electrospinogram recorded from the intrathecal, epidural and cutaneous levels. Electroencephalogr. Clin. Neurophysiol. 44:683–690.
112. Ervin, F.R., and Mark, V.H. 1964. Studies of the human thalamus: Evoked responses. Ann. N.Y. Acad. Sci. 112:81–92.
113. Favale, E., Ratto, S., Leandri, M., and Abbruzzese, M. 1982. Investigations on the nervous mechanisms underlying the somatosensory cervical response in man. J. Neurol. Neurosurg. Psychiatry 45:796–801.
114. Findler, G., and Feinsod, M. 1982. Sensory evoked re-

sponse to electrical stimulation of the trigeminal nerve in humans. J. Neurosurg. 56:545–549.

115. Fine, E.J., and Hallett, M. 1980. Neurophysiological study of subacute combined degeneration. J. Neurol. Sci. 45:331–336.
116. Finley, W.W. 1983. Operant conditioning of the short-latency cervical somatosensory evoked potential in quadriplegics. Exp. Neurol. 81:542–558.
117. Fruhstorfer, H., Guth, H., and Pfaff, U. 1976. Cortical responses evoked by thermal stimuli in man. In: The responsive brain, ed. W.C. McCallum and J.R. Knott, pp. 30–33. Bristol: John Wiley.
118. Fukushima, T., Mayanagi, Y., and Bouchard, G. 1976. Thalamic evoked potentials to somatosensory stimulation in man. Electroencephalogr. Clin. Neurophysiol. 40:481–490.
119. Gandevia, S.C., Burke, D., and McKeon, B. 1984. The projection of muscle afferents from the hand to cerebral cortex in man. Brain 107:1–13.
120. Ganes, T. 1980. Somatosensory conduction times and peripheral, cervical and cortical evoked potentials in patients with cervical spondylosis. J. Neurol. Neurosurg. Psychiatry 43:683–689.
121. Ganes, T. 1980. A study of peripheral, cervical and cortical evoked potentials and afferent conduction times in the somatosensory pathway. Electroencephalogr. Clin. Neurophysiol. 49:446–451.
122. Ganes, T. 1980. Somatosensory evoked responses and central afferent conduction times in patients with multiple sclerosis. J. Neurol. Neurosurg. Psychiatry 43:948–953.
123. Ganes, T. 1982. Synaptic and non-synaptic components of the human cervical evoked response. J. Neurol. Sci. 55:313–326.
124. Ganes, T., and Lundar, T. 1983. The effect of thiopentone on somatosensory evoked responses and EEGs in comatose patients. J. Neurol. Neurosurg. Psychiatry 46:509–514.
125. Ganes, T., and Nakstad, P. 1984. Subcomponents of the cervical evoked response in patients with intracerebral circulatory arrest. J. Neurol. Neurosurg. Psychiatry 47:292–297.
126. Garg, B.P., Markand, O.N., DeMyer, W.E., and Warren, C. 1983. Evoked response studies in patients with adrenoleukodystrophy and heterozygous relatives. Arch. Neurol. 40:356–359.
127. Giblin, D.R. 1964. Somatosensory evoked potentials in healthy subjects and in patients with lesions of the nervous system. Ann. N.Y. Acad. Sci. 112:93–142.
128. Giblin, D.R. 1980. Scalp-recorded somatosensory evoked potentials. In: Electrodiagnosis in clinical neurology, ed. M.J. Aminoff, pp. 414–450. New York: Churchill Livingstone.
129. Gilliat, R.W., and Sears, T.A. 1958. Sensory nerve action potentials in patients with peripheral nerve lesions. J. Neurol. Neurosurg. Psychiatry 21:109–118.
130. Glover, J.L., Worth, R.M., Bendick, P.J., Hall, P.V., and Markand, O.M. 1981. Evoked responses in the diagnosis of thoracic outlet syndrome. Surgery 89:86–93.
131. Goff, W.R., Allison, T., Shapiro, A., and Rosner, B.S. 1966. Cerebral somatosensory responses evoked during sleep in man. Electroencephalogr. Clin. Neurophysiol. 21:1–9.
132. Goff, G.D., Matsumiya, Y., Allison, T., and Goff, W.R.

1977. The scalp topography of human somatosensory and auditory evoked potentials. Electroencephalogr. Clin. Neurophysiol. 42:57–76.
133. Goff, W.R., Rosner, B.S., and Allison, T. 1962. Distribution of cerebral somatosensory evoked responses in normal man. Electroencephalogr. Clin. Neurophysiol. 14:697–713.
134. Goff, W.R., Shaywitz, B.A., Goff, G.D., Reisenauer, M.A., Jasiorkowski, J.G., Venes, J.L., and Rothstein, P.T. 1983. Somatic evoked potential evaluation of cerebral status in Reye syndrome. Electroencephalogr. Clin. Neurophysiol. 55:388–398.
135. Goldie, W.D., Chiappa, K.H., Young, R.R., and Brooks, E.B. 1981. Brainstem auditory and short-latency somatosensory evoked responses in brain death. Neurology 31:248–256.
136. Goto, A., Kosaka, K., Kubota, K., Nakamura, R., and Narabayashi, H. 1968. Thalamic potentials from muscle afferents in the human. Arch. Neurol. 19:301–309.
137. Green, J.B., and Walcoff, M.R. 1982. Evoked potentials in multiple sclerosis. Arch. Neurol. 39:696–697.
138. Green, J.B., Walcoff, M.R., and Lucke, J.F. 1982. Comparison of phenytoin and phenobarbital effects on far-field auditory and somatosensory evoked potential interpeak latencies. Epilepsia 23:417–421.
139. Greenberg, R.P., Becker, D.P., Miller, J.D., and Mayer, D.J. 1977. Evaluation of brain function in severe human head trauma with multimodality evoked potentials. Part 2. Localization of brain dysfunction and correlation with posttraumatic neurological conditions. J. Neurosurg. 47:163–177.
140. Greenberg, R.P., Newlon, P.G., Hyatt, M.S., Narayan, R.K., and Becker, D.P. 1981. Prognostic implications of early multimodality evoked potentials in severely head-injured patients: A prospective study. J. Neurosurg. 55:227–236.
141. Grisolia, J.S., and Wiederholt, W.C. 1980. Short latency somatosensory evoked potentials from radial, median and ulnar nerve stimulation in man. Electroencephalogr. Clin. Neurophysiol. 50:375–381.
142. Grundy, B.L., Nash, C.L., and Brown, R.H. 1981. Arterial pressure manipulation alters spinal cord function during correction of scoliosis. Anesthesiology 54:249–253.
143. Gupta, P.R., and Dorfman, L.J. 1981. Spinal somatosensory conduction in diabetes. Neurology 31:841–845.
144. Haldeman, S., Bradley, W.E., Bhatia, N.N., and Johnson, B.K. 1982. Pudendal evoked responses. Arch. Neurol. 39:280–283.
145. Halliday, A.M. 1967. The electrophysiological study of myoclonus in man. Brain 90:241–284.
146. Halliday, A.M. 1967. Changes in the form of cerebral evoked responses in man associated with various lesions of the nervous system. Electroencephalogr. Clin. Neurophysiol. Suppl. 26:178–192.
147. Halliday, A.M., and Mason, A.A. 1964. The effect of hypnotic anaesthesia on cortical responses. J. Neurol. Neurosurg. Psychiatry 27:300–312.
148. Halliday, A.M., and Wakefield, G.S. 1963. Cerebral evoked potentials in patients with dissociated sensory loss. J. Neurol. Neurosurg. Psychiatry 26:211–219.
149. Hammond, E.J., and Wilder, B.J. 1983. Evoked potentials in olivopontocerebellar atrophy. Arch. Neurol. 40:366–369.

150. Hari, R. 1980. Evoked potentials elicited by long vibrotactile stimuli in the human EEG. Pfluegers Arch. 384:167–170.
151. Hashimoto, T., Tayama, M., Hiura, K., Endo, S., Fukuda, K., Tamura, Y., Mori, A., and Miyao, M. 1983. Short latency somatosensory evoked potential in children. Brain Dev. 5:390–396.
152. Hazemann, P., Olivier, L., and Fischgold, H. 1969. Potentiel évoqué somesthésique ipsilateral, enregistré au niveau du scalp, chez l'homme hémisphérectomisé. C.R. Acad. Sci. [D] (Paris) 268:195–198.
153. Hrbek, A., Hrbková, M., and Lenard, H.G. 1968. Somato-sensory evoked responses in newborn infants. Electroencephalogr. Clin. Neurophysiol. 25:443–448.
154. Hrbek, A., Hrbková, M., and Lenard, H.G. 1969. Somato-sensory, auditory and visual evoked responses in newborn infants during sleep and wakefulness. Electroencephalogr. Clin. Neurophysiol. 26:597–603.
155. Hrbek, A., Karlberg, P., Kjellmer, I., Olsson, T., and Riha, M. 1977. Clinical application of evoked electroencephalographic responses in newborn infants. Perinatal asphyxia. Dev. Med. Child Neurol. 19:34–44.
156. Hrbek, A., Karlberg, P., and Olsson, T. 1973. Development of visual and somatosensory evoked responses in preterm newborn infants. Electroencephalogr. Clin. Neurophysiol. 34:225–232.
157. Hume, A.L., and Cant, B.R. 1978. Conduction time in central somatosensory pathways in man. Electroencephalogr. Clin. Neurophysiol. 45:361–375.
158. Hume, A.L., and Cant, B.R. 1981. Central somatosensory conduction after head injury. Ann. Neurol. 10:411–419.
159. Hume, A.L., Cant, B.R., and Shaw, N.A. 1979. Central somatosensory conduction time in comatose patients. Ann. Neurol. 5:379–384.
160. Hume, A.L., Cant, B.R., Shaw, N.A., and Cowan, J.C. 1982. Central somatosensory conduction time from 10 to 79 years. Electroencephalogr. Clin. Neurophysiol. 54:49–54.
161. Ikuta, T., and Furuta, N. 1982. Sex differences in the human group mean SEP. Electroencephalogr. Clin. Neurophysiol. 54:449–457.
162. Ishiko, N., Hanamori, T., and Murayama, N. 1980. Spatial distribution of somatosensory responses evoked by tapping the tongue and finger in man. Electroencephalogr. Clin. Neurophysiol. 50:1–10.
163. Johnson, D., Jürgens, R., and Kornhuber, H.H. 1980. Somatosensory-evoked potentials and perception of skin velocity. Arch. Psychiatr. Nervenkr. 228:95–100.
164. Johnson, D., Jürgens, R., and Kornhuber, H.H. 1980. Somatosensory-evoked potentials and vibration. Arch. Psychiatr. Nervenkr. 228:101–107.
165. Jones, S.J. 1977. Short latency potentials recorded from the neck and scalp following median nerve stimulation in man. Electroencephalogr. Clin. Neurophysiol. 43:853–863.
166. Jones, S.J. 1981. An "interference" approach to the study of somatosensory evoked potentials in man. Electroencephalogr. Clin. Neurophysiol. 52:517–530.
167. Jones, S.J. 1982. Clinical applications of short-latency evoked potentials. Ann. N.Y. Acad. Sci. 388:369–387.
168. Jones, S.J., Baraitser, M., and Halliday, A.M. 1980. Peripheral and central somatosensory nerve conduction de-

fects in Friedreich's ataxia. J. Neurol. Neurosurg. Psychiatry 43:495–503.
169. Jones, S.J., Carroll, W.M., and Halliday, A.M. 1983. Peripheral and central sensory nerve conduction in Charcot-Marie-Tooth disease and comparison with Friedreich's ataxia. J. Neurol. Sci. 61:135–148.
170. Jones, S.J., Edgar, M.A., and Ransford, A.O. 1982. Sensory nerve conduction in the human spinal cord: Epidural recordings made during scoliosis surgery. J. Neurol. Neurosurg. Psychiatry 45:446–451.
171. Jones, S.J., Parry, W., and Landi, A. 1981. Diagnosis of brachial plexus traction lesions by sensory nerve action potentials and somatosensory evoked potentials. Injury 12:376–382.
172. Jones, S.J., and Small, D.G. 1978. Spinal and sub-cortical evoked potentials following stimulation of the posterior tibial nerve in man. Electroencephalogr. Clin. Neurophysiol. 44:299–306.
173. Jörg, J., Düllberg, W., and Koeppen, S. 1982. Diagnostic value of segmental somatosensory evoked potentials in cases with chronic progressive para- or tetraspastic syndromes. Adv. Neurol. 32:347–358.
174. Josiassen, R.C., Shagass, C., Mancall, E.L., and Roemer, R.A. 1982. Somatosensory evoked potentials in Huntington's disease. Electroencephalogr. Clin. Neurophysiol. 54:483–493.
175. Kakigi, R., and Shibasaki, H. 1983. Scalp topography of the short latency somatosensory evoked potentials following posterior tibial nerve stimulation in man. Electroencephalogr. Clin. Neurophysiol. 56:430–437.
176. Kakigi, R., and Shibasaki, H. 1984. Scalp topography of mechanically and electrically evoked somatosensory potentials in man. Electroencephalogr. Clin. Neurophysiol. 59:44–56.
177. Kelly, D.L., Goldring, S., and O'Leary, J.L. 1965. Averaged somatosensory responses from exposed cortex of man. Arch. Neurol. 13:1–9.
178. Kelly, J.J., Sharbrough, F.W., and Daube, J.R. 1981. A clinical and electrophysiological evaluation of myoclonus. Neurology 31:581–589.
179. Kimura, J., Yamada, T., and Kawamura, H. 1978. Central latencies of somatosensory cerebral evoked potentials. Arch. Neurol. 35:683–688.
180. Kjaer, M. 1980. The value of brain stem auditory, visual and somatosensory evoked potentials and blink reflexes in the diagnosis of multiple sclerosis. Acta Neurol. Scand. 62:220–236.
181. Koshbin, S., and Hallett, M. 1981. Multimodality evoked potentials and blink reflex in multiple sclerosis. Neurology 31:138–144.
182. Kritchevsky, M., and Wiederholt, W.C. 1978. Short-latency somatosensory evoked potentials. Arch. Neurol. 35:706–711.
183. Krumholz, A., Singer, H.S., Niedermeyer, E., Burnite, R., and Harris, K. 1983. Electrophysiological studies in Tourette's syndrome. Ann. Neurol. 14:638–641.
184. Krumholz, A., Weiss, H.D., Goldstein, P.J., and Harris, K.C. 1981. Evoked responses in vitamin B_{12} deficiency. Ann Neurol. 9:407–409.
185. Laget, P., Raimbault, J., d'Allest, A.M., Flores-Guevara, R., Mariani, J., and Thierot-Prevost, G. 1976. La maturation des potentiels évoqués somethésiques (PES) chez l'homme. Electroencephalogr. Clin. Neurophysiol. 40:499–515.

186. La Joie, W.J., Reddy, N.M., and Melvin, J.L. 1982. Somatosensory evoked potentials: Their predictive value in right hemiplegia. Arch. Phys. Med. Rehab. 63:223–226.
187. Landi, A., Copeland, S.A., Wynn Parry, C.B., and Jones, S.J. 1980. The role of somatosensory evoked potentials and nerve conduction studies in the surgical management of brachial plexus injuries. J. Bone Joint Surg. [Br.] 62B:492–496.
188. Larson, S.J., and Sances, A. 1968. The specific somatosensory system and dyskinesia. Arch. Neurol. 18:543–548.
189. Larson, S.J., Sances, A., and Baker, J.B. 1966. Evoked cortical potentials in patients with stroke. Circulation 33/34, Suppl. 2:15–19.
190. Larson, S.J., Sances, A., and Christenson, P.C. 1966. Evoked somatosensory potentials in man. Arch. Neurol. 15:88–93.
191. Larsson, L.E., and Prevec, T.S. 1970. Somatosensory response to mechanical stimulation as recorded in the human EEG. Electroencephalogr. Clin. Neurophysiol. 28:162–172.
192. Lastimosa, A.C.B., Bass, N.H., Stanback, K., and Norvell, E.E. 1982. Lumbar spinal cord and early cortical evoked potentials after tibial nerve stimulation: Effects of stature on normative data. Electroencephalogr. Clin. Neurophysiol. 54:499–507.
193. Lavine, R.A., Buchsbaum, M.S., and Schechter, G. 1980. Human somatosensory evoked responses: Effects of attention and distraction on early components. Physiol. Psychol. 8:405–408.
194. Leandri, M., Favale, E., Ratto, S., and Abbruzzese, M. 1981. Conducted and segmental components of the somatosensory cervical response. J. Neurol. Neurosurg. Psychiatry 44:719–722.
195. Lehmkuhl, D., Dimitrijevic, M.R., and Renouf, F. 1984. Electrophysiological characteristics of lumbosacral evoked potentials in patients with established spinal cord injury. Electroencephalogr. Clin. Neurophysiol. 59:142–155.
196. Lesser, R.P., Koehle, R., and Lueders, H. 1979. Effect of stimulus intensity on short latency somatosensory evoked potentials. Electroencephalogr. Clin. Neurophysiol. 47:377–382.
197. Lesser, R.P., Lueders, H., Hahn, J., and Klem, G. 1981. Early somatosensory potentials evoked by median nerve stimulation: Intraoperative monitoring. Neurology 31:1519–1523.
198. Levy, R., and Behrman, J. 1970. Cortical evoked responses in hysterical hemianaesthesia. Electroencephalogr. Clin. Neurophysiol. 29:400–402.
199. Lewis, E.G., Dustman, R.E., and Beck, E.C. 1970. The effects of alcohol on visual and somato-sensory evoked responses. Electroencephalogr. Clin. Neurophysiol. 28:202–205.
200. Lewis, E.G., Dustman, R.E., and Beck, E.C. 1978. Visual and somatosensory evoked potential characteristics of patients undergoing hemodialysis and kidney transplantation. Electroencephalogr. Clin. Neurophysiol. 44:223–231.
201. Lindsay, K.W., Karlin, J., Kennedy, I., Fry, J., McInnes, A., and Teasdale, G.M. 1981. Evoked potentials in severe head injury: Analysis and relation to outcome. J. Neurol. Neurosurg. Psychiatry 44:796–802.
202. Lüders, H. 1970. The effects of aging on the wave form of the somatosensory cortical evoked potential. Electroencephalogr. Clin. Neurophysiol. 29:450–460.
203. Lueders, H., Dinner, D.S., Lesser, R.P., and Klem, G. 1983. Origin of far-field subcortical evoked potentials to

posterior tibial and median nerve stimulation: A comparative study. Arch. Neurol. 40:93–97.
204. Lueders, H., Gurd, A., Hahn, J., Andrish, J., Weiker, G., and Klem, G. 1982. A new technique for intraoperative monitoring of spinal cord function: Multichannel recording of spinal cord and subcortical evoked potentials. Spine 7:110–115.
205. Lüders, H., Kato, M., and Kuroiwa, Y. 1969. Cortical evoked potentials in hepatolenticular degeneration. Electroencephalogr. Clin. Neurophysiol. 27:425–428.
206. Lueders, H., Lesser, R.P., Hahn, J., Dinner, D.S., and Klem, G. 1983. Cortical somatosensory evoked potentials in response to hand stimulation. J. Neurosurg. 58:885–894.
207. Lütschg, J., Pfenninger, J., Ludin, H.P., and Vassella, F. 1983. Brain-stem auditory evoked potentials and early somatosensory evoked potentials in neurointensively treated comatose children. Am. J. Dis. Child. 137:421–426.
208. Maccabee, P.J., Levine, D.B., Pinkhasov, E.I., and Cracco, R.Q. 1983. Evoked potentials recorded from scalp and spinous processes during spinal column surgery. Electroencephalogr. Clin. Neurophysiol. 56:569–582.
209. Maccabee, P.J., Pinkhasov, E.I., and Cracco, R.Q. 1983. Short latency somatosensory evoked potentials to median nerve stimulation: Effect of low frequency filter. Electroencephalogr. Clin. Neurophysiol. 55:34–44.
210. McCallum, J.E., and Bennett, M.H. 1975. Electrophysiologic monitoring of spinal cord function during intraspinal surgery. Surg. Forum 26:469–471.
211. McGill, K.C., Cummins, K.L., Dorfman, L.J., Berlizot, B.B., Luetkemeyer, K., Nishimura, D.G., and Widrow, B. 1982. On the nature and elimination of stimulus artifact in nerve signals evoked and recorded using surface electrodes. IEEE Trans. Biomed. Eng. 29:129–136.
212. Macon, J.B., and Poletti, C.E. 1982. Conducted somatosensory evoked potentials during spinal surgery. Part 1. Control conduction velocity measurements. J. Neurosurg. 57:349–353.
213. Macon, J.B., Poletti, C.E., Sweet, W.H., Ojemann, R.G., and Zervas, N.T. 1982. Conducted somatosensory evoked potentials during spinal surgery. Part 2. Clinical applications. J. Neurosurg. 57:354–359.
214. Manfredi, M., Bini, G., Cruccu, G., Accornero, N., Berardelli, A., and Medolago, L. 1981. Congenital absence of pain. Arch. Neurol. 38:507–511.
215. Markand, O.N., Dilley, R.S., Moorthy, S.S., and Warren, C. 1984. Monitoring of somatosensory evoked responses during carotid endarterectomy. Arch. Neurol. 41:375–378.
216. Markand, O.N., Garg, B.P., DeMyer, W.E., and Warren, C. 1982. Brain stem auditory, visual and somatosensory evoked potentials in leukodystrophies. Electroencephalogr. Clin. Neurophysiol. 54:39–48.
217. Markand, O.N., Garg, B.P., and Weaver, D.D. 1984. Familial startle disease (hyperexplexia). Electrophysiologic studies. Arch. Neurol. 41:71–74.
218. Maruyama, Y., Shimoji, K., Shimizu, H., Kuribayashi, H., and Fujioka, H. 1982. Human spinal cord potentials evoked by different sources of stimulation and conduction velocities along the cord. J. Neurophysiol. 48:1098–1107.
219. Mastaglia, F.L., Black, J.L., Cala, L.A., and Collins, D.W.K. 1977. Evoked potentials, saccadic velocities, and computerised tomography in diagnosis of multiple sclerosis. Br. Med. J. 1:1315–1317.

220. Mastaglia, F.L., Black, J.L., Edis, R., and Collins, D.W.K. 1978. The contribution of evoked potentials in the functional assessment of the somatosensory pathway. Clin. Exp. Neurol. Proc. Aust. Assoc. Neurol. 15:279–298.
221. Matthews, W.B. 1980. The cervical somatosensory evoked potential in diagnosis. In: Electrodiagnosis in clinical neurology, ed. M.J. Aminoff, pp. 451–467. New York: Churchill Livingstone.
222. Matthews, W.B., Beauchamp, M., and Small, D.G. 1974. Cervical somato-sensory evoked responses in man. Nature 252:230–232.
223. Matthews, W.B., and Esiri, M. 1979. Multiple sclerosis plaque related to abnormal somatosensory evoked potentials. J. Neurol. Neurosurg. Psychiatry 42:940–942.
224. Matthews, W.B., and Small, D.G. 1979. Serial recording of visual and somatosensory evoked potentials in multiple sclerosis. J. Neurol. Sci. 40:11–21.
225. Matthews, W.B., Wattam-Bell, J.R.B., and Pountney, E. 1982. Evoked potentials in the diagnosis of multiple sclerosis: A follow up study. J. Neurol. Neurosurg. Psychiatry 45:303–307.
226. Mauguière, F., Brunon, A.M., Echallier, J.F., and Courjon, J. 1982. Early somatosensory evoked potentials in thalamo-cortical lesions of the lemniscal pathways in humans. Adv. Neurol. 32:321–338.
227. Mauguière, F., and Courjon, J. 1980. Effects of intravenous clonazepam on cortical somatosensory evoked responses (SER) in dyssynergia cerebellaris myoclonica (Ramsey-Hunt syndrome). In: EEG and clinical neurophysiology, Proceedings of the 2nd European congress of EEG and clinical neurophysiology, Salzburg, Austria, September 16–19, 1979, Excerpta Medica International Congress Series No. 526, ed. H. Lechner, and A. Aranibar, pp. 433–444. Amsterdam: Elsevier North-Holland.
228. Mauguière, F., and Courjon, J. 1981. The origins of short-latency somatosensory evoked potentials in humans. Ann. Neurol. 9:607–611.
229. Mauguière, F., Courjon, J., and Schott, B. 1983. Dissociation of early SEP components in unilateral traumatic section of the lower medulla. Ann. Neurol. 13:309–313.
230. Mauguière, F., Desmedt, J.E., and Courjon, J. 1983. Astereognosis and dissociated loss of frontal or parietal components of somatosensory evoked potentials in hemispheric lesions: Detailed correlations with clinical signs and computerized tomographic scanning. Brain 106:271–311.
231. Mauguière, F., Sonnet, M.L., Fischer, C., Chauplannaz, G., Courjon, J., and Schott, B. 1983. A study of far-field somatosensory evoked potentials (SEPs) in two patients with hemianesthesia due to subcortical lesions. Rev. Neurol. 139:141–148.
232. Meyer-Hardting, E., Wiederholt, W.C., and Budnick, B. 1983. Recovery function of short-latency components of the human somatosensory evoked potential. Arch. Neurol. 40:290–293.
233. Miyoshi, S., Lueders, H., Kato, M., and Kuroiwa, Y. 1971. The somatosensory evoked potential in patients with cerebrovascular diseases. Folia Psychiatr. Neurol. Jpn. 25:9–25.
234. Mongia, S.K., and Lundervold, A. 1975. Electrophysiological abnormalities in cases of dystrophia myotonica. Eur. Neurol. 13:360–376.
235. Nakanishi, T., Shimada, Y., Sakuta, M., and Toyokura,

Y. 1978. The initial positive component of the scalp-recorded somatosensory evoked potential in normal subjects and in patients with neurological disorders. Electroencephalogr. Clin. Neurophysiol. 45:26–34.
236. Nakanishi, T., Shimada, V., and Toyokura, Y. 1974. Somatosensory evoked responses to mechanical stimulation in normal subjects and in patients with neurological disorders. J. Neurol. Sci. 21:289–298.
237. Nakanishi, T., Takita, K., and Toyokura, Y. 1973. Somatosensory evoked responses to tactile tap in man. Electroencephalogr. Clin. Neurophysiol. 34:1–6.
238. Nakanishi, T., Tamaki, M., Ozaki, Y., and Arasaki, K. 1983. Origins of short latency somatosensory evoked potentials to median nerve stimulation. Electroencephalogr. Clin. Neurophysiol. 56:74–85.
239. Namerow, N.S., Sclabassi, R.J., and Enns, N.F. 1974. Somatosensory responses to stimulus trains: Normative data. Electroencephalogr. Clin. Neurophysiol. 37:11–21.
240. Nash, C.L., Lorig, R.A., Schatzinger, L.A., and Brown, R.H. 1977. Spinal cord monitoring during operative treatment of the spine. Clin. Orthop. 126:100–105.
241. Niazy, H.M.A., and Lundervold, A. 1982. Correlation of evoked potentials (SEP and VEP), EEG and CT in the diagnosis of brain tumors and cerebrovascular disease. Clin. Electroencephalogr. 13:71–81.
242. Noël, P., and Desmedt, J.E. 1975. Somatosensory cerebral evoked potentials after vascular lesions of the brain-stem and diencephalon. Brain 98:113–128.
243. Noël, P., and Desmedt, J.E. 1980. Cerebral and far-field somatosensory evoked potentials in neurological disorders involving the cervical spinal cord, brainstem, thalamus and cortex. Prog. Clin. Neurophysiol. 7:205–230.
244. Noth, J., Engel, L., Friedemann, H.H., and Lange, H.W. 1984. Evoked potentials in patients with Huntington's disease and their offspring. I. Somatosensory evoked potentials. Electroencephalogr. Clin. Neurophysiol. 59:134–141.
245. Nuwer, M.R., Perlman, S.L., Packwood, J.W., and Kark, R.A.P. 1983. Evoked potential abnormalities in the various inherited ataxias. Ann. Neurol. 13:20–27.
246. Obeso, J.A., Iragui, M.I., Marti-Masso, J.F., Maravi, E., Teijeira, J.M., Carrera, N., and Teijeira, J. 1980. Neurophysiological assessment of alpha pattern coma. J. Neurol. Neurosurg. Psychiatry 43:63–67.
247. Obeso, J.A., Marti-Masso, J.F., and Carrera, N. 1980. Somatosensory evoked potentials: Abnormalities with focal brain lesions remote from the primary sensorimotor area. Electroencephalogr. Clin. Neurophysiol. 49:59–65.
248. Oepen, G., Doerr, M., and Thoden, U. 1981. Visual (VEP) and somatosensory (SSEP) evoked potentials in Huntington's chorea. Electroencephalogr. Clin. Nuerophysiol. 51:666–670.
249. Papakostopoulos, D., Cooper, R., and Crow, H.J. 1974. Cortical potentials evoked by finger displacement in man. Nature 252:582–584.
250. Papakostopoulos, D., Cooper, R., and Crow, H.J. 1975. Inhibition of cortical evoked potentials and sensation by self-initiated movement in man. Nature 258:321–324.
251. Papakostopoulos, D., and Crow, H.J. 1980. Direct recording of the somatosensory evoked potentials from the cerebral cortex of man and the difference between precentral and postcentral potentials. Prog. Clin. Neurophysiol. 7:15–26.
252. Pedersen, L., and Trojaborg, W. 1981. Visual, auditory and somatosensory pathway involvement in hereditary cer-

ebellar ataxia, Friedreich's ataxia and familial spastic paraplegia. Electroencephalogr. Clin. Neurophysiol. 52:283–297.
253. Perot, P.L., and Vera, C.L. 1982. Scalp-recorded somatosensory evoked potentials to stimulation of nerves in the lower extremities and evaluation of patients with spinal cord trauma. Ann. N.Y. Acad. Sci. 388:359–368.
254. Phillips, L.H., and Daube, J.R. 1980. Lumbosacral spinal evoked potentials in humans. Neurology 30:1175–1183.
255. Pratt, H., Politoske, D., and Starr, A. 1980. Mechanically and electrically evoked somatosensory potentials in humans: Effects of stimulus presentation rate. Electroencephalogr. Clin. Neurophysiol. 49:240–249.
256. Pratt, H., and Starr, A. 1981. Mechanically and electrically evoked somatosensory potentials in humans: Scalp and neck distributions of short latency components. Electroencephalogr. Clin. Neurophysiol. 51:138–147.
257. Pratt, H., Starr, A., Amlie, R.N., and Politoske, D. 1979. Mechanically and electrically evoked somatosensory potentials in normal humans. Neurology 29:1236–1244.
258. Prevec, T.S. 1980. Effect of Valium on the somatosensory evoked potentials. In: Clinical uses of cerebral, brainstem and spinal somatosensory evoked potentials, ed. J.E. Desmedt, pp. 311–318. Basel: Karger.
259. Purves, S.J., Low, M.D., Galloway, J., and Reeves, B. 1981. A comparison of visual, brainstem auditory, and somatosensory evoked potentials in multiple sclerosis. Can. J. Neurol. Sci. 8:15–19.
260. Rappaport, M., Hopkins, H.K., Hall, K., and Beleza, T. 1981. Evoked potentials and head injury. 2. Clinical applications. Clin. Electroencephalogr. 12:167–176.
261. Ratto, S., Abbruzzese, M., Abbruzzese, G., and Favale, E. 1983. Surface recording of the spinal ventral root discharge in man: An experimental study. Brain 106:897–909.
262. Raudzens, P.A. 1982. Intraoperative monitoring of evoked potentials. Ann. N.Y. Acad. Sci. 388:308–326.
263. Riffel, B., and Stöhr, M. 1982. Spinal and subcortical somatosensory evoked potentials after stimulation of the tibial nerve. Arch. Psychiatr. Nervenkr. 232:251–263.
264. Riffel, B., Stöhr, M., Petruch, F., Ebensperger, H., and Scheglmann, K. 1982. Somatosensory evoked potentials following tibial nerve stimulation in multiple sclerosis and space-occupying spinal cord diseases. Adv. Neurol. 32:493–500.
265. Ropper, A.H., Miett, T., and Chiappa, K.H. 1982. Absence of evoked potential abnormalities in acute transverse myelopathy. Neurology 32:80–82.
266. Rossini, P.M., Caltagirone, C., David, P., and Macchi, G. 1979. Jakob-Creutzfeldt disease: Analysis of EEG and evoked potentials under basal conditions and neuroactive drugs. Eur. Neurol. 18:269–279.
267. Rossini, P.M., Cracco, R.Q., Cracco, J.B., and House, W.J. 1981. Short latency somatosensory evoked potentials to peroneal nerve stimulation: Scalp topography and the effect of different frequency filters. Electroencephalogr. Clin. Neurophysiol. 52:540–552.
268. Rothwell, J.C., Obeso, J.A., and Marsden, C.D. 1984. On the significance of giant somatosensory evoked potentials in cortical myoclonus. J. Neurol. Neurosurg. Psychiatry 47:33–42.
269. Rowed, D.W., McLean, J.A.G., and Tator, C.H. 1978. Somatosensory evoked potentials in acute spinal cord injury: Prognostic value. Surg. Neurol. 9:203–210.
270. Rumpl, E., Prugger, M., Gerstenbrand, F., Hackl, J.M.,

and Pallua, A. 1983. Central somatosensory conduction time and short latency somatosensory evoked potentials in post-traumatic coma. Electroencephalogr. Clin. Neurophysiol. 56:583–596.
271. Rushton, D.N., Rothwell, J.C., and Craggs, M.D. 1981. Gating of somatosensory evoked potentials during different kinds of movement in man. Brain 104:465–491.
272. Salamy, A. 1978. Commissural transmission: Maturational changes in humans. Science 200:1409–1411.
273. Salar, G., Job, I., and Mingrino, S. 1981. Cortical-evoked responses before and after percutaneous thermocoagulation of the Gasserian ganglion. Appl. Neurophysiol. 44:355–362.
274. Saletu, B., Itil, T.M., and Saletu, M. 1971. Evoked responses after hemispherectomy. Confin. Neurol. 33:221–230.
275. Sances, A., Larson, S.J., Cusick, J.F., Myklebust, J., Ewing, C.L., Jodat, R., Ackmann, J.J., and Walsh, P. 1978. Early somatosensory evoked potentials. Electroencephalogr. Clin. Neurophysiol. 45:505–514.
276. Sauer, M. 1980. Somatosensible Leitungsvermessungen bei neurologischen Systemerkrangkungen. Neurale Muskelatrophien und spinocerebelläre Ataxien. Arch. Psychiatr. Nervenkr. 228:223–242.
277. Scarff, T.B., Dallmann, D.E., and Bunch, W.H. 1981. Dermatomal somatosensory evoked potentials in the diagnosis of lumbar root entrapment. Surg. For. 32:489–491.
278. Sclabassi, R.J., Namerow, N.S., and Enns, N.F. 1974. Somatosensory response to stimulus trains in patients with multiple sclerosis. Electroencephalogr. Clin. Neurophysiol. 37:23–33.
279. Seppäläinen, A.M., Savolainen, K., and Kovola, T. 1981. Changes induced by xylene and alcohol in human evoked potentials. Electroencephalogr. Clin. Neurophysiol. 51:148–155.
280. Serra, C., C'Angellillo, A., Facciolla, D., Romano, F., Rossi, A., Ruocco, A., and Sorrentino, F. 1979. Somatosensory cerebral evoked potentials in uremic polyneuropathy. Acta Neurol. 34:1–14.
281. Seyal, M., Emerson, R.G., and Pedley, T.A. 1983. Spinal and early scalp-recorded components of the somatosensory evoked potential following stimulation of the posterior tibial nerve. Electroencephalogr. Clin. Neurophysiol. 55:320–330.
282. Shagass, C. 1975. EEG and evoked potentials in the psychoses. Res. Publ. Assoc. Res. Nerv. Ment. Dis. 54:101–127.
283. Shagass, C., Roemer, R.A., Straumanis, J.J., and Josiassen, R.C. 1981. Differentiation of depressive and schizophrenic psychoses by evoked potentials. Adv. Biol. Psychiatry 6:173–179.
284. Shagass, C., and Schwartz, M. 1965. Age, personality, and somatosensory cerebral evoked responses. Science 148:1359–1361.
285. Shagass, C., and Straumanis, J.J. 1978. Drugs and human sensory evoked potentials. In: Psychopharmacology: A generation of progress, ed. M.A. Lipton, A. DiMascio, and K.F. Killam, pp. 699–709. New York: Raven Press.
286. Shagass, C., Straumanis, J.J., Roemer, R.A., and Amadeo, M. 1977. Evoked potentials of schizophrenics in several sensory modalities. Biol. Psychiatry 12:221–235.
287. Shibasaki, H., Kakigi, R., Ohnishi, A., and Kuroiwa, Y. 1982. Peripheral and central nerve conduction in subacute myelo-optico-neuropathy. Neurology 32:1186–1189.

288. Shibasaki, H., Kakigi, R., Tsuji, S., Kimura, S., and Kuroiwa, Y. 1982. Spinal and cortical somatosensory evoked potentials in Japanese patients with multiple sclerosis. J. Neurol. Sci. 57:441–453.
289. Shibasaki, H., Ohnishi, A., and Kuroiwa, Y. 1982. Use of SEPs to localize degeneration in a rare polyneuropathy: Studies on polyneuropathy associated with pigmentation, hypertrichosis, edema, and plasma cell dyscrasia. Neurology 12:355–360.
290. Shibasaki, H., Yamashita, Y., and Kuroiwa, Y. 1978. Electroencephalographic studies of myoclonus. Myoclonus-related cortical spikes and high amplitude somatosensory evoked potentials. Brain 101:447–460.
291. Shibasaki, H., Yamashita, Y., and Tsuji, S. 1977. Somatosensory evoked potentials: Diagnostic criteria and abnormalities in cerebral lesions. J. Neurol. Sci. 34:427–439.
292. Shimoji, K., Kano, T., Higashi, H., Morioka, T., and Henschel, E.O. 1972. Evoked spinal electrograms recorded from epidural space in man. J. Appl. Physiol. 33:468–471.
293. Shimoji, K., Shimizu, H., and Maruyama, Y. 1978. Origin of somatosensory evoked responses recorded from the cervical skin surface. J. Neurosurg. 48:980–984.
294. Siivola, J. 1980. Estimation of the brain and spinal cord conduction time in man by means of the somatosensory evoked potentials and F and H responses. J. Neurol. Neurosurg. Psychiatry 43:1103–1111.
295. Siivola, J., Myllylä, V.V., Sulg, I., and Hokkanen, E. 1979. Brachial plexus and radicular neurography in relation to cortical evoked responses. J. Neurol. Neurosurg. Psychiatry 42:1151–1158.
296. Siivola, J., Sulg, I., and Heiskari, M. 1981. Somatosensory evoked potentials in diagnostics of cervical spondylosis and herniated disc. Electroencephalogr. Clin. Neurophysiol. 52:276–282.
297. Simpson, D.M., and Erwin, C.W. 1983. Evoked potential latency change with age suggests differential aging of primary somatosensory cortex. Neurobiol. Aging 4:59–63.
298. Singh, N., Sachdev, K.K., and Brisman, R. 1982. Trigeminal nerve stimulation: Short latency somatosensory evoked potentials. Neurology 32:97–101.
299. Small, D.G., Beauchamp, M., and Matthews, W.B. 1980. Subcortical somatosensory evoked potentials in normal man and in patients with central nervous system lesions. Prog. Clin. Neurophysiol. 7:190–204.
300. Small, M., and Matthews, W.B. 1984. A method of calculating spinal cord transit time from potentials evoked by tibial nerve stimulation in normal subjects and in patients with spinal cord disease. Electroencephalogr. Clin. Neurophysiol. 59:156–164.
301. Small, D.G., Matthews, W.B., and Small, M. 1978. The cervical somatosensory evoked potential (SEP) in the diagnosis of multiple sclerosis. J. Neurol. Sci. 35:211–224.
302. Soininen, K., and Järvilehto, T. 1983. Somatosensory evoked potentials associated with tactile stimulation at detection threshold in man. Electroencephalogr. Clin. Neurophysiol. 56:494–500.
303. Spielholz, N.I., Benjamin, M.V., Engler, G., and Ransohoff, J. 1979. Somatosensory evoked potentials and clinical outcome in spinal cord injury. In: Neural trauma, ed. A.J. Popp et al., pp. 217–222. New York: Raven Press.
304. Starr, A., McKeon, B., Skuse, N., and Burke, D. 1981.

Cerebral potentials evoked by muscle stretch in man. Brain 104:149–166.
305. Stejskal, L., Trávníček, V., Šourek, K., and Kredba, J. 1980. Somatosensory evoked potentials in deep hypothermia. App. Neurophysiol. 43:1–7.
306. Stöhr, M., Buettner, U.W., Riffel, B., and Koletzki, E. 1982. Spinal somatosensory evoked potentials in cervical cord lesions. Electroencephalogr. Clin. Neurophysiol. 54:257–265.
307. Stöhr, M., Buettner, U.W., Wietholter, H., and Riffel, B. 1983. Combined recordings of compound nerve action potentials and spinal cord evoked potentials in differential diagnosis of spinal root lesions. Arch. Psychiatr. Nervenkr. 233:103–110.
308. Stöhr, M., Dichgans, J., Voigt, K., and Buettner, U.W. 1983. The significance of somatosensory evoked potentials for localization of unilateral lesions within the cerebral hemispheres. J. Neurol. Sci. 61:49–63.
309. Stöhr, M., and Petruch, F. 1979. Somatosensory evoked potentials following stimulation of the trigeminal nerve in man. J. Neurol. 220:95–98.
310. Stöhr, M., Petruch, F., and Scheglmann, K. 1981. Somatosensory evoked potentials following trigeminal nerve stimulation in trigeminal neuralgia. Ann. Neurol. 9:63–66.
311. Stohr, P.E., and Goldring, S. 1969. Origin of somatosensory evoked scalp responses in man. J. Neurosurg. 31:117–127.
312. Stowell, H. 1977. Cerebral slow waves related to the perception of pain in man. Brain Res. Bull. 2:23–30.
313. Straumanis, J.J., Shagass, C., and Overton, D.A. 1973. Somatosensory evoked responses in Down syndrome. Arch. Gen. Psychiatry 29:544–549.
314. Strenge, H., and Hedderich, J. 1982. Age-dependent changes in central somatosensory conduction time. Eur. Neurol. 21:270–276.
315. Strenge, H., Tackmann, W., Barth, R., and Sojka-Raytscheff, A. 1980. Central somatosensory conduction time in diagnosis of multiple sclerosis. Eur. Neurol. 19:402–408.
316. Sugioka, H., Tsuyama, N., Hara, T., Nagano, A., Tachibana, S., and Ochiai, N. 1982. Investigation of brachial plexus injuries by intraoperative cortical somatosensory evoked potentials. Arch. Orthop. Trauma Surg. 99:143–151.
317. Sutton, G.G., and Mayer, R.F. 1974. Focal reflex myoclonus. J. Neurol. Neurosurg. Psychiatry 37:207–217.
318. Symon, L. 1980. The relationship between CBF, evoked potentials and the clinical features in cerebral ischaemia. Acta Neurol. Scand. Suppl. 78:175–190.
319. Symon, L., Wang, A.D., Silva, I.E.C.E., and Gentili, F. 1984. Perioperative use of somatosensory evoked responses in aneurysm surgery. J. Neurosurg. 60:269–275.
320. Synek, V.M. 1983. Somatosensory evoked potentials from musculocutaneous nerve in the diagnosis of brachial plexus injuries. J. Neurol. Sci. 61:443–452.
321. Synek, V.M., and Cowan, J.C. 1982. Somatosensory evoked potentials in patients with supraclavicular brachial plexus injuries. Neurology 32:1347–1352.
322. Synek, V.M., and Cowan, J.C. 1983. Saphenous nerve evoked potentials and the assessment of intraabdominal lesions in the femoral nerve. Muscle Nerve 6:453–456.
323. Tackmann, W., Ettlin, T., and Strenge, H. 1982. Multimodality evoked potentials and electrically elicited blink reflex in optic neuritis. J. Neurol. 227:157–163.

324. Tackmann, W., and Radü, E.W. 1983. Observations on the application of electrophysiological methods in the diagnosis of cervical root compressions. Eur. Neurol. 22:397–404.
325. Takahashi, K., and Fujitani, Y. 1970. Somatosensory and visual evoked potentials in hyperthyroidism. Electroencephalogr. Clin. Neurophysiol. 29:551–556.
326. Tamura, K. 1972. Ipsilateral somatosensory evoked responses in man. Folia Psychiatr. Neurol. Jpn. 26:83–94.
327. Terao, A., and Araki, S. 1975. Clinical application of somatosensory cerebral evoked response for the localization and the level diagnosis of neuronal lesions. Folia Psychiatr. Neurol. Jpn. 29:341–354.
328. Thomas, P.K., Jefferys, J.G.R., Smith, I.S., and Loulakakis, D. 1981. Spinal somatosensory evoked potentials in hereditary spastic paraplegia. J. Neurol. Neurosurg. Psychiatry 44:243–246.
329. Thompson, D.S., Woodward, J.B., Ringel, S.P., and Nelson, L.M. 1983. Evoked potential abnormalities in myotonic dystrophy. Electroencephalogr. Clin. Neurophysiol. 56:453–456.
330. Tokuomi, H., Uchino, M., Imamura, S., Yamanaga, H., Nakanishi, R., and Ideta, T. 1982. Minamata disease (organic mercury poisoning): Neuroradiologic and electrophysiologic studies. Neurology 32:1369–1375.
331. Trojaborg, W., Böttcher, J., and Saxtrup, O. 1981. Evoked potentials and immunoglobulin abnormalities in multiple sclerosis. Neurology 31:866–871.
332. Trojaborg, W., and Jørgenson, E.O. 1973. Evoked cortical potentials in patients with "isoelectric" EEGs. Electroencephalogr. Clin. Neurophysiol. 35:301–309.
333. Trojaborg, W., and Petersen, E. 1979. Visual and somatosensory evoked cortical potentials in multiple sclerosis. J. Neurol. Neurosurg. Psychiatry 42:323–330.
334. Tsumoto, T., Hirose, N., Nonaka, S., and Takahashi, M. 1972. Analysis of somatosensory evoked potentials to lateral popliteal nerve stimulation in man. Electroencephalogr. Clin. Neurophysiol. 33:379–388.
335. Tsumoto, T., Hirose, N., Nonaka, S., and Takahashi, M. 1973. Cerebrovascular disease: Changes in somatosensory evoked potentials associated with unilateral lesions. Electroencephalogr. Clin. Neurophysiol. 35:463–473.
336. Vas, G.A., Cracco, J.B., and Cracco, R.Q. 1981. Scalp-recorded short latency cortical and subcortical somatosensory evoked potentials to peroneal nerve stimulation. Electroencephalogr. Clin. Neurophysiol. 52:1–8.
337. Vaughan, H.G. 1975. Electrophysiologic analysis of regional cortical maturation. Biol. Psychiatry 10:513–526.
338. Vaziri, D., Pratt, H., Saiki, J.K., and Starr, A. 1981. Evaluation of somatosensory pathway by short latency evoked potentials in patients with end-stage renal disease maintained on hemodialysis. Int. J. Artif. Organs 4:17–22.
339. Veale, J.L., Mark, R.F., and Rees, S. 1973. Differential sensitivity of motor and sensory fibres in human ulnar nerve. J. Neurol. Neurosurg. Psychiatry 36:75–86.
340. Velasco, F., Velasco, M., Cepeda, C., and Munoz, H. 1980. Wakefulness-sleep modulation of cortical and subcortical somatic evoked potentials. Electroencephalogr. Clin. Neurophysiol. 48:64–72.
341. Vera, C.L., Perot, P.L., and Fountain, E.L. 1983. Scalp recorded somatosensory evoked potentials to posterior tibial nerve stimulation in humans. Electroencephalogr. Clin. Neurophysiol. 56:159–168.
342. Vogel, P., and Vogel, H. 1982. Somatosensory cortical

343. potentials evoked by stimulation of leg nerves: Analysis of normal values and variability; diagnostic significance. J. Neurol. 228:97–111.
343. Walsh, J.C., Garrick, R., Cameron, J., and McLeod, J.G. 1982. Evoked potential changes in clinically definite multiple sclerosis: A two year follow up study. J. Neurol. Neurosurg. Psychiatry 45:494–500.
344. Walsh, J.C., Yiannikas, C., and McLeod, J.G. 1984. Abnormalities of proximal conduction in acute idiopathic polyneuritis: Comparison of short latency evoked potentials and F-waves. J. Neurol. Neurosurg. Psychiatry 47:197–200.
345. Wang, A.D., Cone, J., Symon, L., and Silva, I.E.C.E. 1984. Somatosensory evoked potential monitoring during the management of aneurysmal SAH. J. Neurosurg. 60:264–268.
346. Wiederholt, W.C. 1980. Early components of the somatosensory evoked potential in man, cat and rat. Prog. Clin. Neurophysiol. 7:105–117.
347. Wiederholt, W.C., Meyer-Hardting, E., Budnick, B., and McKeown, K.L. 1982. Stimulating and recording methods used in obtaining short-latency somatosensory evoked potentials (SEPs) in patients with central and peripheral neurologic disorders. Ann. N.Y. Acad. Sci. 388:349–358.
348. Williamson, P.D., Goff, W.R., and Allison, T. 1970. Somato-sensory responses in patients with unilateral cerebral lesions. Electroencephalogr. Clin. Neurophysiol. 28:566–575.
349. Worth, R.M., Markand, O.N., DeRosa, G.P., and Warren, C.H. 1982. Intraoperative somatosensory evoked response monitoring during spinal cord surgery. Adv. Neurol. 32:367–373.
350. Yamada, T., Kayamori, R., Kimura, J., and Beck, D.O. 1984. Topography of somatosensory evoked potentials after stimulation of the median nerve. Electroencephalogr. Clin. Neurophysiol. 59:29–43.
351. Yamada, T., Kimura, J., and Nitz, D.M. 1980. Short latency somatosensory evoked potentials following median nerve stimulation in man. Electroencephalogr. Clin. Neurophysiol. 48:367–376.
352. Yamada, T., Kimura, J., Wilkinson, J.T., and Kayamori, R. 1983. Short- and long-latency median somatosensory evoked potentials: Findings in patients with localized neurological lesions. Arch. Neurol. 40:215–220.
353. Yamada, T., Kimura, J., Young, S., and Powers, M. 1978. Somatosensory-evoked potentials elicited by bilateral stimulation of the median nerve and its clinical application. Neurology 28:218–223.
354. Yamada, T., Machida, M., and Kimura, J. 1982. Far-field somatosensory evoked potentials after stimulation of the tibial nerve. Neurology 32:1151–1158.
355. Yamada, T., Shivapour, E., Wilkinson, T., Kimura, J. 1982. Short- and long-latency somatosensory evoked potentials in multiple sclerosis. Arch. Neurol. 39:88–94.
356. Yiannikas, C., and Walsh, J.C. 1983. Somatosensory evoked responses in the diagnosis of thoracic outlet syndrome. J. Neurol. Neurosurg. Psychiatry 46:234–240.
357. Young, W. 1982. Correlation of somatosensory evoked potentials and neurological findings in spinal cord injury. In: Early management of acute spinal cord injury, ed. C.H. Tator, pp. 153–165. New York: Raven Press.

E

Event-related and other potentials

21. **Event-related potentials, olfactory evoked potentials, and magnetic evoked fields**
    21.1 The $\overline{P300}$
       21.1.1 Methods of producing the $\overline{P300}$
       21.1.2 The normal $\overline{P300}$
          21.1.2.1 Distribution
          21.1.2.2 Effect of age
       21.1.3 The abnormal $\overline{P300}$
    21.2 The contingent negative variation
       21.2.1 Methods of producing the CNV
       21.2.2 The normal CNV
          21.2.2.1 Distribution
          21.2.2.2 The effect of age
       21.2.3 The abnormal CNV
    21.3 Readiness potential *(Bereitschaftspotential)* and other movement-related potentials
    21.4 Potentials preceding speech and accompanying writing
    21.5 Olfactory EPs
    21.6 Magnetic evoked fields

References for part E

21

Event-related potentials, olfactory evoked potentials, and magnetic evoked fields

SUMMARY

21.1. The P300 is a positive peak, with a maximum at the vertex, which occurs about 300 msec after rare or unexpected stimuli that are interspersed within a sequence of more common stimuli. The P300 seems to depend on cognitive processes rather than the physical characteristics of the stimulus. It is therefore called an *event-related potential* (ERP) rather than an *evoked potential*. Attempts have been made to relate abnormalities of the P300 to conditions of altered cognitive functioning.

21.2. The contingent negative variation (CNV) is a negative deflection that develops between a first (warning) stimulus and a constantly following second (imperative) stimulus that requires some response. The CNV appears to depend on the subject's expectancy of the second stimulus and also is an ERP, not an EP.

21.3. The readiness potential *(Bereitschaftspotential)* and other movement-related potentials consist of scalp-recorded peaks that precede and follow actual movements and are ERPs, not EPs.

21.4. Potentials related to speech and writing resemble movement-related potentials.

21.5. Olfactory EPs are elicited by odorous stimuli.

21.6. Magnetic evoked fields are averaged changes of the magnetic field about the head induced by visual, auditory, and somatosensory stimuli.

21.1 THE P̄300

21.1.1 Methods of producing the P̄300

A P̄300 may be elicited in various ways. Usually, two different kinds of stimuli are mixed in one sequence, and the subject is asked to pay attention to one kind, usually the rarer one, and to ignore the other. This is sometimes referred to as the odd-ball paradigm and often implemented with tones of high and low frequency ("beep-boop" stimuli). Responses to each kind of stimulus are recorded between an electrode over the central or posterior head regions and a distant reference electrode and are averaged separately. A P̄300 appears after the AEP to stimulation with the rare stimulus if the subject pays attention to this stimulus (Figure 21.1). These methods require the facility to present two kinds of stimuli, mixed irregularly in different proportions, and to sort the responses to each kind of stimulus in separate averaging channels.

A P̄300 may also be obtained by instructing the subject to guess whether the next stimulus will be visual or auditory[107] or whether it will be a single or double click.[108] The subject may be asked to count or otherwise pay attention to brief presentations of certain visual patterns or colors,[41] words,[28] letters,[13,30] or sounds[26,77] which occur irregularly and infrequently within a series of visual or auditory stimuli from which they differ in some regard. Even the omission of a stimulus in a train of uniform and regular stimuli may produce a P̄300 at a fixed time after the expected stimulus,[50,77,91,100,108,115] clearly indicating that this potential is not evoked by an exogenous stimulus but generated endogenously, or emitted.[115] On the other hand, the simple presentation of an unexpected stimulus unrelated to any instructions may also be followed by a P̄300 which, however, has a shorter latency and a more anterior distribution.[15,105] Thus, although the production of the P̄300 seems to depend on cognitive factors, the precise physiological and psychological mechanisms involved are unclear.[22,31,42]

21.1.2 The normal P̄300

The P̄300 is a positive peak with a latency of 250–600 msec (Figure 21.1). It may be preceded by EP peaks if the recording electrodes are located near the cortical receiving area for the sensory modality of the stimulus used. The latency of the P̄300 increases with the time which the subject needs to distinguish the rare stimulus[65]; the amplitude increases with the rarity of the stimulus[23] and, to some extent, with stimulus intensity.[89] Maneuvers affecting the P̄300, such as manipulations of the subject's attention, may also alter the amplitude of the preceding long-latency peaks of VEPs,[88] AEPs,[21,93,94] and SEPs.[20,58]

21.1.2.1 Distribution

The P̄300 has a wide distribution with a maximum in the parietotemporal areas,[77,101,113] unrelated to the specific sensory areas but probably related to the parietotemporal association cortex and to subcortical structures[116] such as the hippocampus[36,69] or thalamus.[117] Even though the P̄300 to novel stimuli may be located more frontally,[13,15] the distribution of the P$\overline{300}$ generally does not reach as far forward as that of the contingent negative variation.[24]

21.1.2.2 Effect of age

A P$\overline{300}$ to unfamiliar stimuli has been obtained in infants of 3 months.[45] Word stimuli normally become effective at 8–12 years.[95] Between childhood and adulthood, the distribution of the P$\overline{300}$ becomes more restricted to the parietal area[57] and the latency decreases.[14,32] In adults, aging increases the latency,[4,7,32,73,74] decreases the amplitude,[32,73] and causes a forward shift in the distribution[73,74] of the P$\overline{300}$.

Figure 21.1. The normal P$\overline{300}$. Frequent (left) and rare (right) auditory stimuli are intermixed irregularly and the responses are averaged separately. When both stimuli are ignored (top row), they produce similar EPs with N1 and P2 peaks (top left and right). When the rare stimuli are attended to (bottom row), frequent stimuli (bottom left) produce the same peaks as do ignored stimuli, but rare stimuli (bottom right) are followed by an additional positive peak, the P$\overline{300}$ (P3). From Squires and Hecox[104] with permission by the authors and Thieme-Stratton Inc.

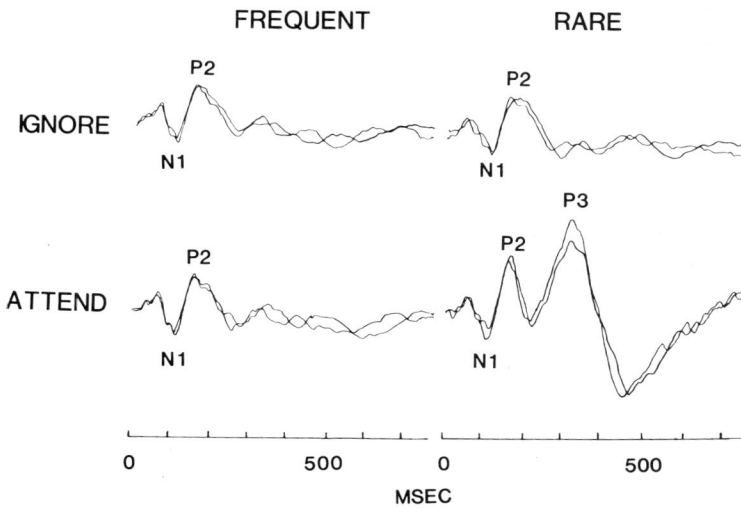

21.1.3 The abnormal $\overline{P300}$

Although the latency of the $\overline{P300}$ varies widely in normal subjects, it has been reported to be prolonged in adult patients with mental retardation,[103] in patients with dementia of various causes,[33,34,76] in patients with chronic renal failure,[12] in patients with Parkinson's disease,[38] in chronic alcoholics shortly after detoxification,[75] and in schizophrenic patients.[76,112]

Changes of $\overline{P300}$ amplitude have been described in various conditions, including frontal lobe lesions,[52,53] hyperactive children treated with methylphenidate,[37,51] increased blood lead levels in children,[71] infantile autism,[67] and schizophrenia,[3,90,112] but these changes are of uncertain diagnostic importance because of the great normal variability of the $\overline{P300}$ amplitude.

21.2 THE CONTINGENT NEGATIVE VARIATION

21.2.1 Methods of producing the CNV

The CNV is usually produced by presenting pairs of stimuli separated by 1–2 sec; the first stimulus serves as a warning signal and the second, or imperative, stimulus as a signal requiring a response.[10] A common example is a click followed after about 1 second by a series of flashes to which the subject is required to push a button. Repetition of the same sequence leads to the buildup of a negative potential which begins after the EP to the first stimulus and ends with the EP to the second stimulus (Figure 21.2). The modality and order of presentation of the stimulus is unimportant: A flash followed by clicks requiring a response is equally effective. However, the mere presentation of paired stimuli without the requirement of a response to the second stimulus does not produce a CNV. Moreover, the amplitude of the CNV decreases if the warning stimulus is not always followed by the imperative stimulus; the CNV usually disappears entirely if less than 70 percent of warning stimuli are followed by imperative stimuli.

If two different warning stimuli are used and the subject is asked to respond to only one of them, a CNV will develop only after that stimulus. This paradigm may be used to test the ability of the subject to discriminate between similar stimuli and to test the effectiveness of the first stimulus. If the intensity of the warning stimulus is reduced below the perceptual threshold, it becomes ineffective in generating a CNV.

The CNV is recorded between electrodes on vertex and ear or mastoid. The averaging epoch should be 2–4 sec long to permit identification of a baseline from which the CNV amplitude can be measured. The CNV is enhanced by averaging 10 or more responses. The technical requirements for eliciting and recording the CNV go beyond those of routine EP recordings. Stimulation requires the ability to generate paired stimuli, usually of different modality. Recording facilities must be able to register very slow, long-lasting electric potential changes and to verify the subject's response to the second stimulus.

21.2.2 The normal CNV

The CNV begins about 400 msec[82] after the warning stimulus and forms a ramp or rectangle having a maximum within about 800 msec of the first stimulus[10]; it has an amplitude of usually up to 50 μV (Figure 21.2). The CNV develops gradually as the paired stimuli are repeated. The speed of development differs between subjects. Anxiety and distraction reduce the CNV amplitude.[64] Because a CNV does not develop if the subject fails to perceive the warning stimulus or to respond to the imperative stimulus,

Figure 21.2. The normal CNV. *(A)* Stimulation with a click produces a LLAEP. *(B)* Stimulation with a train of flashes elicits a VEP. *(C)* Click stimulation followed by flash stimulation produces a sequence of the EPs shown in *A* and *B*. *(D)* Click stimulation followed by flash stimulation gradually leads to the development of a slow negative wave between the click and flash EPs when the subject follows the instruction to press a button terminating the flashes. Recording between vertex and mastoid electrodes. Negativity at the vertex is plotted upward. From Walter et al.,[114] reprinted with permission by the authors and the publisher from Nature 203:380–384, Copyright (c) 1964, Macmillan Journals Limited.

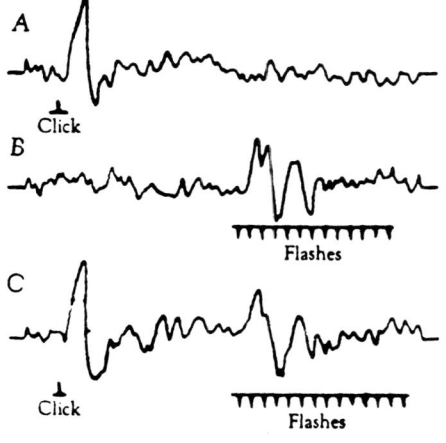
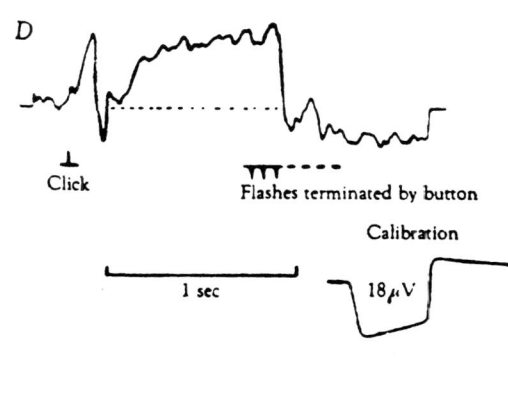

a CNV cannot be elicited in infants or uncooperative subjects.

An increase of the interval between the warning and the imperative stimulus to about 4 sec splits the CNV into two waves that have different distribution and functional characteristics, suggesting that the CNV may consist of a combination of long-latency potentials evoked by the first stimulus and of potentials related to the preparation for a reaction to the second stimulus[60,63,86,92] (21.3).

The name *contingent negative variation* has been applied because this potential is contingent on a stimulus-response sequence, is electrically negative, and is a slow variation from the electric baseline. Because it seems to depend on the expectation by the subject of the second stimulus, it has also been called *expectancy wave*.[114] The methods of producing a CNV suggest that the CNV depends on attention, motivation, effort, preparation for action, and other psychological variables.[23]

21.2.2.1 Distribution
The CNV has a maximum at the vertex. It extends farther frontally than parietally and has a minimum in the occipital and posterior temporal areas.[10]

21.2.2.2 The effect of age
In old age, the CNV has been reported to increase[79] and to depend on the subject's ability to switch attention.[109]

21.2.3 The abnormal CNV

Only few studies have attempted to relate the CNV to neurological and psychiatric diseases.[8,11,62,64,85,111] The CNV has been used in audiological studies to determine the threshold of auditory perception[80,81] or the discrimination of phonemes.[46]

21.3 READINESS POTENTIAL (*BEREITSCHAFTSPOTENTIAL*) AND OTHER MOVEMENT-RELATED POTENTIALS

Computer averaging of the electric activity that precedes a voluntary movement shows a gradually rising negative potential which begins about 1 sec before the movement, has a wide, bilateral distribution with a maximum at the vertex and probably reflects the process of getting ready to move; no such potentials are seen before passive movement (Figure 21.3). Both passive and active movements are followed by several peaks.[18,29,54] Movement-related potentials have now been described in great detail.[44,55,56,59,72,96,97,98] Only a few attempts have so far been made to use these potentials for the study of cerebral disorders, for instance, Parkinson's disease,[17,66,99] cerebellar disorders,[99] and Tourette's syndrome.[68]

Figure 21.3. Normal movement-related potentials. Passive extension of the right middle finger (left half, at time 0) is followed by sequences of peaks having different latencies in various locations (top to bottom). Volitional right middle finger extension (right half, at time 0) is preceded by a widespread, slowly increasing negative deflection and followed by simpler fast waves. Each tracing is the grand average of recordings from four subjects. Recordings from midfrontal, left central, vertex, and left parietal electrodes in reference to linked ear electrodes. Negativity at the scalp electrodes is plotted upward. From Shibasaki et al.[97] with permission by the authors and Elsevier Biomedical Press B.V.

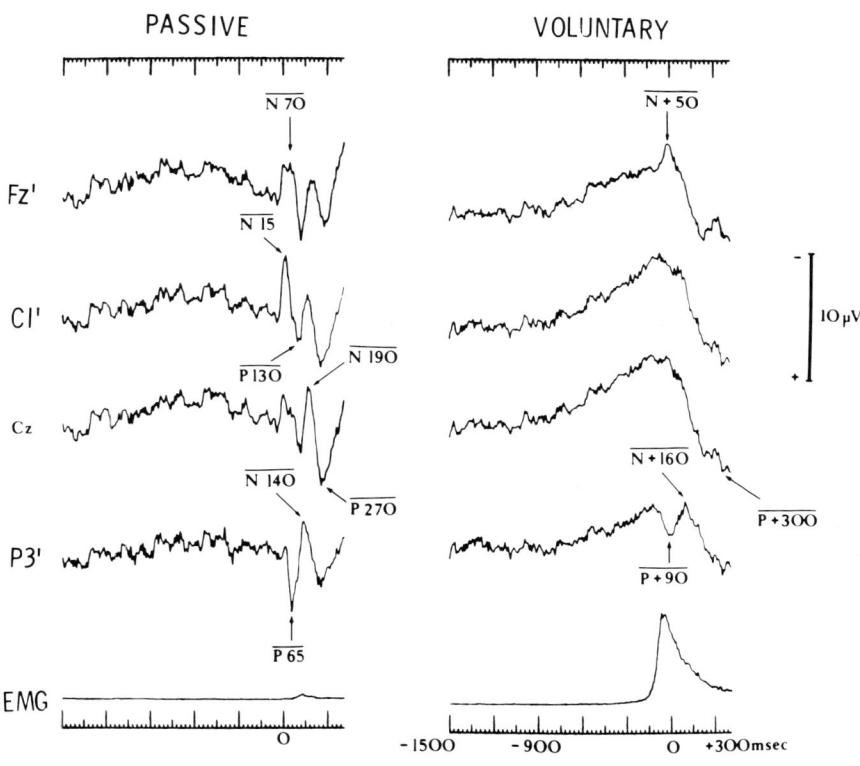

21.4 POTENTIALS PRECEDING SPEECH AND ACCOMPANYING WRITING

Slow negative potentials can be recorded several seconds before the onset of speech. They have a frontal distribution with an asymmetry suggesting a specific relation to speech centers.[27,35,43,61,106] Potentials with a maximum at the vertex and the left motor area appear during writing in right-handed subjects.[47]

21.5 OLFACTORY EPs

Olfactory stimuli elicit EPs of relatively long latency and duration. Precise control of stimulus and recording conditions poses many problems. Even when they are overcome, it is difficult to be certain that EPs to odorous stimuli are due to specific olfactory stimulation rather than to excitation of the nasal somatosensory fibers of the trigeminal nerve.[78,102]

21.6 MAGNETIC EVOKED FIELDS

Visual,[6,110] auditory,[2,25,40,83,84,87] somatosensory,[5,39,48] and cognitive[69] stimuli, and finger,[19,70] foot, and toe[1,16] movements induce changes of the magnetic field about the head that can be recorded and averaged. These magnetic evoked fields generally resemble the corresponding EPs and ERPs and may give some information regarding the localization of the generators in the head.[9,49]

References for part E

1. Antervo, A., Hari, R., Katila, T., Poutanen, T., Seppänen, M., and Tuomisto, T. 1983. Cerebral magnetic fields preceding self-paced plantar flexions of the foot. Acta Neurol. Scand. 68:213–217.
2. Arlinger, S., Elberling, C., Bak, C., Kofoed, B., Lebech, J., and Saermark, K. 1982. Cortical magnetic fields evoked by frequency glides of a continuous tone. Electroencephalogr. Clin. Neurophysiol. 54:642–653.
3. Baribeau-Braun, J., Picton, T.W., and Gosselin, J.Y. 1983. Schizophrenia: A neurophysiological evaluation of abnormal information processing. Science 219:874–876.
4. Beck, E.C., Swanson, C., and Dustman, R.E. 1980. Long latency components of the visually evoked potential in man: Effects of aging. Exp. Aging Res. 6:523–545.
5. Brenner, D., Lipton, J., Kaufman, L., and Williamson, S.J. 1978. Somatically evoked magnetic fields of the human brain. Science 199:81–83.
6. Brenner, D., Williamson, S.J., and Kaufman, L. 1975. Visually evoked magnetic fields of the human brain. Science 190:480–482.
7. Brown, W.S., Marsh, J.T., and LaRue, A. 1983. Exponential electrophysiological aging: P3 latency. Electroencephalogr. Clin. Neurophysiol. 55:277–285.
8. Burian, K., Gestring, G.F., Gloning, K., and Haider, M. 1972. Objective examination of verbal discrimination and comprehension in aphasia using the contingent negative variation. Audiology 11:310–316.
9. Cohen, D., and Cuffin, B.N. 1983. Demonstration of useful differences between magnetencephalogram and electroencephalogram. Electroencephalogr. Clin. Neurophysiol. 56:38–51.
10. Cohen, J. 1969. Very slow brain potentials relating to expectancy: The CNV. In: Average evoked potentials, ed. E. Donchin, and D.B. Lindsley, pp. 143–198. Washington: NASA SP-191.
11. Cohen, J. 1974. Cerebral psychophysiology: The contingent negative variation. In: Bioelectric recording techniques, Part B, Electroencephalography and human brain potentials, ed. R.F. Thompson, and M.M. Patterson, pp. 259–280. New York: Academic Press.
12. Cohen, S.N., Syndulko, K., Rever, B., Kraut, J., Coburn, J., and Tourtellotte, W.W. 1983. Visual evoked potentials and long latency event-related potentials in chronic renal failure. Neurology 33:1219–1222.
13. Courchesne, E. 1978. Changes in P3 waves with event repetition: Long-term effects on scalp distribution and amplitude. Electroencephalogr. Clin. Neurophysiol. 45:754–766.

14. Courchesne, E. 1979. From infancy to adulthood: The neurophysiological correlates of cognition. Prog. Clin. Neurophysiol. 6:224–242.
15. Courchesne, E., Hillyard, S.A., and Galambos, R. 1975. Stimulus novelty, task relevance and the visual evoked potential in man. Electroencephalogr. Clin. Neurophysiol. 39:131–143.
16. Deecke, L., Boschert, J., Weinberg, H., and Brickett, P. 1983. Magnetic fields of the human brain (Bereitschaftsmagnetfeld) preceding voluntary foot and toe movements. Exp. Brain Res. 52:81–86.
17. Deecke, L., Englitz, H.G., Kornhuber, H.H., and Schmitt, G. 1977. Cerebral potentials preceding voluntary movement in patients with bilateral or unilateral Parkinson akinesia. Prog. Clin. Neurophysiol. 1:151–163.
18. Deecke, L., Scheid, P., and Kornhuber, H.H. 1969. Distribution of readiness potential, pre-motion positivity, and motor potential of the human cerebral cortex preceding voluntary finger movements. Exp. Brain Res. 7:158–168.
19. Deecke, L., Weinberg, H., and Brickett, P. 1982. Magnetic fields of the human brain accompanying voluntary movement: Bereitschaftsmagnetfeld. Exp. Brain Res. 48:144–148.
20. Desmedt, J.E., and Robertson, C. 1977. Differential enhancement of early and late components of the cerebral somatosensory evoked potentials during forced-pace cognitive tasks in man. J. Physiol. (Lond.) 271:761–782.
21. Donald, M.M., and Young, M.J. 1982. A time-course analysis of attentional tuning of the auditory evoked response. Exp. Brain Res. 46:357–367.
22. Donchin, E., and Isreal, J.B. 1980. Event-related potentials and psychological theory. In: Motivation, motor and sensory processes of the brain: Electrical potentials, behaviour and clinical use, ed. H.H. Kornhuber and L. Deecke, pp. 697–715. Amsterdam: Elsevier.
23. Donchin, E., Ritter, W., and McCallum, W.C. 1978. Cognitive psychophysiology: The endogenous components of the ERP. In: Event-related brain potentials in man, ed. E. Callaway, pp. 349–411. New York: Academic Press.
24. Donchin, E., Tueting, P., Ritter, W., Kutas, M., and Heffley, E. 1975. On the independence of the CNV and the P300 components of the human averaged evoked potential. Electroencephalogr. Clin. Neurophysiol. 38:449–461.
25. Elberling, C., Bak, C., Kofoed, B., Lebech, J., and Saermark, K. 1982. Auditory magnetic fields from the human cerebral cortex: Location and strength of an equivalent current dipole. Acta Neurol. Scand. 65:553–569.
26. Ford, J.M., and Hillyard, S.A. 1981. Event-related potentials (ERPs) to interruptions of a steady rhythm. Psychophysiology 18:322–328.
27. Fried, I., Ojeman, G.A., and Fetz, E.E. 1981. Language-related potentials specific to human language cortex. Science 212:353–356.
28. Friedman, D., Simson, R., Ritter, W., and Rapin, I. 1975. The late positive component (P300) and information processing in sentences. Electroencephalogr. Clin. Neurophysiol. 38:255–262.
29. Gilden, L., Vaughan, H.G., and Costa, L.D. 1966. Summated human EEG potential with voluntary movement. Electroencephalogr. Clin. Neurophysiol. 20:433–438.
30. Gomer, F.E., Spicuzza, R.J., and O'Donnell, R.D. 1976. Evoked potential correlates of visual item recognition during memory-scanning tasks. Physiol. Psychol. 4:61–65.
31. Goodin, D.S., and Aminoff, M.J. 1984. The relationship

between the evoked potential and brain events in sensory discrimination and motor response. Brain 107:241–251.
32. Goodin, D.S., Squires, K.C., Henderson, B.H., and Starr, A. 1978. Age-related variations in evoked potentials to auditory stimuli in normal human subjects. Electroencephalogr. Clin. Neurophysiol. 44:447–458.
33. Goodin, D.S., Squires, K.C., and Starr, A. 1978. Long latency event-related components of the auditory evoked potential in dementia. Brain 101:635–648.
34. Goodin, D.S., Starr, A., Chippendale, T., and Squires, K.C. 1983. Sequential changes in the P3 component of the auditory evoked potential in confusional states and dementing illnesses. Neurology 33:1215–1218.
35. Grözinger, B., Kornhuber, H.H., and Kriebel, J. 1977. Human cerebral potentials preceding speech production, phonation, and movements of the mouth and tongue, with reference to respiratory and extracerebral potentials. In: Language and cerebral specialization in man: Cerebral ERPs, ed. J.E. Desmedt, pp. 87–103. Basel: Karger.
36. Halgren, E., Squires, N.K., Wilson, C.L., Rohrbaugh, J.W., Babb, T.L., and Crandall, P.H. 1980. Endogenous potentials generated in the human hippocampal formation and amygdala by infrequent events. Science 210:803–805.
37. Halliday, R., Callaway, E., and Rosenthal, J.H. 1984. The visual ERP predicts clinical response to methylphenidate in hyperactive children. Psychophysiology 21:114–121.
38. Hansch, E.C., Syndulko, K., Cohen, S.N., Goldberg, Z.I., Potvin, A.R., and Tourtellotte, W.W. 1982. Cognition in Parkinson disease: An event-related potential perspective. Ann. Neurol. 11:599–607.
39. Hari, R., Hämäläinen, M., Kaukoranta, E., Reinikainen, K., and Teszner, D. 1983. Neuromagnetic responses from the second somatosensory cortex in man. Acta Neurol. Scand. 68:207–212.
40. Hari, R., Kaila, K., Katila, T., Tuomisto, T., and Varpula, T. 1982. Interstimulus interval dependence of the auditory vertex response and its magnetic counterpart: Implications for their neural generation. Electroencephalogr. Clin. Neurophysiol. 54:561–569.
41. Harter, R.M., and Salmon, L.E. 1972. Intra-modality selective attention and evoked cortical potentials to randomly presented patterns. Electroencephalogr. Clin. Neurophysiol. 32:605–613.
42. Hillyard, S.A., and Kutas, M. 1983. Electrophysiology of cognitive processing. Annu. Rev. Psychol. 34:33–61.
43. Hillyard, S.A., and Woods, D.L. 1979. Electrophysiological analysis of human brain function. In: Handbook of behavioral neurobiology, Vol. 2, ed. M. Gazzaniga, pp. 345–378. New York: Plenum Press.
44. Hink, R.F., Kohler, H., Deecke, L., and Kornhuber, H.H. 1982. Risk-taking and the human Bereitschaftspotential. Electroencephalogr. Clin. Neurophysiol. 53:361–373.
45. Hofmann, M.J., and Salapatek, P. 1981. Young infants' event-related potentials (ERPs) to familiar and unfamiliar visual and auditory events in a recognition memory task. Electroencephalogr. Clin. Neurophysiol. 52:405–417.
46. Jacobson, G.P., and Gans, D.P. 1981. An electrophysiological correlate of phonemic discrimination. Audiology 20:480–487.
47. Jung, R., Hufschmidt, A., and Moschallski, W. 1982. Slow brain potentials in writing: Interaction of writing hand and speech dominance in right-handers. Arch. Psychiatr. Nervenkr. 232:305–324.
48. Kaufman, L., Okada, Y., Brenner, D., and Williamson,

S.J. 1981. On the relation between somatic evoked potentials and fields. Intern. J. Neurosci. 15:223–239.
49. Kaufman, L., and Williamson, S.J. 1982. Magnetic location of cortical activity. Ann. N.Y. Acad. Sci. 388:197–213.
50. Klinke, R., Fruhstorfer, H., and Finkenzeller, P. 1968. Evoked responses as a function of external and stored information. Electroencephalogr. Clin. Neurophysiol. 25:119–122.
51. Klorman, R., Salzman, L.F., Bauer, L.O., Coons, H.W., Borgstedt, A.D., and Halpern, W.I. 1983. Effects of two doses of methylphenidate on cross-situational and borderline hyperactive children's evoked potentials. Electroencephalogr. Clin. Neurophysiol. 56:169–185.
52. Knight, R.T. 1984. Decreased response to novel stimuli after prefrontal lesions in man. Electroencephalogr. Clin. Neurophysiol. 59:9–20.
53. Knight, R.T., Hillyard, S.A., Woods, D.L., and Neville, H.J. 1981. The effects of frontal cortex lesions on the event-related potentials during auditory selective attention. Electroencephalogr. Clin. Neurophysiol. 52:571–582.
54. Kornhuber, H.H., and Deecke, L. 1965. Hirnpotentialänderungen bei Willkürbewegungen und passiven Bewegungen des Menschen: Bereitschaftspotential und reafferente Potentiale. Pfluegers Arch. 284:1–17.
55. Kristeva, R., Keller, E., Deecke, L., and Kornhuber, H.H. 1979. Cerebral potentials preceding unilateral and simultaneous bilateral finger movements. Electroencephalogr. Clin. Neurophysiol. 47:229–238.
56. Kurtzberg, D., and Vaughan, H.G. 1982. Topographic analysis of human cortical potentials preceding self-initiated and visually triggered saccades. Brain Res. 243:1–9.
57. Kurtzberg, D., Vaughan, H.G., and Kreuzer, J. 1979. Task-related cortical potentials in children. Prog. Clin. Neurophysiol. 6:216–223.
58. Lavine, R.A., Buchsbaum, M.S., and Schechter, G. 1980. Human somatosensory evoked responses: Effects of attention and distraction on early components. Physiol. Psychol. 8:405–408.
59. Libet, B., Wright, E.W., and Gleason, C.A. 1982. Readiness-potentials preceding unrestricted "spontaneous" vs. pre-planned voluntary acts. Electroencephalogr. Clin. Neurophysiol. 54:322–335.
60. Loveless, N.E., and Sanford, A.J. 1973. The CNV baseline: Considerations of internal consistency of data. Electroencephalogr. Clin. Neurophysiol. Suppl. 33:19–23.
61. McAdam, D.W., and Whittaker, H.A. 1971. Language production: Electroencephalographic localization in the normal human brain. Science 172:499–502.
62. McCallum, W.C., and Cummins, B. 1973. The effects of brain lesions on the contingent negative variation in neurosurgical patients. Electroencephalogr. Clin. Neurophysiol. 35:449–456.
63. McCallum, W.C., and Curry, S.H. 1981. Late slow wave components of auditory evoked potentials: Their cognitive significance and interaction. Electroencephalogr. Clin. Neurophysiol. 51:123–137.
64. McCallum, W.C., and Walter, W.G. 1968. The effects of attention and distraction on the contingent negative variation in normal and neurotic subjects. Electroencephalogr. Clin. Neurophysiol. 25:319–329.
65. McCarthy, G., and Donchin, E. 1981. A metric for thought: A comparison of P300 latency and reaction time. Science 211:77–80.

66. Marsden, C.D. 1982. The mysterious motor function of the basal ganglia: The Robert Wartenberg lecture. Neurology 32:514–539.
67. Novick, B., Vaughan, H.G., Kurtzberg, D., and Simson, R. 1980. An electrophysiologic indication of auditory processing defects in autism. Psychiatry Res. 3:107–114.
68. Obeso, J.A., Rothwell, J.C., and Marsden, C.D. 1981. Simple tics in Gilles de la Tourette's syndrome are not prefaced by a normal premovement EEG potential. J. Neurol. Neurosurg. Psychiatry 44:735–738.
69. Okada, Y.C., Kaufman, L., and Williamson, S.J. 1983. The hippocampal formation as a source of the slow endogenous potentials. Electroencephalogr. Clin. Neurophysiol. 55:417–426.
70. Okada, Y.C., Williamson, S.J., and Kaufman, L. 1982. Magnetic field of the human sensorimotor cortex. Int. J. Neurosci. 17:33–38.
71. Otto, D.A., Benignus, V.A., Muller, K.E., and Barton, C.N. 1981. Effects of age and body lead burden on CNS function in young children. I. Slow cortical potentials. Electroencephalogr. Clin. Neurophysiol. 52:229–239.
72. Papakostopoulos, D. 1980. A no-stimulus, no-response, event-related potential of the human cortex. Electroencephalogr. Clin. Neurophysiol. 48:622–638.
73. Pfefferbaum, A., Ford, J.M., Roth, W.T., and Kopell, B.S. 1980. Age-related changes in auditory event-related potentials. Electroencephalogr. Clin. Neurophysiol. 49:266–276.
74. Pfefferbaum, A., Ford, J.M., Wenegrat, B.G., Roth, W.T., and Kopell, B.S. 1984. Clinical application of the P3 component of event-related potentials. I. Normal aging. Electroencephalogr. Clin. Neurophysiol. 59:85–103.
75. Pfefferbaum, A., Horvath, T.B., Roth, W.T., and Kopell, B.S. 1979. Event-related changes in chronic alcoholics. Electroencephalogr. Clin. Neurophysiol. 47:637–647.
76. Pfefferbaum, A., Wenegrat, B.G., Ford, J.M., Roth, W.T., and Kopell, B.S. 1984. Clinical application of the P3 component of event-related potentials. II. Dementia, depression and schizophrenia. Electroencephalogr. Clin. Neurophysiol. 59:104–124.
77. Picton, T.W., and Hillyard, S.A. 1974. Human auditory evoked potentials. II. Effects of attention. Electroencephalogr. Clin. Neurophysiol. 36:191–199.
78. Plattig, K.H., and Kobal, G. 1979. Spatial and temporal distribution of olfactory evoked potentials and techniques involved in their measurement. In: Event-related potentials in man: Applications and problems, ed. D. Lehmann and E. Callaway, pp. 285–301. New York: Plenum Press.
79. Podlesny, J.A., and Dustman, R.E. 1982. Age effects on heart rate, sustained potential, and P3 responses during reaction-time tasks. Neurobiol. Aging 3:1–9.
80. Prevec, T.S., and Ribarić, K. 1980. Improved contingent negative variation audiometry. Audiology 19:457–468.
81. Prevec, T., Ribarić, K., and Butinar, D. 1984. Contingent negative variation audiometry in children. Audiology 23:114–126.
82. Rebert, C.S., and Knott, J.R. 1970. The vertex non-specific evoked potential and latency of contingent negative variation. Electroencephalogr. Clin. Neurophysiol. 28:561–565.
83. Reite, M., Zimmerman, J.T., Edrich, J., and Zimmerman, J.E. 1982. Auditory evoked magnetic fields: Response amplitude vs. stimulus intensity. Electroencephalogr. Clin. Neurophysiol. 54:147–152.

84. Reite, M., Zimmerman, J.T., and Zimmerman, J.E. 1982. MEG and EEG auditory responses to tone, click and white noise. Electroencephalogr. Clin. Neurophysiol. 53:643–651.
85. Rizzo, P.A., Spadaro, M., Albani, G., and Morocutti, C. 1983. Contingent negative variation and phobic disorders. Neuropsychobiology 9:73–77.
86. Rohrbaugh, J.W., Syndulko, K., Sanquist, T.F., and Lindsley, D.B. 1980. Synthesis of the contingent negative variation brain potential from noncontingent stimulus and motor elements. Science 108:1165–1168.
87. Romani, G.L., Williamson, S.J., and Kaufman, L. 1982. Tonotopic organization of the human auditory cortex. Science 216:1339–1340.
88. Rösler, F. 1981. Event-related brain potentials in a stimulus-discrimination learning paradigm. Psychophysiology 18:447–455.
89. Roth, W.T., Blowers, G.H., Doyle, C.M., and Kopell, B.S. 1982. Auditory stimulus intensity effects on components of the late postive complex. Electroencephalogr. Clin. Neurophysiol. 54:132–146.
90. Roth, W.T., Pfefferbaum, A., Horvath, T.B., Berger, P.A., and Kopell, B.S. 1980. P3 reduction in auditory evoked potentials of schizophrenics. Electroencephalogr. Clin. Neurophysiol. 49:497–505.
91. Ruchkin, D.S., Sutton, S., and Tueting, P. 1975. Emitted and evoked P300 potentials and variation in stimulus probability. Psychophysiology 12:591–595.
92. Sanquist, T.F., Beatty, J.T., and Lindsley, D.B. 1981. Slow potential shifts of human brain during forewarned reaction. Electroencephalogr. Clin. Neurophysiol. 51:639–649.
93. Schwent, V.L., and Hillyard, S.A. 1975. Evoked potential correlates of selective attention with multi-channel auditory inputs. Electroencephalogr. Clin. Neurophysiol. 38:131–138.
94. Schwent, V.L., Hillyard, S.A., and Galambos, R. 1976. Selective attention and the auditory vertex potential. I. Effects of stimulus delivery rate. Electroencephalogr. Clin. Neurophysiol. 40:604–614.
95. Shelburne, S.A. 1973. Visual evoked responses to language stimuli in normal children. Electroencephalogr. Clin. Neurophysiol. 34:135–143.
96. Shibasaki, H., Barrett, G., Halliday, E., and Halliday, A.M. 1980. Components of the movement-related cortical potential and their scalp topography. Electroencephalogr. Clin. Neurophysiol. 49:213–226.
97. Shibasaki, H., Barrett, G., Halliday, E., and Halliday, A.M. 1980. Cortical potentials following voluntary and passive finger movements. Electroencephalogr. Clin. Neurophysiol. 50:201–213.
98. Shibasaki, H., Barrett, G., Halliday, E., and Halliday, A.M. 1981. Cortical potentials associated with voluntary foot movement in man. Electroencephalogr. Clin. Neurophysiol. 52:507–516.
99. Shibasaki, H., Shima, F., and Kuroiwa, Y. 1978. Clinical studies of the movement-related cortical potential (MP) and the relationship between the dentatorubrothalamic pathway and readiness potential (RP). J. Neurol. 219:15–25.
100. Simson, R., Vaughan, H.G., and Ritter, W. 1976. The scalp topography of potentials associated with missing visual or auditory stimuli. Electroencephalogr. Clin. Neurophysiol. 40:33–42.
101. Simson, R., Vaughan, H.G., and Ritter, W. 1977. The scalp topography of potentials in auditory and visual dis-

crimination tasks. Electroencephalogr. Clin. Neurophysiol. 42:528–535.
102. Smith, D.B., Allison, T., Goff, W.R., and Principato, J.J. 1971. Human odorant evoked responses: Effects of trigeminal or olfactory deficit. Electroencephalogr. Clin. Neurophysiol. 30:313–317.
103. Squires, N.K., Galbraith, G.C., and Aine, C.J. 1979. Event related potential assessment of sensory and cognitive deficits in the mentally retarded. In: Event-related potentials in man: Applications and problems, ed. D. Lehmann and E. Callaway, pp. 397–413. New York: Plenum Press.
104. Squires, K.C., and Hecox, K.E. 1983. Electrophysiological evaluation of higher level auditory processing. Semin. Hear. 4:415–432.
105. Squires, N.K., Squires, K.C., and Hillyard, S.A. 1975. Two varieties of long-latency positive waves evoked by unpredictable auditory stimuli in man. Electroencephalogr. Clin. Neurophysiol. 38:387–401.
106. Stuss, D.T., Sarazin, F.F., Leech, E.E., and Picton, T.W. 1983. Event-related potentials during naming and mental rotation. Electroencephalogr. Clin. Neurophysiol. 56:133–146.
107. Sutton, S., Braren, M., Zubin, J., and John, E.R. 1965. Evoked-potential correlates of stimulus uncertainty. Science 150:1187–1188.
108. Sutton, S., Tueting, P., Subin, J., and John, E.R. 1967. Information delivery and the sensory evoked potential. Science 155:1436–1439.
109. Tecce, J.J., Cattanach, L., Yrchik, D.A., Meinbresse, D., and Dessonville, C.L. 1982. CNV rebound and aging. Electroencephalogr. Clin. Neurophysiol. 54:175–186.
110. Teyler, T.J., Cuffin, B.N., and Cohen, D. 1975. The visual evoked magnetencephalogram. Life Sci. 17:683–691.
111. Timsit-Berthier, M., Delaunoy, J., Koninckx, N., and Rousseau, J.C. 1973. Slow potential changes in psychiatry. I. Contingent negative variation. Electroencephalogr. Clin. Neurophysiol. 35:355–361.
112. Tueting, P., and Levit, R.A. 1979. Long-term changes of event-related potentials in normals, depressives and schizophrenics. Prog. Clin. Neurophysiol. 6:265–279.
113. Vaughan, H.G., and Ritter, W. 1970. The sources of auditory evoked responses recorded from the human scalp. Electroencephalogr. Clin. Neurophysiol. 28:360–367.
114. Walter, W.G., Cooper, R., Aldridge, V.J., McCallum, W.C., and Winter, A.L. 1964. Contingent negative variation: An electric sign of sensorimotor association and expectancy in the human brain. Nature 203:380–384.
115. Weinberg, H., Walter, W.G., and Crow, H.J. 1970. Intracerebral events in humans related to real and imaginary stimuli. Electroencephalogr. Clin. Neurophysiol. 29:1–9.
116. Wood, C.C., Allison, T., Goff, W.R., Williamson, P.D., and Spencer, D.D. 1980. On the neural origin of P300 in man. Prog. Brain Res. 54:51–56.
117. Yingling, C.D., and Hosobuchi, Y. 1984. A subcortical correlate of P300 in man. Electroencephalogr. Clin. Neurophysiol. 59:72–76.

Appendix

List of major manufacturers of averagers sold in the United States

This list is incomplete and does not intend to suggest that the equipment of the manufacturers listed is better than other equipment.

Amplaid USA, Inc.
545 West Golf Road
Arlington Heights, IL 60005
(312) 437–2298

Bio-Logic Systems Corp.
1215 Washington Avenue
Wilmette, IL 60091
(312) 256–7060

Cadwell Laboratories, Inc.
4312 West Kennewick Avenue
Kennewick, WA 99336
(509) 735–6481

Disa Dantec Electronics
779 Susquehanna Avenue
Franklin Lakes, NJ 07417
(201) 891–9460

Grass Instrument Company
101 Old Colony Avenue
Quincy, MA 02169
(617) 773–0002

Life-Tech, Inc.
P.O. Box 36221
Houston, TX 77036
(713) 988–6060

Neuro Diagnostics Inc.
111 West Dyer Road
Santa Ana, CA 92707
(714) 556–4439

Nicolet Biomedical Instruments
5225 Verona Road
Madison, WI 53711
(608) 271–3333

Nihon Kohden America, Inc.
1652 Deere Avenue
Irvine, CA 92714
(714) 957–3340

Tracor Analytic, Inc.
1842 Brummel Drive
Elk Grove Village, IL 60007
(312) 364–9100

Index

Numbers in the index refer to pages. Italicized numbers distinguish major from minor items.

EP refers to evoked potentials in general and is listed ahead of other entries.

VEP, checkerboard, refers to the transient VEP to checkerboard pattern reversal. *VEP, checkerboard-on,* refers to the transient VEP to checkerboard pattern appearance. *VEP, flash,* refers to the transient VEP to diffuse light flashes.

Abbreviations are listed on page xiii

Abnormal. *See* Criteria for abnormal
Absence
  EP, 70
  BAEP, 216, *218, 219–221*
  SEP
    arm, *301,* 302
    leg, *330,* 331
  VEP
    checkerboard, 117, *121*
    flash, 140
Accommodation, VEP, steady state, checkerboard, 147
Acoustic nerve action potential, ECochG, 10, *247–249*
Acoustic neurinoma
  BAEP, 224–226
  ECochG, 249
  MLAEP, 241
  SEP, trigeminal, 340
Action potential, nerve fibers, 14
  See also Acoustic nerve action potential; Compound action potential
Active filter, 28, 30, 31
Active recording electrode, 24
Adaptation to fast stimulation, BAEP, 210
  See also Rate-dependent latency shift
Adaptive filter, 34
Addition of EPs, 50–51
Address. *See* Point

Adie's syndrome, SEP
  arm, 304
  leg, 336
Adrenoleukodystrophy
  BAEP, 231
  SEP
    arm, 318
    leg, 336
  VEP, checkerboard, 129
AEP classification, 195
Age effect
  EP, *61,* 66
  BAEP, 206, *208*
  CNV, 372
  LLAEP, 245
  P300, 369
  SEP, arm, 297
  VEP
    checkerboard, 96, *98*
    flash, 138
  See also Maturation
Albinism, VEP, checkerboard, 131, 146
Alcohol, acute effect
  BAEP, 230
  LLAEP, 245
  SEP, arm, 284
  VEP, flash, 142
Alcoholism, chronic
  BAEP, 230
  P300, 370
  VEP
    checkerboard, 129
    flash, 142
Alertness, 41, *61*
Aliasing, 44–45

383

Alpha rhythm, EEG, 61, 149
Alternating current (AC) amplifier, 30
Alzheimer's disease. *See* Senile dementia
Ambient light, VEP, 88, 91, 139
Amblyopia ex anopia
  contrast sensitivity, 153
  ERG, pattern, 159
  VEP
    checkerboard, 133, 148
    checkerboard-on and flash, 146
    flash, 143
    sine grating, 153, 155
Amblyopia
  nutritional, VEP, checkerboard, 133
  toxic, VEP
    checkerboard, 133, 148
    flash, 143
American EEG Society, Guidelines for EP recording, x, 63, 66, 325
  See also International Federation of Societies for Electroencephalography and Clinical Neurophysiology
Aminoglycosides, BAEP, 230
Amitriptylene, LLAEP, 245
Amplifier, 21
Amplifier
  alternating current (AC), 30
  differential, 22–27
  direct current (DC), 30
  discrimination, 23–27
  gain, 27
  noise level, 27
  polarity convention, 23
  sensitivity, 27
Amplitude
  EP, 57–59, 63
  measurement, 22, 54–55
  peak, 57–59
  peak-to-peak, 59
Amplitude ratio
  EP, 59, 63
  abnormal
    EP, 70
    BAEP, 218, 221
    SEP, arm, 302
    VEP, checkerboard, 117, 121
Amyotrophic lateral sclerosis
  BAEP, 230
  SEP, arm, 317
Analog filter, 28
Analog recording, 36
Analog-to-digital conversion, 36, 45
Analysis period
  EP, 36, 41–45
    transient vs. steady-state, 41
  BAEP, 206, 212
  CNV, 370
  ECochG, 246
  LLAEP, 244
  MLAEP, 240
  SEP
    arm, 292, 298
    leg, 323, 327
  VEP, 94, 97, 109
Anatomical correlation, 71
  See also Generator
Anesthesia
  BAEP, 226, 228, 230
  SEP, leg, 338
  VEP, flash, 141
  See also Surgical monitoring
Anode, stimulus, SEP, 285
Anoxic encephalopathy. *See* Postanoxic encephalopathy
Anterior spinal artery infarct, SEP, arm, 308
Antidromic motor conduction, SEP, 286, 290
Anxiety, CNV, 371
Apneic syndrome, BAEP, 231
Arm length. *See* Limb length
Artifact
  EP, 22
  60 Hz, 20, 26
  stimulus
    AEP, 197
    SEP, 288
Artifact rejection
  by computer, 41, 47–48
  by differential amplifier, 25–26
Artificial ear, 198
Artificial pupil, 88, 91
Asphyxic infantile cerebral damage, BAEP, 231
Astigmatism, VEP
  bar grating, 155
  checkerboard, 115
  sine grating, 155
Asymmetry, VEP, checkerboard, half-field, *106*, 114
Attention
  EP, 61
  CNV, 371–372
  LLAEP, 197, *242*
  P300, 368
  SEP, 284
  VEP, 89
Auditory agnosia, MLAEP, 241
Autism. *See* Infantile autism
Average reference electrode, 25
Averager, 38
Averaging, general principle, 36–38
Axonal lesion, 69–70

Background illumination, 88
  See also Ambient light
BAEP, 205–235
Bandpass, bandwidth, 28
  See also Filter settings
Bar grating VEP, 155
Batten's disease, BAEP, 231
Beep-boop stimuli, 368
Bereitschaftspotential, 372
Bias potential, recording electrode, 20

Bifid peak
 EP, 55
 BAEP, 207
 VEP, 97
Bilateral stimulation
 EP, 12
 AEP. See Binaural stimulation
 SEP, 287
 VEP. See Binocular stimulation; Full-field stimulation; Half-field stimulation
Bilobed. See Bifid peak
Bin. See Point
Binaural interaction, 210
Binaural stimulation, 202, 210–211
Binocular stimulation, 103
Binocular vision, VEP
 bar grating, 155
 checkerboard-on, 146
 grid pattern, 157
 moving dot pattern, 157
 sine grating, 155
Bipolar montage, 25
Bipolar recording, 24–25
Birth injury, BAEP, 231
Bit, 36, 45
Bladder stimulus, SEP, 340
Blink VEP, 158
Blood pressure, EP, 61
 See also Surgical monitoring
Blurring, effect on VEP
 checkerboard, 101, 102
 sine grating, 151

Body height, EP, 62
 See also Limb length
Body temperature. See Temperature
Bone conduction. See Hearing loss, conductive; Stimulator, AEP, bone vibrator
Bowel dysfunction, SEP, pudendal, 340
Brachial plexus lesion, SEP, arm, 306
Brain death. See Cerebral death
Brain stem stroke. See Stroke, brain stem
Brain stem tumor. See Tumor, brain stem
Brain tumor. See Tumor, brain
Brightness contrast
 definition, 91
 counterphase modulation, 91–92
 depth modulation, 91, 153
 effect on VEP, checkerboard, 101–102
Broadband click, 199
Brown-Sequard syndrome, SEP, leg, 334

Calibration
 EP, 22
 sound stimulus intensity, 198
Candela/m², 91
Carotid endarterectomy, SEP, arm, 317

Carpal tunnel syndrome, SEP, arm, 306
Cataract, VEP
 checkerboard, 118, 121, *134*
 checkerboard-on, 146
 flash, 143
 laser speckle pattern, 157
Cathode, stimulus, SEP, 285
Central conduction time
 definition, 57
 SEP
  arm, 299, 301, 302
  leg, 324
 See also Central conduction velocity, SEP, leg; Interpeak latency; Peripheral nerve conduction velocity
Central conduction velocity, SEP, leg, 326, 328 *330–331*
Central pontine myelinolysis, BAEP, 230
Central retinal artery occlusion, VEP, flash, 143
Central serous retinopathy, VEP, checkerboard, 134
Cerebellopontine angle tumor, BAEP, 224–226
Cerebral asthenopia, VEP, horizontal line pattern, 157
Cerebral death
 BAEP, 227
 SEP, arm, 315
 VEP, flash, 141

Cerebral infarct. See Stroke, cerebral
Cerebral tumor. See Tumor, brain
Cervical cord injury, SEP, arm, 308
Cervical cord tumor, SEP, arm, 308
Cervical rib, SEP, arm, 304–305
Cervical root lesion, SEP, arm, 306
Cervical SEP, 283, *290–293*
Cervical spondylotic myelopathy, SEP
 arm, 308
 leg, 334
CFPD, 149, 151
Channels, recording. See Number of recording channels
 spatial frequency, 151
Charcot-Marie-Tooth disease
 SEP
  arm, 304
  leg, 336
 VEP, checkerboard, 128
Check size
 measurement, 92
 effect on VEP, 102
Checkerboard flash VEP, 145–146
Checkerboard reversal VEP, transient, 96–134
Checkerboard-on, -off (appearance and disappearance) VEP, 145–146

385

Chiasmal lesions, VEP, checkerboard, 130–131
Chiasmal strategy, VEP, checkerboard, 111–114
　See also Half-field stimulation
Chronic alcoholism. See Alcoholism
Chronic hemodialysis, VEP
　checkerboard, 129
　flash, 142
Chronic renal failure
　BAEP, 228, 230
　P300, 370
　SEP, arm, 304
　VEP
　　checkerboard, 129
　　flash, 142
Chronic vegetative state, SEP, arm, 315
Classification
　EP, 8–12
　AEP, 195
　SEP, 283–284
　VEP, 86–87
Clavicular SEP, 283, *290*
Click stimulus, 198–200
　broadband, 199
　condensation or compression, *201–202*, 207, 210, 247
　filtered, 199
　polarity or phase, *201–202*, 210
　rarefaction, *201–202*, 207, 210, 247

Clinical interpretation
　EP, 66
　BAEP, 219–223
　SEP
　　arm, 302–303
　　leg, 331–333
　VEP, checkerboard, 117–122
　See also Strategy of stimulation and recording
Clioquinol intoxication, SEP, leg, 336
Clip recording electrode, 18
CM, 247, *249*
CNV, 370–372
CNS depressant drugs
　BAEP, 228, 230
　LLAEP, 245
　MLAEP, 240
　SEP, 284
　SEP, arm, 318
　See also Alcohol; Amitriptylene; Diazepam; Enflurane; Halothane, Imipramine; Phenobarbital; Sedation; Thiopental
Cochlear microphonic, 247, *249*
Cochlear sound receptors, 199
Cognitive potential, 368
Cold stimulation, 343
Collapse of outer ear canal, 197
Color VEP, 157–158
Coma, BAEP
　metabolic and toxic encephalopathies 226, *228–229*

　structural lesions, 226–228
　See also Head injury; Cerebral death
Common mode rejection ratio, 26
Common peroneal nerve stimulation
　SEP, 320, 325
　electrode placement, 326
Complicated pregnancy, BAEP, 231
Component
　EP,
　　definition, 7
　　spatial frequency, checkerboard pattern. See Fourier component
Compound action potential, 14
Compression click. See Click stimulus, condensation
Computer
　averaging, 38
　general purpose, 38
　number of channels, 38
　special purpose, 38
　triggering, 39
Condensation click. See Click stimulus, condensation
Conduction time. See Central conduction time
Conduction velocity. See Central conduction velocity; Peripheral nerve conduction velocity
Conductive hearing loss. See Hearing loss, conductive

Congenital insensitivity to pain, SEP arm and tooth pulp, 306
Congenital oculomotor apraxia, VEP, checkerboard, 134
Congenital nystagmus, VEP, flash, 143
Constant current stimulator, 286–287
Contingent negative variation (CNV), 370–372
Contrast, brightness. See Brightness contrast
Contrast sensitivity, sine gratings, 153
Controls, normal, 66–68
Convulsive disorder. See Seizure disorder
Cord. See Spinal cord; Cervical cord; Lumbar; Thoracic
Corneal opacities, VEP, checkerboard, 121, 134
Correlation
　anatomical, 71
　See also Generator
　clinical. See Clinical interpretation
Cortical blindness, VEP
　checkerboard, 132
　flash, 141–142
　steady state
　　checkerboard, 147
　　sine grating, 153

Cortical deafness
  BAEP, 231
  MLAEP, 241
Cortical EP, 8, 14
Counter, number of responses, 40
  See also Number of responses
Counterphase contrast modulation, 91–92
Criteria for abnormal
  EP, 68–70
  BAEP, 206, 218–219
  SEP
    arm, 292, 301–302
    leg, 330–331, 323–324
  VEP, checkerboard, 97, 117
Critical frequency of photic driving (CFPD), 149, 151
Critical fusion frequency, 149, 151
Critical sampling rate, 42
Crossed acoustic response, 250
Crossed asymmetry, VEP, checkerboard, half-field, 114
Cursors, 22, 49–50
Cyclopean visual stimulus, 157
Cycloplegia. See Pupillary dilatation

Decibels
  amplifier gain, 27
  auditory stimulus intensity, 200

Dejerine-Roussy syndrome, SEP, arm, 311
Dejerine-Sottas disease, BAEP, 230
Dementia, P300, 370
  See also Senile dementia
Demyelination
  EP, 68
  VEP, checkerboard, 111
Depression
  BAEP, 230
  LLAEP, 245
  SEP, arm, 318
Derived BAEP, 209
Dermatomal stimulation. See Segmental stimulation
Detached retina, VEP, flash, 143
Developmental disorders, VEP, flash, 142
Dextroamphetamine, LLAEP, 245
Diabetes
  BAEP, 230
  SEP, leg, 336
  VEP, checkerboard, 129
Diabetic ketoacidosis, BAEP, 228
Diamond pattern stimulus, VEP, checkerboard, 93, 103, 124
Diazepam, LLAEP, 245
Differential amplifier, 22–27
Diffuse light effect vs. pattern effect, 102
Digital filter, 28, 31–34

Digital recording, 36
Digitization, 36, 45–46
Dipole, 13
Direct current (DC) amplifier, 30
Disc, magnetic, 51
Discrimination, 23–27
Dispersion, 68, 69
Display gain, 27
Distribution
  EP, 59
  statistical, 67–68
Dominance
  ear, 67
  eye, 67, 99
Dorsal column system, SEP, 284
Dot stimuli
  moving, VEP, 157
  stationary, VEP, 157
Down's syndrome
  BAEP, 231
  SEP, arm, 318
  VEP, flash, 142
Drugs. See Aminoglycosides; Clioquinol; CNS depressant drugs; Dextroamphetamine; Ethambutol; Methylphenidate; Ototoxic drugs; Phenytoin; Quinine
Drusen, optic nerve head, VEP, flash, 143
Dwell time, 36, 42–45
  See also Sampling rate

Dynamic contrast sensitivity, 153
Dynamic random dot correlogram and stereogram, 157
Dyslexia, VEP
  flash, 142
  letters and words, 157
Dyssynergia cerebellaris myoclonica, SEP, arm, 317

Ear canal collapse, 197
Ear dominance, 67
Early AEP, 195, 240
  See also BAEP; Slow brain stem AEP
Earphone, 197
ECG recording electrode, 18
ECochG, 246–249
EEG recording electrode, 17–18
Electric field
  EP, 13, 62
  VEP, checkerboard, 106
Electric response audiometry (ERA), 212, 214–216
  BAEP, 232–235
  CNV, 372
  ECochG, 247
  40 Hz AEP, 242
  FFP, 239
  LLAEP, 245–246
  MLAEP, 241–242
  Sonomotor AEP, 250–251
  See also Hearing loss

Electric safety, 21, 26, 288
Electric stimulus. *See* Stimulus, electric
Electrocerebral silence
  BAEP, 229
  SEP, arm, 315
  VEP, flash, 141
Electrocochleogram (ECochG), 246–249
Electroconvulsive therapy
  BAEP, 230
  VEP, flash, 142
Electrocorticographic recording electrode, 19
Electrode
  recording. *See* Recording electrode
  stimulating. *See* Stimulating electrode
Electrodynamic and electrostatic earphone, 197
Emitted potential, P300, 368
Endarterectomy, carotid, SEP, arm, 317
Endocrine orbitopathy, VEP, checkerboard, 129
Endogenous potential, P300, 368
Enflurane, 228
EP
  definition, 7
  addition, 50–51
  bilateral stimulation, 12
  classification, 8–12
  cortical, 8, 14
  distribution, 59
  far-field, 11
  generator, *12–14*, 62, 374
  hard copy, 51
  inversion, 50–51
  near-field, 11
  polarity, 54
  power spectrum, 28
  replication, 40
  steady-state, 11–12
  subcortical, 8–10, 14
  subtraction, 50–51
  transient, 11
  unilateral stimulation, 12
  variability, 41, *61–62*
Epidural recording, SEP, leg, 325, 338
Epilepsy. *See* Seizure disorder
Epoch. *See* Analysis period
ERA. *See* Electric response audiometry
Erb's point potential, 283, *290*
ERG, 159
  recording electrode, 11, *24*
ERP, 7, 368–374
Ethambutol optic neuropathy, VEP, checkerboard, 134
Event-related potential, 7, *368–374*
Expectancy wave, 372
Exploring electrode, 24
Extratympanic ECochG, 246
Eye dominance, 67, *99*
Eye injury, VEP, flash, 143
Eye stimulation, electric, 158

Facial nerve lesion
  BAEP, 231
  sonomotor AEP, 251
False negative and positive test results, 68
Familial spastic paraplegia, SEP, arm, 317
Familial startle disease, SEP, arm, 317
Far-field recording
  EP, 11
  BAEP, 195
  SEP
    arm, 283, *295*
    leg, 326
  VEP, flash, 136
FFP, 237–239
Field size, VEP, checkerboard, *93, 96*
Filter
  general description, 27–34
  active, 28, 30, 31
  adaptive, 34
  analog, 28
  digital, 28, *31–34*
  high frequency or low-pass, 30–31
  low frequency or high-pass, 28–30
  narrow-band. *See* Narrow-band filter
  notch, 209
  passive, 28, 31
  phase shift, 31
  roll-off, 30, 31
  settings
    BAEP, 206, *211*
    ECochG, 246
    40 Hz AEP, 242
    LLAEP, 244
    MLAEP, 240
    SEP, *288*, 292
    Slow brain stem AEP, 237
    VEP, *94*, 97
  60 Hz, 31
  slope, 30, 31
  time constant, 30
  time varying, 42
  tuned, 52
  Wiener, 42
Filtered click, 199
Fixation point, 93, 99–100
Flash ERG, 159
Flash VEP, 86–87
  steady-state, 149–151
  transient, 136–143
Floating inputs, 26
Focusing, VEP, checkerboard, 118
40 Hz AEP, 242
Fourier analysis. *See* Frequency analysis
Fourier component, checkerboard pattern, 93, 103
Free field acoustic stimulation, 197
Frequency analysis
  EP, 52
  VEP, steady state
    checkerboard, 150–151
    sine grating, 155

Frequency following potential
(FFP), 237–239
Frequency, stimulus. See Stimulus rate
Friedreich's ataxia
  BAEP, 229
  LLAEP, 245
  SEP
    arm, 304, 317
    leg, 336
  VEP
    checkerboard, 127–128
    flash, 141
Frontal lobe lesion
  P300, 370
  SEP, arm, 313
Full-field stimulation, VEP
  checkerboard, 93, 103–106
  flash, 139

Gain, definition, 27
  display, 27
  recording, *27*, 45
Ganzfeld stimulator, VEP, flash, 139
Gaucher's disease, 232
Gaze shift VEP, 158
General purpose computer, 38
Generator
  EP, *12–14*, 62, 374
  acoustic nerve action potential, 247
  BAEP, 212
  CM, 249

FFP, 239
LLAEP, 245
MLAEP, 241
P300, 368
SEP
  arm, 283, 284
    antidromic, 286
    cervical, 293
    clavicular, 290
    scalp, 294
  leg
    low thoracic, 325
    lumbar, 325
slow brain stem AEP, 327
sonomotor AEP, 250
SP, 249
VEP, checkerboard, 106
Glaucoma, ERG, pattern, 159
  contrast sensitivity, 153
  VEP
    checkerboard, 121, *134*
    flash, 142–143
    sine grating, 152
    steady state, checkerboard, 148
    steady state, sine grating, 153–155
Gold foil ERG electrode, 19
Grid pattern VEP, 157
Ground electrode
  EP, 21, 26
  AEP, 203
  SEP, 288, 298
  VEP, checkerboard, 106
Guidelines for EP recordings,

American EEG Society, x, 63, 66, 325
See also Recommendations for EP recording
Guillain-Barré syndrome, SEP, arm, 304

Habituation, EP, 61
Half-field stimulation
  method, 93
  VEP, checkerboard
    normal, 103, 106–109
    abnormal, 121, 130, 132
Hallervorden-Spatz disease, BAEP, 231
Halothane, BAEP, 228
Handedness, EP, 67
Hard copy, 51
Head injury
  BAEP, 226
  LLAEP, 245
  SEP, arm, 315
  VEP
    checkerboard, 129
    flash, *141*, 143
Heart rate, EP, 61
Hearing and BAEP, 216, 232
Hearing level (HL), 200
Hearing loss
  general, 209
  BAEP, 214, *232–235*
  FFP, 239
  LLAEP, 245–246
  MLAEP, 241–242

Sonomotor AEP, 251
See also Electric response audiometry
conductive
  BAEP, 222, 223, *233*
  ECochG, 247
neural
  BAEP, 233–234
  ECochG, 249
nonorganic. See Nonorganic hearing loss
sensorineural, BAEP, 223, *233–235*
sensorineural, ECochG, 249
sensory
  BAEP, 233–234
  ECochG, 249
Hearing threshold. See Electric response audiometry
Heat stimulation, SEP, 343
Heating, effect on EP. See Temperature
Height. See Body height; Limb length
Hemifacial spasm, BAEP, 230, 231
Hemi-field stimulation. See Half-field stimulation
Hemiretina stimulation, 93, 103
Hemispherectomy
  VEP, checkerboard-on, 146
  SEP, arm, 313–315
Hemodialysis. See Chronic hemodialysis

Hemorrhage, intracerebral,
    BAEP, 226
Hepatic encephalopathy
    BAEP, 228–229
    SEP, arm, 318
Hereditary cerebellar ataxia
    BAEP, 229
    SEP, arm, 317
    VEP, checkerboard, 127–128
Hereditary optic neuropathy.
    See Leber's hereditary
        optic neuropathy
Hereditary pressure-sensitive
    neuropathy, 306
Hereditary sensory neuropathy,
    BAEP, 230
Hereditary spastic ataxia, VEP,
    checkerboard, 127–128
Hereditary spastic paraplegia,
    VEP
    flash, 141
    checkerboard, 128
Heredity, EP, 62
Herpes zoster radiculitis, SEP,
    arm, 306
Hertz (Hz), definition, 17
High frequency filter, 30–31
High-pass filter. *See* Low frequency filter
Hodgkin's disease of spinal column, SEP, leg, 334
Horizontal line pattern VEP,
    157
Huntington's disease
    BAEP, 230
    SEP
        arm, 317
        leg, 338
    VEP
        flash, 142
        checkerboard, 132
Hydrocephalus, VEP, flash, 141
Hydromyelia, SEP, arm, 308
Hyperactive child
    LLAEP, 245
    P300, 370
    VEP, flash, 142
Hypercapnia, BAEP, 230
Hyperglycinemia
    BAEP, 230
    VEP, flash, 142
Hyperkinetic child. *See* Hyperactive child
Hyperthermia. *See* Temperature
Hyperthyroidism
    SEP, arm, 318
    VEP, flash, 142
Hypesthesia, SEP, 297
    arm, 310, 311
    leg, 326
Hypnotic anesthesia, SEP
    arm, 318
    arm, touch stimulus, 341
Hypnotic hearing loss, LLAEP,
    246
Hypothermia
    BAEP, 229
    SEP, 284
Hypothyroidism, VEP, flash, 142
Hypoxia, BAEP, 230
Hysterical anesthesia, SEP,
    arm, 318

Hysterical blindness, VEP,
    checkerboard, 130

Idiopathic central serous retinopathy, VEP, checkerboard, 134
Imipramine, BAEP, 228
Impedance
    input, 22
    recording electrode, 19–20
Impedance audiometry, 233
Imperative stimulus, CNV, 370
Inactive recording electrode, 24
Increased intracranial pressure,
    VEP, flash, 141
Indifferent recording electrode,
    24
Infantile autism
    BAEP, 232
    P300, 370
Infantile neuraxonal dystrophy,
    VEP, checkerboard,
    132
Infarct. *See* Stroke
Input board, 21–22
Input 1, input 2, *23*, 54, 64
Input impedance, 22
Intensity, stimulus. *See* Stimulus
    intensity
Interaural latency difference
    AEP, 69
    BAEP, 218, 222
Interference artifact, 26
Interindividual variability, 62

Interlaboratory variability, 61
International Federation of Societies for Electroencephalography and Clinical Neurophysiology (IFSECN), recommendations for EPs, x
    See also American EEG Society
International 10-20 System of
    EEG recording electrodes, 21
Interocular latency difference,
    VEP, 69
    checkerboard, *117*, 118
Interpeak latency (IPL)
    EP, *56–57*, 63
    BAEP, *218*, *222*, 233
    See also Central conduction time
Interpretation. *See* Clinical
    interpretation
Intracerebral hemorrhage,
    BAEP, 226
Intracerebral recording electrode, 19
Intraindividual variability, EP,
    61
Intraventricular hemorrhage,
    BAEP, 231
Intramedullary brain stem tumor, BAEP, 226
    See also Tumor, brain stem
Intrathecal SEP recording, 235
Inversion of EP, 50–51

Ischemic optic neuropathy, VEP
  checkerboard, 127
  flash, 141
Isopotential lines, 13

Jakob-Creutzfeldt disease
  BAEP, 230
  SEP, arm, 317
  VEP, flash, 142
Jet electrode, ERG, 19
Joint movement stimulus, SEP, 342
Joint position sense, SEP, leg, 334
  See also Posterior column system

Kinky hair disease, VEP, flash, 142
Korsakoff's psychosis, VEP, checkerboard, 129

Labeling, EP records, 63–64
Labyrinthitis, BAEP, 230
Laser, VEP
  sine grating, 151
    steady-state, 155
  speckle pattern, 157
Late AEP, 195
  See also Long-latency AEP

Latency
  EP
    interpeak (IPL). See Interpeak latency; Central conduction time; Central conduction velocity, SEP, leg; Peripheral nerve conduction velocity
    peak, 54–56, 63, 68–69
    steady-state, 57
  BAEP
    interaural. See Interaural latency difference
    interpeak. See Interpeak latency (IPL), BAEP
    peak, 219, 221–222, 232
  SEP
    interpeak. See Central conduction time
    peak, arm
      cervical, 293
      clavicular, 290
      far-field, 295
      side-to-side, 274
      scalp, 294–295
    peak, leg
      popliteal fossa, 320
      lumbar and low thoracic, 320–324
      scalp, 325
    VEP, checkerboard, 96–97, 106, 117, 118
      interocular difference, *117*, 118
Latency difference, lateral, 69

  See also Interaural latency difference; Interocular latency difference; Lateral occipital amplitude ratio
Latency-intensity curve, BAEP, 215, 219, 223, *232–233*
Latency shift, BAEP, rate-dependent, 209–210, 219, 222
Lateral femoral cutaneous nerve stimulation, SEP, 326
Lateral medullary syndrome of Wallenberg. See Wallenberg's lateral medullary syndrome
Lateral occipital amplitude ratio, VEP, checkerboard, 121
Lead poisoning, P300, 370
Learning disorder, VEP, flash, 142
Leber's hereditary optic neuropathy, VEP
  checkerboard, 128–129
  flash, 141
Left-right latency difference. See Latency difference, lateral
Leigh's disease, BAEP, 229, 230
Letter stimuli, VEP, 157
Leukodystrophies
  BAEP, 231
  SEP, arm, 318
  VEP

  checkerboard, 129
  flash, 141
Light emitting diodes in visual stimulators, VEP
  checkerboard, 100
  flash, 139
Limb length, SEP
  arm, 297
  leg, 320, *326*, 328
Lipid storage diseases, VEP, flash, 141
LLAEP, 242–246
Lobectomy, occipital, VEP, checkerboard, 132
Localization of lesions, EP, 71
Locked-in syndrome
  BAEP, 230–231
  SEP, arm, 311
Logon stimulus, 199–200
Long-latency AEP (LLAEP), 242–246
Loudspeaker stimulus, 197
Low frequency filter, 28–30
Low-pass filter. See High frequency filter
Low thoracic SEP, 283, *320–325*
Lumbar SEP, 283, *320–325*
Lumbosacral root lesion, SEP, leg, 334
Luminance
  VEP stimulus, 89
  effect on VEP
    checkerboard, 102
    in multiple sclerosis, 124
    flash, 139

Macular degeneration, VEP, steady-state, checkerboard, 148
Macular light spot VEP, 155–157
Magnetic disc, 51
Magnetic evoked field, 374
Magnetic tape, 51–52
Maple syrup urine disease, BAEP, 229
Masking noise
  nonstimulated ear, 202
  stimulated ear, 209
Maturation
  EP, *61*, 66
  BAEP, 208
  LLAEP, 245
  MLAEP, 241
  P300, 369
  SEP
    arm, 295–296
    leg, 326
  VEP
    bar grating, 155
    checkerboard reversal, 98
    checkerboard-on and flash, 146
    flash, 137–138
    steady-state
      checkerboard, 146–147
      sine grating, 153
Maxwellian system, VEP, 91, *101*
Medial lemniscal system, SEP, 284

Median nerve stimulation, SEP, 297–298
Ménière's disease
  BAEP, 230, 235
  ECochG, 247, 249
Meningioma, cerebellopontine angle, BAEP, 224–226
Meningitis, BAEP, 232
Meningoencephalitis, BAEP, 226
Menke's kinky hair disease, VEP, flash, 142
Mental retardation
  BAEP, *231*, 232
  P300, 370
Metachromatic leukodystrophy
  BAEP, 231
  SEP, arm, 318
  VEP, checkerboard, 129
Methylphenidate
  P300, 370
  VEP, flash, 142
Middle latency AEP (MLAEP), 195, *240–242*
Minamata disease, SEP, arm, 318
Minimal brain damage, BAEP, 231–232
Miosis, VEP, checkerboard, 118, *134*
  See also Pupillary size
Mixed nerve stimulation, SEP, 286
MLAEP, 195, *240–242*
Monaural stimulation, 202, 210–211

Monocular stimulation, 89, 103
Monopolar recording, 24
Montage
  AEP, *202–203*, 206
  bipolar, 25
  referential, 25
  SEP
    arm, 292
    leg, 320, 323
  VEP
    checkerboard
      full-field stimulation, 96, 106
      half-field stimulation, 97, 106–109
    flash, 139
Motor threshold, SEP, 286
Movement-related potential, 372
Moving random dot stereogram, 157
Multiple sclerosis
  EP, 7, 69
  BAEP, 223
  contrast sensitivity, 153
  ECochG, 247
  ERG, pattern, 159
  LLAEP, 245
  MLAEP, 241
  SEP
    arm, 308–311
    leg, 334
    trigeminal, 340
  Sonomotor AEP, 251
  VEP
    checkerboard, 122–127
    checkerboard-on, 146

    color stimuli, 158
    flash, 140
    macular light spots, 155–156
    scotopic, 157
    sine grating, 152
    steady-state
      checkerboard, 147
      flash, 150
      flash, CFPD, 150
      sine grating, 153
Muscle stretch SEP, 343
Musculocotaneous nerve stimulation, 298
Mydriasis. *See* Pupillary dilatation
Myoclonus, SEP, arm, 317
Myoclonus epilepsy
  SEP, arm, 317
  VEP
    checkerboard, 132
    flash, 142
Myotonic dystrophy
  BAEP, 230
  SEP
    arm, 318
    leg, 336
  VEP, checkerboard, 132

N1
  LLAEP, 244
  VEP, checkerboard, 97
N2
  LLAEP, 244
  VEP, checkerboard, 97

N9, SEP, arm, 290, 293
N11, SEP, arm, 293
N13, SEP, arm, 293
N18, SEP, arm, 295
N20, SEP, arm, 294, 295
N75, VEP, checkerboard, 97
N105, VEP, checkerboard, 106
N145, VEP, checkerboard, 97
N_a, MLAEP, 240
N_b, MLAEP, 240
N_o, MLAEP, 240
NAP. See Nerve action potential
Narcolepsy
 BAEP, 230
 LLAEP, 245
Narrow-band filter
 EP, 52
 VEP
  checkerboard, steady-state, 147–148
  sine grating, 155
Near-field EP, 11
Neck SEP, 283, 290–293
Needle recording electrode, 18, 338, 340
 See also Recording electrode, ECochG
Needle stimulating electrode, 18, 285
Nerve action potential (NAP), 14
 acoustic nerve, ECochG, 10, 247–249
 See also Sensory nerve action potential

Neurinoma, cerebellopontine angle, BAEP, 224–226
Neurosyphilis, VEP, checkerboard, 128
Noise
 definition, 36
 residual, 41, 55
 variability, 41
Noise level, amplifier, 27
Noise reduction
 by averaging, 36, 40–41
 by filtering, 27
Nomenclature, 54, 59–61
Nonorganic hearing loss
 BAEP, 232
 LLAEP, 246
Nonparametric statistical tests, 67–68
Nonspecific EP, 59
Normal controls, 66–68
Normal hearing level (nHL), 200–201
Normal statistical distribution, 67
Normative data, 66–68
Notch filtered masking noise, 209
Number of recording channels
 EP, 38
 BAEP, 206, 212
 SEP, 288, 292
 VEP, 94
  checkerboard, full-field, 96, 104–106
  checkerboard, half-field, 96, 106–109

Number of responses
 EP, 40–41
 BAEP, 206, 212
 CNV, 370
 counter, 40
 ECochG, 246
 LLAEP, 244
 MLAEP, 240
 SEP
  arm, 292, 298
  leg, 323, 327
 VEP, checkerboard, 97, 109
Numbness. See Hypesthesia
Nutritional amblyopia, VEP, checkerboard, 133
Nyqvist frequency, 42

Occipital lobectomy, 132
Ocular dominance, 67, 99
Odd-ball paradigm, 368
Old age effect. See Age effect
Olfactory EP, 374
Olivopontocerebellar degeneration
 BAEP, 229
 SEP
  arm, 317
  leg, 336
 VEP, checkerboard, 128
Omitted stimulus, P300, 368
Ophthalmological strategy, VEP, checkerboard, 114–115
Optic nerve hypoplasia, VEP, flash, 143

Optic neuritis. See Retrobulbar neuritis; Parainfectious optic neuritis
Optic neuropathy
 toxic, VEP, flash, 141
 traumatic. See Traumatic optic neuropathy
Ototoxic drugs, ECochG, 247
Outer ear canal collapse, 197

P1
 LLAEP, 241, 244
 VEP, checkerboard, 96
P2
 LLAEP, 244
 VEP, checkerboard, 98
P3
 LLAEP, 244
 See also P300
P9, SEP, arm, 295
P11, SEP, arm, 295
P13, SEP, arm, 295
P13/14, SEP, arm, 295
P14, SEP, arm, 295
P27-N35, SEP, leg, common peroneal nerve stimulation, 325
P37-N45, SEP, leg, posterior tibial nerve stimulation, 325
P75, VEP, checkerboard, 106
P100, VEP, checkerboard, 96
P135, VEP, checkerboard, 106
P300

cognitive, 368–370
LLAEP, 244
SEP, arm, 294–295
P_a, MLAEP, 240
P_o, MLAEP, 240
Pain stimulation, 343
Palatal myoclonus, BAEP, 230
Palinopsia, VEP, checkerboard, 132
Papilledema, VEP
  checkerboard, 127
  flash, steady-state, 151
Paradoxical distribution
  SEP, leg, 325
  VEP, checkerboard, 106
Parainfectious optic neuritis, VEP, checkerboard, 129
Parasagittal tumor, SEP, leg, 338
Parietal infarct, SEP, arm, 311
Parietal tumor, SEP, arm, 311–312
Parkinson's disease
  BAEP, 229
  P300, 370
  readiness potential, 372
  VEP
    checkerboard, 132
    flash, 142
    sine grating, 152
Passive filter, 28, 31
Pattern appearance and disappearance
  definition, 87
  VEP, 145–146
  TV stimulator, 99

Pattern element size, 92
  See also Check size
Pattern ERG, 159
Pattern orientation. See Diamond pattern stimulus
Pattern reversal, definition, 87
  TV stimulator, 99
  VEP 96–134
    steady state, 146–149
Pattern shift, definition, 87
  stimulator, 100
Pattern VEP, definition, 86–87
  versus luminance VEP 102
Patterned light flashes
  definition, 87
  VEP, 145–146
    steady-state, 146–149
Patterned mirror stimulator, VEP, 101
Peak
  definition, 7
  bifid, bilobed, 55
Peak amplitude, 57–59
Peak equivalent sound pressure level (peSPL), 201
Peak identification, 54–55
Peak separation. See Interpeak latency
Peak-to-peak amplitude, 59, 63
Pelizaeus-Merzbacher disease
  BAEP, 231
  SEP, arm, 318
  VEP, checkerboard, 129
Percentile method, 68
Perinatal asphyxia, SEP, arm, 317

Perinatal disorders, VEP, checkerboard, 142
Peripheral nerve conduction velocity
  SEP
    arm, 299, 301
    leg, 328, 330, 331
  SNAP, 340
  See also Sensory nerve conduction velocity
Peripheral nerve lesion, SEP
  arm, 304
  leg, 333
Pernicious anemia, VEP, checkerboard, 129
  See also Subacute combined degeneration; Vitamin B_{12} deficiency
PF potential, 283, 320
Phase
  click stimulus, 201–202, 210
  checkerboard reversal stimulus, 96, 101
Phase lag, EP, steady state, 57
Phase shift, filter, 31
Phenobarbital, SEP, 284
Phenylketonuria
  BAEP, 229
  VEP
    checkerboard, 129
    checkerboard-on, 146
    flash, 142
Phenytoin
  BAEP, 230
  SEP, 284
Photographic copy, 51

Photometer, 90
Photopic VEP, 157
Photosensitive epilepsy, VEP
  checkerboard, 132
  flash, 142
  steady-state, flash, 151
Photosensitive paper copy, 51
Piezoelectric earphone, 197
Pip stimulus, 199–200
Pituitary tumor, VEP, checkerboard, 130–131
Plus-minus average, 55
Point, 36, 42–45
Polarity
  EP, 54
  click stimulus, 201–202, 210
  electric shock stimulus, 285
  pattern reversal stimulus, 96
Polarity convention
  EP, 23–24, 54, 63–64
  AEP, 203
  VEP, 94
Polarization, recording electrodes, 20
Polarized light stimulator, VEP, 101
Polyneuropathy, SEP
  arm, 304
  leg, 336
  See also Peripheral nerve conduction velocity; Peripheral nerve lesion
Pontine glioma, BAEP, 226
Pontomesencephalic stroke, BAEP, 226

Popliteal fossa (PF) potential, 283, *320*
Position sense, SEP, leg, 334
   See also Posterior column system
Postanoxic encephalopathy
   BAEP, 226
   LLAEP, 245
   VEP, flash, 141
Postanoxic myoclonus, SEP, arm, 317
Postauricular AEP, 250
Postchiasmal. *See* Retrochiasmal
Postconcussion syndrome
   BAEP, 231
   VEP
      checkerboard, 129
      flash, 141
Posterior column system, SEP, 284
   See also Position sense
Posterior tibial nerve stimulation
   SEP, 320, *325*
   electrode placement, 326
Poststimulus triggering, 39
Postsynaptic potential, 14, 69
Posttraumatic encephalopathy, BAEP, 226
Power spectral analysis, definition, 52
   BAEP, 211
   filter settings, 28
   VEP, steady-state
      checkerboard, 148
      flash, 151
      sine grating, 155

Prechiasmal lesions, VEP, checkerboard, 122–130
Prechiasmal strategy, VEP, checkerboard, 109–111
   See also Full-field stimulation
Precocious puberty, BAEP, 230
Pregnancy complications, BAEP, 231
Presbyacousis, BAEP, 235
Prestimulus triggering, 39
Primary, EP, 59
Probability ellipse, 68
Progressive supranuclear palsy, BAEP, 230
Pseudotumor cerebri, VEP, checkerboard, 132
Psychiatric disorders
   SEP, arm, 318
   VEP, checkerboard, 142
   See also Depression, Schizophrenia
Ptosis, VEP, checkerboard, 118
Pudendal SEP, 340
Pupillary dilatation, VEP, 88, 91
   See also Artificial pupil
Pupillary size
   VEP, 91
      checkerboard, 99
      flash, 138

Quinine amplyopia, VEP
   checkerboard, 134
   flash, 143

Radial nerve stimulation, SEP, 298
Radiculopathy, SEP
   arm, 306
   leg, 334
Ramsey Hunt syndrome, SEP, arm, 317
Random dot VEP, 157
Rarefaction click. *See* Click stimulus, rarefaction
Rate, stimulus. *See* Stimulus rate
Rate-dependent latency shift, BAEP, 209–210, 219, 222
Readiness potential, 372
Reading disability. *See* Dyslexia
Recommendations for EP recording, of the International Federation of Societies for Electroencephalography and Clinical Neurophysiology (IFSECN), x
   See also Guidelines for EP recordings
Recording channels. *See* Number of recording channels
Recording electrode
   active, 24
   application, 17–18
   average reference, 25
   bias potential, 20
   clip, 18
   ECG, 18

   ECochG, 18
   EEG, 17–18
   electrocorticographic, 19
   ERG, 18–19
   exploring, 24
   impedance, 19–20
   inactive, 24
   indifferent, 24
   intracerebral, 19
   material, 19
   needle, *18*, 338, 340
   placement
      International 10–20 System, 21
      AEP, 202
      BAEP, 206, 211
      ECochG, 246
      LLAEP, 242–244
      MLAEP, 240
      P300, 368
      SEP, 287–288
         arm, 292
         arm, cervical, 290–293
         arm, clavicular, 290
         arm, scalp, 294–295
         leg, 323
         leg, interspinal ligament, 338
         leg, low thoracic, 320
         leg, lumbar, 320
         leg, popliteal fossa, 320
         leg, scalp, 325
         leg, thoracic, 320
   sonomotor AEP, 249–250
   VEP, 94

checkerboard, full-field, 96, 103–106
checkerboard, half-field, 96, 106–109
flash, 139
polarization, 20
reference, 24
resistance, 19
silver-silver chloride, 19
Recording gain, 27, 45
Recording montage. See Montage
Recruitment, 233
Reference electrode, 24
Referential montage, 25
Referential recording, 24
Refraction, VEP
  bar grating, 155
  checkerboard, 132
  checkerboard-on, 146
  steady state, checkerboard, 147–148
  See also Visual acuity
Refractive error, VEP, 88
  checkerboard, 101, 115, 122, 132
  dots, 157
Rejection of artifact. See Artifact rejection
Renal failure. See Chronic renal failure
Replication, EP, 40
  See also Number of responses
Report, EP, 71–72
Residual noise, 41, 55

Resistance of recording electrode, 19
Resolution, EP amplitude, 45
Respiratory distress syndrome, BAEP, 231
Response
  definition, 7
  number averaged. See Number of responses
Retinal detachment, VEP, flash, 143
Retinitis pigmentosa, VEP, flash, 143
Retinopathy, VEP, checkerboard, 134
Retrobulbar neuritis
  BAEP, 223
  contrast sensitivity, 153
  ERG, pattern, 159
  SEP, arm, 310
  VEP
    checkerboard, 122
    flash, 141–142
    macular light spot, 156
    steady-state checkerboard, 147
    flash, 150
Retrochiasmal lesions, VEP, checkerboard, 131–132
Retrochiasmal strategy, VEP, checkerboard, 111–114
  See also Half-field stimulation
Reye syndrome, SEP, arm, 315
Rhythmical afterdischarge, VEP, flash, 136

Roll-off slope, filter, 30, 31
Root lesion. See Radiculopathy

Saltatory conduction, 68
Sampling rate
  EP, 36, 42–45
  BAEP, 206, 212
  critical, 42
  SEP, 327
  VEP, 94
Saphenous nerve stimulation, SEP, 326
Sarcoidosis
  SEP, trigeminal, 340
  VEP, checkerboard, 129
Scalp, EP, 10
Scalp AEP. See LLAEP; MLAEP
Scalp SEP
  arm stimulation, 283, 294–295
  leg stimulation, 325–326
Scalp VEP. See Cortical EP
Schizophrenia
  BAEP, 230
  LLAEP, 245
  P300, 370
  SEP, arm, 318
  VEP, checkerboard-on, 146
Scotopic VEP, 157
Secondary EP, 59
Sedation
  AEP, 197
  MLAEP, 240

SEP, 284
  See also CNS depressant drugs
Seeing and VEP, 115
Segmental stimulation, SEP, 285
  arm, 298
  leg, 327, 328, 334
  touch, 341
Seizure disorder, VEP
  checkerboard, 132
  flash, 142
Seizures, BAEP, 230
Senile dementia
  BAEP, 229
  P300, 370
  VEP
    checkerboard, 131
    flash, 142
Sensitivity, definition, 27
Sensorineural hearing loss. See Hearing loss, sensorineural
Sensory level (SL), 201
Sensory loss. See Hypesthesia
Sensory nerve action potential (SNAP), 340
Sensory nerve conduction velocity, 340
Sensory nerve stimulation, SEP, 286, 298
Sensory threshold, SEP, 286
SEP classification, 283–284
Sex
  EP, 62, 66
  BAEP, 206, 208

SEP, arm, 292, *297*
VEP, checkerboard, 96, *99*
Sexual dysfunction, pudendal SEP, 340
Shift of gaze, VEP, 158
Signal, definition, 36
Signal enhancement. *See* Noise reduction
Signal, variability, 41
Signal-to-noise ratio, 40
Silver-silver chloride recording electrode, 19
Sine grating
  stimulus, 91–93, 151
  VEP
    steady-state, 153–155
    transient, 151–152
60 Hz artifact, 20, 26
60 Hz filter, 31
Sleep
  EP, 41, 61
  AEPs, 197
  LLAEP, 242
  MLAEP, 240
  SEP, 284
Sleep apnea, BAEP, 231
Slide projector stimulator, VEP, 101
Slope, filter roll-off, 30, 31
Slow brain stem AEP, 237
Smoking, VEP, flash, 142
Smoothing, 48–49
SNAP, 340
SN_{10}, 237
Somatomotor SEP, 343
Sonomotor AEP, 249–251

Sound pressure level (SPL), 201
SP, 247, *249*
Spatial frequency
  definition, 92–93
  channels, 151
Spatial sine wave, square wave, 91, 151
Special purpose computer, 38
Specific EP, 59
Spectral analysis. *See* Power spectral analysis
Speech-related potentials, 374
Spinal cord compression, SEP, leg, 334–336
Spinal cord conduction velocity, SEP, leg, 326, 328
Spinal cord injury, SEP, leg, 334
Spinocerebellar degeneration, BAEP, 230
  See also Friedreich's ataxia; Hereditary cerebellar ataxia
Spondylotic myelopathy, SEP
  arm, 306
  leg, 334
Spondylotic radiculopathy. *See* Radiculopathy
Square size. *See* Check size
Static contrast sensitivity, 153
Stationary dot VEP, 157
Statistical distribution of normal values, 67–68
Steady-state EP
  description, 11–12
  analysis period, 41

latency, 57
triggering, 39
Steady-state AEP. *See* FFP; 40 Hz AEP
Steady-state SEP, arm, 287, 310–311
Steady-state VEP
  checkerboard reversal, checkerboard-on, -off, flash, 146–149
  flash, 149–151
  sine grating, 153–155
Stereopsis. *See* Binocular vision
Stimulator
  AEP, 197–198
    bone vibrator, *197–198*, 233
    click, 199
    logon, 199–200
    loudspeaker, 197
    tone pip, burst, 199–200
  ECochG, bone vibrator, 246, 247
  SEP, constant current, 286–287
  VEP
    checkerboard, 99–101
    flash, 138–139
    sine grating, 151
Stimulus
  anode, SEP, 285
  bar grating, VEP, 155
  beep-boop, P300, 368
  bilateral. *See* Bilateral stimulation
  bladder, SEP, 340

blinking, VEP, 158
cathode, SEP, 285
checkerboard, VEP, 91–93
click, AEP. *See* Click stimulus
cold, SEP, 343
color, VEP, 157–158
constant current, 286–287
dermatomal, SEP. *See* Segmental stimulation
dot, VEP. *See* Dot stimuli
electric
  eye, 158
  mixed nerve, SEP, 286
  sensory nerve, SEP, 286
  tooth pulp, 306
filtered click, AEP, 199
flash, VEP, 89–91
full-field, 93
ganzfeld, VEP, 139
gaze shift, VEP, 158
half-field, 93
heat, SEP, 343
hemiretina, 93
imperative, CNV, 370
isolated, 285
joint movement, SEP, 342
laser, VEP. *See* Laser VEP
macular light spot, VEP, 155
maximum, EP, 16
Maxwellian, VEP, 91, *101*
muscle stretch, SEP, 343
omitted, P300, 368
pain, SEP, 343
photopic, VEP, 157
pudendal, SEP, 340

random dot, VEP, 157
scotopic, VEP, 157
segmental, SEP. *See* Segmental stimulation
shift of gaze, VEP, 158
sine grating, VEP, 91–93, 151
temperature, SEP, 343
threshold, EP, 16
tone burst, pip, AEP, 199–200
tongue, SEP, 341
tooth pulp, SEP, 306, *343*
touch, SEP, 341
trigeminal, SEP, 340
unilateral, 12
    See also Monaural stimulation; Monocular stimulation; Half-field stimulation
vibrotactile, SEP, 341
warm, SEP, 343
warning, CNV, 370
Stimulus artifact. *See* Artifact, stimulus
Stimulus duration
  EP, 16
  AEP, 199
  BAEP, 209
  LLAEP, 242
  SEP, 287
    arm, 292, *298*
    leg, 327
Stimulus electrode
  SEP, 285
    arm, 292, *297–298*

leg, 326–327
needle, 18, 285
SNAP, 340
Stimulus field size, VEP, checkerboard, *93,* 96
Stimulus intensity
  EP, 16
  AEP, 200–201
  BAEP, 206, *210*
  P300, 368
  SEP, 286
    arm, 292, *298*
    leg, 323, *327*
  VEP, 98
    checkerboard, 102
    flash, 139
Stimulus isolation, SEP, 285
Stimulus phase, VEP, checkerboard, 96, *101*
Stimulus polarity
  AEP, click stimulus, *201–202,* 210
  SEP, 285
Stimulus rate
  EP, 16–17
  AEP, 200
  BAEP, 206, *209–210*
  40 Hz AEP, 242
  FFP, 237
  LLAEP, 242
  MLAEP, 240
  SEP, 287
    arm, 292, *298*
    leg, 327
    touch, 341
  sonomotor, 249

VEP, 98
  checkerboard, 96, *101*
  flash, 139
Strategy of stimulation and recording
  EP, 71
  BAEP, 212–216
  SEP
    arm, 299
    leg, 327–328
  VEP, checkerboard, 109–115
  See also Clinical interpretation
Stroboscopic flash stimulator, 138
Stroke
  brain stem
    BAEP, 226, *230–231*
    LLAEP, 245
    SEP
      arm, 311
      trigeminal, 340
  cerebral
    SEP, arm, 311–313
    VEP
      checkerboard, 131–132
      flash, 142
      steady-state, checkerboard, 147
      steady-state, flash, 151
Subacute combined degeneration of the spinal cord,
  SEP
    arm, 308
    leg, 336
  VEP, checkerboard, 129

  See also Vitamin B$_{12}$ deficiency
Subacute myelo-optico-neuropathy, SEP, leg, 336
Subacute sclerosing panencephalitis (SSPE), BAEP, 230, 231
Subarachnoid hemorrhage, SEP, arm, 315
Subcortical EP, 8–10, 14
Subjective contrast sensitivity, sine grating, 153
Subtraction of EPs, 50–51
Sudden hearing loss, ECochG, 247
Sudden infant death syndrome (SIDS), 231
Summating potential (SP), ECochG, 247, *249*
Superficial peroneal nerve stimulation, SEP, 327
Supratentorial lesion, BAEP, 230
Sural nerve stimulation, SEP, 327
Surgical monitoring
  BAEP, 226
  SEP
    arm, 317
    leg, 336
  VEP, flash, 140–141
Sweep, 36, 41
Sweep counter, 40
Sweep length. *See* Analysis period
Synaptic transmission, 69
Syringomyelia, SEP, arm, 308

Tabes dorsalis
  SEP
    arm, 304
    leg, 336–338
  VEP, checkerboard, 128
Tachistoscope, 101
Tape, magnetic, 51–52
Television (TV) pattern stimulator, 99–100
Temperature
  BAEP, *208–209*, 229
  SEP, 284
    arm 292, *297*
    leg, 320, *326*
  VEP, checkerboard, 124
Temperature stimulus, SEP, 343
Temporal dispersion, 68, 69
10-20 System of EEG electrode placement, 21
Terminology, 59–61
Texture contrast, VEP, 157
Thalamic lesion, SEP, arm, 311
Thalamic syndrome of Dejerine-Roussy, 311
Thiopental
  BAEP, 228
  SEP, 284
Thoracic outlet syndrome, 304–305
Thoracic SEP, 283, *320–325*
Threshold
  EP, 16
  Acoustic nerve action potential, 247

BAEP, 215, 219, *232*
  See also Electric response audiometry
  hearing, 200, *201*
  SEP, 286
  VEP, 89
Time constant, 30
Time varying filter, 42
Tolerance limits, 67–68
Toluene, BAEP, 230
Tone burst stimulus, AEP, 200
Tone pip stimulus, AEP, 199–200
Tongue stimulation, SEP, 341
Tooth pulp stimulation, SEP, 306, *343*
Touch stimulation, SEP, 341
Tourette's syndrome
  BAEP, 230
  readiness potential, 372
  SEP, arm, 318
  VEP, checkerboard, 132
Toxic amblyopia, VEP
  checkerboard, 133–134, 148
  flash, 143
  steady-state, checkerboard, 148
Toxic optic neuropathy, VEP, flash, 141
Transient EP, 11
  triggering, 39
Transtympanic ECochG, 246
Transverse myelitis
  BAEP, 223
  SEP, arm, 311
  VEP, checkerboard, 124

Traumatic optic neuropathy, VEP
  checkerboard, 129
  flash, 141, *143*
  steady-state, checkerboard, 147
Trigeminal SEP, 340
Trigeminal neuralgia
  BAEP, 231
  SEP, trigeminal, 340
Triggering, 39
Trolard, 91
Tumor, brain
  LLAEP, 245
  SEP
    arm, 311–312
    leg, 338
  VEP
    checkerboard, *127*, 130–131, 132
    flash, 141, 142
  steady-state
    checkerboard, 147
    flash, 151
  brain stem
    BAEP, 224–226
    SEP
      arm, 311
      trigeminal, 340
    cerebellopontine angle, 224–226
    cervical cord, SEP, arm, 308
    spinal cord, SEP, leg, 334
Tuned filter, 52
  See also Narrow-band filter

TV pattern stimulator, 99–100
Twins, EP, 62

Ulnar nerve stimulation, SEP, 298
Uncrossed asymmetry, VEP, checkerboard, half-field, 114
Unilateral stimulation
  EP, 12
  AEP. See Monaural stimulation
  SEP, 287
  VEP. See Monocular stimulation; Half-field stimulation
Unipolar recording, 24
Upper and lower hemiretina stimulation, 93, 103
Uremic encephalopathy. See Chronic renal failure
Uremic peripheral neuropathy, SEP, arm, 304
Urinary bladder SEP, 340

Variability
  EP, 41, *61–62*
  interindividual, 62
  interlaboratory, 61
  intraindividual, 61
  noise, 41
  signal, 41

Vascular malformation BAEP, 231
VEP classification, 86–87
Vestibular neuronitis, BAEP, 230
Vibrotactile stimulation, SEP, 341
Vision and VEP, checkerboard, 115
Visual acuity, VEP
   checkerboard, *96, 99,* 101, 115, *132*
   dot pattern, 157
   See also Refraction
Visual angle, 92

Visuogram, 153
Vitamin B_{12} deficiency
   BAEP, 230
   SEP, arm, 308
   See also Subacute combined degeneration

Wallenberg's lateral medullary syndrome
   BAEP, 231
   SEP, arm, 311
Warm stimulus, SEP, 343

Warning stimulus, CNV, 370
Wave, definition, 7
Waveshape, *59, 70*
Weber's peduncular syndrome, SEP, arm, 311
Wernicke's encephalopathy, BAEP, 230
White noise, 202
Wiener filter, 42
Wilson's disease
   BAEP, 230

SEP, arm, 318
VEP, flash, 142
Word length, computer, *45,* 46
Word stimuli
   P300, 369
   VEP, 157
Writing-related potentials, 374

XY plot, 51